Quaternary Alloys
Based on II-VI
Semiconductors

Quaternary Alloys Based on II-VI Semiconductors

Vasyl Tomashyk
V. Ye. Lashkaryov Institute of Semiconductor
Physics of NAS of Ukraine, Kyiv

CRC Press
Taylor & Francis Group
Boca Raton London New York

CRC Press is an imprint of the
Taylor & Francis Group, an informa business

CRC Press
Taylor & Francis Group
6000 Broken Sound Parkway NW, Suite 300
Boca Raton, FL 33487-2742

First issued in paperback 2019

ISBN-13: 978-1-4822-5296-5 (hbk)
ISBN-13: 978-0-367-37826-4 (pbk)

Library of Congress Cataloging-in-Publication Data

Tomashyk, Vasyl, author.
　　Quaternary alloys based on II-VI semiconductors / Vasyl Tomashyk.
　　　　pages cm
　　Includes bibliographical references and index.
　　ISBN 978-1-4822-5296-5 (hardcover : alk. paper) 1.
　　Chromium-cobalt-nickel-molybdenum alloys. 2. Semiconductors. I. Title.

　　TN693.A45T66 2015
　　621.3815'20284--dc23　　　　　　　　　　　　　　　　　　　　　　2014033170

Visit the Taylor & Francis Web site at
http://www.taylorandfrancis.com

and the CRC Press Web site at
http://www.crcpress.com

Contents

Contents   xiii

Index ...

Preface

Applications of II–VI semiconductors in various branches of modern technology (micro- and optoelectronics; photovoltaics; partly as photoresistors and transistors; for blue lasers and LEDs; in solar cells, infrared detectors, and infrared imaging sensors; as electro-optic modulators, components of microwave generators, and solid-state x-ray and γ-ray detectors that can operate at room temperature; and so on) are well known. As they are used in various devices, the search for new semiconductor materials and the improvement of existing materials are important fields of study in materials science. Doping, in which impurities are introduced, is a common method of modifying the physical and chemical properties of a semiconductor. Quaternary alloys permit a simultaneous adjustment of band gap and lattice constant, increasing radiant efficiency at a wide range of wavelengths. Phase diagrams are visual representations of the state of a material as a function of temperature, pressure, and concentrations of the constituent components serve as a basic map for the choice of alloys, their understanding, processing, and development.

Though ternary phase diagrams based on zinc, cadmium, and mercury chalcogenides were collected and published in the handbook *Ternary Alloys Based on II–VI Semiconductor Compounds*, by V. Tomashyk, P. Feychuk, and L. Shcherbak (Taylor & Francis, 2013), data pertaining to diagrams of quaternary systems based on these semiconductor compounds are preferentially dispersed in the scientific literature. This reference book is intended to illustrate the up-to-date experimental and theoretical information about phase relations in based on II–VI semiconductor systems with four components. The book critically evaluates many industrially significant systems presented in the form of two-dimensional section for the condensed phases.

All materials are classified according to the periodic table groups of their constituent atoms, that is, possible combinations of Zn, Cd, and Hg with chalcogens S, Se, and Te and additional components in the order of their group number. This information is divided into nine chapters.

Each quaternary database description contains brief information in the following order: the diagram type, possible phase transformations and physical–chemical interactions of the components, thermodynamic characteristics, and methods of equilibrium investigation and samples preparation.

Most of the figures have been presented in their original form, although some have been modified. If the published data varied essentially, several versions have been presented for comparison. The content of system components is mostly indicated in mol.%. If the original phase diagram is given in mass.%, the axis with content in mol.% is also provided after each such figure.

The book will be helpful for researchers in industrial and national laboratories, universities, and graduate students majoring in materials science and engineering. It will also be suitable for phase relations researchers, inorganic chemists, and semiconductor physicists.

1 Systems Based on ZnS

1.1 ZINC–HYDROGEN–CARBON–SULFUR

Some compounds are formed in this quaternary system.

[MeZnSMe]$_x$ or C$_2$H$_6$SZn, methylzinc methyl sulfide. This compound was precipitated with some gas evolution, when methanethiol was slowly condensed into dimethylzinc in hexane (Coates and Ridley 1965). When all the thiol had been added, the mixture was allowed to warm to room temperature and more gas was evolved.

[Zn(MeS)$_2$]$_n$ or C$_2$H$_6$S$_2$Zn, (methanethiolato)zinc. This compound could be prepared by a reaction of ZnCl$_2$ with NaMeS: ZnCl$_2$ (27 mmol) was dissolved in H$_2$O (100 mL) (Osakada and Yamamoto 1991). A small amount of undissolved solid was removed by filtration. Aqueous NaMeS (15%) (27 mL, *ca.* 4.1 g NaMeS) was added dropwise to the solution at room temperature to cause immediate precipitation of a white solid. The solid was separated from the solution by centrifugation followed by careful decantation of the solution from the mixture. Addition of H$_2$O (100 mL) to the solid with stirring followed by the centrifugation and decantation procedure was repeated until no Cl$^-$ was detected in the washings. Drying the resulting white solid under vacuum gave C$_2$H$_6$S$_2$Zn as a white powder.

(Methanthiolato)zinc can be also obtained by reaction of ZnEt$_2$ with MeSH (Osakada and Yamamoto 1987, 1991). A hexane solution (10 vol.%) of ZnEt$_2$ (19 mmol) in a Schlenk flask equipped with a dry ice condenser and a dropping funnel was cooled at −30°C under Ar atmosphere. MeSH (46 mmol) was added dropwise from the dropping funnel. A white solid was precipitated immediately. The temperature of the mixture was raised gradually to room temperature. After the reaction mixture was maintained for several hours at that temperature, the white solid was filtered out with a glass tube equipped with a glass frit and washed with hexane several times. Drying the solid under vacuum gave C$_2$H$_6$S$_2$Zn.

Thermolysis of this compound at 260°C gives β-ZnS accompanied by evolution of Me$_2$S (Osakada and Yamamoto 1987, 1991).

[MeZnSPr^n]$_x$ or C$_4$H$_{10}$SZn, methylzinc n-propyl sulfide. This compound was precipitated when n-propylthiol in hexane was added to dimethylzinc (also in hexane) at −80°C and the mixture allowed warming to room temperature (Coates and Ridley 1965).

[Zn(EtS)$_2$]$_n$ or C$_4$H$_{10}$S$_2$Zn, (ethanethiolato)zinc. To obtain this compound, LiEtS (6.6 mmol) in ethanol (10 mL) was added slowly under a N$_2$ atmosphere to an ethanol (20 mL) solution of ZnI$_2$ (3.3 mmol) in a Schlenk flask (Osakada and Yamamoto 1991). The resulting solution became milky on stirring. Allowing the reaction mixture to settle overnight resulted in the separation of C$_4$H$_{10}$S$_2$Zn as a white solid, which was filtered out, washed with H$_2$O until I$^-$ was not detected in the washings, and dried in vacuo.

Thermolysis of this compound gives β-ZnS.

1

[Zn(Pr^{i}S)$_2$]$_n$ or C$_6$H$_{14}$S$_2$Zn, (i-propylthiolato)zinc. To obtain this compound, LiPr^{i}S (6.6 mmol) in ethanol (10 mL) was added slowly under a N$_2$ atmosphere to an ethanol (20 mL) solution of ZnI$_2$ (3.3 mmol) in a Schlenk flask (Osakada and Yamamoto 1991). The resulting solution became milky on stirring. Allowing the reaction mixture to settle overnight resulted in the separation of C$_6$H$_{14}$S$_2$Zn as a white solid, which was filtered out, washed with H$_2$O until I$^-$ was not detected in the washings, and dried in vacuo.

Thermolysis of this compound gives β-ZnS.

MeZnSPh or C$_7$H$_8$SZn, methylzinc phenyl sulfide. This compound was precipitated when thiophenol in hexane was added to dimethylzinc (also in hexane) at −80°C and the mixture allowed warming to room temperature (Coates and Ridley 1965).

Zn(Bu^{n}S)$_2$ or C$_8$H$_{18}$S$_2$Zn, bis(n-butylthiolato)zinc. This compound could be prepared electrochemically using a simple cell: Pt(−)/CH$_3$CN + n-C$_4$H$_9$SH/Zn(+) (Said and Tuck 1982). The anode was a piece of Zn, approximately 2 cm^2 area, suspended on a Pt wire. When the electrolysis began, a white solid formed at the anode almost immediately. The resulted solid was collected, washed several times with acetonitrile and petroleum ether, and dried. C$_8$H$_{18}$S$_2$Zn is air stable.

Zn(Bu^{t}S)$_2$ or C$_8$H$_{18}$S$_2$Zn, bis(t-butylthiolato)zinc. This compound could be prepared by the same procedure as for bis(n-butylthiolato)zinc using t-C$_4$H$_9$SH or (t-C$_4$H$_9$S)$_2$ instead of n-C$_4$H$_9$SH (Said and Tuck 1982). This compound is air stable.

Zn(PhS)$_2$ or C$_{12}$H$_{10}$S$_2$Zn, bis(phenylthiolato)zinc. This compound could be prepared by the same procedure as for bis(n-butylthiolato)zinc using C$_6$H$_5$SH instead of n-C$_4$H$_9$SH (Said and Tuck 1982). White solid was formed at the anode within a few hours after the electrolysis was started. This compound is air stable.

Bis(phenylthiolato)zinc could be also obtained by another procedure (Brennan et al. 1990). Thiophenol (21.8 mmol) was added to diethylzinc (10.8 mmol) in toluene (20 mL). There was immediate gas evolution, and a white precipitate is formed. After 2 h, heptane (10 mL) was added to the mixture, and the white solid was collected by filtration and washed with heptane (10 mL).

[Zn(S$_2$C$_2$Ph$_2$)]$_x$ or C$_{14}$H$_{10}$S$_2$Zn. C$_{60}$H$_{52}$S$_8$Zn$_2$ was heated to 250°C in a small vacuum sublimation apparatus to yield a pale yellow sublimate identified as cis-bis(methylthio) stilbene (Zhang et al.1991). The pale red-brown nonvolatile residue was identified as C$_{14}$H$_{10}$S$_2$Zn. This compound melts at 300°C.

C$_{14}$H$_{10}$S$_6$Zn, bis(dithiobenzoato)dithiozinc. This was obtained by direct photolysis of degassed CH$_3$Cl solution containing sulfur and zinc dithiobenzoate (Fackler et al. 1972).

C$_{14}$H$_{10}$S$_6$Zn, bis(trithioperoxybenzoato)zinc. This compound crystallizes in the monoclinic structure with the lattice parameters $a = 2065.4 \pm 0.1$, $b = 407.8 \pm 0.1$, $c = 2130.3 \pm 0.1$ pm, and $β = 112°30 \pm 5'$ and calculation and experimental densities 1.75 and 1.75 ± 0.1 g·cm^{-3} (Bonamico et al. 1971).

C$_{16}$H$_{14}$S$_4$Zn, bis(p-dithiotoluato)zinc. To obtain this compound, perthiotoluatozinc (2 g) was dissolved in 150 mL of xylene, and 3 g (excess) of triphenylphosphine was added (Fackler et al. 1968). The solution was brought to reflux and filtered. Upon addition of 50 mL of n-pentane and cooling to room temperature, a white oily substance separated (a mixture of Ph$_3$P and Ph$_3$PS). The solution was decanted from the oil and an additional 50 mL of n-pentane was added. After removing some of the n-pentane under vacuum, yellow needles are formed. They were separated from the solution, washed with methanol, and dried in a stream of N$_2$. C$_{16}$H$_{14}$S$_4$Zn melts at 206°C–208°C.

C$_{16}$H$_{14}$S$_6$Zn, bis(perhiotoluato)zinc. To obtain this compound, p-tolualdehyde (0.1 mol) was added to a mixture of 25 mL of (NH$_4$)$_2$S solution, 3.0 g of sulfur, and 20 mL of tetrahydrofuran (Fackler et al. 1968). The mixture was refluxed for 10 min with frequent agitation. The resulting dark red solution was diluted with 100 mL of distilled water and extracted with 200 mL of ether to remove unreacted aldehyde and oxidation products. The aqueous solution was filtered to remove a small amount of sulfur and added directly to an aqueous solution of ZnCl$_2$ (10 g in 100 mL of H$_2$O). An insoluble orange precipitate is immediately formed, which was isolated and washed with MeOH. The crude precipitate was dissolved in tetrahydrofuran, filtered to separate the zinc hydroxide contaminant, and evaporated to dryness in a hood. Orange-red crystals of C$_{16}$H$_{14}$S$_6$Zn were isolated. The product was recrystallized from CS$_2$ by adding absolute EtOH and cooling. Large orange-red needles were produced. This compound melts at 192°C–193°C.

Zn(SC$_6$H$_2$Me$_3$-2,4,6)$_2$ or C$_{18}$H$_{22}$S$_2$Zn. This compound was obtained by a pyrolysis of Zn[N(SiMe$_3$)$_2$]$_2$ with 2,4,6-Me$_3$C$_6$H$_2$SH as polymeric solids that dissolved only in coordinating solvents (Bochmann et al. 1991). It melts at a temperature higher than 295°C (Bochmann and Webb 1991).

The experiment was carried out under Ar atmosphere using standard vacuumline techniques. Crystals suitable for x-ray diffraction (XRD) could not be grown (Bochmann et al. 1991).

Zn[S$_2$CC$_6$H$_4$-p-(i-C$_3$H$_7$)]$_2$ or C$_{20}$H$_{22}$S$_4$Zn, bis(p-dithiocumato)zinc. To obtain this compound, a solution of triphenylphosphine in CHCl$_3$ in slight excess of stoichiometric requirements was added to a hot CHCl$_3$ solution of Zn[S$_2$CC$_6$H$_4$-p-(i-C$_3$H$_7$)]$_2$S$_2$ (Fackler et al. 1972). After cooling the solution with cold water, anhydrous ether was added and crystals precipitated out. Recrystallization was accomplished by dissolving the compound in a minimum amount of CHCl$_3$ and adding n-pentane. This compound melts at 169°C–170°C.

Zn[S$_2$CC$_6$H$_4$-p-(i-C$_3$H$_7$)]$_2$S$_2$ or C$_{20}$H$_{22}$S$_6$Zn, bis(p-perthiocumato)zinc. To obtain this compound, sulfur (7 g) and 50 mL of (NH$_4$)$_2$S (25% solution) in 50 mL of tetrahydrofuran were added to a solution of p-isopropylbenzaldehyde (0.2 mol) in tetrahydrofuran (Fackler et al. 1972). After the reaction mixture was boiled for 3 min, ice was added to cool it to room temperature. A solution of 15 g of ZnCl$_2$ in 100 mL of H$_2$O was added to the obtained mixture, and water was added to bring the total volume to approximately 500 mL. The red oil that formed was separated from aqueous layer and triturated with MeOH, which caused solidification. The MeOH was decanted, and the orange-red precipitate was washed with 400 mL of 10 vol.% HCl solution and several portions of MeOH. Upon drying in the air, C$_{20}$H$_{22}$S$_6$Zn was obtained. The product was recrystallized several times from CHCl$_3$–n-pentane, yielding small red-orange crystals, which melt at 121°C–123°C. Crystals of C$_{20}$H$_{22}$S$_6$Zn were grown from saturated chloroform–methanol (20:80 volume ratio) solution by slow evaporation and obtained as large, well-faceted prismatic plates.

This compound crystallizes in the monoclinic structure with the lattice parameters $a = 2526.2 \pm 0.6$, $b = 806.5 \pm 0.2$, $c = 2104.7 \pm 0.5$ pm, and $\beta = 147.49° \pm 0.02°$ and calculation and experimental densities 1.49 and 1.49 ± 0.01 g·cm^{-3} (Fackler et al. 1972).

[MeZnSPr^i]$_6$ or C$_{24}$H$_{60}$S$_6$Zn$_6$, methylzinc i-propyl sulfide. This compound was crystallized when the solution of i-propylthiol and dimethylzinc in hexane was cooled after evaporation of some hexane (Coates and Ridley 1965). It melts over the range 90°C–105°C with previous shrinking at 75°C and under reduced pressure evolved ZnMe$_2$ at 95°C. The experiment was carried out under Ar atmosphere using standard vacuum-line techniques.

[MeZnSBu^t]$_5$ or C$_{25}$H$_{60}$S$_5$Zn$_5$, pentameric methylzinc t-butyl sulfide. This compound crystallizes from hexane as colorless needles (Coates and Ridley 1965, Adamson et al. 1982). It crystallizes in the monoclinic structure with the lattice parameters $a = 959 \pm 2$, $b = 3904 \pm 6$, $c = 1213 \pm 2$ pm, and $\beta = 117.13° \pm 0.17°$ (Adamson et al. 1982) [$a = 1202$, $b = 3880$, $c = 961$ pm, and $\beta = 117°30'$ (Coates and Ridley 1965)] and calculation and experimental densities 1.38 and 1.39 ± 0.02 g·cm^{-3} (Coates and Ridley 1965, Adamson et al. 1982).

[EtZnSBu^t]$_5$ or C$_{30}$H$_{75}$S$_5$Zn$_5$, ethylzinc t-butyl sulfide. This compound was crystallized from the solution of t-butylthiol and diethylzinc in hexane at –70°C (Coates and Ridley 1965).

[MeZnSPr^i]$_8$ or C$_{32}$H$_{80}$S$_8$Zn$_8$, methylzinc i-propyl sulfide. This compound was prepared by the reaction of ZnMe$_2$ with an equimolar quantity of propan-2-thiol (Adamson and Shearer 1969). It crystallizes in the tetragonal structure with the lattice parameters $a = 1361$ and $c = 1516$ pm. The crystals are air sensitive and deteriorate rapidly on exposure to x-radiation.

Zn(SC$_6$H$_2$Bu$^t_3$-2,4,6)$_2$ or C$_{36}$H$_{58}$S$_2$Zn, bis(2,4,6-tri-t-butylbenzenethiolato)zinc. To obtain this compound, tri-t-butylbenzenethiol (3.59 mmol) and light petroleum (20 mL) are injected into a 100 mL, three-necked flask, equipped with a magnetic stirring bar and connected to the inert gas supply of a vacuum line via a stopcock adaptor (Bochmann et al. 1991, 1997). To the stirred solution is added via syringe Zn[N(SiMe$_3$)$_2$]$_2$ (1.71 mmol). The mixture is heated gently with hair dryer and allowed to stir at room temperature for 1 h. The white precipitate is filtered off, washed with light petroleum (2 × 10 mL), dissolved in warm toluene (*ca.* 10 mL), and recrystallized at room temperature to give fine colorless crystals, which sublimes at 225°C (0.13 Pa) (Bochmann et al. 1991, 1997, Bochmann and Webb 1991).

Crystals of C$_{36}$H$_{58}$S$_2$Zn were grown by sublimation at 200°C (1.3 Pa) (Bochmann et al. 1993). They crystallize in the triclinic structure with the lattice parameters $a = 1033.3 \pm 0.5$, $b = 1122.6 \pm 0.9$, $c = 3226.6 \pm 3.0$ pm, $\alpha = 81.29° \pm 0.02°$, $\beta = 86.12° \pm 0.02°$, and $\gamma = 86.07° \pm 0.02°$ and calculated density 1.118 g·cm^{-3}. The molecule of this compound is dimmer (C$_{72}$H$_{116}$S$_4$Zn$_2$).

Zn(SC$_6$H$_2$Bu$^t_3$-2,4,6)$_2$(C$_4$H$_8$S) or C$_{40}$H$_{66}$S$_3$Zn. To obtain this compound, 1.1 mmol of tetrahydrothiophene (C$_4$H$_8$S) was added to a suspension of Zn(SC$_6$H$_2$Bu$^t_3$-2,4,6)$_2$ (0.16 mmol) in 10 mL of petroleum ether (Bochmann et al. 1994). The starting material dissolved within a few second, followed by the precipitation of the white solid.

The mixture was stirred for 1 h and the precipitate was filtered off and recrystallized from hot toluene with a few drops of tetrahydrothiophene added to give colorless needles of $C_{40}H_{66}S_3Zn$, which sublimes slowly at 150°C (1.3 Pa).

$Zn_2[Ph(SCH_3)C = C(S)Ph]_4$ or $C_{60}H_{52}S_8Zn_2$, zinc bis[1-(methylthio)-cis-stilbene-2-thiolate]. To obtain this compound, $ZnCl_2$ (1.47 mmol) was dissolved in 20 mL of deionized water to form a clear solution, which was added, drop by drop, to 20 mL of an ethanol solution of 3.1 mmol of 1-(methylthio)-cis-stilbene-2-thiol, $Ph(SCH_3)$ $C = C(SH)Ph$ (Zhang et al.1991). 1 N solution of $NaOCH_3$ in CH_3OH was slowly added drop by drop to the resulting cloudy suspension until the appearance of a white precipitate. The supernatant was decanted after 30 min, and the product was collected by filtration. After consecutive washing with 3×10 mL each of ethanol and n-hexane, the white solid was dried under vacuum. It melts at 221°C with decomposition.

According to the data of Zhang et al. (1991), $C_{60}H_{52}S_8Zn_2$ crystallizes in the triclinic structure with the lattice parameters $a = 850.25 \pm 0.14$, $b = 1607.0 \pm 0.5$, and $c = 2168.3 \pm 0.7$ pm and experimental density 1.382 g·cm^{-3}, but for such structure, angles α, β, and γ could not be equal to 90°.

$Zn_{10}S_{16}Ph_{12}$ or $C_{72}H_{60}S_{16}Zn_{10}$. Bulk samples of this compound were prepared by heating $(Me_4N)_2[S_4Zn_{10}(SPh)_{16}]$ in quartz tubes attached to a high vacuum line (Farneth et al. 1992). Samples were heated at 10°C/min to 300°C and held there for 15 min, all in dynamic vacuum.

1.2 ZINC–HYDROGEN–OXYGEN–SULFUR

$Zn(OH)_2ZnSO_4$ quaternary compound is formed in this system. It crystallizes in the orthorhombic structure with the lattice parameters $a = 493.7 \pm 0.5$, $b = 623.1 \pm 0.6$, and $c = 1437.6 \pm 1.5$ pm and calculation and experimental densities 3.925 and 3.930 g·cm^{-3}, respectively (Iitaka von et al. 1962). This compound was obtained at the interaction of ZnO and $ZnSO_47H_2O$ at 300°C during 48–72 h.

1.3 ZINC–LITHIUM–SODIUM–SULFUR

$ZnS–Li_2S–Na_2S$. The phase diagram is not constructed. The $NaLiZnS_2$ quaternary compound is formed in this system, which crystallizes in the trigonal structure with the lattice parameters $a = 397.11 \pm 0.03$ and $c = 671.86 \pm 0.12$ pm and calculated density 2.885 g·cm^{-3} (Deng et al. 2007). $NaLiZnS_2$ has been synthesized by the reaction of Zn and $Li_2S/S/Na_2S$ flux at 500°C.

1.4 ZINC–LITHIUM–POTASSIUM–SULFUR

$ZnS–Li_2S–K_2S$. The phase diagram is not constructed. The $LiKZnS_2$ quaternary compound is formed in this system, which crystallizes in the tetragonal structure with the lattice parameters $a = 660.0 \pm 0.5$ and $c = 1557.0 \pm 0.5$ pm [$a = 399.8 \pm 0.2$ and $c = 1326.2 \pm 0.6$ pm (Schmitz and Bronger 1987)] and calculation and experimental densities 3.43 and 3.18 g·cm^{-3}, respectively (Nicholas et al. 1983). $LiKZnS_2$ was obtained by the interaction of Li_2CO_3, K_2CO_3, and ZnO or ZnS at 850°C in the H_2S flow (Nicholas et al. 1983) or by a reaction of alkali carbonates with Zn in a stream of H_2S at 800°C.

1.5   ZINC-LITHIUM-RUBIDIUM-SULFUR

ZnS–Li$_2$S–Rb$_2$S. The phase diagram is not constructed. The LiRbZnS$_2$ quaternary compound is formed in this system, which crystallizes in the tetragonal structure with the lattice parameters $a = 404.6 \pm 0.2$ and $c = 1353.7 \pm 0.5$ pm (Schmitz and Bronger 1987). LiRbZnS$_2$ was synthesized by a reaction of alkali carbonates with Zn in a stream of H$_2$S at 800°C.

1.6   ZINC-LITHIUM-CESIUM-SULFUR

ZnS–Li$_2$S–Cs$_2$S. The phase diagram is not constructed. The LiCsZnS$_2$ quaternary compound is formed in this system, which crystallizes in the tetragonal structure with the lattice parameters $a = 409.2 \pm 0.4$ and $c = 1397.4 \pm 0.8$ pm (Schmitz and Bronger 1987). LiCsZnS$_2$ was synthesized by a reaction of alkali carbonates with Zn in a stream of H$_2$S at 800°C.

1.7   ZINC-LITHIUM-TIN-SULFUR

ZnS–Li$_2$S–SnS$_2$. The phase diagram is not constructed. The Li$_2$ZnSnS$_4$ quaternary compound is formed in this system, which melts at 919°C and has calculated density 3.173 g·cm^{-3} and energy gap 2.87 eV (Lekse et al. 2008). Its structure ($a = 637.28 \pm 0.13$, $b = 672.86 \pm 0.13$, and $c = 796.21 \pm 0.16$ pm) has been solved from a pseudo-merohedrally twinned crystal. Li$_2$ZnSnS$_4$ was synthesized using Li$_2$S, Zn, Sn, and S mixture, which was heated to 700°C in 12 h and then held at this temperature for 96 h.

1.8   ZINC-SODIUM-COPPER-SULFUR

ZnS–Na$_2$S–Cu. The phase diagram is not constructed. At 1100°C–1150°C, the next reactions take place in this system (Shishkin et al. 1968):

$$ZnS + 2Cu = Cu_2S + Zn$$

$$2Cu + Na_2S = Cu_2S + 2Na$$

1.9   ZINC-SODIUM-OXYGEN-SULFUR

ZnS–Na$_2$S–ZnO. This system was investigated according to 9 polythermal section from the Na$_2$S corner (Maslov et al. 1972). Some of these sections are represented in Figure 1.1. The liquidus surface of the ZnS–Na$_2$S–ZnO quasiternary system includes seven fields of primary crystallization (Figure 1.2). The 1 and 2 regions are divided by the eutectic line (dotted line). This line was not determined experimentally but its existence is caused by the existence of two eutectic points (e_4 in the ZnS–ZnO system and E_1 on the Ee eutectic line) and the existence of low-temperature isotherm deflection in the region of radial section (55.6 mol.% ZnS + 44.4 mol.% ZnO)–Na$_2$S (Figure 1.1c).

Regions 1 and 2 are the projections of primary crystallization fields of solid solutions based on ZnS and ZnO, respectively. The 2ZnS·Na$_2$S ternary compound primarily crystallizes in region 3 (the beginning of crystallization at 880°C; 23.8 mol.% Na$_2$S and 76.2 mol.% ZnS). This compound crystallizes along the p_1E peritectic line that borders with region 1.

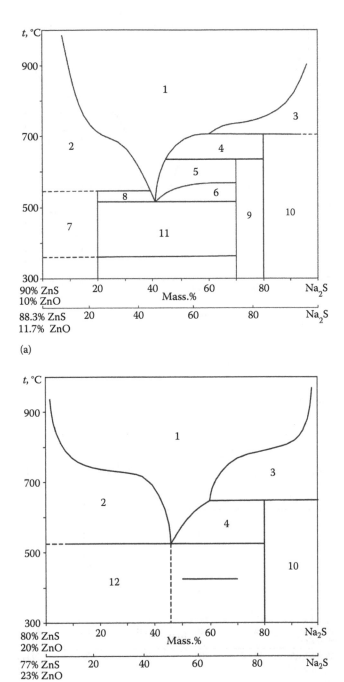

FIGURE 1.1 Polythermal sections of the ZnS–Na$_2$S–ZnO quasiternary system: (a) (88.3 mol.% ZnS + 11.7 mol.% ZnO) – Na$_2$S; (b) (77 mol.% ZnS + 23 mol.% ZnO) – Na$_2$S.   (*Continued*)

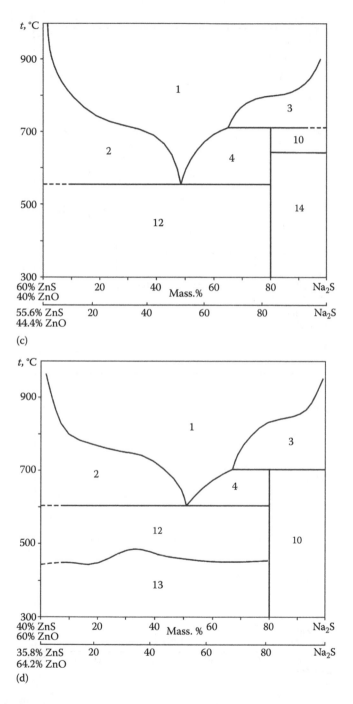

FIGURE 1.1 (*Continued*) Polythermal sections of the ZnS–Na$_2$S–ZnO quasiternary system: (c) (55.6 mol.% ZnS + 44.4 mol.% ZnO) – Na$_2$S; (d) (35.8 mol.% ZnS + 64.2 mol.% ZnO) – Na$_2$S.

(*Continued*)

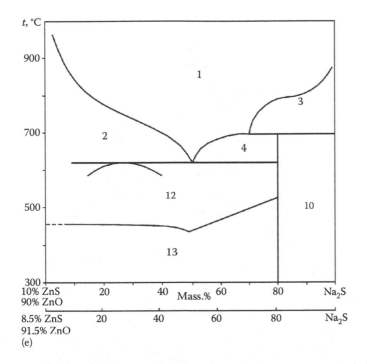

FIGURE 1.1 (Continued) Polythermal sections of the ZnS–Na₂S–ZnO quasiternary system: (e) (8.5 mol.% ZnS + 91.5 mol.% ZnO) – Na₂S; 1, L; 2, L + ZnS + ZnO; 3, L + Na₂S; 4, L + γ-ZnS · 4Na₂S; 5, L + γ-ZnS · 3Na₂S; 6, L + β-ZnS · 3Na₂S + α-ZnS · 3Na₂S; 7, ZnS + ZnO + β-4ZnS·Na₂S; 8, L + β-4ZnS·Na₂S; 9, γ-ZnS · 3Na₂S + β-ZnS · 4Na₂S; 10, Na₂S + γ-ZnS · 4Na₂S; 11, α-4ZnS·Na₂S + α-ZnS · 3Na₂S; 12, ZnO + ZnS + β-ZnS · 4Na₂S; 13, ZnO + ZnS + α-ZnS · 4Na₂S; 14, Na₂S + β-ZnS · 4Na₂S. (From Maslov, V.I. et al., *Deposited in VINITI*, № 4652-72Dep. 1972.)

The $3Na_2S \cdot ZnS$ ternary compound crystallizes in region 4 (the beginning of crystallization at 670°C; 74.4 mol.% Na_2S and 25.6 mol.% ZnS) along the p_2E peritectic line. Its crystallization ends on the Ee eutectic line.

Region 5 characterizes the projection of $ZnS \cdot 4Na_2S$ primary crystallization. The temperatures of the ingot primary crystallizations, which are located on the p_3p_4 peritectic line, are within the interval from 600°C (Figure 1.1a) to 810°C in the $ZnS–Na_2S$ and 715°C in the $ZnO–Na_2S$ systems. In addition, the temperatures of the ingot primary crystallizations from the (88.3 mol.% ZnS + 11.7 mol.% ZnO) – Na_2S to (8.5 mol.% ZnS + 91.5 mol.% ZnO) – Na_2S section sometimes increase and sometimes decrease in a narrow interval (20°C–30°C) but on average increase in the direction to $ZnO–Na_2S$ system and are equal to 640°C–665°C–685°C. The decrease of the crystallization temperature is explained by the existence of Na_2O in the system.

Region 6 is the projection of eutectic mixture crystallization. The eutectic crystallization begins at 660°C and ends in the ternary peritectic point at 580°C.

The E point is a quaternary eutectic, which crystallizes on the bases of the $2ZnS \cdot Na_2S$, $ZnS \cdot 3Na_2S$, and $ZnS \cdot 4Na_2S$ ternary compounds and solid solutions based on ZnS, and has a minimum melting temperature (470°C).

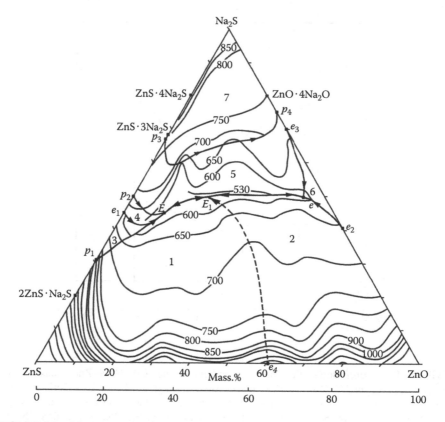

FIGURE 1.2 Liquidus surface of the ZnS–Na$_2$S–ZnO quasiternary system. (From Maslov, V.I. et al., *Deposited in VINITI*, № 4652-72Dep. 1972.)

Region 7 is the projection of Na$_2$S primary crystallization from the liquid phase at 700°C–1180°C (Maslov et al. 1972).

1.10 ZINC–POTASSIUM–COPPER–SULFUR

ZnS–K$_2$S–Cu$_2$S. The phase diagram is not constructed. The KCuZnS$_2$ quaternary compound is formed in this system. It melts at 915°C and crystallizes in the tetragonal structure with the lattice parameters a = 392.2 and c = 1326.9 pm [a = 392.2 ± 0.3 and c = 1308.5 ± 0.3 pm (Mouallem-Bahout et al. 2001)] and calculated density 3.78 g·cm^{-3} (Savel'eva and Gromilov 1996).

KCuZnS$_2$ was obtained by sulfidation of Cu$_2$O + ZnO + KHCO$_3$ mixture in the necessary stoichiometry (Cu/Zn = 1:1) with 50% excess of KHCO$_3$ at 900°C ± 50°C in the CS$_2$ flow (Savel'eva and Gromilov 1996, Mouallem-Bahout et al. 2001).

1.11 ZINC–POTASSIUM–GALLIUM–SULFUR

ZnS–K$_2$S–Ga$_2$S$_3$. The phase diagram is not constructed. The KZn$_4$Ga$_5$S$_{12}$ quaternary compound is formed in this system, which crystallizes in the hexagonal structure with the lattice parameters a = 1339.17 ± 0.15 and c = 910.01 ± 0.17 pm

(a = 830.62 ± 0.07 and α = 107.453° ± 0.005° in rhombohedral settings) and calculated density 3.644 g·cm^{-3} (Schwer et al. 1993). In order to produce $KZn_4Ga_5S_{12}$, $ZnGa_2S_4$ was solved in KBr in molar ratio 1:9 in quartz-glass ampoule, which was sealed under vacuum and heated to 900°C. After 1 day, the furnace was switched off and the sample cooled down slowly. The reguli were treated with hot water to remove the unreacted halide.

1.12   ZINC–POTASSIUM–GERMANIUM–SULFUR

ZnS–K$_2$S–GeS$_2$. The phase diagram is not constructed. The $K_{10}Zn_4Ge_4S_{17}$ quaternary compound is formed in this system. It melts congruently at 656°C and crystallizes in the cubic structure with the lattice parameter a = 981.7 ± 0.4 pm, calculated density 2.612 g·cm^{-3}, and energy gap 3.16 eV (Palchik et al. 2004). $K_{10}Zn_4Ge_4S_{17}$ was synthesized from a mixture of Ge (0.5 mmol), Zn (0.5 mmol), K$_2$S (2.5 mmol), and S (6 mmol). The reagents were mixed, sealed in an evacuated silica tube, and heated at 500°C for 4 days. This was followed by cooling to room temperature at 5°C/h. The excess flux was removed with methanol to reveal yellow-brown transparent rectangular crystals that are air sensitive.

The glass-forming region in this ternary system is shown in Figure 1.3 (Imaoka and Jamadzaki 1967).

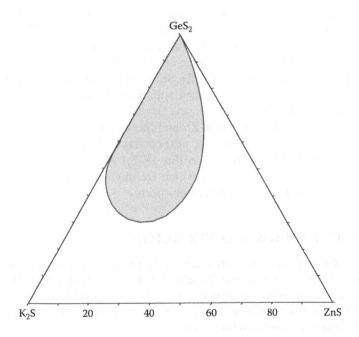

FIGURE 1.3   The glass-forming region in the ZnS–K$_2$S–GeS$_2$ quasiternary system. (From Imaoka, M. and Jamadzaki, T. [in Japanese], *Monthly J. Inst. Industr. Sci. Univ. Tokyo*, 19(9), 261, 1967.)

1.13 ZINC–POTASSIUM–TIN–SULFUR

ZnS–K$_2$S–SnS$_2$. The phase diagram is not constructed. K$_2$ZnSn$_3$S$_8$, K$_6$Zn$_4$Sn$_5$S$_{17}$, and K$_{10}$Zn$_4$Sn$_4$S$_{17}$ quaternary compounds are formed in this system. The first of them crystallizes in two polymorph modifications. α-K$_2$ZnSn$_3$S$_8$ crystallizes in the triclinic structure with the lattice parameters a = 736.56 ± 0.15, b = 799.96 ± 0.16, c = 1531.9 ± 0.3, α = 95.04° ± 0.03°, β = 91.64° ± 0.03°, and γ = 115.76° ± 0.03°; calculated density 3.11 g·cm^{-3}; and energy gap 2.30 eV. β-K$_2$ZnSn$_3$S$_8$ crystallizes in the cubic structure with the lattice parameter a = 1312.6 ± 0.2, calculated density 3.331 g·cm^{-3}, and energy gap 2.55 eV. K$_2$ZnSn$_3$S$_8$ melts at 445°C and its glassy form is characterized by the energy gap of 2.15 eV (Fard and Kanatzidis 2012).

K$_6$Zn$_4$Sn$_5$S$_{17}$ crystallizes in the tetragonal structure with the lattice parameters a = 1374.25 ± 0.07 and c = 972.72 ± 0.05 pm (Manos et al. 2005) [a = 1379.6 ± 0.2 and c = 958.0 ± 0.2 pm (Kanatzidis et al. 1997a, b)], energy gap 2.94 eV [≈ 2.87 eV (Manos et al. 2005)], and calculated density 2.977 g·cm^{-3} (Kanatzidis et al. 1997a,b).

K$_{10}$Zn$_4$Sn$_4$S$_{17}$ melts congruently at 820°C and crystallizes in the cubic structure with the lattice parameter a = 1992.3 ± 0.3 pm, calculated density 2.809 g·cm^{-3}, and energy gap 3.1–3.2 eV (Palchik et al. 2003, 2004).

α-K$_2$ZnSn$_3$S$_8$ is formed through the slow cooling (~1°C/min) of a melt K$_2$CO$_3$ + Zn + Sn + S from 700°C to room temperature (Fard and Kanatzidis 2012). However, water quenching of the same melt led to amorphous glassy K$_2$ZnSn$_3$S$_8$, which upon annealing below its melting point crystallizes to metastable single-phase β-K$_2$ZnSn$_3$S$_8$. Glassy K$_2$ZnSn$_3$S$_8$ shows a glass transition at ~220°C, followed by a sharp, single-step crystallization at 398°C and melting at 468°C.

A mixture of Sn, Zn, K$_2$S, and S was sealed under vacuum in a silica tube, heated (≈40°C/h) to 400°C for 92–96 h, and then cooled to 25°C at a rate of 4°C/h–6°C/h to obtain yellowish-white crystals of K$_6$Zn$_4$Sn$_5$S$_{17}$ (Kanatzidis et al. 1997a,b, Manos et al. 2005). The product must be washed with degassed dimethyl formamide and dried with acetone and ether.

K$_{10}$Zn$_4$Sn$_4$S$_{17}$ forms by reaction of Zn and Sn with K$_2$S flux in Zn:Sn:K$_2$S:S = 1:1:3:12 ratio at 500°C for 60 h with the next cooling to room temperature at 5°C/h (Palchik et al. 2003, 2004). The excess flux (K$_2$S$_x$) was removed with MeOH to reveal yellowish-white cubic crystals, which are moderately air stable (for a few days) and need to be stored under inert atmosphere.

1.14 ZINC–RUBIDIUM–COPPER–SULFUR

ZnS–Rb$_2$S–Cu$_2$S. The phase diagram is not constructed. The RbCuZnS$_2$ quaternary compound is formed in this system. It melts at 970°C and crystallizes in the tetragonal structure with the lattice parameters a = 395.8 and c = 1359.5 pm [a = 396.0 ± 0.2 and c = 1310.5 ± 0.4 pm (Mouallem-Bahout et al. 2001)] and calculated density 4.34 g·cm^{-3} (Savel'eva and Gromilov 1996).

RbCuZnS$_2$ was obtained by sulfidation of Cu$_2$O + ZnO + Rb$_2$CO$_3$ mixture in the necessary stoichiometry (Cu/Zn = 1:1) with 50% excess of Rb$_2$CO$_3$ at 900°C ± 50°C in the CS$_2$ flow (Savel'eva and Gromilov 1996).

1.15   ZINC–RUBIDIUM–GALLIUM–SULFUR

ZnS–Rb$_2$S–Ga$_2$S$_3$. The phase diagram is not constructed. The RbZn$_4$Ga$_5$S$_{12}$ quaternary compound is formed in this system, which crystallizes in the hexagonal structure with the lattice parameters a = 1342.10 ± 0.32 and c = 911.50 ± 0.46 pm (a = 832.08 ± 0.11 and α = 107.433° ± 0.009° in rhombohedral settings) and calculated density 3.785 g·cm^{-3} (Schwer et al. 1993). This compound was obtained if ZnGa$_2$S$_4$ was kept in RbBr at a temperature above 150°C of its melting point for 1 day and then cooled down slowly.

1.16   ZINC–RUBIDIUM–TIN–SULFUR

ZnS–Rb$_2$S–SnS$_2$. The phase diagram is not constructed. The Rb$_2$ZnSn$_2$S$_6$ quaternary compound is formed in this system. It crystallizes in the monoclinic structure with the lattice parameters a = 687.3 ± 0.2, b = 1345.6 ± 0.4, c = 728.5 ± 0.2 pm, and β = 113.12° ± 0.02° at 183 K; energy gap 3.00 eV; and calculated density 3.569 g·cm^{-3} (Kanatzidis et al. 1997a,b).

The reaction of Sn, Zn, Rb$_2$S, and S at 400°C for 4 days upon a cooling rate of 4°C/h afforded brownish-orange crystals of Rb$_2$ZnSn$_2$S$_6$ (Kanatzidis et al. 1997a,b). The product must be washed with degassed dimethyl formamide and dried with acetone and ether.

1.17   ZINC–CESIUM–COPPER–SULFUR

ZnS–Cs$_2$S–Cu$_2$S. The phase diagram is not constructed. The CsCuZnS$_2$ quaternary compound is formed in this system. It melts at 985°C and crystallizes in the tetragonal structure with the lattice parameters a = 401.2 and c = 1408.2 pm and calculated density 4.78 g·cm^{-3} (Savel'eva and Gromilov 1996). CsCuZnS$_2$ was obtained by sulfidation of Cu$_2$O + ZnO + Cs$_2$CO$_3$ mixture in the necessary stoichiometry (Cu/Zn = 1:1) with 50% excess of Cs$_2$CO$_3$ at 900°C ± 50°C in the CS$_2$ flow.

1.18   ZINC–CESIUM–YTTERBIUM–SULFUR

ZnS–Cs$_2$S–Yb$_2$S$_3$. The phase diagram is not constructed. The CsZnYbS$_3$ quaternary compound is formed in this system. It crystallizes in the orthorhombic structure with the lattice parameters a = 395.43 ± 0.05, b = 1527.6 ± 0.2, and c = 1044.0 ± 0.1 pm at 153 ± 2 K; energy gap 2.61 eV for (010) face; and calculated density 4.924 g·cm^{-3} (Mitchell et al. 2004). CsZnYbS$_3$ was synthesized using the mixture of Cs$_2$S$_3$, Zn, S, and CsI. The sample was heated to 900°C in 48 h, kept at this temperature for 96 h, and cooled to room temperature in 148 h.

1.19   ZINC–CESIUM–GERMANIUM–SULFUR

ZnS–Cs$_2$S–GeS$_2$. The phase diagram is not constructed. The Cs$_2$ZnGe$_3$S$_8$ quaternary compound is formed in this system. It crystallizes in the monoclinic structure with the lattice parameters a = 726.24 ± 0.04, b = 1698.01 ± 0.09, c = 1262.31 ± 0.08 pm, and $\beta = 97.706° \pm 0.005°$; calculated density 3.468 g·cm^{-3}; and energy gap 3.32 eV (Morris et al. 2013). To obtain $Cs_2ZnGe_3S_8$, a mixture of Cs_2S, Zn, Ge, and S in the appropriate ratios was added to a fused silica tube inside a nitrogen-filled glove box. The tube was then evacuated, flame-sealed, and heated gently at first in a weak methane–oxygen flame to start the reaction. It was then heated more strongly until a glowing liquid was obtained for a total of ~2 min. The tube was allowed to cool in air at a rate of ~200°C/min.

1.20   ZINC–COPPER–ALUMINUM–SULFUR

ZnS–CuAlS$_2$. The phase diagram is not constructed. Solid solutions with sphalerite structure are formed within the interval of 68–100 mol.% ZnS at the annealing temperature of 1000°C (Donohue and Bierstadt 1974). The lattice parameters of solid solutions increase up to 1 mol.% CuAlS$_2$ and then decrease. At 33 mol.% CuAlS$_2$, a new phase is formed in this system that crystallizes in the tetragonal structure with the lattice parameters $a = 533.2$ and $c = 1042.9$ pm.

According to the data of Robbins and Miksovsky (1972), the lattice parameters change linearly within the homogeneity region. The chalcopyrite–sphalerite transformation takes place at approximately 65 mol.% ZnS. Quaternary phases were not found in this system.

It was indicated (Gallardo 1992) that it is possible to predict the value at which the chalcopyrite phase in this system disappears (x_c) and $T_c(x)$ behavior for the $Zn_{2x}(CuAl)_{1-x}S_2$ alloys for which a chalcopyrite–sphalerite-like disordering transition with extensive terminal solid solution formation exists.

1.21   ZINC–COPPER–GALLIUM–SULFUR

ZnS–CuGaS$_2$. The phase diagram is not constructed. The CuGaS$_2$ ternary compound crystallizes in the chalcopyrite-type structure ($a = 534$ and $c = 1047$ pm), which is closed to the sphalerite-type structure (Hahn et al. 1955). The structure similarity facilitates the solid solution formation. The sphalerite structure conserves in this system up to 40 mol.% CuGaS$_2$ and the lattice parameter linearly decreases from 540.4 pm for ZnS to 534.0 pm for the solid solution, containing 40 mol.% CuGaS$_2$ (Apple 1958). The chalcopyrite–sphalerite transformation takes place at approximately 33 mol.% ZnS for the ingots annealing at 800°C. The solubility of CuGaS$_2$ in ZnS is equal to 33.3 mol.% at 975°C. Quaternary phases were not found in this system. It was indicated (Gallardo 1992) that it is possible to predict the value at which the chalcopyrite phase in this system disappears (x_c) and $T_c(x)$ behavior for the $Zn_{2x}(CuGa)_{1-x}S_2$ alloys for which a chalcopyrite–sphalerite-like disordering transition with extensive terminal solid solution formation exists.

The addition of small amount of CuGaS$_2$ increases the hardness and thermal stability of ZnS (Do et al. 1992). Solid solutions in which 5.3 and 10.3 mol.% of ZnS are replaced by CuGaS$_2$ begin to decompose at temperatures higher than both end members.

The ingots of this system were obtained from ZnS and CuGaS$_2$ in the H$_2$S atmosphere at 900°C for 1 h (Apple 1958). Single crystals (ZnS)$_{1-x}$(CuGaS$_2$)$_x$ at $x = 0.053$

and 0.103 were grown by chemical vapor transport using iodine as the transport agent (Do et al. 1992). The charge zone was kept at 850°C and the growth zone at 810°C. The transport process was carried out for 1 week.

1.22 ZINC–COPPER–INDIUM–SULFUR

ZnS–CuInS$_2$. Continuous solid solutions are formed between both isostructural modifications (hexagonal and cubic) of ZnS and CuInS$_2$ (Figure 1.4) (Parasyuk et al. 2003). The solid solution range of the tetragonal CuInS$_2$ modifications extends from 0 to 29 mol.% ZnS, while that of the cubic modification of ZnS from 41 to 100 mol.% ZnS. The obtained results confirmed the data of Robbins and Miksovsky (1972) and Sombuthawee et al. (1978) where it was found that ZnS and CuInS$_2$ form a solid solution series with a broad miscibility gap at room temperature. The same results have been obtained in Schorr et al. (2005) and Schorr and Wagner (2005). Wagner et al. (2005) obtained more accurate information about the two-phase field in this system, where cubic and tetragonal mixed crystals coexist. It was shown that this field covers at room temperature the compositional range $0.10 \leq x \leq 0.41$ for the alloys $2(ZnS)_x(CuInS_2)_{1-x}$, and these data are included in Figure 1.4. It was indicated (Gallardo 1992) that it is possible to predict the value at which the chalcopyrite phase in this system disappears (x_c) and $T_c(x)$ behavior for the $Zn_{2x}(CuIn)_{1-x}S_2$ alloys for which a chalcopyrite–sphalerite-like disordering transition with extensive terminal solid solution formation exists.

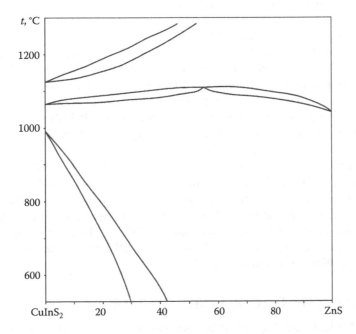

FIGURE 1.4 Phase diagram of the ZnS–CuInS$_2$ system. (From Parasyuk, O.V. et al., *J. Alloys Compd.*, 348(1–2), 57, 2003; Wagner, G. et al., *J. Cryst. Growth*, 283(3–4), 356, 2005.)

In the single-phase regions, the lattice parameters follow Vegard's rule (Schorr et al. 2005, Schorr and Wagner 2005). The concentration dependence of the wurtzite phase lattice parameters has small deviation from linearity (Sombuthawee et al. 1978).

According to the data of Parthé et al. (1969), the $CuZn_2InS_4$ quaternary compound is formed in this system. It crystallizes in the cubic structure of sphalerite type with the lattice parameter $a = 547.5 \pm 0.5$ pm, but this compound can be one of the solid solution compositions.

Powder sample $Zn_{2x}(CuIn)_{1-x}S_2$ alloys were prepared by solid-state reaction of the elements at 950°C and annealed with cooling rates between 2°C/h and 42°C/h in the entire composition range (Schorr et al. 2005, Schorr and Wagner 2005). Powder samples with $0.4 < x < 1.0$ were also prepared by the directional freezing method (Schorr et al. 2005). The alloys were homogenized at 900°C for 240 h or at 860°C for 120 h (Wagner et al. 2005) [they were annealed for 500 h at 600°C, and then they were quenched in cold water (Parasyuk et al. 2003); at 310°C–1330°C from 15 min to 1 month (Sombuthawee et al. 1978)]. The $CuZn_2InS_4$ compound was obtained by the heating of the mixture from chemical elements at 600°C–800°C (Parthé et al. 1969).

This system was investigated through differential thermal analysis (DTA), XRD, neutron powder diffraction, diffraction using synchrotron radiation, and HRTEM with EDX (Parasyuk et al. 2003, Schorr et al. 2005, Schorr and Wagner 2005, Wagner et al. 2005).

1.23   ZINC–COPPER–SILICON–SULFUR

$ZnS–Cu_2S–SiS_2$. The phase diagram is not constructed. The Cu_2ZnSiS_4 quaternary compound is formed in this system. It melts incongruently at 1123°C ± 5°C (Schäfer and Nitsche 1977) [according to the data of Yao et al. (1987), it decomposes at 620°C ± 10°C] and crystallizes in the orthorhombic structure with the lattice parameters $a = 743.74 \pm 0.01$, $b = 640.01 \pm 0.01$, and $c = 613.94 \pm 0.01$ pm (Rosmus and Aitken 2011) [$a = 744$, $b = 639$, and $c = 613$ pm (Schleich and Wold 1977); $a = 740$, $b = 640$, and $c = 608$ pm (Nitsche et al. 1967); $a = 743.5$, $b = 639.6$, and $c = 613.5$ pm (Schäfer and Nitsche 1974, 1977); $a = 743.6 \pm 0.1$, $b = 639.8 \pm 0.1$, and $c = 613.7 \pm 0.2$ pm (Yao et al. 1987)]; calculation and experimental densities 3.97 [3.964 (Rosmus and Aitken 2011)] and 3.94 g·cm^{-3}, respectively; and energy gap $E_g = 3.25$ eV [$E_g = 3.04 \pm 0.2$ eV (Yao et al. 1987)] (Schleich and Wold 1977).

To obtain Cu_2ZnSiS_4, stoichiometric ratios of the elements were heated under a vacuum to 1000°C in 12 h, held at this temperature for 168 h, and then cooled at 7.5°C/h to room temperature (Rosmus and Aitken 2011). Its single crystals have been prepared via high-temperature solid-state synthesis (Rosmus and Aitken 2011) or by the horizontal gradient freezing (Matsuhita et al. 2005) or by chemical vapor transport using I_2 as the transport agent (Nitsche et al. 1967, Schäfer and Nitsche 1974, Schleich and Wold 1977, Yao et al. 1987).

1.24   ZINC–COPPER–GERMANIUM–SULFUR

$ZnS–Cu_2GeS_3$. Phase diagram of this system is shown in Figure 1.5 (Parasyuk et al. 2005). The peritectic point is at 43 mol.% ZnS. The thermal effects at 810°C that could correspond to the phase transition of Cu_2ZnGeS_4 from tetragonal to

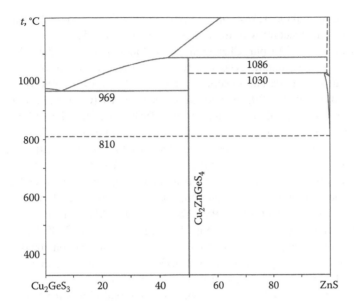

FIGURE 1.5 Phase diagram of the ZnS–Cu₂GeS₃ system. (From Parasyuk, O.V. et al., *J. Alloys Compd.*, 397(1–2), 85, 2005.)

orthorhombic structure were not observed. The eutectic point coordinates are 969°C and 6 mol.% ZnS. The homogeneity range of the Cu_2ZnGeS_4 is narrow and the solid solubility in the components is less than 2 mol.%.

There is one intermediate quaternary phase Cu_2ZnGeS_4 in this system that melts incongruently at 1086°C [1120°C (Matsuhita et al. 2005); 1107°C ± 5°C (Schäfer and Nitsche 1977), decomposes at 620°C ± 10°C (Yao et al. 1987)], and crystallizes in the tetragonal structure with the lattice parameters $a = 534.127 \pm 0.009$ and $c = 1050.90 \pm 0.02$ pm [$a = 534.1$ and $c = 1049.2$ pm (Schäfer and Nitsche 1977); $a = 542.9$ and $c = 1084.7$ pm (Das et al. 2013)] and calculated density 4.3561 ± 0.0002 g·cm⁻³ (Parasyuk et al. 2005).

This compound can also crystallize in the orthorhombic structure with the lattice parameters $a = 750.6 \pm 0.3$, $b = 647.6 \pm 0.4$, and $c = 618.9 \pm 0.2$ pm (Yao et al. 1987) [$a = 750$, $b = 648$, and $c = 618$ pm (Schleich and Wold 1977); $a = 747$, $b = 645$, and $c = 612$ pm (Nitsche et al. 1967); $a = 750.4$, $b = 647.4$, and $c = 618.5$ (Schäfer and Nitsche 1974, 1977, Guen et al. 1979, Guen and Glaunsinger 1980)]; calculation and experimental densities 4.35 and 4.37 g·cm⁻³ (Schleich and Wold 1977) [4.34 g·cm⁻³ (Guen et al. 1979, Guen and Glaunsinger 1980)], respectively; and energy gap $E_g = 2.2$ eV (Tsuji et al. 2010) [$E_g = 2.04 \pm 0.2$ eV (Yao et al. 1987); $E_g = 2.1$ eV (Schleich and Wold 1977)].

The structures of two polymorphs of Cu_2ZnGeS_4 have been determined by Moodie and Whitefield (1986). The first of them is a tetragonal polymorph with the lattice parameters $a = 527$ and $c = 1054$ pm. The second polymorph has a pseudo-rhombohedral structure and can be described in terms of triply primitive cell with orthogonal axes $a = 3660$, $b = 655$, and $c = 752$ pm.

Because the energy between kesterite- and stannite-type structures is small, both structures can coexist in synthesized samples of Cu_2ZnGeS_4. According to the first-principles calculations, kesterite-type structure of this compound is characterized by the tetragonal structure with the lattice parameters $a = 526.4$ and $c = 1084.3$ pm [$a = 535.8$ and $c = 1064.1$ pm (Chen et al. 2010b)] and energy gap 2.43 eV (Chen and Ravindra 2013), and stannite-type structure is characterized also by the tetragonal structure with the lattice parameters $a = 532.8$ and $c = 1074.1$ pm [$a = 533.3$ and $c = 1074.1$ pm (Chen et al. 2010b)] and energy gap 2.14 eV (Chen and Ravindra 2013). According to the calculation of Chen et al. (2010b), wurtzite-derived polytypes of kesterite (orthorhombic structure with the lattice parameters $a = 754.4$, $b = 651.9$, and $c = 622.6$ pm) and stannite (orthorhombic structure with the lattice parameters $a = 750.3$, $b = 654.7$, and $c = 622.6$ pm) could exist for Cu_2ZnGeS_4 compound.

Cu_2ZnGeS_4 was obtained by the chemical vapor reactions using I_2 as the transport agent (Nitsche et al. 1967, Schäfer and Nitsche 1974, Schleich and Wold 1977, Guen et al. 1979, Guen and Glaunsinger 1980, Yao et al. 1987, Matsuhita et al. 2005) or via vertical or horizontal gradient freezing method (Matsuhita et al. 2005, Das et al. 2013). The powder of this compound was synthesized by solid-state reaction: the starting materials ZnS, Cu_2S, and GeS_2 were mixed with 15% excess amount of ZnS and GeS_2 (Tsuji et al. 2010). The mixture was sealed in quartz ampoule tube in vacuo and heat-treated at 550°C–650°C for 10 h. Cu_2ZnGeS_4 synthesized at 650°C was the low-temperature phase of stannite type. A wurtzite-type high-temperature phase was not confirmed.

The alloys were annealed at 400°C during 500 h and this system was investigated through DTA, XRD, and SEM (Parasyuk et al. 2005).

ZnS–Cu_2S–GeS_2. The isothermal section of this system at 400°C is shown in Figure 1.6 (Parasyuk et al. 2005). The alloys were annealed at 400°C during 500 h. This system was investigated through DTA, XRD, and SEM.

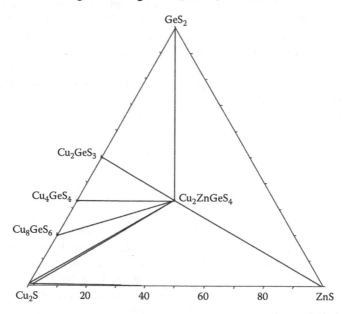

FIGURE 1.6 Isothermal section of the ZnS–Cu_2S–GeS_2 system at 400°C. (From Parasyuk, O.V. et al., *J. Alloys Compd.*, 397(1–2), 85, 2005.)

1.25 ZINC–COPPER–TIN–SULFUR

ZnS–Cu$_2$SnS$_3$. This system is a quasibinary section of the quasiternary system ZnS–Cu$_2$S–SnS$_2$ (Figure 1.7) (Olekseyuk et al. 2000). The eutectic, which melts at 850°C and contains 5 mol.% ZnS, forms between Cu$_2$SnS$_3$ and Cu$_2$ZnSnS$_4$ compounds. Solid solubility based on this section components at 400°C is insignificant (less than 1 mol.% based on Cu$_2$SnS$_3$ and less than 2 mol.% based on ZnS). Solid solution based on sphalerite modification of ZnS tends to increase with increasing temperature. The highest solubility (10 mol.% Cu$_2$SnS$_3$) takes place at the peritectic temperature. Its transition to wurtzite modification is accompanied with the peritectic process at 1080°C.

According to the data of Moh (1975), the solubility of ZnS in Cu$_2$ZnSnS$_4$ reaches a maximum of ≈10.4 mol.% (≈2.5 mass.%) at ≈825°C. Maximum solubility of Cu$_2$ZnSnS$_4$ in ZnS takes place at 972°C ± 4°C and reaches ≈12 mol.% (≈38 mass.%). At higher temperatures, the solubility decreases to ≈3.1 mol.% (≈12.5 mass.%) at 1050°C where sphalerite inverts to wurtzite, which takes only ≈1.4 mol.% (≈6 mass.%) Cu$_2$ZnSnS$_4$ in the solid solution.

The Cu$_2$ZnSnS$_4$ quaternary compound is formed in this system. It melts incongruently at 980°C [990°C (Matsuhita et al. 2005); 982 ± 5°C (Schäfer and Nitsche 1977)], possesses narrow homogeneity region (Olekseyuk et al. 2000), and crystallizes in the tetragonal structure with the lattice parameters $a = 541.98 ± 0.03$ and $c = 1089.2 ± 0.5$ pm (Olekseyuk et al. 2000) [$a = 542.7$ and $c = 1085.4$ pm for kesterite-type crystal structure (Kheraj et al. 2013); $a = 545.5$ and $c = 1088$ pm (Nagaoka et al. 2011, 2012); $a = 542.1 ± 0.1$ and $c = 1081.9 ± 0.2$ pm and $a = 542.90 ± 0.02$ and $c = 1083.40 ± 0.5$ pm for two different methods of Cu$_2$ZnSnS$_4$ single crystals

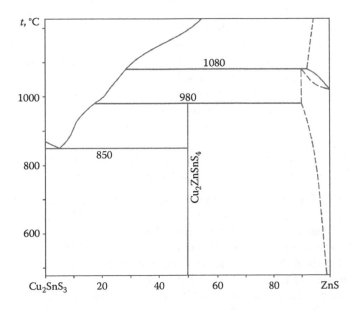

FIGURE 1.7   Phase diagram of the ZnS–Cu$_2$SnS$_3$ system. (From Olekseyuk, I. et al., *Visnyk L'viv. un-tu. Ser. khim.* (39), 48, 2000.)

obtaining (Colombara et al. 2013); $a = 543$ and $c = 1081$ pm (Matsuhita et al. 2005); $a = 543.5$ and $c = 1084.3$ pm (Guen et al. 1979, Guen and Glaunsinger 1980); $a = 543$ and $c = 1083$ pm (Nitsche et al. 1967); $a = 542.7$ and $c = 1084.8$ pm (Schäfer and Nitsche 1974, 1977); $a = 543.6$ and $c = 1085$ pm (Hahn and Schulze 1965)]; calculation and experimental densities 4.56 and 4.60 g·cm^{-3} (Guen et al. 1979) [4.57 and 4.61 g·cm^{-3} (Guen and Glaunsinger 1980); 4.55 and 4.49 g·cm^{-3} (Hahn and Schulze 1965)], respectively; and energy gap 1.45 eV (Kheraj et al. 2013) [1.4 eV (Tsuji et al. 2010); 1.50 eV (Chen et al. 2009, Persson 2010); 1.39 eV (Matsuhita et al. 2005)].

Using the first-principles density functional method, it was shown that the low-energy crystal structure of Cu_2ZnSnS_4 is a kesterite-type structure (Chen et al. 2009). However, the stannite or partially disordered kesterite structure can also exist in synthesized samples due to the small energy cost. According to the first-principles calculations, kesterite-type structure of this compound is characterized by the tetragonal structure with the lattice parameters $a = 539.3$ and $c = 1079.7$ pm [$a = 532.6$ and $c = 1066.3$ pm (Persson 2010)] and calculated density 4.6484 g·cm^{-3}, and stannite-type structure is characterized also by the tetragonal structure with the lattice parameters $a = 539.5$ and $c = 1078.4$ pm [$a = 532.5$ and $c = 1062.9$ pm (Persson 2010)] and calculated density 4.6494 g·cm^{-3} (Bensalem et al. 2014).

The existence of the quaternary compound $Cu_4ZnSn_2S_7$ was not confirmed (Olekseyuk et al. 2000).

This compound was obtained by the chemical vapor reactions (Nitsche et al. 1967, Schäfer and Nitsche 1974, Guen et al. 1979, Guen and Glaunsinger 1980), or by sintering of binary sulfide or constituent elements mixtures (Hahn and Schulze 1965, Kheraj et al. 2013), or by the horizontal gradient freezing (Matsuhita et al. 2005). Cu_2ZnSnS_4 could be also obtained if an aqueous solution (150 mL) of $Zn(NO_3)_2 \cdot 6H_2O$ with 10%–15% excess amounts (about 0.029 mol/L) and $SnCl_4 \cdot 5H_2O$ (0.025 mol/L) was purged with N_2 with the next addition of CuCl into the mixture solution (Tsuji et al. 2010). Cu_2ZnSnS_4 is precipitated with bubbling with a H_2S gas into the mixed solution containing $Zn(NO_3)_2$, $SnCl_4$, and CuCl.

Single crystals of Cu_2ZnSnS_4 have been produced within sealed quartz ampoules via the chemical vapor transport technique using I_2 as the transport agent (Colombara et al. 2013) or by the traveling heater method, where the Sn solvent was used (Nagaoka et al. 2011, 2012, 2013).

The alloys were annealed at 400°C during 250 h and this system was investigated through DTA, metallography, and XRD (Olekseyuk et al. 2000).

ZnS–Cu$_2$Sn$_4$S$_9$. The phase relations in this system are shown in Figure 1.8 (Olekseyuk et al. 2004). The liquidus consists of lines that correspond to the fields of primary crystallization of the solid solutions based on α- and β-ZnS, SnS$_2$, and Cu_2ZnSnS_4. The secondary crystallization L + Cu_2SnS_3 + (SnS$_2$) and L + (SnS$_2$) + (Cu$_2$ZnSnS$_4$) ends at 762°C in a ternary eutectic reaction. The secondary crystallization of the binary peritectic L + (β-ZnS) + (Cu$_2$ZnSnS$_4$) ends at the temperature of the ternary peritectic reaction (790°C). In the subsolidus region of the diagram, the change of the phase equilibria occurs, which is caused by the formation of

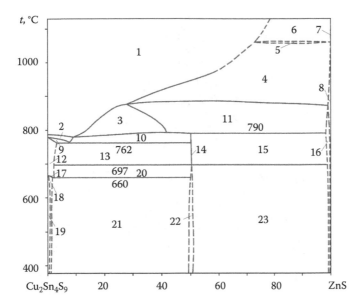

FIGURE 1.8 Phase diagram of ZnS–Cu₂Sn₄S₉ system: 1, L; 2, L + Cu₂SnS₃; 3, L + Cu₂ZnSnS₄; 4, L + α-ZnS; 5, L + β-ZnS + α-ZnS; 6, L + β-ZnS; 7, β-ZnS; 8, α-ZnS; 9, L + Cu₂SnS₃ + SnS₂; 10, L + SnS₂ + Cu₂ZnSnS₄; 11, L + α-ZnS + Cu₂ZnSnS₄; 12, Cu₂SnS₃ + SnS₂; 13, Cu₂ZnSnS₄ + Cu₂SnS₃ + SnS₂; 14, Cu₂ZnSnS₄ + SnS₂; 15, α-ZnS + Cu₂ZnSnS₄ + SnS₂; 16, α-ZnS + Cu₂ZnSnS₄; 17, Cu₂SnS₃ + Cu₂Sn₄S₉ + SnS₂; 18, Cu₂Sn₄S₉ + SnS₂; 19, Cu₂Sn₄S₉; 20, Cu₂SnS₃ + Cu₂ZnSn₃S₈ + SnS₂; 21, Cu₂Sn₄S₉ + Cu₂ZnSn₃S₈ + SnS₂; 22, Cu₂ZnSn₃S₈ + SnS₂; and 23, Cu₂ZnSn₃S₈ + α-ZnS + SnS₂. (From Olekseyuk, I.D. et al., *J. Alloys Compd.*, 368(1–2), 135, 2004.)

Cu₂Sn₄S₉ on the side and of Cu₂ZnSn₃S₈ within the concentration triangle. This system was investigated through DTA, metallography, and XRD (Olekseyuk et al. 2004).

ZnS–Cu₂S–SnS₂. The liquidus surface projection of this quasiternary system consists of seven fields of primary crystallization (Figure 1.9) (Olekseyuk et al. 2004). ZnS has the largest area of the primary crystallization field. This causes the elongated form of the primary crystallization fields of Cu₂SnS₃, SnS₂, and Cu₂ZnSnS₄ along the Cu₂S–SnS₂ side. The fields of the primary crystallization are separated by 16 monovariant lines and 15 invariants points of which 8 correspond to the binary and 7 to the ternary invariant reactions.

The isothermal section of the ZnS–Cu₂S–SnS₂ system at 400°C is given in Figure 1.10 (Olekseyuk et al. 2004). Cu₂ZnSnS₄ and Cu₂ZnSn₃S₈ quaternary compounds are formed in this system. Cu₂ZnSn₃S₈ melts incongruently at 700°C and crystallizes in the tetragonal structure with the lattice parameters $a = 543.5 \pm 0.1$ and $c = 1082.5 \pm 0.6$ pm.

The stable phases and their tie line connections in this system at 600°C have been illustrated schematically by Moh (1975), but this author indicated the formation of Cu₂Sn₃S₇ ternary compounds instead of Cu₂Sn₄S₉.

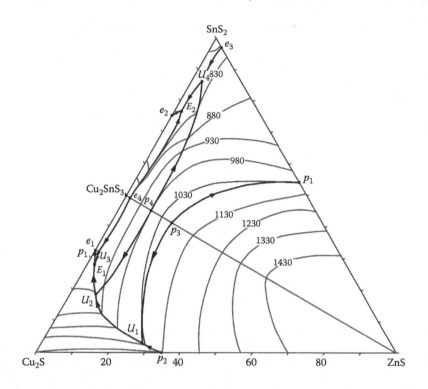

FIGURE 1.9 Liquidus surface of ZnS–Cu_2S–SnS_2 system. (From Olekseyuk, I.D. et al., *J. Alloys Compd.*, 368(1–2), 135, 2004.)

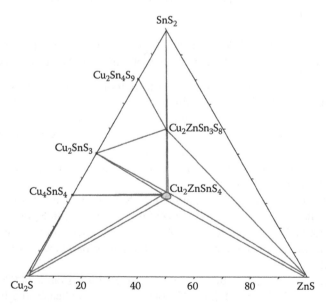

FIGURE 1.10 Isothermal section of ZnS–Cu_2S–SnS_2 system at 400°C. (From Olekseyuk, I.D. et al., *J. Alloys Compd.*, 368(1–2), 135, 2004.)

1.26 ZINC–COPPER–LEAD–SULFUR

In the Zn–Cu–S ternary system, a two-phase region consisting of the L_1 alloy phase rich in Cu and the L_2 matte phase exists in a considerably wide composition range from 0 to about 20 mol.% Zn (Figure 1.11a) (Surapunt et al. 1995, 1996). When 0.8, 1.5, or 2.4 mol.% Pb is added, the range of two-liquid region does not change with the lead content (Figure 1.11b through d).

ZnS–Cu$_{1.8}$S–PbS. The liquidus surface of this quasiternary system (Figure 1.12) was constructed according to three polythermal sections (Figure 1.13), which are traced from ZnS corner (Kopylov et al. 1976). It includes three fields of primary crystallization of binary sulfides, and the field of ZnS primary crystallization occupies the most part of the liquidus surface. The phase diagram of this system belongs to the diagrams with four-phase transition equilibrium, which is preceded by one eutectic and one peritectic three-phase equilibria. The border lines characterized the eutectic crystallization of ZnS and PbS and the peritectic interaction L + ZnS ⇔ (Cu$_{1.8}$S) is situated near the Cu$_{2-x}$S–PbS quasibinary system and converges in the U transition point. The crystallization in this systems ends by the three-phase eutectic equilibrium L ⇔ (Cu$_{1.8}$S) + PbS. The most fusible part of the ZnS–Cu$_{1.8}$S–PbS quasiternary system adjoins to the region of the binary eutectic in the Cu$_{2-x}$S–PbS system at 560°C.

At 1200°C in the ZnS–Cu$_2$S–PbS system, the solubility of ZnS in the liquid phase is around 10 mol.% ZnS in the low composition range of PbS (Figure 1.14) (Surapunt

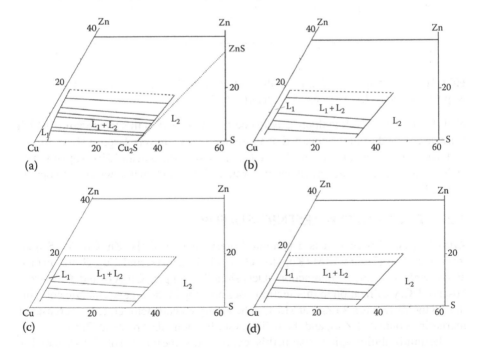

FIGURE 1.11 Phase relations in the Cu corner of the Zn–Cu–Pb–S system at 1200°C at (a) 0, (b) 0.8, (c) 1.5, and (d) 2.4 mol. Pb. (From Surapunt, S. et al., *Metal Rev. MMIJ*, 12(2), 84, 1995; Surapunt, S. et al., *Shigen-to-Sozai*, 112(1), 56, 1996.)

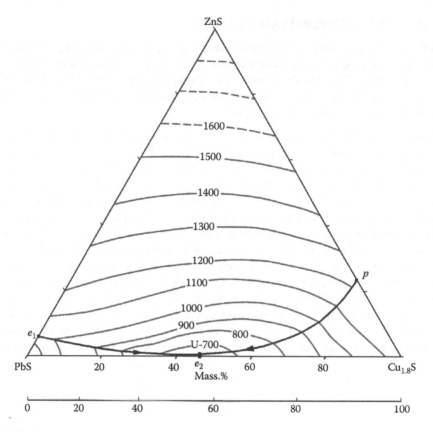

FIGURE 1.12 Liquidus surface of the $ZnS–Cu_{1.8}S–PbS$ quasiternary system. (From Kopylov, N.I. et al., *Izv. AN SSSR. Ser. Metally*, (6), 80, 1976.)

et al. 1995, 1996). The Raoultian activity coefficients of ZnS at this temperature, which were determined from the obtained phase relation data, are very large at about 10.

Quaternary compounds are not formed in this system (Strohfeldt 1936, Kopylov et al. 1976). This system was investigated through DTA and metallography (Kopylov et al. 1976).

1.27   ZINC–COPPER–ARSENIC–SULFUR

$ZnS–Cu_3As$. This section is a nonquasibinary section of the Zn–Cu–As–S quaternary system (Figure 1.15) (Kopylov et al. 1984). Cu_5As_2 and $Cu_{2-x}S$ are formed at the ZnS and Cu_3As interaction. Occasionally, copper, Cu_2As, and CuS were observed. Crystallization in this section does not stop in the binary eutectic containing 7.8 mol.% (3 mass.%) ZnS at 815°C. The binary eutectic temperature decreases at increasing content of ZnS, and the solid crystallization takes place at 705°C.

The immiscibility region exists in this section within the interval of 37.4–73.9 mol.% (18–51 mass.%) ZnS and above 1130°C. Thermal effects at 1000°C correspond to the polymorphous transformation of ZnS. This system was investigated through DTA, metallography, and local XRD.

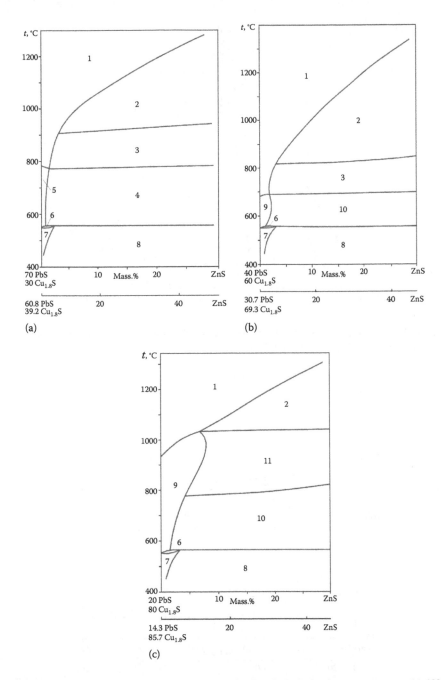

FIGURE 1.13 Polythermal sections of the ZnS–Cu$_{1.8}$S–PbS quasiternary system: (a) (60.8 mol.% PbS + 39.2 mol.% Cu$_{2-x}$S) – ZnS; (b) (30.7 mol.% PbS + 69.3 mol.% Cu$_{2-x}$S) – ZnS; (c) (14.3 mol.% PbS + 85.7 mol.% Cu$_{2-x}$S) – ZnS; 1, L; 2, L + α-ZnS; 3, L + β-ZnS; 4, L + β-ZnS + PbS; 5, L + PbS; 6, L + Cu$_{1.8}$S + PbS; 7, Cu$_{1.8}$S + PbS; 8, Cu$_{1.8}$S + β-ZnS + PbS; 9, L + Cu$_{1.8}$S; 10, L+ Cu$_{1.8}$S + β-ZnS; and 11, L+ Cu$_{1.8}$S + α-ZnS. (From Kopylov, N.I. et al., *Izv. AN SSSR. Ser. Metally,* (6), 80, 1976.)

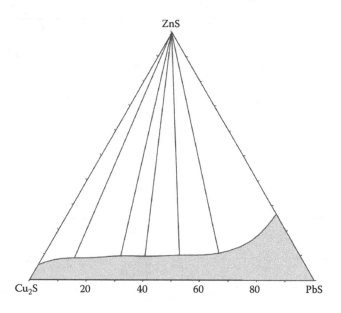

FIGURE 1.14 Isothermal sections of the ZnS–Cu₂S–PbS system at 1200°C. (From Surapunt, S. et al., *Metal Rev. MMIJ*, 12(2), 84, 1995; Surapunt, S. et al., *Shigen-to-Sozai*, 112(1), 56, 1996.)

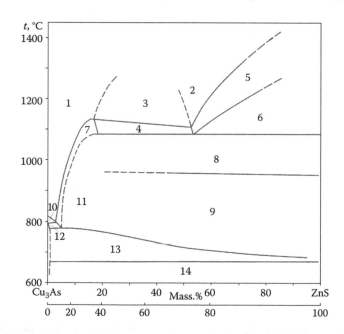

FIGURE 1.15 Phase relations in the ZnS–Cu₃As system: 1, L_2; 2, L_1; 3, $L_1 + L_2$; 4, $L_1 + L_2$ + β-ZnS; 5, L_1 + β-ZnS; 6, L_1 + β-ZnS + Cu₂S(CuS); 7, L_2 + β-ZnS; 8, L_2 + β-ZnS + Cu₂S(CuS); 9, L_2 + α-ZnS + Cu₂S(CuS); 10, L_2 + Cu₃As; 11, L; 12, L_2 + Cu₃As + α-ZnS; 13, L + Cu₃As + Cu₂S(CuS) + α-ZnS; and 14, Cu₃As + Cu₅As₂ + Cu₂S(CuS) + α-ZnS. (From Kopylov, N.I. et al., *Zhurn. neorgan. khimii*, 29(6), 1621, 1984.)

ZnS–Cu$_2$S–As$_2$S$_3$. The phase diagram is not constructed. The Cu$_6$Zn$_3$As$_4$S$_{12}$ quaternary compound is formed in this system. It crystallizes in the hexagonal structure with the lattice parameters $a = 1344 \pm 1.5$ and $c = 917 \pm 1$ pm and calculated density 4.38 g·cm^{-3} (Novatski 1982).

1.28   ZINC–COPPER–IRON–SULFUR

This quaternary system embraces many important sulfide minerals, which are primary sources for the extraction of Zn and Cd (Raghavan 2004a).

In the Zn–Cu–S ternary system, a two-phase region consisting of the L$_1$ alloy phase rich in Cu and the L$_2$ matte phase exists in a considerably wide composition range from 0 to about 20 mol.% Zn (Figure 1.16a) (Surapunt et al. 1995, Raghavan 2004a). The presence of Fe increases significantly the width of the liquid miscibility gap (Figure 1.16b through d).

Wiggins and Craig (1980) measured the Cu content of sphalerite, which is dependent upon temperature and the FeS activity, in this quaternary system between 500°C and 800°C and found that α-ZnS dissolves 5 mass.% Cu, when it coexists with pyrrhotite (Fe$_{1-x}$S) and intermediate solid solutions (*iss*), and 7 mass.%, when it coexists with bornite (Cu$_5$FeS$_4$). The solid solution of CuS in Fe-bearing α-ZnS slightly reduces the unit cell dimension. In contrast, naturally occurring sphalerites contain less than 0.5 mass.% Cu. This is attributed, at least partially,

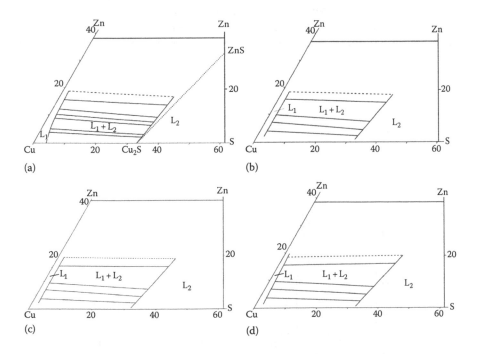

FIGURE 1.16   Phase relations in the Cu corner of the Zn–Cu–Fe–S system at 1200°C at (a) 0, (b) 3.4, (c) 4.3, and (d) 5.5 mol. Fe. (From Surapunt, S. et al., *Metal Rev. MMIJ*, 12(2), 84, 1995; Surapunt, S. et al., *Shigen-to-Sozai*, 112(1), 56, 1996.)

to the precipitation from high-temperature solution during cooling, as evidenced by the presence of inclusions of chalcopyrite in sphalerite. These data are included in the review of Raghavan (2004a).

Phase relations have been determined among sphalerite, pyrite (FeS_2), pyrrhotite, and *iss* from 439°C to 725°C at 10^2 kPa and 353°C to 760°C at $5 \cdot 10^2$ MPa (Hutchison and Scott 1981, Raghavan 2004a). The extents of solid solutions in the four-phase assemblage are described as functions of temperature (in K) by the following equations:

$$\log \text{mol.\% ZnS in } iss \ (10^2 \text{ kPa}) = 2.795 - 1.632T^{-1}$$

$$\log \text{mol.\% ZnS in } iss \ (5 \cdot 10^2 \text{ MPa}) = 2.659 - 1.603T^{-1}$$

$$\log \text{mol.\% CuS in } Fe_{1-x}S \ (10^2 \text{ kPa}) = 2.546 - 1.866T^{-1}$$

$$\log \text{mol.\% CuS in sphalerite} \ (10^2 \text{ kPa}) = 4.202 - 3.735T^{-1}$$

$$\log \text{mol.\% CuS in sphalerite} \ (5 \cdot 10^2 \text{ MPa}) = 4.084 - 3.791T^{-1}$$

It was found that in a certain temperature range, the FeS content of Cu-bearing sphalerite is a function only of the pressure and is given by the equation p (Pa) = 37.30 − 32.10 log (mol.% FeS) (Hutchison and Scott 1981, Raghavan 2004a). The temperature-independent range is illustrated in Figure 1.17 (Hutchison and Scott 1981). The deviation from this range is seen above 600°C and $5 \cdot 10^2$ MPa.

Sphalerites and sulfides in selected reaction assemblages of the Zn–Cu–Fe–S system were crystallized in molten salt fluxes contained in evacuated capsules at

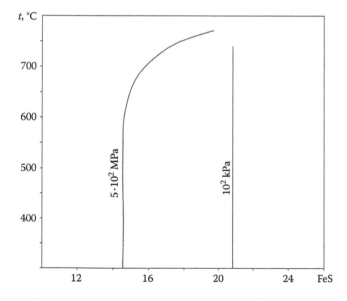

FIGURE 1.17   Zn–Cu–Fe–S composition of sphalerite as a function of temperature and pressure. (From Hutchison, M.N. and Scott, S.D., *Econ. Geol.*, 76(1), 143, 1981.)

temperatures between ~250°C and ~535°C (Lusk and Calde 2004). The combined compositional data for ZnS indicate large increase in the mol.% FeS content and Fe–Cu ratio with decreasing sulfur fugacity at given temperatures. FeS_2 and associated Cu–Fe sulfides show parallel increases for Fe/Cu and Fe/Zn ratios. These compositional changes occur over a range of S fugacity–temperature conditions that include many sulfide deposits. The equation $\log f(S_2) = 11.01 - 9.49(1000/T) + [0.187 - 0.252(1000/T)](mol.\% \ FeS) + [0.35 - 0.2(1000/K)]$ (mol.% CuS in sphalerite) can be applied to sphalerites that have equilibrated with buffer assemblages, containing FeS_2, or FeS_2 and $Fe_{1-x}S$, at temperature between 250°C and 550°C at 10^2 kPa. The results indicate relatively uniform composition for idaite $(Cu_{5.5}FeS_{6.5})$ over 452°C–323°C range. Respective mean composition of $(Cu_{5.35}Zn_{0.09})FeS_{6.45}$ is indicated. The idaites of this system have constant Zn contents of ~1 mass.% over this temperature range.

ZnS–CuFeS₂. Figure 1.18 shows the phase relations in the ZnS–$CuFeS_2$ system above 300°C (Moh 1975). The α–β-transition temperature of $CuFeS_2$ is lowered by nearly 50°C to ~500°C. At this temperature, α-$CuFeS_2$ takes ≈7.3 mol.% (≈4 mass.%) ZnS in the solid solutions, whereas β-$CuFeS_2$ in α-ZnS just exceeds 0.53 mol.% (1 mass.%) and increases very slowly to reach not even 1.6 mol.% (3 mass.%) at 800°C.

Peritectic conditions were found to occur at 872°C ± 4°C and at 1032°C ± 7°C: at 872°C ± 4°C, the maximum α-ZnS content in $CuFeS_2$ is 28.5 mol.% (17.5 mass.%), and α-ZnS, when in equilibrium with $CuFeS_2$, takes ≤1.6 mol.% (≤3 mass.%) $CuFeS_2$ in solid solution, whereas the composition of coexisting liquid is ≈23.5 mol.% (≈14 mass.%) ZnS. At 1032°C ± 7°C, the solubility of $CuFeS_2$ in α-ZnS is 1 mol.% (2 mass.%) or less,

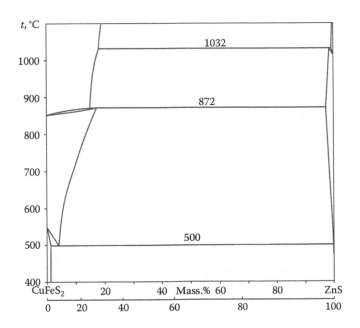

FIGURE 1.18 Phase diagram of ZnS–CuFeS₂ system. (From Moh, G.H., *Chem. Erde*, 34(1), 1, 1975.)

β-ZnS dissolves not even 0.53 mol.% (1 mass.%) CuFeS$_2$, and the composition of coexisting liquid is ≈27.8 ± 1.9 mol.% (≈17 ± 1 mass.%) ZnS (Moh 1975).

The existence of CuFeS$_2$ solid-phase solubility in α-ZnS has been noted for the first time by Buerger (1934), and later, it was indicated (Kullerud 1955) that at 600°C, α-ZnS dissolves 40 mol.% β-CuFeS$_2$.

ZnS–CuS–FeS. The compositional fields of chalcopyrite (CuFeS$_2$), *iss*, and ZnS were determined at temperatures between 300°C and 500°C (Kojima and Sugaki 1985). At 300°C, stable assemblages of CuFeS$_2$ + FeS$_2$ + Fe$_{1-x}$S + ZnS and CuFeS$_2$ + *iss* + Fe$_{1-x}$S + ZnS exist (Figure 1.19a). Also, a stable tie line between Cu$_5$FeS$_4$ and FeS$_2$ commonly found in natural ores occurs.

The stable univariant phase assemblages in the central portion of the system at both 400°C and 500°C are covellite (CuS) + nukundamite [(Cu,Fe)$_4$S$_4$] + FeS$_2$ + ZnS, CuS + Cu$_5$FeS$_4$ + (Cu,Fe)$_4$S$_4$ + ZnS, Cu$_5$FeS$_4$ + (Cu,Fe)$_4$S$_4$ + FeS$_2$ + ZnS, Cu$_5$FeS$_4$ +

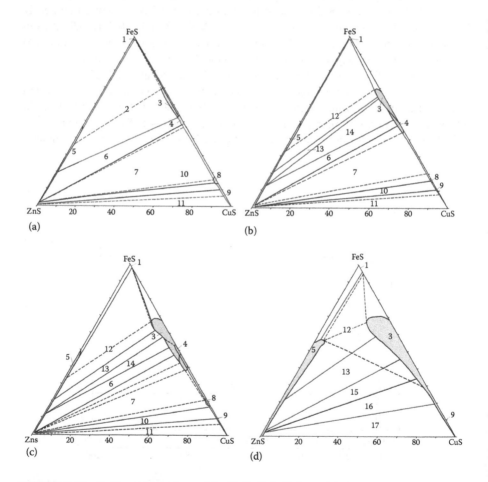

FIGURE 1.19 Isothermal sections of the ZnS–CuS–FeS at (a) 300°C (49 MPa), (b) 400°C (49 MPa), (c) 500°C (49 MPa), (d) 600°C. *(Continued)*

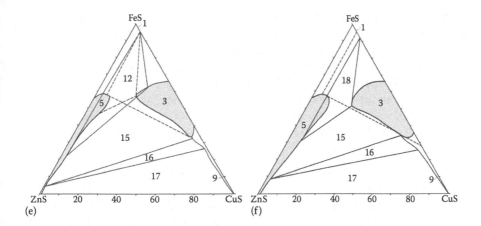

FIGURE 1.19 (Continued) Isothermal sections of the ZnS–CuS–FeS at (e) 700°C, and (f) 800°C [84Koj, 85Koj]. Solid and broken lines represent S-rich assemblages including FeS_2 and S-deficient assemblages, respectively: 1, $Fe_{1-x}S$; 2, $CuFeS_2 + FeS_2 + Fe_{1-x}S + ZnS$; 3, iss, 4; $CuFeS_2$; 5, ZnS; 6, $CuFeS_2 + FeS_2 + ZnS$; 7, $CuFeS_2 + Cu_5FeS_4 + FeS_2 + ZnS$; 8, $(Cu,Fe)_4S_4$; 9, Cu_5FeS_4; 10, $Cu_5FeS_4 + FeS_2 + ZnS$; 11, $CuS + (Cu,Fe)_4S_4 + FeS_2 + ZnS$; 12, $iss + Fe_{1-x}S + FeS_2 + ZnS$; 13, $iss + FeS_2 + ZnS$; 14, $CuFeS_2 + iss + FeS_2 + ZnS$; 15, $iss + ZnS + S_{liq}$; 16, $Cu_5FeS_4 + iss + ZnS + S_{liq}$; 17, $Cu_5FeS_4 + ZnS + S_{liq}$; 18, $iss + Fe_{1-x}S + ZnS + S_{liq}$. (From Kojima, S. and Sugaki, A., *Mineral. J.*, 12(1), 15, 1984; Kojima, S. and Sugaki, A., *Econ. Geol.*, 80(1), 158, 1985.)

$CuFeS_2 + FeS_2 + ZnS$, $Cu_5FeS_4 + CuFeS_2 + iss + ZnS$, $CuFeS_2 + iss + FeS_2 + ZnS$ and $iss + FeS_2 + Fe_{1-x}S + ZnS$ (Figure 1.19b, c) (Kojima and Sugaki 1985).

$CuFeS_2$ has a very small solid solution field close to the stoichiometric composition and dissolves small amounts of Zn (<0.9 at.%) at all three temperatures. *Iss* has an extensive solid solution field that dissolves up to 3.3 at.% Zn at 500°C and 1.2 at.% Zn at 300°C, with a slight reduction in solubility due to pressure. The solid solution area of *iss* becomes narrower as the temperature decreases and becomes richer in Cu as the pressure increases. Solid solution based on ZnS contains a maximum of about 2.4 mol.% CuS at the three temperatures and has no relation to FeS content and sulfur fugacity (Kojima and Sugaki 1985). At 500°C, additional univariant assemblages of $CuS + Cu_5FeS_4 + ZnS + S_{liq}$, $(Cu,Fe)_4S_4 + Cu_5FeS_4 + ZnS + S_{liq}$, and $(Cu,Fe)_4S_4 + FeS_2 + ZnS + S_{liq}$ should exist; however, they were not found experimentally because reactions among phases were very slow at this temperature (Kojima and Sugaki 1984).

Two univariant assemblages of $iss + FeS_2 + ZnS + S_{liq}$ and $iss + FeS_2 + Fe_{1-x}S + ZnS$ become stable at 600°C (Figure 1.19d), but the former assemblage was not confirmed at 700°C (Figure 1.19e) (Kojima and Sugaki 1984).

At 800°C, five univariant phase assemblages are stably present as follows: $Cu_5FeS_4 + iss + ZnS + S_{liq}$, $iss + Fe_{1-x}S + ZnS + S_{liq}$, $Cu_5FeS_4 + Fe_{1-x}S + Fe + ZnS$, $Cu_5FeS_4 + Fe + Cu + ZnS$, and $Cu_5FeS_4 + iss + Fe_{1-x}S$ (Figure 1.19f) (Kojima and Sugaki 1984).

The ingots were annealing 10–22 days at 800°C, 20–135 days at 700°C, 30–148 days at 600°C, and 100–214 days at 500°C (Kojima and Sugaki 1984). After annealing to equilibration, all samples were quenched in ice water.

Phase equilibria in the ZnS–CuS–FeS system between 500°C and 800°C were studied by the evacuated silica glass tube method (Kojima and Sugaki 1984). Phase relations in this system were studied between 300°C and 500°C using both the thermal gradient transport method and the isothermal in situ recrystallization method under hydrothermal conditions (Kojima and Sugaki 1985).

All these data are included in the review of Raghavan (2004a).

Isothermal sections of this system at 400°C and 800°C were also constructed by Shima et al. (1982), but these sections differ only slightly from those presented in Kojima and Sugaki (1984, 1985).

ZnS–Cu$_{1.8}$S–FeS. The liquidus surface of this quasiternary system (Figure 1.20) was constructed according to four polythermic sections (Figure 1.21), which are traced from ZnS corner (Novoselov 1955, Kopylov and Toguzov 1975). It includes three fields of primary crystallization of FeS and solid solutions based on ZnS and Cu$_{1.8}$S. The field of primary crystallization of solid solutions based on ZnS occupies the most part of the liquidus surface. The phase diagram belongs to the diagrams with four-phase peritectic equilibrium (P_1, 940°C), which is preceded by one eutectic (e_2P_1 line) and one peritectic (pP_1 line) three-phase equilibria. The crystallization in these systems ends in the e_1 binary eutectic of the Cu$_2$S–FeS quasibinary system (Kopylov and Toguzov 1975).

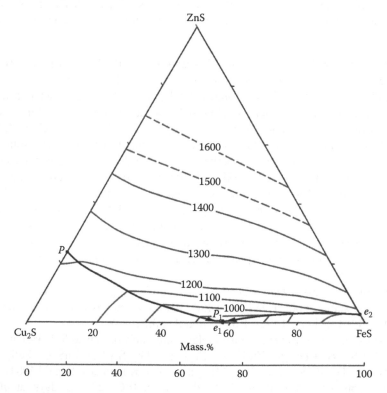

FIGURE 1.20 Liquidus surface of the ZnS–Cu$_{1.8}$S–FeS quasiternary system. (From Novoselov, S.S., *Tsvet. metally*, (3), 15, 1955.)

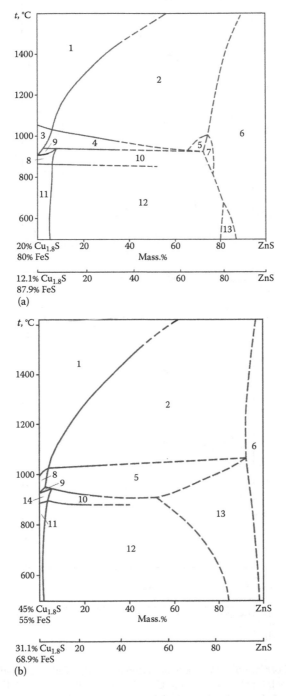

FIGURE 1.21 Polythermal sections of the ZnS–Cu$_{1.8}$S–FeS quasiternary system: (a) (87.9 mol.% FeS + 12.1 mol.% Cu$_{2-x}$S) – ZnS; (b) (68.9 mol.% FeS + 31.1 mol.% Cu$_{2-x}$S) – ZnS.   (*Continued*)

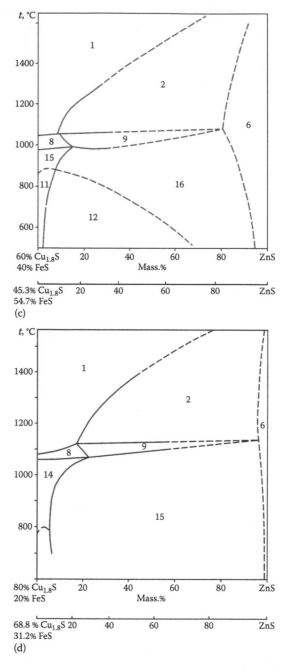

FIGURE 1.21 (Continued) Polythermal sections of the ZnS–Cu$_{1.8}$S–FeS quasiternary system: (c) (54.7 mol.% FeS + 45.3 mol.% Cu$_{2-x}$S) – ZnS; (d) (31.2 mol.% FeS + 68.8 mol.% Cu$_{2-x}$S) – ZnS: 1, L; 2, L + ZnS; 3, L + FeS; 4, L + FeS + ZnS; 5, L + ZnS + δ; 6, ZnS; 7, ZnS + δ; 8, L + δ; 9, L + FeS + δ; 10, FeS + ZnS + δ; 11, CuFeS$_2$ + FeS + δ; 12, CuFeS$_2$ + FeS + ZnS + δ; 13, FeS + FeS; 14, FeS + δ; 15, δ; 16, ZnS + δ. (From Novoselov, S.S., *Tsvet. metally*, (3), 15, 1955.)

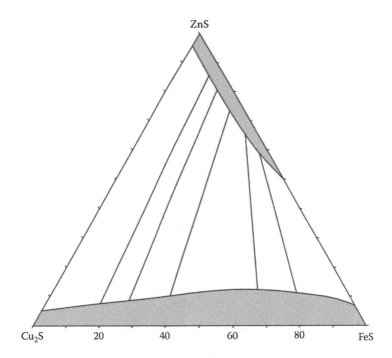

FIGURE 1.22 Isothermal section of the $ZnS–Cu_2S–FeS$ system at 1200°C. (From Surapunt, S. et al., *Metal Rev. MMIJ*, 12(2), 84, 1995; Surapunt, S. et al., *Shigen-to-Sozai*, 112(1), 56, 1996.)

Isothermal section of the $ZnS–Cu_2S–FeS$ quasiternary system at 1200°C is shown in Figure 1.22 (Surapunt et al. 1995, 1996, Raghavan 2004a). β-ZnS is the only solid sulfide at this temperature. It dissolves up to 48 mol.% FeS. The liquid phase is present along the $Cu_2S–FeS$ system, which is in equilibrium with the previously mentioned solid solution.

Nesterov and Ponomarev (1960) determined the ZnS crystallization region at 1200°C–1400°C using measurement of vapor pressure (Figure 1.23).

This system was investigated through DTA and metallography (Novoselov 1955).

1.29 ZINC–SILVER–ALUMINUM–SULFUR

ZnS–AgAlS$_2$. The phase diagram is not constructed. There were investigated phase transformations in the solid state at temperatures below 1020°C (Figure 1.24) (Robbins and Miksovsky 1972). At 1000°C, the chalcopyrite structure exists up to 20 mol.% ZnS, and at more than 20 mol.% ZnS, two phases are in this system: the phase with wurtzite and the phase with chalcopyrite structures. The ingots, containing 83–86 mol.% ZnS and annealing at 800°C, crystallize in the wurtzite structure and the mixtures, containing 86–91 mol.% ZnS, have the phases with wurtzite and sphalerite structures. Only one phase with sphalerite structure exists within the interval of 91–100 mol.% ZnS. The ingots, containing 83–100 mol.% ZnS and annealing at 1000°C, crystallize in the wurtzite structure.

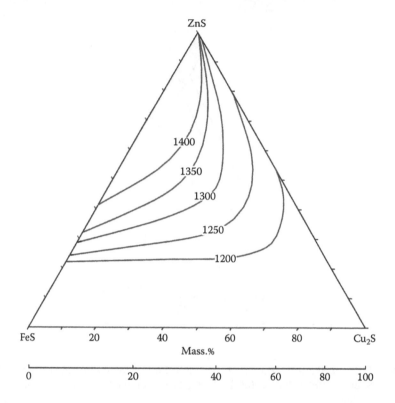

FIGURE 1.23 The ZnS crystallization region at 1200°C–1400°C in the ZnS–Cu$_{1.8}$S–FeS quasiternary system. (From Nesterov, V.N. and Ponomarev, V.D., *Izv. AN KazSSR. Ser. metallurgii, obogashchenia i ogneuporov*, 3(9), 77, 1960.)

According to the data of Parthé et al. (1969), the AgZn$_2$AlS$_4$ quaternary compound is formed in this system. It crystallizes in the hexagonal structure of wurtzite type with the lattice parameters $a = 384.6 \pm 0.5$ and $c = 631.3 \pm 0.5$ pm, but this compound can be one of the solid solution compositions.

This system was investigated through XRD (Robbins and Miksovsky 1972). The AgZn$_2$AlS$_4$ compound was obtained by the heating of the mixture from chemical elements at 600°C–800°C (Parthé et al. 1969).

1.30 ZINC–SILVER–GALLIUM–SULFUR

ZnS–AgGaS$_2$. The phase diagram is a eutectic type (Figure 1.25) (Olekseyuk et al. 2001). The coordinates of the eutectic point are 18 mol.% ZnS and 960°C. The solid solubility of the initial compounds of this system is insignificant at 600°C. The presence of the wurtzite phase at this temperature, which was reported by Robbins and Miksovsky (1972), has not been revealed. The solid solutions based on α-ZnS has an extension of 91–100 mol.% ZnS. The homogeneity region of the solid solution based on AgGaS$_2$ is less than 2 mol.% ZnS. The solid solution based on β-ZnS exceeds 70 mol.% AgGaS$_2$ at the eutectic temperature. These solid solutions decompose by the eutectoid reaction at 690°C.

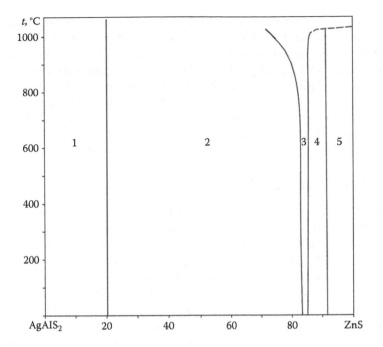

FIGURE 1.24 Phase relations in the ZnS–AgAlS$_2$ system at temperatures below 1020°C: 1, chalcopyrite; 2, chalcopyrite + wurtzite; 3, wurtzite; 4, wurtzite + sphalerite; 5, sphalerite. (From Robbins, M. and Miksovsky, M.A., *J. Solid State Chem.*, 5(3), 462, 1972.)

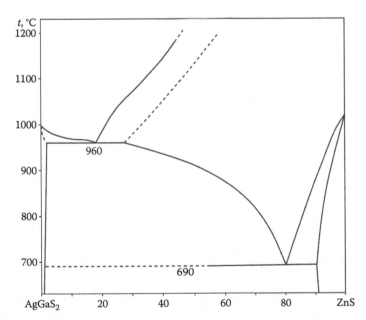

FIGURE 1.25 Phase diagram of the ZnS–AgGaS$_2$ system. (From Olekseyuk, I.D. et al., *J. Alloys Compd.*, 325(1–2), 204, 2001.)

According to the data of Apple (1958), the ingots of the $ZnS-AgGaS_2$ system crystallize in the cubic structure at 5 mol.% $AgGaS_2$. The phases with the cubic and hexagonal structures exist in this system within the interval of 5–20 mol.% $AgGaS_2$, and at more than 20 mol.% $AgGaS_2$, only one phase with hexagonal structure crystallizes. The solubility of $AgGaS_2$ in ZnS at 900°C is within the interval of 5–10 mol.%.

This system was investigated using DTA and XRD. All samples were annealed at 600°C during 500 h and then quenched in cold water (Olekseyuk et al. 2001, Robbins and Miksovsky 1972). The ingots were obtained by the heating of the ZnS + $AgGaS_2$ mixtures at 900°C for 1 h in the H_2S atmosphere (Apple 1958).

1.31   ZINC–SILVER–INDIUM–SULFUR

$ZnS-AgInS_2$. This system is not quasibinary (Figure 1.26) due to incongruent melting of $AgInS_2$ (Olekseyuk et al. 2001). The liquidus consists of the primary crystallization curve of $AgIn_5S_8$ and the primary crystallization of solid solution based on β-ZnS. The solubility of α-ZnS in α-$AgInS_2$ is smaller than 2 mol.% and the solubility of α-$AgInS_2$ in α-ZnS is 3 mol.% at 600°C. The phase with wurtzite structure exists in the range 62–88 mol.% ZnS [66.67–85.5 mol.% ZnS (Lambrecht 1972)] at this temperature. The extent of the solid solutions is much larger than determined by Robbins and Miksovsky (1972). Longer annealing than that used in Robbins and Miksovsky (1972) proved to be necessary for reaching equilibrium in the samples.

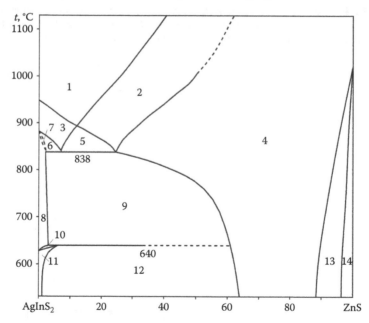

FIGURE 1.26   Phase diagram of the $ZnS-AgInS_2$ system: 1, L; 2, L + (β-ZnS); 3, L + $AgIn_5S_8$; 4, (β-ZnS); 5, L + $AgIn_5S_8$ + (β-ZnS); 6, L + $AgIn_5S_8$ + (β-$AgInS_2$); 7, $AgIn_5S_8$ + (β-$AgInS_2$); 8, (β-$AgInS_2$); 9, (β-$AgInS_2$) + (β-ZnS); 10, (β-$AgInS_2$) + (α-$AgInS_2$); 11, (α-$AgInS_2$); 12, (α-$AgInS_2$) + (β-ZnS); 13, (α-ZnS) + (β-ZnS); 14, (β-ZnS). (From Olekseyuk, I.D. et al., *J. Alloys Compd.*, 325(1–2), 204, 2001.)

According to the data of Parthé et al. (1969), the $AgZn_2InS_4$ quaternary compound is formed in this system. It crystallizes in the hexagonal structure of wurtzite type with the lattice parameters $a = 394.4 \pm 0.5$ and $c = 645.9 \pm 0.5$ pm, but this compound can be one of the solid solution compositions.

This system was investigated using DTA and XRD. All samples were annealed at 600°C during 500 h and then quenched in cold water (Olekseyuk et al. 2001, Robbins and Miksovsky 1972). The $AgZn_2InS_4$ compound was obtained by the heating of the mixture from chemical elements at 600°C–800°C (Parthé et al. 1969).

1.32 ZINC–SILVER–SILICON–SULFUR

$ZnS–Ag_8SiS_6$. The phase diagram is a eutectic type (Figure 1.27) (Piskach et al. 2006). The eutectic composition and temperature are 47 ± 2 mol.% ZnS and 934°C, respectively. The solubility of ZnS in Ag_8SiS_6 at the eutectic temperature is 42 mol.%. The crystallization of the solid solutions occurs in a narrow temperature range that does not exceed 20°C. These solid solutions are characterized by the polymorph transformation at 238°C, and solid solutions based on ZnS have the transformation sphalerite–wurtzite at 1037°C. This system was investigated by DTA, XRD, and metallography.

$ZnS–Ag_2S–SiS_2$. The phase diagram is not constructed. The quaternary compound Ag_2ZnSiS_4 that crystallizes in the monoclinic structure with the lattice parameters $a = 640.52 \pm 0.01$, $b = 654.84 \pm 0.01$, $c = 793.40 \pm 0.01$ pm, and $\beta = 90.455° \pm 0.001°$ and calculated density 4.366 g·cm^{-3} was found in this quasiternary system (Brunetta et al. 2012). This compound is a direct bandgap semiconductor with $E_g = 3.28$ eV.

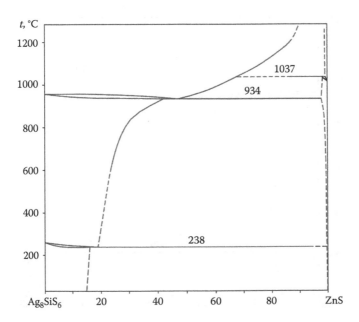

FIGURE 1.27 The $ZnS–Ag_8SiS_6$ phase diagram. (From Piskach, L.V. et al., *J. Alloys Compd.*, 421(1–2), 98, 2006.)

Single crystals of Ag_2ZnSiS_4 were produced by heating the powder mixture of Zn, Ag, Si, and S in stoichiometric ratio in a fused silica tube under a vacuum up to 800°C over 12 h and held at that temperature for 96 h. After that, the sample was slow-cooled to 500°C at 5°C/h (60 h) and then allowed to cool radiatively to ambient temperature (Brunetta et al. 2012).

1.33   ZINC–SILVER–GERMANIUM–SULFUR

$ZnS–Ag_8GeS_6$. The phase diagram is a eutectic type (Figure 1.28) (Piskach et al. 2006). The eutectic composition and temperature are 42 ± 2 mol.% ZnS and 928°C, respectively. The solubility of ZnS in Ag_8GeS_6 at the eutectic temperature is 38 mol.%. The crystallization of the solid solutions occurs in a narrow temperature range that does not exceed 20°C. These solid solutions are characterized by the polymorph transformation at 226°C and solid solutions based on ZnS have the transformation sphalerite–wurtzite at 1041°C. This system was investigated by DTA, XRD, and metallography.

$ZnS–Ag_2S–GeS_2$. The isothermal section of this system at room temperature is shown in Figure 1.29 (Parasyuk et al. 2010). The quaternary compound Ag_2ZnGeS_4 that crystallizes in the tetragonal structure with the lattice parameters $a = 574.996 ± 0.009$ and $c = 1034.34 ± 0.03$ pm and calculated density 4.6799 ± 0.0003 g·cm^{-3} was found in this quasiternary system. The energy gap of this compound is equal to 2.5 eV (Tsuji et al. 2010).

The glass-formation region in the $ZnS–Ag_2S–GeS_2$ quasiternary system is localized along the $Ag_2S–GeS_2$ quasibinary system and the maximum amount of ZnS in the glass phase is 12 mol.% (Figure 1.30) (Parasyuk et al. 2010). It narrows with

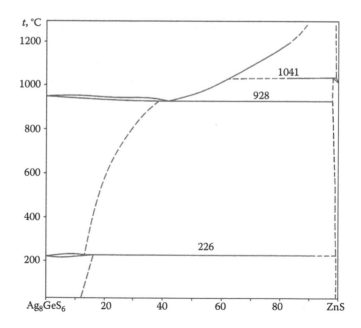

FIGURE 1.28 The $ZnS–Ag_8GeS_6$ phase diagram. (From Piskach, L.V. et al., *J. Alloys Compd.*, 421(1–2), 98, 2006.)

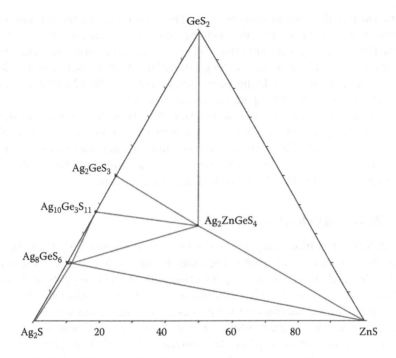

FIGURE 1.29 Isothermal section of $ZnS–Ag_2S–GeS_2$ quasiternary system at room temperature. (From Parasyuk, O.V. et al., *J. Alloys Compd.*, 500(1), 26, 2010.)

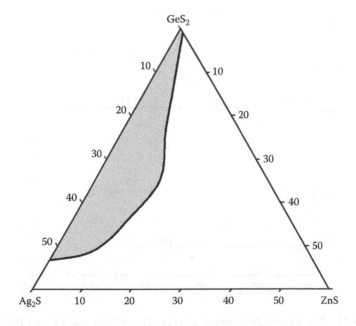

FIGURE 1.30 The glass-formation region in the $ZnS–Ag_2S–GeS_2$ quasiternary system. (From Parasyuk, O.V. et al., *J. Alloys Compd.*, 500(1), 26, 2010.)

higher content of the glass-forming element. The glass-formation temperature varies only narrowly for obtained glasses, likely because the amounts of ZnS that can be sustained by the glassy state are quite moderate. The glass-formation temperature values increase slightly from ~280°C for glasses of the Ag_2S–GeS_2 system to ~300°C for the glasses containing the highest concentration of ZnS (10–12 mol.%). The variations of the crystallization temperature are likewise minor.

The powder of Ag_2ZnGeS_4 was synthesized by solid-state reaction: the starting materials ZnS, Ag_2S, and GeS_2 were mixed with 15% excess amount of ZnS and GeS_2 (Tsuji et al. 2010). The mixture was sealed in quartz ampoule tube in vacuo and heat-treated at 550°C–650°C for 10 h. This system was investigated by DTA and the alloys were annealed at 400°C for 800 h (Parasyuk et al. 2010).

1.34   ZINC–SILVER–TIN–SULFUR

ZnS–Ag_8SnS_6. The phase diagram is a eutectic type (Figure 1.31) (Dudchak and Piskach 2001, Piskach et al. 2006). The eutectic composition and temperature are 32 mol.% ZnS and 814°C, respectively. The solubility of ZnS in Ag_8SiS_6 at the eutectic temperature is 24 mol.%. The crystallization of the solid solutions occurs in a narrow temperature range that does not exceed 20°C. These solid solutions are characterized by the polymorph transformation at 169°C and solid solutions based on ZnS have the transformation sphalerite–wurtzite at 1039°C. The solubility of ZnS in Ag_8SnS_6 is not higher than 5 mol.%. This system was investigated by DTA, XRD, and metallography and the alloys were annealed at 400°C for 250 h.

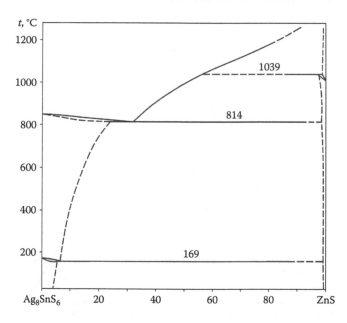

FIGURE 1.31   The ZnS–Ag_8SnS_6 phase diagram. (From Dudchak, I.V. and Piskach, L.V., *Nauk. visnyk Volyn. derzh. un-tu. Khim. nauky*, (6), 59, 2001; Piskach, L.V. et al., *J. Alloys Compd.*, 421(1–2), 98, 2006.)

ZnS–Ag$_2$S–SnS$_2$. The phase diagram is not constructed. Quaternary compound Ag$_2$ZnSnS$_4$ with energy gap 2.0 eV was found in this quasiternary system (Tsuji et al. 2010). To obtain this compound, a mixed aqueous solution (150 mL) of Zn(NO$_3$)$_2 \cdot$ 6H$_2$O with 10%–15% excess amounts (about 0.029 mol/L) and SnCl$_4 \cdot$ 5H$_2$O (0.025 mol/L) was bubbled with a H$_2$S gas. Then the aqueous solution containing 7.5 mmol of AgNO$_3$ was slowly added into the mixed solution.

1.35   ZINC–SILVER–PHOSPHORUS–SULFUR

ZnS–Ag$_2$S–P$_2$S$_5$. The phase diagram is not constructed. The ZnAgPS$_4$ quaternary compound is formed in this system, which crystallizes in the orthorhombic structure with the lattice parameters a = 1250.2, b = 759.9, and c = 606.96 pm and calculated density 3.829 g$\cdot$cm^{-3} (Toffoli et al. 1985).

1.36   ZINC–SILVER–SELENIUM–SULFUR

ZnS–Ag$_2$Se. The phase diagram is a eutectic type (Figure 1.32) (Trishchuk et al. 1985). The eutectic composition and temperature are 8 mol.% ZnS and 862°C, respectively. The solubility of ZnS in Ag$_2$Se reaches 6 mol.% and decreases at decreasing temperature (at 840°C and 800°C, it reaches 5 and 2.5 mol.%, respectively). This system was investigated through DTA and metallography.

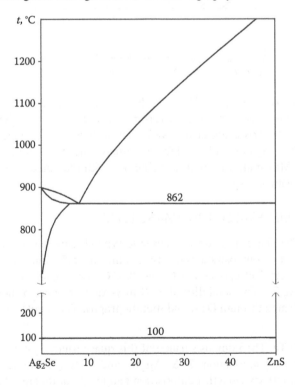

FIGURE 1.32   The ZnS–Ag$_2$Se phase diagram. (From Trishchuk, L.I. et al., *Izv. AN SSSR. Neorgan. materialy*, 21(2), 210, 1985.)

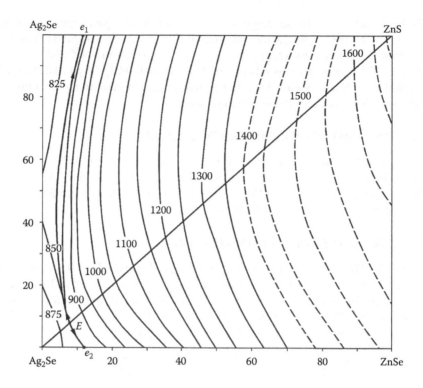

FIGURE 1.33 Liquidus surface of the ZnS + Ag$_2$Se ⇔ ZnSe + Ag$_2$S ternary mutual system. (From Trishchuk, L.I. et al., *Izv. AN SSSR. Neorgan. materialy*, 21(2), 210, 1985.)

ZnS + Ag$_2$Se ⇔ ZnSe + Ag$_2$Se. The liquidus surface of this ternary mutual system includes two fields of primary crystallization: ZnS$_x$Se$_{1-x}$ and Ag$_2$S$_x$Se$_{1-x}$ solid solutions (Figure 1.33) (Trishchuk et al. 1985). These fields are divided by the e_1Ee_2 line of secondary crystallization which is situated near the Ag$_2$S–Ag$_2$Se quasibinary system. Maximum on this line coincides with the eutectic point (E) in the ZnS–Ag$_2$Se quasibinary system.

1.37 ZINC–SILVER–TELLURIUM–SULFUR

ZnS–Ag$_2$Te. The phase diagram is a eutectic type (Figure 1.34) (Trishchuk et al. 1992). The eutectic composition and temperature are 6.5 mol.% ZnS and 945°C, respectively. Thermal effects at 145°C and 802°C correspond to the phase transformations of Ag$_2$Te. The solubility of ZnS in γ-Ag$_2$Te reaches 5 mol.%. This system was investigated through DTA and metallography. The ingots were annealed at 920°C for 20 h.

ZnS–Ag$_2$Te–ZnTe. The liquidus surface of this quasiternary system includes three fields of primary crystallization: ZnS, Ag$_2$Te, and ZnTe (Figure 1.35) (Trishchuk et al. 1992). The ternary eutectic composition and temperature are 69 mol.% Ag$_2$Te, 28 mol.% ZnTe, 3 mol.% ZnS, and 865°C, respectively. This system was investigated through DTA and metallography. The ingots were annealed at 920°C for 20 h.

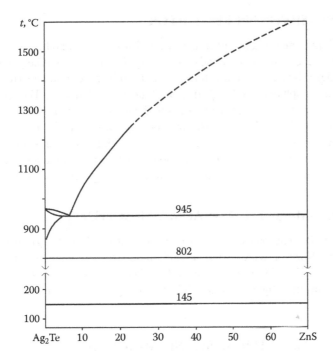

FIGURE 1.34 The ZnS–Ag$_2$Te phase diagram. (From Trishchuk, L.I. et al., *Izv. AN SSSR. Neorgan. materialy*, 28(4), 735, 1992.)

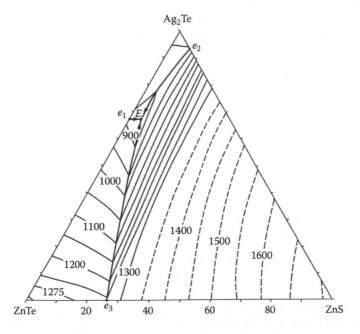

FIGURE 1.35 Liquidus surface of the ZnS–Ag$_2$Te–ZnTe quasiternary system. (From Trishchuk, L.I. et al., *Izv. AN SSSR. Neorgan. materialy*, 28(4), 735, 1992.)

1.38 ZINC–SILVER–NICKEL–SULFUR

Starting with pure metals and sulfur in evacuated silica tubes and heating up to 900°C led to several Ag-based compounds in Zn–Ag–Ni–S system (Garg and Ganguli 2000). Powder diffraction studies showed either a thiospinel-related structure (cubic, $a \sim 1030$ pm) or sphalerite-related structure (cubic, $a \sim 541$ pm). The spinel-related phases were found to crystallize from melts in the temperature range of 670°C–850°C while sphalerite-related phases crystallized between 850°C and 900°C. Single crystals have been obtained by slow cooling. These materials are interesting with respect to their ion exchange/intercalation of the monovalent ions suitable for use as cathode materials, and some of these compounds have interesting electro-optical properties.

1.39 ZINC–MAGNESIUM–CADMIUM–SULFUR

ZnS–MgS–CdS. The phase diagram is not constructed. The regions of the solid solution crystallization with hexagonal and cubic structure were determined in this system (Figure 1.36) (Dmitrenko et al. 1990). These solid solutions are not regular. This system was investigated through XRD. The ingots were annealed at 800°C and 900°C for 2 h.

1.40 ZINC–MAGNESIUM–SELENIUM–SULFUR

ZnS + MgSe ⇔ ZnSe + MgS. Binodal and spinodal isotherms in this system were calculated in a regular solution approximation by Sorokin et al. (2000). Instability and immiscibility regions at typical growth temperatures are isolated from the sides of a composition square diagram and cover a wide range of compositions (Figure 1.37). The spinodal decomposition of $Zn_{1-x}Mg_xSe_{1-y}S_y$ alloys occurs along the ZnSe–MgS

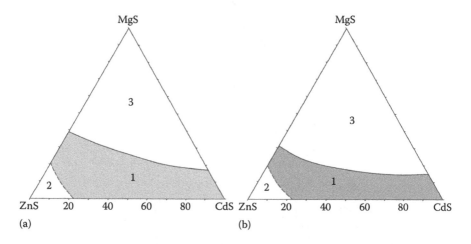

FIGURE 1.36 Arrangement of the phase regions in the ZnS–MgS–CdS quasiternary system at (a) 900°C and (b) 800°C: 1, solid solutions with wurtzite structure; 2, wurtzite + sphalerite mixtures; 3, wurtzite + MgS mixtures. (From Dmitrenko, A.O. et al., *Izv. AN SSSR. Neorgan. materialy*, 26(12), 2483, 1990.)

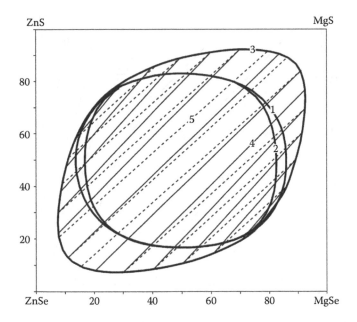

FIGURE 1.37 Instability and immiscibility regions in $Zn_{1-x}Mg_xSe_{1-y}S_y$ alloys at 300°C: 1, chemical spinodal; 2, coherent spinodal; 3, binodal isotherms; 4, nodes connecting compositions of equilibrium phases; 5, directions of coherent decomposition. (From Sorokin, V.S. et al., *J. Cryst. Growth*, 214–215, 130, 2000.)

diagonal. The spinodal and bimodal isotherms calculated with and without taking account of strain effects are practically identical. The calculated critical temperature of $Zn_{1-x}Mg_xSe_{1-y}S_y$ alloys $T_c \approx 537$°C. Calculations give a critical temperature of these alloys' coherent decomposition $T_c \approx 529$°C.

The composition dependence of the energy bandgap of the $Zn_{1-x}Mg_xSe_{1-y}S_y$ solid solutions has been calculated using the full-potential linearized augmented plane wave method within the density functional theory (Figure 1.38) (Hassan El Haj et al. 2007b).

1.41 ZINC–MAGNESIUM–TELLURIUM–SULFUR

ZnS + MgTe ⇔ ZnTe + MgS. The calculation bandgap energies of the $Zn_{1-x}Mg_xTe_{1-y}S_y$ solid solutions depend nonlinearly on the compositions, especially at high values of *x* and *y* (Figure 1.39) (Hassan El Haj et al. 2007a). A nonlinear behavior of the lattice constant and bulk modulus dependence on *x* and *y* has been also observed.

1.42 ZINC–MAGNESIUM–FLUORINE–SULFUR

ZnS–MgF₂. The phase diagram is not constructed. The solid solution thin films containing up to 30 mol.% ZnS in MgF_2 and MgF_2 in ZnS were obtained at simultaneous evaporation of ZnS and MgF_2 at different temperatures (Vankar et al. 1979). Both solid solutions exist in the middle composition region. Lattice parameters of ZnS increase linearly at the MgF_2 addition and the lattice parameters of MgF_2 linearly decrease at the ZnS addition.

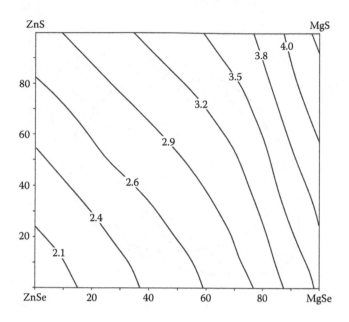

FIGURE 1.38 Composition dependence of the calculated energy bandgap (eV) of the $Zn_{1-x}Mg_xSe_{1-y}S_y$ solid solutions. (From Hassan El Haj, F. et al., *Mater. Lett.*, 61(4–5), 1178, 2007b.)

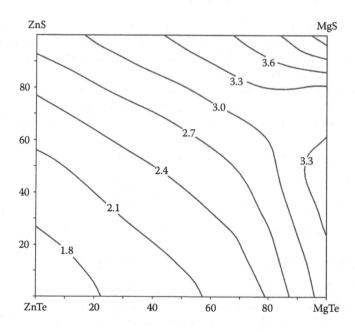

FIGURE 1.39 Composition dependence of the calculated energy bandgap (eV) of the $Zn_{1-x}Mg_xS_yTe_{1-y}$ quaternary alloys. (From Hassan El Haj, F. and Akbarzadeh, H., *Alloys Compd.*, 433(1–2), 306, 2007a.)

1.43 ZINC–CALCIUM–OXYGEN–SULFUR

ZnS–CaO. This section is a nonquasibinary section of the ZnS + CaO ⇔ ZnO + CaS ternary mutual system (Figure 1.40) (Yarygin et al. 1975, Toguzov et al. 1979). The crystallization temperature of ZnS decreases sharply at the CaO addition and reaches the minimum at 16.2 mol.% (10 mass.%) CaO. The further addition of CaO leads to an increase of the primary crystallization temperature and it achieves 1350°C at approximately 63.5 mol.% (50 mass.%) CaO.

The crystallization order can be presented by the next scheme. ZnS primarily crystallizes from the melt at up to 16.2 mol.% (10 mass.%) CaO and calcium oxide primarily crystallizes at the more CaO concentrations. Zinc calcium oxysulfide secondarily crystallizes from the melt, containing more than 42.7 mol.% (30 mass.%) CaO. Complete crystallization of the melts takes place at 1205°C and 1190°C by the crystallization of zinc calcium oxysulfide with earlier crystallizable phases (Yarygin et al. 1975, Toguzov et al. 1979).

According to the data of Yarygin et al. (1977), Petrova et al. (2003), Gulyaeva et al. (2006), and Sambrook et al. (2007), ZnCaOS quaternary compound is formed in the ZnS–CaO system. It melts at 1281°C (Gulyaeva et al. 2006) and crystallizes in the hexagonal structure with the lattice parameters $a = 375.726 \pm 0.003$ and $c = 1140.13 \pm 0.01$ pm (Sambrook et al. 2007) [$a = 375.547 \pm 0.001$ and $c = 1140.14 \pm 0.05$ pm (Petrova et al. 2003)], energy gap $E_g = 3.7 \pm 0.1$ eV (Sambrook et al. 2007), and experimental and calculation densities 3.53 and 3.66 [3.56 and 3.73 (Yarygin et al. 1977)] g·cm⁻³,

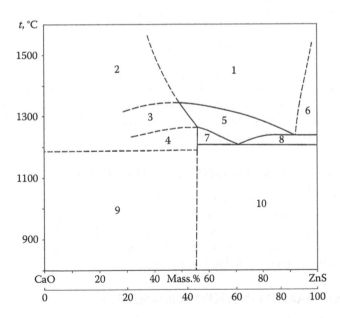

FIGURE 1.40 Phase relations in the ZnS–CaO system: 1, L; 2, L + CaO; 3, L + CaO + CaS; 4, L + CaO + zinc calcium oxysulfide; 5, L + CaO + CaS + ZnS; 6, L + ZnS; 7, L + CaS + zinc calcium oxysulfide; 8, L + CaS + ZnS; 9, CaO + CaS + zinc calcium oxysulfide; and 10, CaS + ZnS + zinc calcium oxysulfide. (From Toguzov, M.Z. et al., *Zhurn. neorgan. khimii*, 24(12), 3354, 1979.)

respectively (Petrova et al. 2003). Temperature dependence of the lattice parameters up to 700°C could be expressed by the next equations (Gulyaeva et al. 2006): $a = 374.2 + 4.57 \cdot 10^{-3}T$ and $c = 1135.1 + 15.13 \cdot 10^{-3}T$ pm, where T is temperature in K. Coefficients of thermal expansion for ZnCaOS are equal to $\alpha_a = 1.25 \cdot 10^{-5}$ K^{-1} and $\alpha_c = 1.30 \cdot 10^{-5}$ K^{-1} (Gulyaeva et al. 2006). Yarygin et al. (1977) noted that this compound crystallizes in the hexagonal structure but with other lattice parameters ($a = 648 \pm 0.003$ and $c = 378$ pm).

This system was investigated through DTA, metallography, and local XRD (Yarygin et al. 1975, 1977, Toguzov et al. 1979, Petrova et al. 2003, Gulyaeva et al. 2006). ZnCaOS was synthesized by the annealing of the ZnS and CaO equimolar mixture at 1000°C–1200°C in the He atmosphere during 3 h (Petrova et al. 2003, Gulyaeva et al. 2006) or by the melting of ZnS and CaO mixture in an inert atmosphere at 1250°C with the next slow cooling up to 400°C (Yarygin et al. 1977). This compound is stable at the heating up to 1100°C and only at the heating up to 1290°C ZnS and CaO were determined (Gulyaeva et al. 2006). Heating of ZnCaOS higher than 700°C at $p(O_2) = 10$ Pa leads to the formation of CaS (Gulyaeva et al. 2006).

1.44   ZINC–BARIUM–LANTHANUM–SULFUR

ZnS–BaS–La$_2$S$_3$. The phase diagram is not constructed. ZnBaLa$_2$S$_5$ quaternary compound is formed in this system (Wakeshima and Hinatsu 2001). It crystallizes in the tetragonal structure with the lattice parameters $a = 798.23 \pm 0.01$ and $c = 1367.08 \pm 0.02$ pm. ZnBaLa$_2$S$_5$ was synthesized by a solid-state reaction of the ZnS, BaS, and La$_2$S$_3$ stoichiometry mixture at 950°C for 2 days.

1.45   ZINC–BARIUM–CERIUM–SULFUR

ZnS–BaS–Ce$_2$S$_3$. The phase diagram is not constructed. ZnBaCe$_2$S$_5$ quaternary compound is formed in this system (Wakeshima and Hinatsu 2001). It crystallizes in the tetragonal structure with the lattice parameters $a = 791.02 \pm 0.01$ and $c = 1365.79 \pm 0.02$ pm. ZnBaCe$_2$S$_5$ was synthesized by a solid-state reaction of the ZnS, BaS, and Ce$_2$S$_3$ stoichiometry mixture at 950°C for 2 days.

1.46   ZINC–BARIUM–PRASEODYMIUM–SULFUR

ZnS–BaS–Pr$_2$S$_3$. The phase diagram is not constructed. ZnBaPr$_2$S$_5$ quaternary compound is formed in this system (Wakeshima and Hinatsu 2001). It crystallizes in the tetragonal structure with the lattice parameters $a = 787.19 \pm 0.01$ and $c = 1363.27 \pm 0.02$ pm. ZnBaPr$_2$S$_5$ was synthesized by a solid-state reaction of the ZnS, BaS, and Pr$_2$S$_3$ stoichiometry mixture at 950°C for 2 days.

1.47   ZINC–BARIUM–NEODYMIUM–SULFUR

ZnS–BaS–Nd$_2$S$_3$. The phase diagram is not constructed. ZnBaNd$_2$S$_5$ quaternary compound is formed in this system (Wakeshima and Hinatsu 2001). It crystallizes in the tetragonal structure with the lattice parameters $a = 783.94 \pm 0.01$ and $c = 1361.31 \pm 0.01$ pm. ZnBaNd$_2$S$_5$ was synthesized by a solid-state reaction of the ZnS, BaS, and Nd$_2$S$_3$ stoichiometry mixture at 950°C for 2 days.

1.48 ZINC–BARIUM–TIN–SULFUR

ZnS–BaS–SnS$_2$. The phase diagram is not constructed. The BaZnSnS$_4$ quaternary compound is formed in this system. It crystallizes in the rhombohedral structure with the lattice parameters $a = 2196$, $b = 2150$, and $c = 1270$ pm and calculation and experimental densities 3.99 and 3.84 g·cm^{-3}, respectively (Teske 1980). This compound was obtained by the solid-state reaction at 700°C–840°C.

1.49 ZINC–BARIUM–OXYGEN–SULFUR

ZnS–BaO. The phase diagram is not constructed. The ZnBaOS quaternary compound is formed in this system. It crystallizes in the orthorhombic structure with the lattice parameters $a = 396.19 \pm 0.02$, $b = 1285.41 \pm 0.07$, and $c = 611.75 \pm 0.04$ pm and energy gap $E_g = 3.9 \pm 0.3$ eV (Broadley et al. 2005).

1.50 ZINC–CADMIUM–MERCURY–SULFUR

ZnS–CdS–HgS. Figure 1.41 shows the position of the calculated miscibility gap in this quasiternary system at 200°C and 400°C (Ohtani et al. 1992). Mercury sulfide tends to conserve the cubic structure of the Zn$_x$Cd$_y$Hg$_{1-x-y}$S solid solutions providing that Cd/Hg ratio cannot exceed some limit (Wachtel 1960).

The mineral saukovite (Hg$_{0.7}$Cd$_{0.2}$Zn$_{0.1}$S) is formed in this system (Vasil'ev 1966, 1968). It crystallizes in the cubic structure with the lattice parameter $a = 579.9 \pm 0.1$ pm and calculation and experimental densities 6.83 and 6.72 $\pm$ 0.09 [6.86 and 6.80 $\pm$ 0.04 (Vasil'ev 1968)] g·cm^{-3}, respectively (Vasil'ev 1966).

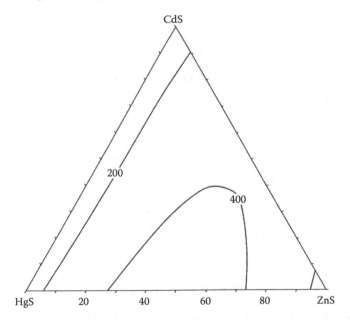

FIGURE 1.41 Calculated miscibility gap in the ZnS–CdS–HgS quasiternary system. (From Ohtani, H. et al., *J. Alloys Compd.*, 182(1), 103, 1992.)

1.51   ZINC–CADMIUM–SILICON–SULFUR

ZnS–CdS–SiS$_2$. The phase diagram is not constructed. Solid solutions are formed in this system along the section $Cd_4SiS_6-(ZnS)_{0.8}(SiS_2)_{0.2}$ within the interval from 75 to 100 mol.% Cd_4SiS_6 (Dubrovin et al. 1989). $(Zn_xCd_{1-x})_4SiS_6$ single crystals at x = 0.05, 0.13, 0.18, and 0.25 were grown using chemical transport reactions with iodine as a transport agent. Lattice parameters of obtained solid solutions change versus composition according to Vegard's rule.

1.52   ZINC–CADMIUM–GERMANIUM–SULFUR

ZnS–CdS–GeS$_2$. The phase diagram is not constructed. It was determined that the $(Zn_xCd_{1-x})_4GeS_6$ ingots at $0 < x \leq 0.25$ are the solid solutions (Dubrovin et al. 1991). The lattice parameters of these solid solutions change linearly with composition.

This system was investigated through XRD. The ingots were annealed at 600°C for 500 h. Single crystals of the $(Zn_xCd_{1-x})_4GeS_6$ solid solutions with x = 0.03, 0.08, 0.14, 0.2, and 0.25 were obtained by the chemical transport reactions (Dubrovin et al. 1991).

1.53   ZINC–CADMIUM–OXYGEN–SULFUR

ZnS–CdSO$_4$. The phase diagram is not constructed. Using DTA and determining of the phase composition at the selective dissolution of oxides, sulfates, and sulfides within the interval of 600°C–625°C, it was determined that the next interaction takes place (Kochkin et al. 1963):

$$ZnS + CdSO_4 = CdS + ZnO + SO_2$$

At 780°C–960°C, other interaction can occur too:

$$CdS + CdSO_4 = 2Cd + 2SO_2$$

The phase composition was determined by the polarography analysis of metals in the obtained solutions (Kochkin et al. 1963).

CdS–ZnSO$_4$. The phase diagram is not constructed. At 530°C–600°C, the next reaction takes place in this system (Kochkin et al. 1963):

$$CdS + 4ZnSO_4 = 4ZnO + CdSO_4 + 4SO_2$$

Cadmium sulfate forming in this reaction decomposes partially with CdO and SO_2 formation. At 700°C–750°C, CdS, CdO, and $CdSO_4$ interact with one another according to the other reactions:

$$CdS + 2CdO = 3Cd + SO_2$$

$$CdS + CdSO_4 = 2Cd + 2SO_2$$

The interaction in this system was investigated through DTA. The phase composition was determined by the polarography analysis of metals in the obtained solutions (Kochkin et al. 1963).

1.54 ZINC–CADMIUM–SELENIUM–SULFUR

ZnS–CdSe. The phase diagram is shown in Figure 1.42 (Tomashik et al. 1981). The sphalerite–wurtzite phase transformation takes place in the solid state, the temperature of which decreases with increasing CdSe concentration. The concentration dependence of the lattice parameters of forming solid solutions does not change linearly: there is a negative deviation from Vegard's law. This system was investigated through DTA and XRD.

ZnSe–CdS. This section is a nonquasibinary section of the ZnS + CdSe ⇔ ZnSe + CdS ternary mutual system (Figure 1.43) (Tomashik et al. 1979, Tomashik and Mizetskaya 1980). Using XRD, it was determined that solid solutions with wurtzite and sphalerite structure are formed in this system (Fischer and Paff 1962). Energy gap of the forming solid solutions has intermediate values in comparison with the energy gap of ZnSe and CdS. Cadmium selenide is not formed at the interaction of ZnSe and CdS.

The ingots were annealed at 1000°C for 12 h (Fischer and Paff 1962).

ZnS + CdSe ⇔ ZnSe + CdS. The liquidus and solidus surfaces of this system are shown in Figure 1.44 (Tomashik et al. 1979, Tomashik and Mizetskaya 1980, Tomashik 1981). Solid solutions are formed in the ZnS + CdSe ⇔ ZnSe + CdS ternary mutual system over the entire range of concentrations.

Calculated miscibility gap in this system at 200°C–800°C is shown in Figure 1.45 (Ohtani et al. 1992). The position of the miscibility gap indicates that the phase

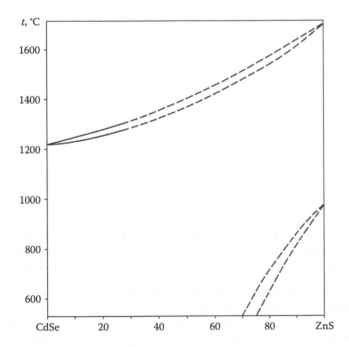

FIGURE 1.42 The ZnS–CdSe phase diagram. (From Tomashik, V.N., *Izv. AN SSSR. Neorgan. materialy*, 17(1), 17, 1981.)

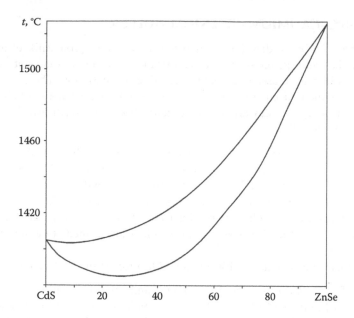

FIGURE 1.43 Phase relations in the ZnSe–CdS system. (From Tomashik, V.N. and Mizetskaya, I.B., *Izv. AN SSSR. Neorgan. materialy*, 17(6), 1116, 1981.)

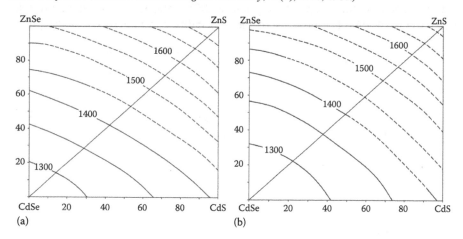

FIGURE 1.44 (a) Liquidus and (b) solidus surfaces of the ZnS + CdSe ⇔ ZnSe + CdS ternary mutual system. (From Tomashik, V.N. et al., *Izv. AN SSSR. Neorgan. materialy*, 15(2), 202, 1979; Tomashik, V.N. and Mizetskaya, I.B., *Izv. AN SSSR. Neorgan. materialy*, 16(4), 608, 1980; Tomashik, V.N., *Izv. AN SSSR. Neorgan. materialy*, 17(6), 1116, 1981.)

separation occurs toward the ZnS and CdSe ends, which have the highest and lowest melting temperatures, respectively, among the four compounds.

$Zn_xCd_{1-x}Se_yS_{1-y}$ solid solutions with high Zn content crystallize in the sphalerite-type structure; others crystallize in the wurtzite-type structure (Hotje et al. 2003). Molar volumes derived from experimentally determined lattice constants can be given in form of a plane as a function of the composition of the mixed phases.

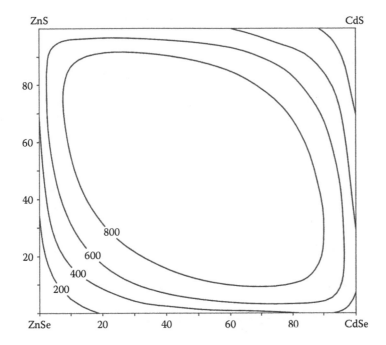

FIGURE 1.45 Calculated miscibility gap in the ZnS + CdSe $\Leftrightarrow$ ZnSe + CdS quasiternary system. (From Ohtani, H. et al., *J. Alloys Compd.*, 182(1), 103, 1992.)

The free standard enthalpy of the cation exchange reaction and inreaction parameters of the cations in the solid solutions $Zn_xCd_{1-x}S$ and $Zn_xCd_{1-x}Se$ were calculated from the cation concentrations at a pseudophase boundary in the system ZnS + CdSe $\Leftrightarrow$ ZnSe + CdS (Jacobi and Leute 1977). The positive values of the interaction parameters correspond to an ordering tendency of the cation sublattice in these solid solutions.

This system was investigated through DTA and using mathematical simulation of experiment and extrapolation method (Tomashik et al. 1979, Tomashik and Mizetskaya 1980, Tomashik 1981, Hotje et al. 2003). $Zn_xCd_{1-x}Se_yS_{1-y}$ solid solutions have been prepared by the transport chemical reactions in the presence of iodine vapor as a mineralizator gas (Hotje et al. 2003).

1.55 ZINC–CADMIUM–TELLURIUM–SULFUR

ZnS–CdTe. The phase diagram is a eutectic type (Figure 1.46) (Tomashik et al. 1978). The eutectic composition and temperature are 14 mol.% ZnS and 1079°C, respectively. The polymorphous transformation of solid solutions based on ZnS takes place at 1040°C ± 10°C. The solubility of ZnS in CdTe at 600°C, 800°C, and 1000°C reaches 5.5, 8, and 12 mol.% and the solubility of CdTe in ZnS at the same temperatures is equal to 1.5, 3.5, and 5 mol.%, respectively.

This system was investigated through DTA, metallography, and XRD. The ingots were annealed at 600°C for 200 h and at 800°C and 1000°C for 100 h (Tomashik et al. 1978).

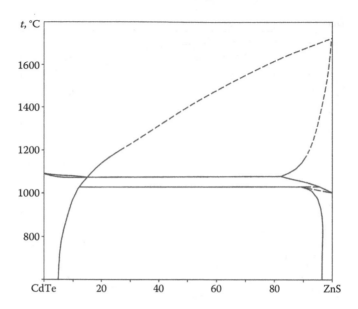

FIGURE 1.46 The ZnS–CdTe phase diagram. (From Tomashik, V.N. et al., *Izv. AN SSSR. Neorgan. materialy*, 14(10), 1838, 1978.)

ZnTe–CdS. This system is a nonquasibinary system of the Zn–Cd–Te–S quaternary system (Fischer and Paff 1962, Tomashik et al. 1978). $Zn_xCd_{1-x}S$ and CdS_xTe_{1-x} solid solutions and CdTe are formed at the interaction of ZnTe and CdS.

ZnS + CdTe ⇔ ZnTe + CdS. Two fields of primary crystallization exist on the liquidus surface of this system (Figure 1.47) (Tomashik et al. 1978). The field of primary crystallization of the solid solutions based on ZnS occupies the most part of this liquidus surface. The minimum liquidus temperature is in the ZnS–CdTe quasibinary system and corresponds to 1079°C.

Isothermal section of the ZnS + CdTe ⇔ ZnTe + CdS ternary mutual system at 740°C is shown in Figure 1.48 (Odin et al. 2005b).

Calculated miscibility gap in this system at 800°C–1400°C is shown in Figure 1.49 (Ohtani et al. 1992). The position of the miscibility gap indicates that the phase separation occurs toward the ZnS and CdTe ends, which have the highest and lowest melting temperatures, respectively, among the four compounds.

This system was investigated through DTA, metallography, and XRD (Tomashik et al. 1978, Odin et al. 2005b). The ingots were annealed at 740°C for 720 h (Odin et al. 2005b).

1.56 ZINC–CADMIUM–FLUORINE–SULFUR

CdS–ZnF₂–CdF₂. The phase diagram is not constructed. The $Zn_2Cd_2F_6S$ quaternary compound is formed in this system (Pannetier et al. 1972). It crystallizes in the cubic structure with the lattice parameter $a = 1067.0 \pm 0.2$ pm. This compound was obtained by the annealing of mixtures of binary compounds at 700°C for 4 days or by their quick heating up to 1000°C during 0.5 h.

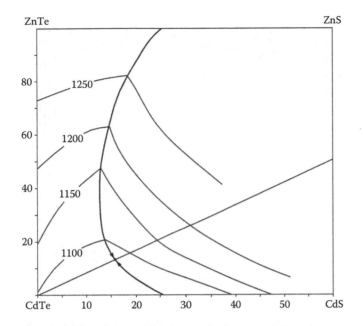

FIGURE 1.47   A part of liquidus surface of the ZnS + CdTe ⇔ ZnTe + CdS ternary mutual system. (From Tomashik, V.N. et al., *Izv. AN SSSR. Neorgan. materialy*, 14(10), 1838, 1978.)

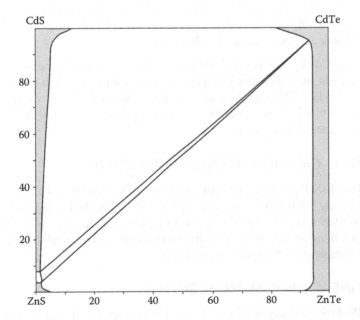

FIGURE 1.48   Isothermal section of the ZnS + CdTe ⇔ ZnTe + CdS ternary mutual system at 740°C. (From Odin, I.N. et al., *Zhurn. neorgan. khimii*, 50(6), 1018, 2005b.)

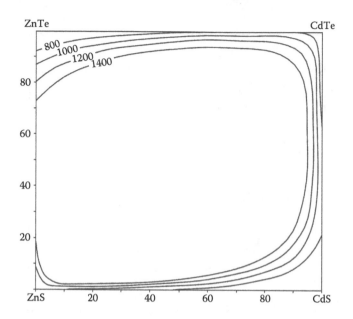

FIGURE 1.49 Calculated miscibility gap in the ZnS + CdTe ⇔ ZnTe + CdS quasiternary system. (From Ohtani, H. et al., *J. Alloys Compd.*, 182(1), 103, 1992.)

1.57   ZINC–CADMIUM–CHLORINE–SULFUR

ZnS–CdCl$_2$. The phase diagram is not constructed. The annealing of ZnS films in the CdCl$_2$ vapor leads to the formation of Zn$_x$Cd$_{1-x}$S solid solutions with $0 \leq x \leq 1$ (Kashina et al. 1986). The ZnS films were annealed at 400°C, 450°C, and 500°C in the CdCl$_2$ vapor and using the inert atmosphere.

CdS–ZnCl$_2$–CdCl$_2$. The phase diagram is not constructed. Using oriented zone recrystallization and radioactive isotopes, the equilibrium coefficient of ZnCl$_2$ distribution in the CdS–CdCl$_2$ eutectic was determined: $k_0 = 1.77$ (Leonov and Chunarev 1977). As $k_0 > 0$, it is obvious that the next exchange reaction CdS + ZnCl$_2$ = ZnS + CdCl$_2$ takes place in this system.

1.58   ZINC–CADMIUM–MANGANESE–SULFUR

ZnS–CdS–MnS. Phase relations in this system at 500°C and 0.1 GPa (1000 atm) are shown in Figure 1.50 (Tauson and Chernyshev 1981). The isothermal section of the ZnS–CdS–MnS system at 900°C is given in Figure 1.51 (Kröger 1940). Practically all solid solutions crystallize in the wurtzite structure, and only solid solutions with high ZnS content show the sphalerite structure.

1.59   ZINC–CADMIUM–IRON–SULFUR

ZnS–CdS–FeS. Cell edges of the (Cd + Fe)-bearing sphalerites in this system should be linear functions of their compositions, and the following relation should hold: a (A) = 5.4093 + 0.000456x + 0.00424y, where x and y are the FeS and CdS

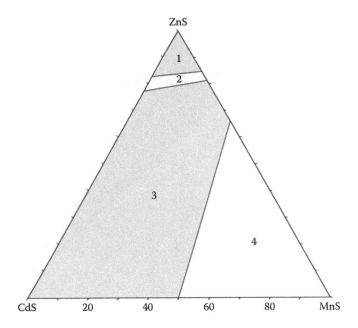

FIGURE 1.50 Phase relations in the ZnS–CdS–MnS quasiternary system at 500°C and 0.1 GPa
(1000 atm): 1, sphalerite; 2, sphalerite + wurtzite; 3, wurtzite; 4, wurtzite + MnS. (From Tauson,
V.L. and Chernyshev, L.V. *Experimental Investigations on Zinc Sulfide Crystal Chemistry and
Geochemistry* [in Russian], Nauka Publish., Novosibirsk, Russia, 1981, 190pp.)

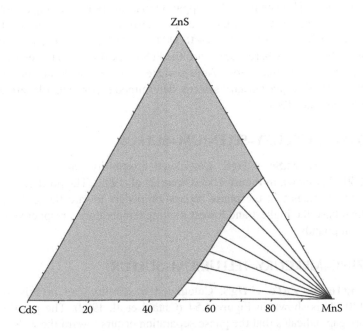

FIGURE 1.51 Isothermal section of the ZnS–CdS–MnS quasiternary system at 900°C.
(From Kröger, F.A., *Z. Kristallogr.*, A102(1), 132, 1940.)

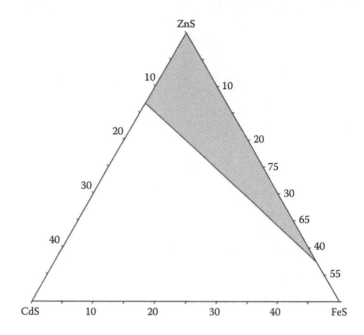

FIGURE 1.52 Approximate limits of sphalerite solid solutions in the ZnS–CdS–FeS qua-siternary system, determined experimentally. (From Skinner, B.J., *Am. Mineral.*, 46(11–12), 1399, 1961.)

contents in mol.% (Skinner 1961). It is apparent that functions defining *a* and *c* of the solid solutions with wurtzite structure can be derived by simple addition from the quasibinary system, so *a* (A) = 3.8230 + 0.000490*x* + 0.003124*y* and *c* (A) = 6.2565 + 0.000886*x* + 0.00455*y*, where *x* and *y* are the FeS and CdS contents, respectively, in mol.% (Skinner and Bethke 1961). Approximate limits of sphalerite solid solutions in the ZnS–CdS–FeS quasiternary system, determined experimentally, are shown in Figure 1.52 (Skinner 1961).

1.60 ZINC–MERCURY–SELENIUM–SULFUR

ZnS + HgSe ⇔ ZnSe + HgS. Calculated miscibility gap in this system at 400°C–800°C is shown in Figure 1.53 (Ohtani et al. 1992). The position of the miscibility gap indicates that the phase separation occurs toward the ZnS and HgSe ends, which have the highest and lowest melting temperatures, respectively, among the four compounds.

1.61 ZINC–MERCURY–TELLURIUM–SULFUR

ZnS + HgTe ⇔ ZnTe + HgS. Calculated miscibility gap in this system at 800°C–1200°C is shown in Figure 1.54 (Ohtani et al. 1992). The position of the miscibility gap indicates that the phase separation occurs toward the ZnS and HgTe ends, which have the highest and lowest melting temperatures, respectively, among the four compounds.

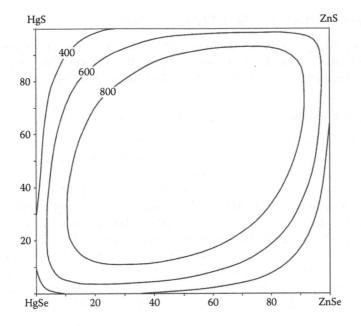

FIGURE 1.53 Calculated miscibility gap in the ZnS + HgSe ⇔ ZnSe + HgS quasiternary system. (From Ohtani, H. et al., *J. Alloys Compd.*, 182(1), 103, 1992.)

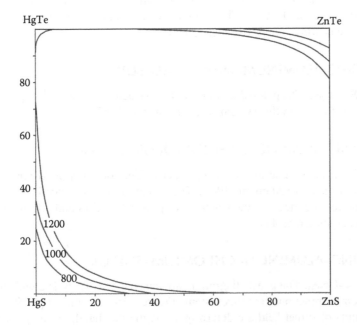

FIGURE 1.54 Calculated miscibility gap in the ZnS + HgTe ⇔ ZnTe + HgS quasiternary system. (From Ohtani, H. et al., *J. Alloys Compd.*, 182(1), 103, 1992.)

1.62 ZINC–MERCURY–FLUORINE–SULFUR

ZnS–ZnF$_2$–HgF$_2$. The phase diagram is not constructed. The Zn$_2$Hg$_2$SF$_6$ quaternary compound is formed in this system. It crystallizes in the cubic structure with the lattice parameter a = 1078.01 ± 0.26 pm (Bernard et al. 1975).

1.63 ZINC–MERCURY–CHLORINE–SULFUR

ZnS–HgS–HgCl$_2$. The phase diagram is not constructed. The ZnHg$_3$S$_2$Cl$_4$ quaternary compound is formed in this system. It crystallizes in the hexagonal structure with the lattice parameters a = 725.7 ± 0.4 and c = 1084.8 ± 0.8 pm, calculated density 5.861 g·cm^{-3}, and optical bandgap 2.65 eV (Chen et al. 2010a).

This compound was prepared from the solid-state reaction of HgCl$_2$, ZnS, and HgS. The starting materials were loaded into a silica tube. The tube was flame-sealed under a 133.3 Pa atmosphere and subsequently placed into a furnace. It was heated to 200°C for 6 h from room temperature and kept for 24 h, then heated to 450°C for 6 h and kept for 15 days, followed by cooling to 100°C at a rate of 5°C/h to promote crystal growth, then cooled to 35°C for 5 h, followed by power-off. It has been investigated through thermogravimetry-DTA, XRD, and SEM with EDX (Chen et al. 2010a).

1.64 ZINC–BORON–OXYGEN–SULFUR

ZnS–ZnO–B$_2$O$_3$. The phase diagram is not constructed. The Zn$_4$(BO$_2$)$_6$S quaternary compound (mineral sodalite) is formed in this system. It melts incongruently at 1070°C and crystallizes in the cubic structure with the lattice parameter a = 763.5 ± 0.2 pm (Fouassier et al. 1970). This compound was obtained from the binary compound by the heating at approximately 500°C.

1.65 ZINC–ALUMINUM–INDIUM–SULFUR

ZnS–Al$_2$S$_3$–In$_2$S$_3$. The phase diagram is not constructed. The Zn$_3$AlInS$_6$ quaternary compound is formed in this system (Radautsan et al. 1987).

1.66 ZINC–ALUMINUM–PHOSPHORUS–SULFUR

ZnS–AlP. The phase diagram is not constructed. The solubility of AlP in ZnS is not higher than 1 mol.% (Addamiano 1960). This system was investigated through XRD. The ingots were obtained by the sintering of powderlike ZnS and AlP at 900°C for 24 h in the argon atmosphere.

1.67 ZINC–ALUMINUM–CHROMIUM–SULFUR

ZnS–Al$_2$S$_3$–Cr$_2$S$_3$. The phase diagram is not constructed. There is a wide homogeneity region of spinel phase in this system (Figure 1.55) (Kovaliv et al. 1981).

The limits of spinel field are determined, on the one hand, by the possibility of the existence of cation-deficient phases in the quasiternary system and, on the other hand, by the limit of chromium concentration in nonstoichiometric solid solutions.

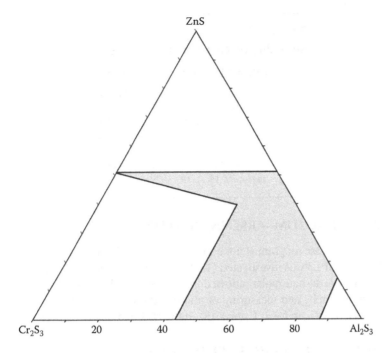

FIGURE 1.55 The limits of spinel solid solutions in the ZnS–Al_2S_3–Cr_2S_3. (From Kovaliv, V.I. et al., *Izv. AN SSSR. Neorgan. materialy*, 17(11), 2084, 1981.)

1.68 ZINC–GALLIUM–INDIUM–SULFUR

ZnS–Ga_2S_3–In_2S_3. The phase diagram is not constructed. The Zn_3GaInS_6 and $ZnGaInS_4$ quaternary compounds are formed in this system (Radautsan et al. 1985, 1987, Sugiyama et al. 2005). Zn_3GaInS_6 crystallizes in the hexagonal structure with the lattice parameters $a = 1540$ and $c = 1240$ pm and energy gap $E_g = 3.26$ eV (Radautsan et al. 1985, 1987). $ZnGaInS_4$ has both a defect chalcopyrite (the lattice parameters $a = 530$ and $c = 1050$ pm) and a layered structure below 920°C, and only the layered structure exists between 920°C and 1020°C (Sugiyama et al. 2005). The bandgap of the layered $ZnGaInS_4$ was estimated to be approximately 2.6 eV and it is possibly a direct bandgap. The layered structure of $ZnGaInS_4$ was grown using the normal freezing method.

1.69 ZINC–GALLIUM–PHOSPHORUS–SULFUR

ZnS–GaP. The phase diagram is not constructed. Solid solutions with sphalerite structure over the entire range of concentrations are formed in this system (Yim 1969, Sonomura et al. 1973, Voitsehovski and Panchenko 1975, 1977). The lattice parameters of these solid solutions change linearly with composition (Yim 1969).

This system was investigated through XRD. The single crystals of $(ZnS)_x(GaP)_{1-x}$ solid solutions were obtained by the chemical transport reactions and by the crystallization from the solutions in the melts of Zn, Ga, and Sn (Yim 1969, Sonomura et al. 1973, Voitsehovski and Panchenko 1975, 1977).

TABLE 1.1
Solubility of Zn + S (cm⁻³) in InAs

t, °C	Zn/S = 3:1	Zn/S = 1:1	Zn/S = 1:3
700	$2.0 \cdot 10^{20}$	$6.1 \cdot 10^{20}$	$3.2 \cdot 10^{20}$
800	$2.3 \cdot 10^{20}$	$7.9 \cdot 10^{20}$	$4.3 \cdot 10^{20}$
850	$3.5 \cdot 10^{20}$	$15.8 \cdot 10^{20}$	$9.4 \cdot 10^{20}$
900	$5.8 \cdot 10^{20}$	$22 \cdot 10^{20}$	$11.5 \cdot 10^{20}$

Source: Glazov, V.M. et al., *Izv. AN SSSR. Neorgan. materialy*, 15(3), 390, 1979.

1.70   ZINC–GALLIUM–ARSENIC–SULFUR

ZnS–InAs. The phase diagram is not constructed. The solubility of Zn + S in InAs at Zn/S = 3:1, 1:1, and 1:3 was investigated (Table 1.1) (Glazov et al. 1979). Maximum solubility takes place at the equimolar ratio of doping elements. This system was investigated through metallography and measuring of microhardness. The ingots were annealed at 700°C, 800°C, 850°C, and 900°C for 1200, 1000, 800, and 500 h, respectively.

1.71   ZINC–GALLIUM–TELLURIUM–SULFUR

ZnS–Ga₂Te₃. The phase diagram is a eutectic type (Figure 1.56) (Odin et al. 2005a). The eutectic composition and temperature are 8 ± 2 mol.% ZnS and 774°C ± 5°C, respectively. Phase transformation of the solid solutions based on ZnS takes place at 1055°C.

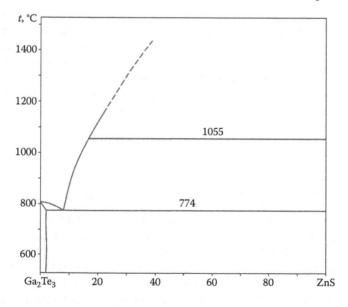

FIGURE 1.56   The ZnS–Ga₂Te₃ phase diagram. (From Odin, I.N. et al., *Zhurn. neorgan. khimii*, 50(5), 848, 2005a.)

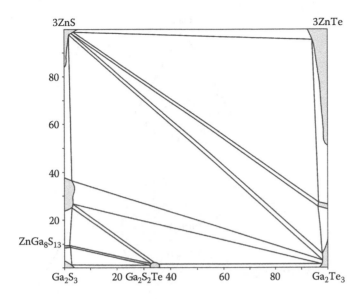

FIGURE 1.57 Isothermal section of the $3ZnS + Ga_2Te_3 \Leftrightarrow 3ZnTe + Ga_2S_3$ ternary mutual system at 720°C. (From Odin, I.N. et al., *Zhurn. neorgan. khimii*, 50(5), 848, 2005a.)

This system was investigated through DTA, metallography, and XRD. The ingots were annealed at 720°C for 1000 h with the next cooling in the ice water.

$3ZnS + Ga_2Te_3 \Leftrightarrow 3ZnTe + Ga_2S_3$. The isothermal section of the $3ZnS + Ga_2Te_3 \Leftrightarrow 3ZnTe + Ga_2S_3$ ternary mutual system at 720°C is shown in Figure 1.57 (Odin et al. 2005a). Stable quaternary compounds were not found in this system.

1.72 ZINC–GALLIUM–IRON–SULFUR

In this system, Ga solubility in wurtzite at 800°C and 900°C is more than 15.6 and 21.4 at.%, respectively, and in sphalerite at 800°C is more than 27.4 at.% (Ueno and Scott 1991). The cell parameters of Ga-bearing sphalerite and Ga-bearing wurtzite decrease dramatically with increasing Ga content. Besides wurtzite and sphalerite solid solutions in the central portion of the Zn–Ga–Fe–S system, other solid solutions are phases V, U, W, and X (Ueno and Scott 2002). Phase V extends into the tetrahedron: its maximum Fe content is 12.5 at.% at 900°C and 11.9 at.% at 800°C. Phase U does not contain Fe, and phases Z and W do not contain Zn. GaS, Ga_2S_3, and phase X have very small area of solid solution in the Zn–Ga–Fe–S tetrahedron. The solid solution between the three monosulfides (with the structure of sphalerite, wurtzite, or both) extends over a large region, reaching a maximum of 28.6 at.% Fe and 28.2 at.% Ga at 900°C and 26.4 at.% Fe and 28.2 at.% Ga at 800°C (Ueno and Scott 2002, Raghavan 2004b). The ingots were annealed at 800°C for 26–56 days and at 900°C for 13–41 days (Ueno and Scott 1991).

ZnS–GaS–FeS. In the projection diagrams of this system at 800°C (Figure 1.58), sphalerite is the only stable form and extends as a solid solution from the ZnS end

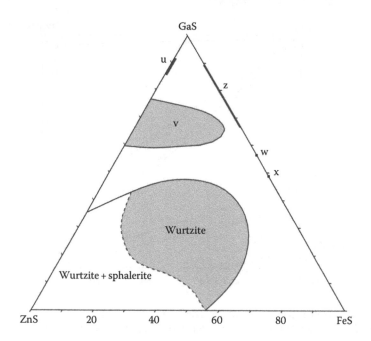

FIGURE 1.58   Isothermal section of the ZnS–FeS–GaS system at 800°C. (From Ueno, T. and Scott, S.D., *Can. Mineral.*, 34(5), 949, 1996.)

up to ~55 mol.% FeS (Ueno and Scott 1996, Raghavan 2004b). Addition of GaS results in the appearance of the sphalerite + wurtzite two-phase mixture. Further addition of FeS stabilizes the wurtzite form. Phases Z, W, and X along the GaS–FeS side do not dissolve any ZnS. The phase U along the ZnS–GaS side does not dissolve any FeS. The phase V along the ZnS–GaS side, however, dissolves up to about 29 mol.% FeS.

The phase distribution at 900°C (Figure 1.59) is somewhat similar to that at 800°C, except that the wurtzite phase field is somewhat larger and the sphalerite + wurtzite two-phase field reappears along the ZnS–GaS side between 30 and 55 mol.% GaS (Ueno and Scott 1996, Raghavan 2004b). There are limitations about the quasiternary sections constructed by Ueno and Scott (1996). Some phases shown in Figures 1.58 and 1.59 lie on the Ga_2S_3–FeS join or the ZnS–Ga_2S_3 join. On the ZnS–GaS–FeS plane, they are expected to be in equilibrium with other phases. In view of this, the constructed sections are not strictly quasiternary (Raghavan 2004b). The ingots were annealed at 800°C for 28–47 days and at 900°C for 13–24 days (Ueno and Scott 1996).

1.73   ZINC–INDIUM–SELENIUM–SULFUR

ZnS–In_2S_3–In_2Se_3. The phase diagram is not constructed. The $Zn_{1.25}In_{2.5}S_3Se_2$ quaternary compound is formed in this system (Haeuseler et al. 1988). It crystallizes in the trigonal structure with the lattice parameters $a = 392.7$ and $c = 3183$ pm. Single crystals of this compound were obtained by the chemical transport reactions.

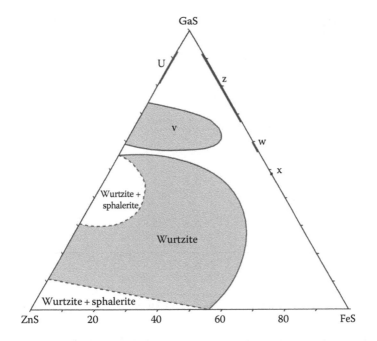

FIGURE 1.59 Isothermal section of the ZnS–FeS–GaS system at 900°C. (From Ueno, T. and Scott, S.D., *Can. Mineral.*, 34(5), 949, 1996.)

1.74   ZINC–INDIUM–TELLURIUM–SULFUR

ZnS–In$_2$Te$_3$. The phase diagram is a eutectic type (Figure 1.60) (Odin et al. 2005b). The eutectic composition and temperature are 7 ± 2 mol.% ZnS and 675°C ± 5°C, respectively. Phase transformation of the solid solutions based on ZnS takes place at 1053°C. This system was investigated through DTA, metallography, and XRD. The ingots were annealed at 740°C for 720 h with the next cooling in the ice water.

ZnS–ZnTe–In$_2$Te$_3$. The isothermal section of the ZnS–ZnTe–In$_2$Te$_3$ quasiternary system at 740°C is shown on Figure 1.61 (Odin et al. 2005b).

1.75   ZINC–LANTHANUM–OXYGEN–SULFUR

ZnS–La$_2$O$_2$S. The phase diagram is not constructed. According to the data of Baranov et al. (1996), this system is a quasibinary system of the Zn–La–O–S quaternary system. New phases were not found in the ZnS–La$_2$O$_2$S system. This system was investigated through XRD. The ingots were annealed at 730°C–830°C for 250 h.

1.76   ZINC–CARBON–FLUOR–SULFUR

Zn(SC$_6$F$_5$)$_2$ or **C$_{12}$F$_{10}$S$_2$Zn** quaternary compound exists in this system. To obtain this compound, approximately 4 mmol of ZnO was stirred magnetically with a MeOH solution of a slight stoichiometric excess of pentafluorothiophenol until ZnO had dissolved (Peach 1968). The methanol–water was fractionated off and the product washed and dried. C$_{12}$F$_{10}$S$_2$Zn melts at a temperature higher than 350°C.

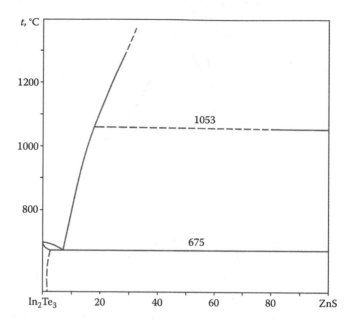

FIGURE 1.60   The ZnS–In$_2$Te$_3$ phase diagram. (From Odin, I.N. et al., *Zhurn. neorgan. khimii*, 50(6), 1018, 2005b.)

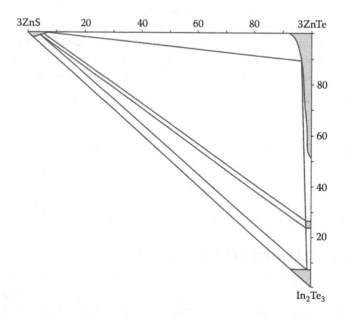

FIGURE 1.61   Isothermal section of the ZnS–ZnTe–In$_2$Te$_3$ quasiternary system at 740°C. (From Odin, I.N. et al., *Zhurn. neorgan. khimii*, 50(6), 1018, 2005b.)

1.77 ZINC–CARBON–CHLORINE–SULFUR

$Zn(SC_6Cl_5)_2$ or $C_{12}Cl_{10}S_2Zn$ quaternary compound exists in this system. This compound could be obtained if a suspension of C_6Cl_5SH and solution of ZnO in 95% ethanol was stirred at room temperature (Lucas and Peach 1969, 1970). $C_{12}Cl_{10}S_2Zn$ was isolated after removal of the solvent. It melts at a temperature higher than 300°C.

1.78 ZINC–TIN–TELLURIUM–SULFUR

ZnS–SnTe. The phase diagram is a eutectic type (Figure 1.62) (Dubrovin et al. 1983). The eutectic composition and temperature are 5 mol.% ZnS and 785°C, respectively. The solubility of ZnS in SnTe at the eutectic temperature is equal to 3 and 2 mol.% at 400°C, and the solubility of SnTe in ZnS is not higher than 1 mol.%. This system was investigated through DTA, metallography, XRD, and measuring of microhardness. The ingots were annealed at 700°C and 400°C for 120 and 250 h, respectively.

1.79 ZINC–LEAD–OXYGEN–SULFUR

ZnS–PbSO₄. The phase diagram is not constructed. Research carried out on the interaction between ZnS and $PbSO_4$ showed that the process starts at 530°C (Malinowski 1992, Malinowski et al. 1996). It was found that a few reactions occur at the same time, including transformation of ZnS into ZnO and PbS formation. The secondary $PbSO_4$ containing the sulfur from ZnS also appears. The PbS is

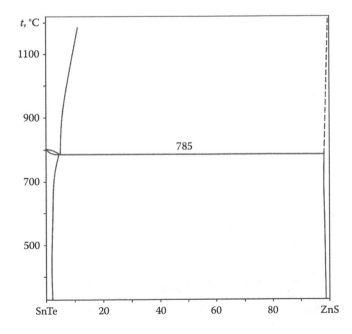

FIGURE 1.62 The ZnS–SnTe phase diagram. (From Dubrovin, I.V. et al., *Izv. AN SSSR. Neorgan. materialy*, 19(11), 1816, 1983.)

derived primarily from $PbSO_4$, and some of it contains the sulfur from ZnS also. The presence of $PbO \cdot PbSO_4$ in the products is possible in the case of a low ZnS content (Malinowski 1992). The dependence of the preparation mass reduction and the degree of change of ZnS into ZnO have been determined. This system was investigated through DTA, differential thermogravimetry (DTG), thermogravimetry (TG), XRD, and chemical analysis (Malinowski 1992, Malinowski et al. 1996).

ZnS–PbO. The phase diagram is not constructed. Using DTA, XRD, and chemical analysis, it was determined that annealing of ZnS + PbO mixtures leads to the formation of ZnO and Pb (Suleimanov et al. 1974a,b). This indicates that ZnS, PbO, and forming products interact according to the next reactions:

$$ZnS + PbO = ZnO + PbS$$

$$PbS + 2PbO = 3Pb + SO_2$$

1.80 ZINC–LEAD–SELENIUM–SULFUR

ZnS–PbSe. The phase diagram is a eutectic type (Figure 1.63) (Oleinik et al. 1982). The eutectic composition and temperature are 18 mol.% ZnS and 1019°C, respectively. Mutual solubility of ZnS and PbSe is not higher than 1 mol.%. This system was investigated through DTA, metallography, and XRD.

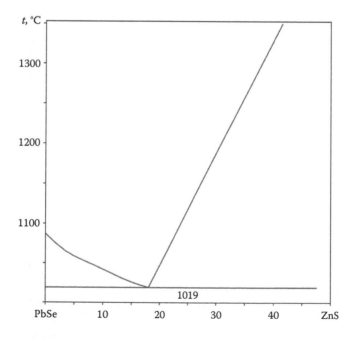

FIGURE 1.63 The ZnS–PbSe phase diagram. (From Oleinik, G.S. et al., *Izv. AN SSSR. Neorgan. materialy*, 18(5), 873, 1982.)

1.81 ZINC–LEAD–IRON–SULFUR

ZnS–PbS–FeS. A part of the liquidus surface of this quasiternary system (Figure 1.64) was constructed according to six polythermal sections, three of which are presented in Figure 1.65 (Avetisian and Gnatyshenko 1956). It includes three fields of primary crystallization. The ternary eutectic composition and temperature are 4.2 mol.% ZnS + 54.9 mol.% FeS + 40.9 mol.% PbS (2.7 mass.% ZnS + 32.1 mass.% FeS + 65.2 mass.% PbS), and 820°C, respectively. This system was investigated through DTA, metallography, and chemical analysis.

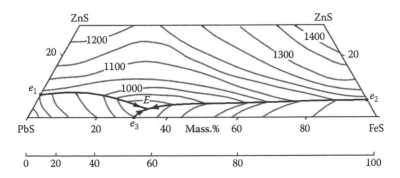

FIGURE 1.64 A part of liquidus surface of the ZnS–PbS–FeS quasiternary system. (From Avetisian, H.K. and Gnatyshenko, G.I., *Izv. AN KazSSR. Ser. gorn. dela, stroimaterialov i metallurgii*, 6, 11, 1956.)

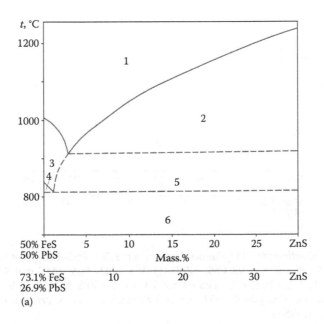

FIGURE 1.65 Polythermal sections of the ZnS–PbS–FeS quasiternary system: (a) (73.1 mol.% FeS + 26.9 mol.% PbS) – ZnS. *(Continued)*

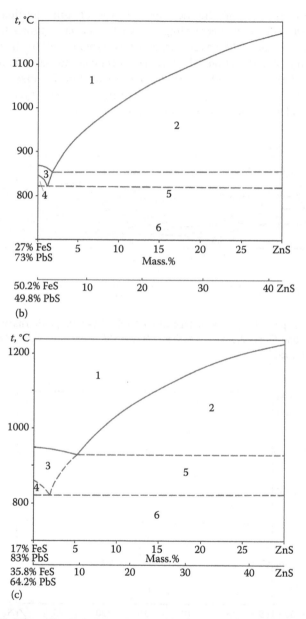

FIGURE 1.65 (*Continued*) Polythermal sections of the ZnS–PbS–FeS quasiternary system: (b) (50.2 mol.% FeS + 49.8 mol.% PbS) – ZnS. (c) (35.8 mol.% FeS + 64.2 mol.% PbS) – ZnS: 1, L; 2, L + ZnS; 3, L + FeS; 4, L + FeS + PbS; 5, L + FeS + ZnS; 6, PbS + FeS + ZnS. (From Avetisian, H.K. and Gnatyshenko, G.I., *Izv. AN KazSSR. Ser. gorn. dela, stroimaterialov i metallurgii*, 6, 11, 1956.)

1.82 ZINC–NITROGEN–PHOSPHORUS–SULFUR

ZnS–Zn$_3$N$_2$–P$_3$N$_5$. The phase diagram is not constructed. The Zn$_8$[P$_{12}$N$_{24}$]S$_2$ [Zn$_4$P$_6$N$_{12}$S (Ronis et al. 1988, 1989)] quaternary compound is formed in this system (Wester and Schnick 1996). It crystallizes in the cubic structure with the lattice parameter $a = 822.66 \pm 0.02$ [$a = 820.4 \pm 0.4$ (Ronis et al. 1988, 1989)] pm and calculated density 3.86 g·cm^{-3} (Wester and Schnick 1996).

Zn$_8$[P$_{12}$N$_{24}$]S$_2$ was obtained by the reaction of HPN$_2$ with ZnS at 750°C (Wester and Schnick 1996) or by the interaction of ZnS and P$_2$S$_5$ in ammonia at 900°C–970°C (Ronis et al. 1988, 1989): 4ZnS + 3P$_2$S$_5$ + 12NH$_3$ = Zn$_4$P$_6$N$_{12}$S.

1.83 ZINC–BISMUTH–TELLURIUM–SULFUR

ZnS–Bi$_2$Te$_3$. The phase diagram is a eutectic type (Figure 1.66) (Odin 1996). The eutectic composition and temperature are 5 ± 1 mol.% ZnS and 575°C $\pm$ 4°C, respectively. Thermal effects at 1050°C correspond to the polymorphous transformation of ZnS at the liquid presence. The solubility of ZnS in Bi$_2$Te$_3$ reaches 3 mol.% and the solubility of Bi$_2$Te$_3$ in ZnS is insignificant. This system was investigated through DTA, metallography, and XRD. The ingots were annealed at temperatures 20°C below the temperatures of nonvariant equilibria with liquidus presence for 1000 h.

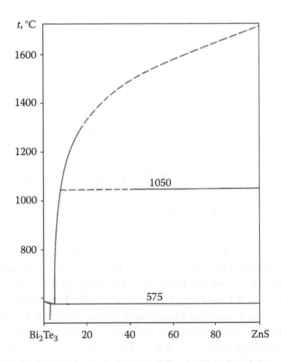

FIGURE 1.66 The ZnS–Bi$_2$Te$_3$ phase diagram. (From Odin, I.N., *Zhurn. neorgan. khimii*, 41(6), 941, 1996.)

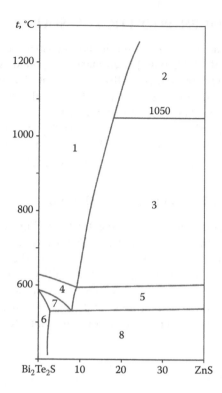

FIGURE 1.67 Phase relations in the ZnS–Bi$_2$Te$_2$S system: 1, L; 2, L + β-ZnS; 3, L + α-ZnS; 4, L + γ; 5, L + γ + α-ZnS; 6, γ + δ; 7, L + γ + δ; and 8, γ + δ + α-ZnS. (From Odin, I.N., *Zhurn. neorgan. khimii*, 41(6), 941, 1996.)

ZnS–Bi$_2$Te$_2$S. This section is a nonquasibinary section of the Zn–Bi–Te–S quaternary system (Figure 1.67) (Odin 1996). The phase based on Bi$_2$Te$_2$S crystallizes from the Bi$_2$Te$_2$S-rich side. Thermal effects at 1050°C correspond to the polymorphous transformation of ZnS at the liquid presence. The solubility of ZnS in Bi$_2$Te$_3$ reaches 3 mol.% and the solubility of Bi$_2$Te$_3$ in ZnS is insignificant. This system was investigated through DTA, metallography, and XRD. The ingots were annealed at temperatures 20°C below the temperatures of nonvariant equilibria with liquidus presence for 1000 h.

3ZnS + Bi$_2$Te$_3$ ⇔ 3ZnTe + Bi$_2$S$_3$. The field of ZnS primary crystallization occupies the most part of the liquidus surface of this ternary mutual system (Figure 1.68) (Odin 1996). There are also the fields of primary crystallization of ZnTe, Bi$_2$S$_3$, Bi$_2$Te$_2$S (γ-phase), and δ-phase (the phase based on Bi$_2$Te$_3$). Three ternary eutectics are in this system. Three- and four-phase equilibria in the 3ZnS + Bi$_2$Te$_3$ ⇔ 3ZnTe + Bi$_2$S$_3$ ternary mutual system are given in Table 1.2. This system was investigated through DTA, metallography, and XRD. The ingots were annealed at temperatures 20°C below the temperatures of nonvariant equilibria with liquidus presence for 1000 h.

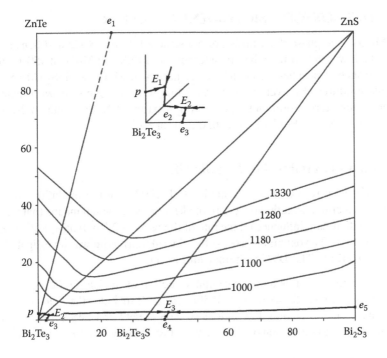

FIGURE 1.68 Liquidus surface of the $3ZnS + Bi_2Te_3 \Leftrightarrow 3ZnTe + Bi_2S_3$ ternary mutual system. (From Odin, I.N., *Zhurn. neorgan. khimii*, 41(6), 941, 1996.)

TABLE 1.2
Three- and Four-Phase Equilibria in the $3ZnS + Bi_2Te_3 \Leftrightarrow$ $3ZnTe + Bi_2S_3$ Ternary Mutual System

Symbol	Reaction	t, °C
e_1	$L \Leftrightarrow \beta\text{-}ZnS + ZnTe$	1260
e_2	$L \Leftrightarrow \alpha\text{-}ZnS + (Bi_2Te_3)$	582
e_3	$L \Leftrightarrow (Bi_2Te_3) + (Bi_2Te_2S)$	581
e_4	$L \Leftrightarrow (Bi_2S_3) + (Bi_2Te_2S)$	622
e_5	$L \Leftrightarrow (Bi_2S_3) + \alpha\text{-}ZnS$	774
e_6	$L \Leftrightarrow (Bi_2Te_2S) + \alpha\text{-}ZnS$	617
p	$L + ZnTe \Leftrightarrow (Bi_2Te_3)$	600
E_1	$L \Leftrightarrow \alpha\text{-}ZnS + ZnTe + (Bi_2Te_3)$	575
E_2	$L \Leftrightarrow \alpha\text{-}ZnS + (Bi_2Te_2S) + (Bi_2Te_3)$	571
E_3	$L \Leftrightarrow \alpha\text{-}ZnS + (Bi_2S_3) + (Bi_2Te_2S)$	615

Source: Odin, I.N., *Zhurn. neorgan. khimii*, 41(6), 941, 1996.

1.84 ZINC–OXYGEN–MOLYBDENUM–SULFUR

ZnS–MoO₃. The phase diagram is not constructed. Using XRD and density measurements, it was determined that the interaction of ZnS and MoO_3 at 450°C–664°C in the nitrogen atmosphere leads to the formation of MoO_2 + MoS_2 (Mo_2S_3) hard separable mixtures (Bhuiyan et al. 1979). MoO_2 contents increase with increasing reaction temperature. Addition to the starting mixture of NH_4Cl leads to the formation of pure MoS_2 at 500°C for 2 h with output of 89.1%.

1.85 ZINC–OXYGEN–IRON–SULFUR

The experimental results show (Živković et al. 1998) that the oxidation process of (Zn, Fe)S (mineral marmatite) occurs in the kinetic field, which means that temperature has the dominant influence on the rate of the process 2(Zn, Fe)S + $4.5O_2$ = ZnO + $ZnFe_2O_4$ + $2SO_2$. The oxidation process of mineral marmatite was investigated using simultaneous DTA–TG–DTG technique and XRD.

ZnS–FeO. This section is a nonquasibinary section of the Zn–O–Fe–S quaternary system (Figure 1.69) (Toguzov et al. 1980a, 1982). It includes the fields of FeO and ZnS primary crystallization, the regions of ZnS + FeO and FeS + FeO secondary crystallization, and the region of ZnS + FeO + FeS ternary eutectic. The solubility of

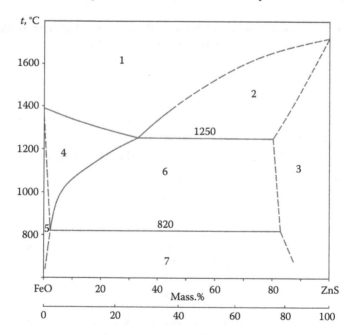

FIGURE 1.69 Phase relations in the ZnS–FeO system: 1, L; 2, L + ZnS; 3, ZnS; 4, L + FeO; 5, FeO; 6, L + ZnS + FeO; and 7, FeO + FeS + ZnS. (From Toguzov, M.Z. et al., *Zhurn. neorgan. khimii*, 25(10), 2863, 1980a; Toguzov, M.Z. et al., Phase equilibria in the Cu₂S–FeS–ZnS–ZnO system [in Russian], in: *Sul'fid. rasplavy tiazh. met.*, Nauka Publish., 1982, p. 122.)

ZnS in liquid FeO increases from 24.9 to 42.4 mol.% (31–50 mass.%) with temperature increasing from 1250°C to 1400°C. Liquidus lines have the common point at
33 mol.% ZnS and 1250°C. The crystallization ends in the ternary eutectic at 920°C.
Its projection on the ZnS–FeO section corresponds to 98.5 mol.% (98 mass.%) FeO
and 1.5 mol.% (2 mass.%) ZnS. At 930°C, the solubility of ZnS in FeO is not higher
than 0.74 mol.% (1 mass.%) and the solubility of FeO in ZnS at 1250°C reaches
24.1 mol.% (19 mass.%).

This system was investigated through DTA, metallography, and local XRD
(Toguzov et al. 1980a, 1982).

ZnS–FeO–FeS. The liquidus surface of this system (Figure 1.70) was constructed according to four polythermal sections from the ZnS corner (Figure 1.71)
(Toguzov et al. 1980b, 1982). It includes three fields of primary crystallization of
ZnS, FeO, and FeS. The field of ZnS primary crystallization occupies the most
part of the liquidus surface. The composition and temperature of ternary eutectic
are 40.9 mol.% (36.0 mass.%) FeO, 57.0 mol.% (61.5 mass.%) FeS, 2.1 mol.%
(2.5 mass.%) ZnS, and 920°C, respectively. This system was investigated through
DTA and metallography.

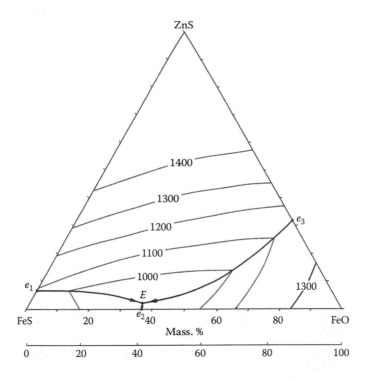

FIGURE 1.70 Liquidus surface of the ZnS–FeO–FeS quasiternary system. (From Toguzov,
M.Z. et al., *Zhurn. neorgan. khimii*, 25(10), 2873, 1980b; Toguzov, M.Z. et al., Phase equilib-
ria in the Cu₂S–FeS–ZnS–ZnO system [in Russian], in: *Sul'fid. rasplavy tiazh. met.*, Nauka
Publish., 1982, p. 122.)

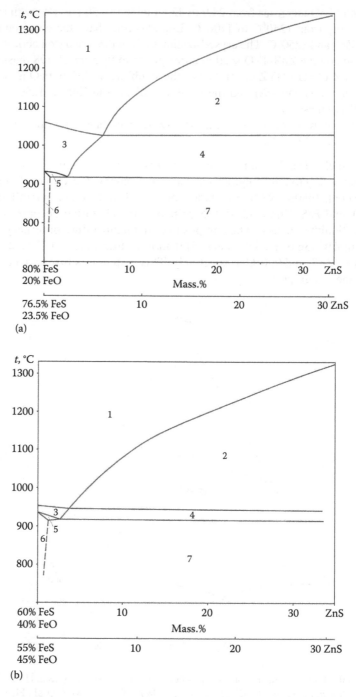

FIGURE 1.71 Polythermal sections of the ZnS–FeO–FeS quasiternary system: (a) (80 mass.% FeS + 20 mass.% FeO) – ZnS; (b) (60 mass.% FeS + 40 mass.% FeO) – ZnS.

(*Continued*)

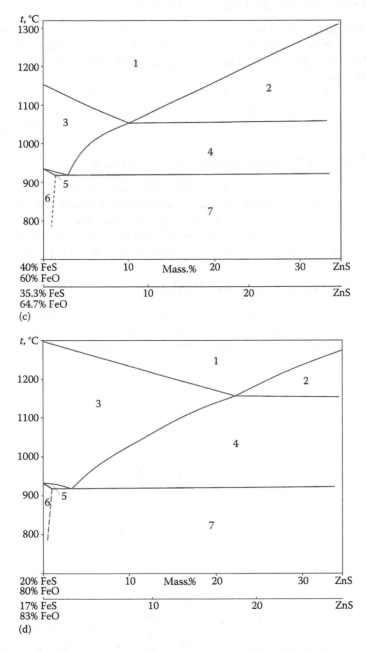

FIGURE 1.71 (*Continued*) Polythermal sections of the ZnS–FeO–FeS quasiternary system: (c) (40 mass.% FeS + 60 mass.% FeO) – ZnS; (d) (20 mass.% FeS + 80 mass.% FeO) – ZnS: 1, L; 2, L + ZnS; 3, L + FeO; 4, L + FeO + ZnS; 5, L + FeO + FeS; 6, FeO_xS_{1-x}; and 7, FeS + FeO + ZnS. (From Toguzov, M.Z. et al., Phase equilibria in the Cu_2S–FeS–ZnS–ZnO system [in Russian], in: *Sul'fid. rasplavy tiazh. met.*, Nauka Publish., 1982, p. 122.)

1.86   ZINC–SELENIUM–TELLURIUM–SULFUR

ZnS–ZnSe–ZnTe. The liquidus surface of this quasiternary system was calculated using mathematical planning and experimental results of the investigation of the ZnS–ZnTe, ZnS–ZnSe, and ZnSe–ZnTe quasibinary system (Figure 1.72) (Tomashik 1990).

Figure 1.73 shows the position of the calculated miscibility gap in the ZnS–ZnSe–ZnTe system at 600°C–1200°C (Ohtani et al. 1992). It is seen that the surface of the miscibility gap extends from the ZnS–ZnTe quasibinary system into the ternary.

1.87   ZINC–TELLURIUM–MANGANESE–SULFUR

ZnS–MnTe. $(ZnS)_{1-x}(MnTe)_x$ powder samples in the composition range $0 < x \le 0.25$ have been prepared by the simple solid-state reaction method (Rao et al. 2009). All samples were polycrystalline with wurtzite structure. Lattice parameters of these samples increase with x linearly, obeying Vegard's law. Their bandgap decreases with the increase of x. Metal–semiconductor transition was observed in all $(ZnS)_{1-x}(MnTe)_x$ samples.

1.88   ZINC–MANGANESE–IRON–SULFUR

ZnS–MnS–FeS. The isothermal sections of this system at 350°C (Shima et al. 1982) and 800°C (Knitter and Binnewies 2000) are shown in Figures 1.74 and 1.75. MnS content in sphalerite is 4 mol.% at 800°C and increases with descending temperature:

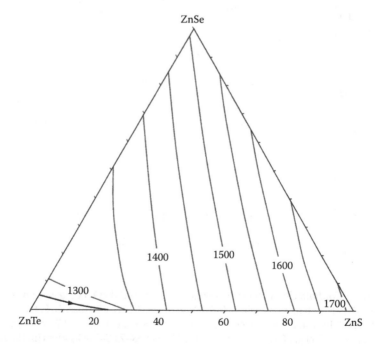

FIGURE 1.72 Liquidus surface of the ZnS–ZnSe–ZnTe quasiternary system. (From Tomashik, V.N., *Izv. AN SSSR. Neorgan. materialy*, 26(11), 2422, 1990.)

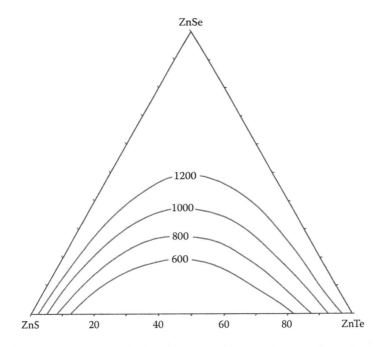

FIGURE 1.73 Calculated miscibility gap in the ZnS–ZnSe–ZnTe quasiternary system. (From Ohtani, H. et al., *J. Alloys Compd.*, 182(1), 103, 1992.)

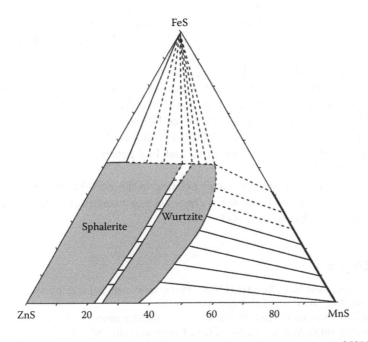

FIGURE 1.74 Isothermal section of the ZnS–MnS–FeS quasiternary system at 350°C. (From Shima, H. et al., *J. Jpn. Assoc. Mineral. Petrol. Econ. Geol.*, 77(Spec. Issue 3), 271, 1982.)

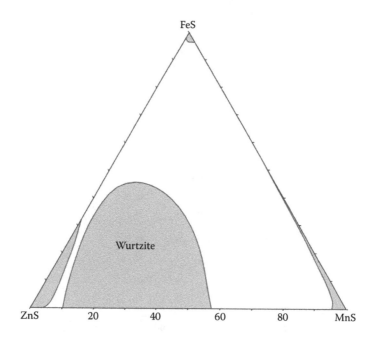

FIGURE 1.75 Isothermal section of the ZnS–MnS–FeS quasiternary system at 800°C. (From Knitter, St. and Binnewies, M., *Z. anorg. und allg. Chem.*, 626(11), 2335, 2000.)

it reaches 23–25 mol.% at 350°C (Shima et al. 1982). The Mn solubility seems to be independent of Fe content in sphalerite. Mn-bearing wurtzite is stable at a temperature as low as 350°C with a considerable wide solid solution area in the central part of the ZnS–MnS–FeS quasiternary system.

Cell edges of the (Fe + Mn)-bearing sphalerites in this system should be linear functions of their compositions, and the following relation should hold: a (A) = 5.4093 + 0.00202x + 0.000456y, where x and y are the MnS and FeS contents in mol.% (Skinner 1961). The functions used to calculate the cell dimensions of the ternary wurtzites are a (A) = 3.8230 + 0.001628x + 0.000490y and c (A) = 6.2565 + 0.002089x + 0.000886y, where x and y are the MnS and FeS contents, respectively, in mol.% (Skinner and Bethke 1961).

Single crystals of all solid solutions existing in the ZnS–MnS–FeS quasiternary system could be prepared by means of chemical vapor transport using I_2 as transport agent (Knitter and Binnewies 2000).

REFERENCES

Adamson G.W., Bell N.A., Shearer H.M.M. Pentameric methylzinc tert-butylsulphide, *Acta Crystallogr.*, **B38**(2),462–465 (1982).

Adamson G.W., Shearer H.M.M. The crystal and molecular structure of octameric methylzinc isopropylsulphide, *J. Chem. Soc. Chem. Commun.*, (16), 897–898 (1969).

Addamiano A. Some observations on the system ZnS–AlP, *J. Electrochem. Soc.*, **107**(12), 1006–1007 (1960).

Apple E.F. Investigations in the CuGaS$_2$–ZnS and AgGaS$_2$–ZnS systems, *J. Electrochem. Soc.*, **105**(5), 251–255 (1958).

Avetisian H.K., Gnatyshenko G.I. Thermal and metallographic investigation of PbS–ZnS–FeS system [in Russian], *Izv. AN KazSSR. Ser. gorn. dela, stroimaterialov i metallurgii*, **6**, 11–25 (1956).

Baranov I.Yu., Dolgih V.A., Popovkin B.A. Investigation of phase correlations in the La$_2$O$_2$X–MX (M = Zn, Cd, Hg; X = S, Se) systems [in Russian], *Zhurn. neorgan. khimii*, **41**(11), 1916–1919 (1996).

Bensalem S., Chegaar M., Maouche D., Bouhemadou A. Theoretical study of structural, elastic and thermodynamic properties of CZTX (X = S and Se) alloys, *J. Alloys Compd.*, **589**, 137–142 (2014).

Bernard D., Pannetier J., Lucas J. Hg$_2$M$_2$F$_6$S et Hg$_2$M$_2$F$_6$O: Deux nouvelles familles de pyrochlores contenant de mercure et des métaux M de transition divalents, *J. Solid State Chem.*, **14**(4), 328–334 (1975).

Bhuiyan N.H., Ahmed A., Begum S. Synthesis of molybdenum (IV) sulphide. Part II. Reaction of molybdenum (VI) oxide with (a) zinc sulphide and (b) cadmium sulphide in the presence of ammonium chloride, *J. Chem. Technol. Biotechnol.*, **29**(3), 169–174 (1979).

Bochmann M., Bwembya G., Grinter R., Lu J., Webb K.J., Williamson D.J., Hursthouse M.B., Mazid M. Three-coordinate thiolato complexes of zinc: Solution and solid-state structures and EHMO analysis of the bonding pattern of [Zn(S-*t*-Bu$_3$C$_6$H$_2$-2,4,6)$_2$]$_2$, *Inorg. Chem.*, **32**(5), 532–537 (1993).

Bochmann M., Bwembya G.C., Grinter R., Powell A.K., Webb K.J., Hursthouse M.B., Malik K.M.A., Mazid M.A. Synthesis of low-coordinate chalcogenolato complexes of zinc with O, N, S, and P donor ligands. Molecular and crystal structures of Zn(S-t-Bu$_3$C$_6$H$_2$-2,4,6)$_2$(L) (L = NC$_5$H$_3$Me$_2$-2,6, PMePh$_2$), Zn(Se-t-Bu$_3$C$_6$H$_2$-2,4,6)$_2$(OSC$_4$H$_8$) and Zn(S-t-Bu$_3$C$_6$H$_2$-2,4,6)$_2$(N-methylimidazole)$_2$, *Inorg. Chem.*, **33**(10), 2290–2296 (1994).

Bochmann M., Bwembya G., Webb K.J., Malik M.A., Walsh J.R., O'Brien P. Arene chalcogenolato complexes of zinc and cadmium, *Inorg. Synthes.*, **31**, 19–24 (1997).

Bochmann M., Webb K. Novel precursors for the deposition of II-VI semiconductor films, *Mater. Res. Symp. Proc.*, **204**, 149–154 (1991).

Bochmann M., Webb K., Hursthouse M.B., Mazid M. Sterically hindered chalcogenolato complexes. Mono- and di-meric thiolates and selenolates of zinc and cadmium; structure of [{Cd(SeC$_6$H$_2$Bu$_3^t$-2,4,6)$_2$}$_2$], the first three-co-ordinate cadmium–selenium complex, *J. Chem. Soc. Dalton Trans.*, (9), 2317–2323 (1991).

Bonamico M., Dessy G., Fares V., Scaramuzza L. Crystal and molecular structures of nickel(II) and zinc(II) bis(trithioperoxybenzoates), *J. Chem. Soc. A: Inorg. Phys. Theor.*, 3191–3195 (1971).

Brennan J.G., Siegrist T., Carroll P.J., Stuczynski S.M., Reynders P., Brus L.E., Steigerwald M.L. Bulk and nanostructure group II-VI compounds from molecular organometallic precursors, *Chem. Mater.*, **2**(4), 403–409 (1990).

Broadley S., Gál Z.A., Corà F., Smura C.F., Clarke S.J. Vertex-linked ZnO$_2$S$_2$ tetrahedra in the oxysulfide BaZnOS: A new coordination environment for zinc in a condensed solid, *Inorg. Chem.*, **44**(24), 9092–9096 (2005).

Brunetta C.D., Karuppannan B., Rosmus K.A., Aitken J.A. The crystal and electronic band structure of the diamond-like semiconductor Ag$_2$ZnSiS$_4$, *J. Alloys Compd.*, **516**, 65–72 (2012).

Buerger N.W. The unmixing of chalcopyrite from sphalerite, *Am. Mineral.*, **19**(11), 525–530 (1934).

Chen D., Ravindra N.M. Electronic and optical properties of Cu$_2$ZnGeX$_4$ (X = S, Se and Te) quaternary semiconductors, *J. Alloys Compd.*, **579**, 468–472 (2013).

Chen S., Gong X.G., Walsh A., Wei S.-H. Crystal and electronic band structure of Cu$_2$ZnSnX$_4$ (X = S and Se) photovoltaic absorbers: First-principles insights, *Appl. Phys. Lett.*, **94**(4), 041903_1–041903_3 (2009).

Chen S., Walsh A., Luo Y., Yang J.-H., Gong X.G., Wei S.-H. Wurtzite-derived polytypes of kesterite and stannite quaternary chalcogenide semiconductors, *Phys. Rev. B*, **82**(19), 195203_1–195203_8 (2010b).

Chen W.-T., Kuang H.-M., Chen H.-L. Solid-state synthesis, crystal structures and properties of two novel metal sulfur chlorides—$Zn_6S_5Cl_2$ and $Hg_3ZnS_2Cl_4$, *J. Solid State Chem.*, **183**(10), 2411–2415 (2010a).

Coates G.E., Ridley D. Alkoxy-, thio-, and amino-derivatives of methylzinc, *J. Chem. Soc.*, 1870–1877 (1965).

Colombara D., Delsante S., Borzone G., Mitchels J.M., Thomas L.H., Mendis B.G., Cummings C.Y., Marken F., Peter L.M. Crystal growth of Cu_2ZnSnS_4 solar cell absorber by chemically vapour transport with I_2, *J. Cryst. Growth*, **364**, 101–110 (2013).

Das S., Krishna R.M., Ma S., Mandal K.C. Single phase polycrystalline Cu_2ZnSnS_4 grown by vertical gradient freeze technique, *J. Cryst. Growth*, **381**, 148–152 (2013).

Deng B., Chan G.H., Huang F.Q., Gray D.L., Ellis D.E., Van Duyne R.P., Ibers J.A. Synthesis, structure, optical properties and electronic structure of $NaLiCdS_2$, *J. Solid State Chem.*, **180**(2), 759–764 (2007).

Dmitrenko A.O., Abramova L.V., Bukesov S.A. Solid solutions in the ZnS–CdS–MgS system [in Russian], *Izv. AN SSSR. Neorgan. materialy*, **26**(12), 2483–2487 (1990).

Do Y.R., Kershaw R., Dwight K., Wold A. The crystal growth and characterization of the solid solutions $(ZnS)_{1-z}(CuGaS_2)_x$, *J. Solid State Chem.*, **96**(2), 360–365 (1992).

Donohue P.C., Bierstadt P.E. Cell dimensions and electrical properties of the solid solutions: $(1-x)ZnS/xCuAlS_2$ and $(1-x)ZnSe/xCuAlSe_2$ where $x = 0$ to 0.33, *J. Electrochem. Soc.*, **121**(3), 327–329 (1974).

Dubrovin I.V., Budionnaya L.D., Mizetskaya I.B., Sharkina E.V. Phase equilibria in the SnTe–ZnS system [in Russian], *Izv. AN SSSR. Neorgan. materialy*, **19**(11), 1816–1819 (1983).

Dubrovin I.V., Budionnaya L.D., Oleynik N.D., Sharkina E.V. Interaction of Cd_4SiS_6 with ZnS and SiS_2 along the section $Cd_4SiS_6–(ZnS)_{0.8}(SiS_2)_{0.2}$ [in Russian], *Izv. AN SSSR. Neorgan. materialy*, **25**(5), 722–725 (1989).

Dubrovin I.V., Budionnaya L.D., Sharkina E.V. Interaction of Cd_4GeS_6 with ZnS and GeS_2 according to the $Cd_4GeS_6–(ZnS)_{0.8}(GeS_2)_{0.2}$ section of CdS–ZnS–GeS₂ system [in Russian], *Izv. AN SSSR. Neorgan. materialy*, **27**(2), 244–247 (1991).

Dudchak I.V., Piskach L.V. Phase equilibria along the section Ag_8SnS_6–ZnS [in Ukrainian], *Nauk. visnyk Volyn. derzh. un-tu. Khim. nauky*, (6), 59–60 (2001).

Fackler J.P., Coucouvanis, Jr. D., Fetchin J.A. Seidel W.C. Sulfur chelates. VIII. Oxidative addition of sulfur to dithioaryl acid complexes of nickel(II) and zinc(II), *J. Am. Chem. Soc.*, **90**(11), 2784–2788 (1968).

Fackler J.P., Fetchin, Jr. J.A., Fries D.C. Sulfur chelates. XV. Sulfur addition and abstraction reactions of dithioaryl acid complexes of zinc(II), nickel(II), palladium(II), and platinum(II) and the X-ray crystal structures of bis(trithioperoxycumato)zinc(II) and dithiocumato(trithioperoxycumato)nickel(II), *J. Am. Chem. Soc.*, **94**(21), 7323–7333 (1972).

Fard Z.H., Kanatzidis M.G. Phase-change materials exhibiting tristability: Interconverting forms of crystalline α-, β-, and glassy $K_2ZnSn_3S_8$, *Inorg. Chem.*, **51**(15), 7963–7965 (2012).

Farneth W.E., Herron N., Wang Y. Bulk semiconductors from molecular solids: A mechanistic investigation, *Chem. Mater.*, **4**(4), 916–922 (1992).

Fischer A.G., Paff R.J. Solubility of ZnSe and ZnTe in CdS, *J. Phys. Chem. Solids*, **23**(10), 1479–1480 (1962).

Fouassier C., Levasseur A., Joubert J.C., Muller J., Hagenmuller P. Les systèmes B_2O_3–MO–MS. Boracites M – S (M = Mg, Mn, Fe, Cd) et sodalites M – S (M = Co, Zn), *Z. anorg. und allg. Chem.*, **375**(2), 202–208 (1970).

Gallardo P.G. $Hg_{2x}(CuIn)_{1-x}Se_2$ alloys: Phase diagram and lattice parameter values, *Phys. Status Solidi (a)*, **130**(1), 39–44 (1992).

Garg G., Ganguli A.K. Phase relations in Ag–Ni–Zn–S and Cu–Si–Ni–Sn–S systems, *Proc. Indian Acad. Sci. Chem. Sci.*, **112**(3), 374 (2000).

Glazov V.M., Kiselev A.N., Shvedkov E.I. Solubility and donor-acceptor interaction in InAs doping by S, Se and Zn [in Russian], *Izv. AN SSSR. Neorgan. materialy*, **15**(3), 390–394 (1979).

Guen L., Glaunsinger W.S. Electrical, magnetic and EPR studies of the quaternary chalcogenides $Cu_2A^{II}B^{IV}X_4$ prepared by iodine transport, *J. Solid State Chem.*, **35**(1), 10–21 (1980).

Guen L., Glaunsinger W.S., Wold A. Physical properties of the quaternary chalcogenides $Cu_2^1B^{II}C^{IV}X_4$ (B^{II} = Zn, Mn, Fe, Co; C^{IV} = Si, Ge, Sn; X = S, Se), *Mater. Res. Bull.*, **14**(4), 463–467 (1979).

Gulyaeva R.I., Selivanov E.N., Vershinin A.D., Chumarev V.M. Thermal expansion of CaZnSO [in Russian], *Neorgan. materially*, **42**(8), 990–993 (2006).

Haeuseler H., Cansiz A., Nimmrich M., Jung M. Materials with layered structures: Crystal structure of $Zn_{1.25}In_{2.5}S_3Se_2$, a new polytype of $Zn_2In_2S_5$ and the isotypic compounds $Cd_{0.5}Ga_2InS_5$ and $Hg_{0.8}Ga_{1.6}In_{1.2}S_5$, *J. Solid State Chem.*, **74**(1), 171–175 (1988).

Hahn H., Frank G., Klingler W., Störger A.D., Störger G. Über ternäre Chalkogenide des Aluminiums, Galliums und Indiums mit Zink, Cadmium und Quecksilber, *Z. anorg. und allg. Chem.*, **279**(5/6), 241–270 (1955).

Hahn H., Schulze H. Über quaternäre Chalkogenide des Germanium und Zinns, *Naturwissenschaften*, **52**(14), 426 (1965).

Hassan El Haj F., Akbarzadeh H. First-principles investigation of wide-gap quaternary alloys $Zn_{1-x}Mg_xTe_{1-y}S_y$, *J. Alloys Compd.*, **433**(1–2), 306–312 (2007a).

Hassan El Haj F., Bleybel A., Hijazi A., Alaeddine A., Beydoun B., Zoaeter M. Structural and electronic properties of $Zn_{1-x}Mg_xS_ySe_{1-y}$ alloys, *Mater. Lett.*, **61**(4–5), 1178–1182 (2007b).

Hotje U., Rose C., Binnewies M. Lattice constants and molar volume in the system ZnS, ZnSe, CdS, CdSe, *Solid State Sci.*, **5**(9), 1259–1262 (2003).

Hutchison M.N., Scott S.D. Sphalerite geobarometry in the Cu–Fe–Zn–S system, *Econ. Geol.*, **76**(1), 143–153 (1981).

Iitaka von Y., Oswald H.R., Locchi S. Die Kristallstruktur von Zink-hydroxidsulfat I, $Zn(OH)_2ZnSO_4$, *Acta Crystallogr.*, **15**(6), 559–563 (1962).

Imaoka M., Jamadzaki T. [in Japanese], *Monthly J. Inst. Industr. Sci. Univ. Tokyo*, **19**(9), 261–262 (1967).

Jacobi W., Leute V. Berechnung thermodynamischer Daten fuer das quasiternaere System (Zn, Cd)(S, Se) aus Randskonzzentrationen an einer Psedophasengrenze, *Z. Phys. Chem. (BRD)*, **104**(1–3), 31–40 (1977).

Kanatzidis M.G., Liao J.H., Marking G.A. Alkali metal quaternary chalcogenides and process for the preparation of thereof, USA Patent 5 614 128. Appl. № 606565. Filed 26.02.96; Data of Patent 25.03.1997 (1997a).

Kanatzidis M.G., Liao J.H., Marking G.A. Alkali metal quaternary chalcogenides and process for the preparation of thereof, USA Patent 5 618 471. Appl. № 606886. Filed 26.02.96; Data of Patent 08.04.1997 (1997b).

Kashina I.A., Kul'sha A.M., Lomako V.M., Mellikov E.Ya. Investigation of kinetics of $Zn_xCd_{1-x}S$ solid solution formation in the ZnS substrates [in Russian], *Zhurn. neorgan. khimii*, **31**(12), 3101–3105 (1986).

Kheraj V., Patel K.K., Patel S.J., Shah D.V. Synthesis and characterization of copper zinc tin sulphide (CZTS) compound for absorber material in solar-cells, *J. Cryst. Growth*, **362**, 174–177 (2013).

Knitter St., Binnewies M. Der Chemische Transport von Mischkristallen im System FeS/MnS/ZnS, *Z. anorg. und allg. Chem.*, **626**(11), 2335–2339 (2000).

Kochkin V.P., Chizhikov D.M., Rumiantsev Yu.V. To the question on the chemical reactions in the $CdSO_4$–ZnS and $ZnSO_4$–CdS systems [in Russian]. In: *Physico-khimicheskii analiz*. Novosibirsk, Russia: Izd-vo Sib. otd. AN SSSR, pp. 55–62 (1963).

Kojima S., Sugaki A. Phase relations in the central portion of the Cu–Fe–Zn–S system between 800 and 500°C, *Mineral. J.*, **12**(1), 15–28 (1984).

Kojima S., Sugaki A. Phase relations in the Cu–Fe–Zn–S system between 500 and 300°C under hydrothermal conditions, *Econ. Geol.*, **80**(1), 158–171 (1985).

Kopylov N.I., Toguzov M.Z. The $Cu_{1.8}$S–FeS–ZnS system [in Russian], *Zhurn. neorgan. khimii*, **20**(9), 2562–2565 (1975).

Kopylov N.I., Toguzov M.Z., Kudriavtseva S.I., Matsenko Yu.A., Smailov S.D. The Cu_3As–ZnS system [in Russian], *Zhurn. neorgan. khimii*, **29**(6), 1621–1623 (1984).

Kopylov N.I., Toguzov M.Z., Yarygin V.I. Investigation of the PbS–$Cu_{1.8}$S–ZnS system [in Russian], *Izv. AN SSSR. Ser. Metally*, (6), 80–83 (1976).

Kovaliv V.I., Kesler Ya.A., Gordeev I.V., Tretiakov Yu.D., Mel'nik P.M. Spinel homogeneity region in the Al_2S_3–Cr_2S_3 system and crystal chemistry of cation deficient thiospinels [in Russian], *Izv. AN SSSR. Neorgan. materialy*, **17**(11), 2084–2088 (1981).

Kröger F.A. Solid solutions in the ternary system ZnS–CdS–MnS, *Z. Kristallogr.*, **A102**(1), 132–135 (1940).

Kullerud G. $CuFeS_2$–ZnS, *Carnegie Inst. Washington Yearbook*, **55**, 180–181 (1955–1956).

Lambrecht, Jr. V.G. Crystal growth by vapor transport of new wurtzite structure materials in the system $AgInS_2$–ZnS, *Mater. Res. Bull.*, **7**(12), 1411–1415 (1972).

Lekse J.W., Leverett B.M., Lake C.H., Aitken J.A. Synthesis, physicochemical characterization and crystallographic twinning of Li_2ZnSnS_4, *J. Solid State Chem.*, **181**(12), 3217–3222 (2008).

Leonov V.V., Chunarev E.N. Distribution of potassium and zinc chlorides at the crystallization of eutectics of cadmium chalcogenides with its chloride [in Russian]. In: *Physico-khim. Processy v geterogennyh systemah*. Krasnoyarsk, Russia: Krasnoyar. Un-t Publish., pp. 59–64 (1977).

Lucas C.R., Peach M.E. Metal derivatives of pentachlorothiophenol, *Inorg. Nucl. Chem. Lett.*, **5**(2), 73–76 (1969).

Lucas C.R., Peach M.E. Reactions of pentachlorothiophenol. I. Preparation of some simple metallic and non-metallic derivatives, *Can. J. Chem.*, 48(12), 1869–1875 (1970).

Lusk J., Calder B.O.E. The composition of sphalerite and associated sulfides in reactions of the Cu–Fe–Zn–S, Fe–Zn–S and Cu–Fe–S systems at 1 bar and temperatures between 250 and 535°C, *Chem. Geol.*, **203**(3–4), 319–345 (2004).

Malinowski C. Phenomenologic analysis of chemical process in the systems $ZnSO_4$–ZnS and $PbSO_4$–ZnS [in Polish], *Zesz. nauk. AGH im. Stanisława Staszika. Met. i odlew.*: [Monogr.], (141), 1–120 (1992).

Malinowski C., Malinowska K., Małecki S. Analysis of the chemical processes occurring in the system $PbSO_4$–ZnS, *Thermochim Acta*, **275**(1), 117–130 (1996).

Manos M.J., Iyer R.G., Quarez E., Liao J.H., Kanatzidis M.G. {$Sn[Zn_4Sn_4S_{17}]^{6-}$}: A robust open framework based on metal-linked supertetrahedral $[Zn_4Sn_4S_{17}]^{10-}$ clusters possessing ion-exchange properties, *Angew. Chem. Int. Ed. Engl.*, **44**(23), 3552–3555 (2005).

Maslov V.I., Polyviannyi I.R., Ivakina L.P., Turisbekov M.T. Investigation of ZnS–Na_2S–ZnO melting diagram [in Russian]. *Deposited in VINITI*, № 4652-72Dep (1972).

Matsuhita H., Ichikawa T., Katsui A. Structural, thermodynamic and optical properties of Cu_2–II–IV–VI_4 quaternary compounds. *J. Mater. Sci.*, **40**(8), 2003–2005 (2005).

Mitchell K., Huang F.Q., Caspi E.N., McFarland A.D., Haynes Ch.L., Somers R.C., Jorgensen J.D., Van Duyne R.P., Ibers J.A. Synthesis, structure and selected physical properties of $CsLnMnSe_3$ (Ln = Sm, Gd, Tb, Dy, Ho, Er, Tm, Yb, Y) and $AYbZnQ_3$ (A = Rb, Cs; Q = S, Se, Te), *Inorg. Chem.*, **43**(3), 1082–1089 (2004).

Moh G.H. Tin-containing mineral systems. Part II: Phase relations and mineral assemblages in the Cu–Fe–Zn–Sn–S system, *Chem. Erde*, **34**(1), 1–61 (1975).

Moodie A.F., Whitefield H.J. Determination of the structure of Cu_2ZnGeS_4 polymorphs by lattice imaging and convergent-beam electron diffraction, *Acta Crystallogr.*, **B42**(3), 236–247 (1986).

Morris C.D., Li H., Jin H., Malliakas C.D., Peters J.A., Trikalitis P.N., Freeman A.J., Wessels B.W., Kanatzidis M.G. $Cs_2M^{II}M_3^{IV}Q_8$ (Q = S, Se, Te): An extensive family of layered semiconductors with diverse band gaps, *Chem. Mater.*, **25**(16), 3344–3356 (2013).

Mouallem-Bahout M., Peña O., Carel C., Ouammou A., Retat M. Obtaining and properties of the $ACuZnS_2$ (A = K, Rb) and $BaKCu_3ZnS_4$ sulfides with low-symmetry structure of the $ThCr_2Si_2$-type [in Russian], *Zhurn. neorgan. khimii*, **46**(5), 742–747 (2001).

Nagaoka A., Yoshino K., Taniguchi H., Taniyama T., Kakimoto K., Miyake H. Growth and characterization of Cu_2ZnSnS_4 single crystals, *Phys. Status Solidi (a)*, **210**(7), 1328–1331 (2013).

Nagaoka A., Yoshino K., Taniguchi H., Taniyama T., Miyake H. Growth of Cu_2ZnSnS_4 single crystal by travelling heater method, *Jpn. J. Appl. Phys.*, **50**(12), 128001_1–128001_2 (2011).

Nagaoka A., Yoshino K., Taniguchi H., Taniyama T., Miyake H. Preparation of Cu_2ZnSnS_4 single crystals from Sn solutions, *J. Cryst. Growth*, **341**(1), 38–41 (2012).

Nesterov V.N., Ponomarev V.D. Vapor pressure of zinc sulfide in the $ZnS–FeS–Cu_2S$ [in Russian]. *Izv. AN KazSSR. Ser. metallurgii, obogashchenia i ogneuporov*, **3**(9), 77–87 (1960).

Nicholas D.M., Waite M.S., Ho T.M. Crystal data on some potassium lithium metal sulphides ($KLiMS_2$, M being Zn, Mn or Fe), *J. Appl. Crystallogr.*, **16**(1), 141–142 (1983).

Nitsche R., Sargent D.F., Wild P. Crystal growth of quaternary $1_2 2 4 6_4$ chalcogenides by iodine vapor transport, *J. Cryst. Growth*, **1**(1), 52–53 (1967).

Novatski V. Isotypie of aktashite $Cu_6Hg_3As_4S_{12}$ and novatskite $Cu_6Zn_3As_4S_{12}$ [in Russian], *Kristallografiya*, **27**(1), 49–50 (1982).

Novoselov S.S. Influence of zinc sulfide on the properties of copper matte [in Russian], *Tsvet. metally*, (3), 15–20 (1955).

Odin I.N. Physico-chemical analysis of ternary and ternary mutual systems, containing cadmium, zinc, silicon, bismuth chalcogenides and properties of ingots in these systems [in Russian], *Zhurn. neorgan. khimii*, **41**(6), 941–953 (1996).

Odin I.N., Rubina M.E., Grigor'yeva A.V., Kozlovskiy V.F. $Zn_3Te_3 + Ga_2S_3 = Zn_3S_3 + Ga_2Te_3$ mutual system [in Russian], *Zhurn. neorgan. khimii*, **50**(5), 848–850 (2005a).

Odin I.N., Visitskiy E.V., Rubina M.E. *T-x-y* phase diagrams of the CdS + ZnTe ⇔ ZnS + CdTe, $CdS–CdIn_2S_4$, $ZnS–In_2Te_3–ZnTe$ [in Russian], *Zhurn. neorgan. khimii*, **50**(6), 1018–1023 (2005b).

Ohtani H., Kojima K., Ishida K., Nishizawa T. Miscibility gap in II-VI semiconductor systems, *J. Alloys Compd.*, **182**(1), 103–114 (1992).

Oleinik G.S., Mizetski P.A., Nizkova A.I. Interaction between lead and zinc chalcogenides [in Russian], *Izv. AN SSSR. Neorgan. materialy*, **18**(5), 873–874 (1982).

Olekseyuk I.D., Dudchak I.V., Piskach L.V. Phase equilibria in the $Cu_2S–ZnS–SnS_2$ system, *J. Alloys Compd.*, **368**(1–2), 135–143 (2004).

Olekseyuk I.D., Halka V.O., Parasyuk O.V., Voronyuk S.V. Phase equilibria in the $AgGaS_2–ZnS$ and $AgInS_2–ZnS$ systems, *J. Alloys Compd.*, **325**(1–2), 204–209 (2001).

Olekseyuk I., Marchuk O., Dudchak I., Parasyuk O., Piskach L. Phase equilibria in the $Cu_2SnS_3–Zn(Hg)S$ systems, *Visnyk L'viv. un-tu. Ser. khim*, (39), 48–52 (2000).

Osakada K., Yamamoto T. Formation of ZnS and CdS by thermolysis of homoleptic thiolato compounds $[M(SMe)_2]_n$ (M = Zn, Cd), *J. Chem. Soc. Chem. Commun.*, (14), 1117–1118 (1987).

Osakada K., Yamamoto T. Preparation of ZnS and CdS by thermal degradation of (methanethiolato)zinc and -cadmium complexes, $[M(SMe)_2]_n$ (M = Zn, Cd), *Inorg. Chem.*, **30**(10), 2328–2332 (1991).

Palchik O., Iyer R.G., Canlas C.G., Weliky D.P., Kanatzidis M.G. $K_{10}M_4M'_4S_{17}$ (M = Mn, Fe, Co, Zn; M' = Sn, Ge) and $Cs_{10}Cd_4Sn_4S_{17}$: Compounds with a discrete supertetrahedral cluster, *Z. anorg. und allg. Chemie*, **630**(13–14), 2237–2247 (2004).

Palchik O., Iyer R.G., Liao J.H., Kanatzidis M.G. $K_{10}M_4Sn_4S_{17}$ (M = Mn, Fe, Co, Zn): Soluble quaternary sulfides with the discrete $[M_4Sn_4S_{17}]^{10-}$ supertetrahedral clusters, *Inorg. Chem.*, **42**(17), 5052–5054 (2003).

Pannetier J., Calaga Y., Lucas J. Nouvelles pyrochlores $Cd_2M_2F_6S$ (M^{II} = Mn, Fe, Co, Ni, Cu, Zn), *Mater. Res. Bull.*, **7**(1), 57–62 (1972).

Parasyuk O.V., Fedorchuk A.O., Kogut Yu.M., Piskach L.V., Olekseyuk I.D. The Ag_2S–ZnS–GeS_2 system: Phase diagram, glass-formation region and crystal structure of Ag_2ZnGeS_4, *J. Alloys Compd.*, **500**(1), 26–29 (2010).

Parasyuk O.V., Piskach L.V., Romanyuk Y.E., Olekseyuk I.D., Zaremba V.I., Pekhnyo V.I. Phase relations in the quasibinary Cu_2GeS_3–ZnS and quasiternary Cu_2S–Zn(Cd)S–GeS_2 systems and crystal structure of Cu_2ZnGeS_4, *J. Alloys Compd.*, **397**(1–2), 85–94 (2005).

Parasyuk O.V., Voronyuk S.V., Gulay L.D., Davidyuk G.Ye., Halka V.O. Phase diagram of the $CuInS_2$–ZnS system and some physical properties of solid solutions phases, *J. Alloys Compd.*, **348**(1–2), 57–64 (2003).

Parthé E., Yvon K., Deitch R.H. The crystal structure of Cu_2CdGeS_4 and other quaternary normal tetrahedral structure compounds, *Acta Crystallogr. B*, **25**(6), 1164–1174 (1969).

Peach M.E. Some reactions of pentafluorothiophenol. Preparation of some pentafluoro-phenylthio metal derivatives, *Can. J. Chem.*, **46**(16), 2699–2706 (1968).

Persson C. Electronic and optical properties of Cu_2ZnSnS_4 and $Cu_2ZnSnSe_4$, *J. Appl. Phys.*, **107**(5), 053710_1–053710_8 (2010).

Petrova S.A., Mar'evich V.P., Zakharov R.G., Selivanov E.N., Chumarev V.M., Udoeva L.Yu. Crystal structure of a zinc-calcium oxysulfide [in Russian], *Dokl. AN*, **393**(1), 52–56 (2003).

Piskach L.V., Parasyuk O.V., Olekseyuk I.D., Romanyuk Y.E., Volkov S.V., Pekhnyo V.I. Interaction of argyrodite family compounds with the chalcogenides of II-b elements, *J. Alloys Compd.*, **421**(1–2), 98–104 (2006).

Radautsan S.I., Raylyan V.Ya., Tsiulyanu I.I., Zhitar V.F., Marcus M.M., Moldovyan N.A. New Zn_3InGaS_6 phase and its main properties, *Progr. Cryst. Growth Charact.*, **10**(1–4), 397–401 (1985).

Radautsan S.I., Tsiulianu I.I., Lialikova R.Yu., Moldovyan N.A., Zhitar' V.F., Markus M.M. Polymorphous transformations in $Zn_3In_2(Ga,Al)S_6$ [in Russian], *Izv. AN SSSR. Neorgan. materialy*, **23**(5), 852–854 (1987).

Raghavan V. Cu–Fe–S–Zn (Copper – Iron – Sulfur – Zinc), *J. Phase Equilib. Diff.*, **25**(5), 465–470 (2004a).

Raghavan V. Fe–Ga–S–Zn (Iron – Gallium – Sulfur – Zinc), *J. Phase Equilib. Diff.*, **25**(4), 384–387 (2004b).

Rao N.M., Krishnaiah G., Sambasivam S., Reddy K.P., Reddy D.R., Reddy B.K., Xu C.N. Structural, optical and electrical properties of luminescent $(ZnS)_{1-x}(MnTe)_x$ powders, *J. Alloys Compd.*, **468**(1–2), 360–364 (2009).

Robbins M., Miksovsky M.A. Preparation of and phase relationships in systems of the type ZnS–$M^IM^{III}S_2$ where M^I = Cu, Ag and M^{III} = In, Ga, Al, *J. Solid State Chem.*, **5**(3), 462–466 (1972).

Ronis Ya.V., Krasnikov V.V., Bondars B.Ya., Vitola A.A., Miller T.N. X-ray investigation of the metal-phosphorus thionitrides $Me_4P_6N_{12}S$ (Me – Mg, Zn, Mn, Fe, Co) [in Russian], *Izv. AN LatvSSR. Ser. khim.*, (2), 139–144 (1989).

Ronis Ya.V., Vitola A.A., Miller T.N. Synthesis and properties of $Me_4P_6N_{12}S$ (Me – Zn, Cd, Mg) thionitrides, *Tr. Mosk. khim.-tehnol. in-ta*, (150), 12–14 (1988).

Rosmus K.A., Aitken J.A. Cu_2ZnSiS_4, *Acta Crystallogr. E*, **67**(4), i28 (2011).

Said F.F., Tuck D.G. The direct electrochemical synthesis of some thiolates of zinc, cadmium and mercury, *Inorg. Chim. Acta*, **59**, 1–4 (1982).

Sambrook T., Smura C.F., Clarke S.J., Ok K.M., Halasyamani P.S. Structure and physical properties of the polar oxysulfide CaZnOS, *Inorg. Chem.*, **46**(7), 2571–2574 (2007).

Savel'eva M.V., Gromilov S.A. The $MCuXS_2$ and $M_2Cu_3FeS_4$ (M = K, Rb, Cs; X = Mn, Zn) new ternary sulfides [in Russian], *Zhurn. neorgan. khimii*, **41**(9), 1423–1426 (1996).

Schäfer W., Nitsche R. Tetrahedral quaternary chalcogenides of the type Cu_2–II–IV–$S_4(Se_4)$, *Mater. Res. Bull.*, **9**(5), 645–654 (1974).

Schäfer W., Nitsche R. Zur Systematik tetraedrischer Verbindungen vom Typ $Cu_2Me^{II}Me^{IV}$ Me_4^{VI} (Stannite und Wurtzstannite), Z. Kristallogr., 145(5–6), 356–370 (1977).

Schleich D.M., Wold A. Optical and electrical properties of quaternary chalcogenides, Mater. Res. Bull., 12(2), 111–114 (1977).

Schmitz D., Bronger W. Zur Kristallchemie einiger Alkalithiomanganate und -zinkate. Mit einer Bemerkung zum $ThCr_2Si_2$-Typ, Z. anorg. und allg. Chem., 553(10), 248–260 (1987).

Schorr S., Tovar M., Sheptyakov D., Keller L., Geandier G. Crystal structure and cation distribution in the solid solution series $2(ZnX)–CuInX_2(X = S, Se, Te)$, J. Phys. Chem. Solids, 66(11), 1961–1965 (2005).

Schorr S., Wagner G. Structure and phase relations of the $Zn_{2x}(CuIn)_{1-x}S_2$ solid solution series, J. Alloys Compd., 396(1–2), 202–207 (2005).

Schwer H., Keller E., Krämer V. Crystal structure and twinning of $KCd_4Ga_5S_{12}$ and some isotypic $AB_4C_5X_{12}$ compounds, Z. Kristallogr., 204(Pt. 2), 203–213 (1993).

Shima H., Ueno H., Nakamura Y. Synthesis and phase studies on sphalerite solid solution— The systems Cu–Fe–Zn–S and Mn–Fe–Zn–S, J. Jpn. Assoc. Mineral. Petrol. Econ. Geol., 77(Spec. Issue 3), 271–280 (1982).

Shishkin V.I., Polyviannyi I.R., Demchenko R.S. The dissolution of lead and copper and phase diagram in the $Na_2S–ZnS$ system [in Russian], Vestn. AN KazSSR, (9), 20–30 (1968).

Skinner B.J. Unit-cell edges of natural and synthetic sphalerites, Am. Mineral., 46(11–12), 1399–1411 (1961).

Skinner B.J., Bethke P.M. The relationship between unit-cell edges and composition of synthetic wurtzites, Am. Mineral., 46(11–12), 1382–1398 (1961).

Sombuthawee C., Bonsall S.B., Hummel F.A. Phase equilibria in the systems $ZnS–MnS$, $ZnS–CuInS_2$ and $MnS–CuInS_2$, J. Solid State Chem., 25(4), 391–399 (1978).

Sonomura H., Uragaki T., Miyauchi T. Synthesis and some properties of solid solutions in the GaP–ZnS and GaP–ZnSe pseudobinary systems, Jpn. J. Appl. Phys., 12(7), 968–973 (1973).

Sorokin V.S., Sorokin S.V., Kaygorodov V.A., Ivanov S.V. Instability and immiscibility regions in $Mg_xZn_{1-x}S_ySe_{1-y}$ alloys, J. Cryst. Growth, 214–215, 130–134 (2000).

Strohfeldt E. Beitraege zu den Systemen Kupfer–Zink–Schwefel und Bleisulfid– Kupfersulfuer–Zinksulfid, Metall und Erz., 33(21), 561–572 (1936).

Sugiyama M., Kinoshita M., Nakanishi H. Growth of $ZnGaInS_4$ by normal freezing method, J. Phys. Chem. Solids, 66(11), 2127–2129 (2005).

Suleimanov T.Zh., Onaev I.A., Kozhahmetov S.M. Interaction rate of lead oxide and zinc sulfide [in Russian], Metallurgia i obogashchenie, Alma-Ata, 9, 77–80 (1974b).

Suleimanov T.Zh., Onaev I.A., Kozhahmetov S.M., Kairbayeva Z.K. Thermographic investigation of the PbO–ZnS system [in Russian], Metallurgia i obogashchenie, Alma-Ata, 9, 74–76 (1974a).

Surapunt S., Nyamai C.M., Hino M., Itagaki K. Phase relations and distribution of minor elements in the Cu–Zn–S, Cu–Zn–Fe–S and Cu–Zn–Pb–S systems at 1473 K, Metal Rev. MMIJ, 12(2), 84–97 (1995).

Surapunt S., Nyamai C.M., Hino M., Itagaki K. Phase relations in the Cu–Zn–S, Cu–Zn–Fe–S and Cu–Zn–Pb–S systems at 1473 K [in Japan], Shigen-to-Sozai, 112(1), 56–60 (1996).

Tauson V.L., Chernyshev L.V. Experimental Investigations on Zinc Sulfide Crystal Chemistry and Geochemistry [in Russian], Novosibirsk, Nauka Publish., 190 pp. (1981).

Teske Chr.L. Darstellung, Kristallstrukturdaten und Eigenschaften der quaternären Thiostannate(IV) $BaZnSnS_4$ und $BaMnSnS_4$, Z. Naturforsch. B, 35(4), 509–510 (1980).

Toffoli P., Rouland J.C., Khodadad P., Rodier N. Structure du tétrathiophosphate (V) de zinc et d'argent, $ZnAgPS_4$, Acta Crystallogr., C41(5), 645–647 (1985).

Toguzov M.Z., Kopylov N.I., Minkevich S.M. Phase equilibria in the Fe–Zn–S–O system [in Russian], Zhurn. neorgan. khimii, 25(10), 2863–2865 (1980a).

Toguzov M.Z., Kopylov N.I., Sychev A.P., Minkevich S.M. The FeS–ZnS–FeO system [in Russian], Zhurn. neorgan. khimii, 25(10), 2873–2875 (1980b).

Toguzov M.Z., Kopylov N.I., Sychev A.P., Yarygin V.I., Minkevich S.M. Phase equilibria in the Cu_2S–FeS–ZnS–ZnO system [in Russian]. In: *Sul'fid. rasplavy tiazh. met.*, M. Nauka Publish., pp. 122–127 (1982).

Toguzov M.Z., Kopylov N.I., Yarygin V.I., Minkevich S.M. The Cu_2S–ZnS–CaO system [in Russian], *Zhurn. neorgan. khimii*, **24**(12), 3354–3357 (1979).

Tomashik V.N. The physico-chemical interaction in the Cd, Zn ‖ S, Se ternary mutual system [in Russian], *Izv. AN SSSR. Neorgan. materialy*, **17**(6), 1116–1117 (1981).

Tomashik V.N. The ZnTe–ZnSe–ZnS ternary system [in Russian], *Izv. AN SSSR. Neorgan. materialy*, **26**(11), 2422–2423 (1990).

Tomashik V.N., Mizetskaya I.B. Phase equilibria in the CdSe–ZnSe–CdS system [in Russian], *Izv. AN SSSR. Neorgan. materialy*, **16**(4), 608–610 (1980).

Tomashik V.N., Oleinik G.S., Mizetskaya I.B. The CdTe–ZnS system [in Russian], *Izv. AN SSSR. Neorgan. materialy*, **14**(10), 1838–1840 (1978).

Tomashik V.N., Oleinik G.S., Mizetskaya I.B. Investigation of interaction in the CdSe + ZnS ⇔ CdS + ZnSe ternary mutual system [in Russian], *Izv. AN SSSR. Neorgan. materialy*, **15**(2), 202–204 (1979).

Tomashik V.N., Oleinik G.S., Mizetskaya I.B., Novitskaya G.N. The CdSe–ZnS system [in Russian], *Izv. AN SSSR. Neorgan. materialy*, **17**(1), 17–19 (1981).

Trishchuk L.I., Oleinik G.S., Mizetskaya I.B. The Ag_2S+ZnSe ⇔ Ag_2Se+ZnS ternary mutual system [in Russian], *Izv. AN SSSR. Neorgan. materialy*, **21**(2), 210–213 (1985).

Trishchuk L.I., Oleinik G.S., Mizetskaya I.B. Liquidus surface of the Ag_2Te–ZnTe–ZnS ternary system [in Russian], *Izv. AN SSSR. Neorgan. materialy*, **28**(4), 735–737 (1992).

Tsuji I., Shimodaira Y., Kato H., Kobayashi H., Akihiko K.A. Novel stannite-type complex sulfide photocatalysts A_2^I–Zn–A^{IV}–S_4 (A^I = Cu and Ag; A^{IV} = Sn and Ge) for hydrogen evolution under visible-light irradiation, *Chem. Mater.*, **22**(4), 1402–1409 (2010).

Ueno T., Scott S.D. Solubility of gallium in sphalerite and wurtzite at 800°C and 900°C, *Can. Mineral.*, **29**(1), 143–148 (1991).

Ueno T., Scott S.D. Inversion between sphalerite- and wurtzite-type structures in the system Zn–Fe–Ga–S, *Can. Mineral.*, **34**(5), 949–958 (1996).

Ueno T., Scott S.D. Phase equilibria in the system Zn–Fe–Ga–S at 900° and 800°C, *Can. Mineral.*, **40**(2), 563–570 (2002).

Vankar V.D., Pandya D.K., Chopra K.L. Structural and optical properties of ZnS–MgF_2 alloy films, *Thin Solid Films*, **59**(1), 43–49 (1979).

Vasil'ev V.I. Saukovite—The new zinc-cadmium containing mercury sulfide [in Russian], *Dokl. AN SSSR*, **168**(1), 182–185 (1966).

Vasil'ev V.I. New ore minerals of mercury depositions of the Mountain Altai and their paragenesis [in Russian]. In: *Voprosy metallogenii rtuti.*: Nauka Publish., pp. 111–129 (1968).

Voitsehovski A.V., Panchenko L.B. About obtaining of $(GaP)_x(ZnS)_{1-x}$ solid solution single crystals [in Russian]. In: *Fizika tverdogo tela*. Kiev, Ukraine: Kiev. ped. in-t, pp. 24–26 (1975).

Voitsehovski A.V., Panchenko L.B. Microstructural investigation of crystals of the GaP–ZnS system [in Russian], *Izv. AN SSSR. Neorgan. materialy*, **13**(1), 160–161 (1977).

Wachtel A. (Zn,Hg)S and (Zn,Cd,Hg)S electroluminescent phosphors, *J. Electrochem. Soc.*, **107**(8), 682–688 (1960).

Wagner G., Fleicher F., Schorr S. Extension of the two-phase field in the system $2(ZnS)_x(CuInS_2)_{1-x}$ and structural relationship between the tetragonal and cubic phase, *J. Cryst. Growth*, **283**(3–4), 356–366 (2005).

Wakeshima M., Hinatsu Y. Crystal structures and magnetic properties of new quaternary sulfides $BaLn_2MS_5$ (Ln = La, Ce, Pr, Nd; M = Co, Zn) and $BaNd_2MnS_5$, *J. Solid State Chem.*, **159**(1), 163–169 (2001).

Wester F., Schnick W. Nitrido-sodalithe: III. Synthese, Struktur und Eigenschaften von $Zn_8[P_{12}N_{14}]X_2$ mit X = O, S, Se, Te, *Z. anorg. und allg. Chem.*, **622**(8), 1281–1286 (1996).

Wiggins L.B., Craig J.R. Reconnaissance of the Cu–Fe–Zn–S system; sphalerite phase relationships, *Econ. Geol.*, **75**(5), 742–751 (1980).

Yao G.-Q., Shen H.-S., Honig E.D., Kershaw R., Dwight K., Wold A. Preparation and characterization of the quaternary chalcogenides $Cu_2B(II)C(IV)X_4[B(II) = Zn, Cd;C(IV) = Si, Ge; X = S, Se]$, *Solid State Ionics*, **24**(3), 249–252 (1987).

Yarygin V.I., Kopylov N.I., Novoselova V.N., Larin V.F., Pestunova N.P. Influence of calcium oxide on fusibility of iron, copper, zinc and lead sulfides [in Russian], *Izv. AN SSSR. Ser. Metally*, (6), 64–68 (1975).

Yarygin V.I., Shokarev M.M., Mamaev V.E., Kolganov I.M., Kopylov N.I. About new oxide-sulfide type phases [in Russian], *Izv. AN SSSR. Ser. Metally*, (2), 104–107 (1977).

Yim M.F. Solid solutions in the pseudobinary (III-V)–(II-VI) systems and their optical energy gap, *J. Appl. Phys.*, **40**(6), 2617–2623 (1969).

Zhang C., Chadha R., Reddy H.K., Schrauzer G.N. Pentacoordinate zinc: Synthesis and structures of bis[1-(methylthio)-*cis*-stilbene-2-thiolato]zinc and of its adducts with mono- and bidentate nitrogen bases, *Inorg. Chem.*, **30**(20), 3865–3869 (1991).

Živković Ž., Živković D., Grujičić D., Savović V. Kinetics of the oxidation process in the system Zn–Fe–S–O, *Thermochim. Acta*, **315**(1), 33–37 (1998).

2 Systems Based on ZnSe

2.1 ZINC–HYDROGEN–CARBON–SELENIUM

Some compounds are formed in this quaternary system.

MeZnSePh or C_7H_8SeZn, methylzinc phenyl selenide. This compound was precipitated when selenophenol in hexane was added to dimethylzinc (also in hexane) at $-80°C$ and the mixture allowed warming to room temperature (Coates and Ridley 1965).

Zn(PhSe)₂ or $C_{12}H_{10}Se_2Zn$, bis(phenylselenolato)zinc. This compound could be obtained by the next procedure (Brennan et al. 1990). Selenophenol (21.6 mmol) was added to diethylzinc (8.37 mmol) in toluene (20 mL). Gas evolution was immediate, and a white precipitate began to form after approximately one-fourth of the selenophenol was added. The mixture was stirred at room temperature for 6 h, after which heptane (20 mL) was added. The precipitate was collected by filtration, washed with heptane (10 mL), and dried under vacuum to give a white solid. It crystallizes in the orthorhombic structure and melts at the temperature higher than $300°C$.

MeZn(SeC₆H₂Pr$_3^i$-2,4,6) or $C_{16}H_{26}SeZn$. To obtain this compound, $HSeC_6H_2Me_3$-2,4,6 (8.65 mmol) was added to a solution of $ZnMe_2$ (10.2 mmol) in light petroleum (10 mL) at $-78°C$ (Bochmann et al. 1992). On warming the clear yellow solution to room temperature, a gas evolved and a white precipitate formed, which dissolved on stirring for another 2 h. Cooling to $-20°C$ produced colorless crystals of $C_{16}H_{26}SeZn$. This compound was obtained as microcrystalline precipitates, which melts at $158°C$. It is sensitive to oxidation. All operations were carried out under Ar using standard vacuum-line techniques.

EtZn(SeC₆H₂Pr$_3^i$-2,4,6) or $C_{17}H_{28}SeZn$. The same procedure as for obtaining $C_{16}H_{26}SeZn$ was used to prepare $C_{17}H_{28}SeZn$ using $ZnEt_2$ instead of $ZnMe_2$ (Bochmann et al. 1992). This compound gives well-formed crystals, which melts at $110°C$. It is sensitive to oxidation.

Pr^{n}Zn(SeC₆H₂Pr$_3^i$-2,4,6) and Pr^{i}Zn(SeC₆H₂ Pr$_3^i$-2,4,6) or $C_{18}H_{30}SeZn$. The same procedure as for obtaining $C_{16}H_{26}SeZn$ was used to prepare $C_{17}H_{28}SeZn$ using $ZnPr_2^n$ or $ZnPr_2^i$ instead of $ZnMe_2$ (Bochmann et al. 1992). Pr^{n}Zn(SeC₆H₂Pr$_3^i$-2,4,6) was obtained as microcrystalline precipitates, which melts at $118°C$, and Pr^{i}Zn(SeC₆H₂ Pr$_3^i$-2,4,6) gives well-formed crystals, which melts at $110°C$. Both compounds are sensitive to oxidation.

Zn(SeC₆H₂Me₃-2,4,6)₂ or $C_{18}H_{22}Se_2Zn$. This compound was obtained by a pyrolysis of $Zn[N(SiMe_3)_2]_2$ with 2,4,6-$Me_3C_6H_2SeH$ as polymeric solids, which dissolved only in coordinating solvents (Bochmann et al. 1991). It melts at $275°C$ (Bochmann and Webb 1991).

The experiment was carried out under Ar atmosphere using standard vacuum-line techniques. Crystals suitable for x-ray diffraction (XRD) could not be grown (Bochmann et al. 1991).

Zn(SeC$_6$H$_2$Bu$_3^t$-2,4,6)$_2$ or C$_{36}$H$_{58}$Se$_2$Zn. To obtain this compound, tri-t-butylbenzeneselenol (3.16 mmol) and light petroleum (20 mL) are injected into a 100 mL, three-necked flask, equipped with a magnetic stirring bar and connected to the inert gas supply of a vacuum line via a stopcock adaptor (Bochmann et al. 1991, 1997). Zn[N(SiMe$_3$)$_2$]$_2$ (1.71 mmol) is added via syringe to the stirred solution. The mixture is heated gently with hair dryer and allowed to stir at room temperature for 1 h. The white precipitate is filtered off, washed with light petroleum (2 × 10 mL), dissolved in warm toluene (ca. 10 mL), and recrystallized at room temperature. More products can be recovered from the mother liquor by adding light petroleum (20 mL) and refluxing for 2–3 h followed by filtration. This compound melts at 210°C with decomposition, sublimes at 220°C (0.13 Pa), and decomposes in refluxing toluene to give ZnSe (Bochmann et al. 1991, 1997, Bochmann and Webb 1991).

2.2   ZINC–HYDROGEN–NITROGEN–SELENIUM

ZnSe–N$_2$H$_4$. The phase diagram is not constructed. The ZnSe·N$_2$H$_4$ quaternary compound is formed in this system (Chaus et al. 1983). It crystallizes in the hexagonal structure with the lattice parameters $a = 1654 \pm 6$ and $c = 1114 \pm 5$ pm and calculation and experimental densities 2.660 and 2.630 g·cm^{-3}, respectively. This compound decomposes at 250°C forming N$_2$H$_4$ and hexagonal ZnSe.

2.3   ZINC–HYDROGEN–OXYGEN–SELENIUM

ZnSeO$_3$·H$_2$O, ZnSeO$_3$·2H$_2$O, Zn$_3$(SeO$_3$)$_3$·H$_2$O, Zn(HSeO$_3$)$_2$·2H$_2$O, and Zn(HSeO$_3$)$_2$·4H$_2$O quaternary compounds are formed in this quaternary system. ZnSeO$_3$ exists in two polymorphs (Markovskiy and Sapozhnikov 1960): α-ZnSeO$_3$ is stable up to 285°C and is characterized by the experimental density 3.94 ± 0.01 g·cm^{-3}, and β-ZnSeO$_3$ has experimental density 4.71 ± 0.02 g·cm^{-3}.

ZnSeO$_3$·H$_2$O crystallizes in the monoclinic structure with the lattice parameters $a = 477.9 \pm 0.1$, $b = 1319.4 \pm 0.5$, $c = 570.1 \pm 0.1$ pm, and $\beta = 90.84° \pm 0.02°$ and calculated density 3.886 [3.830 ± 0.005 (Markovskiy and Sapozhnikov 1960)] g·cm^{-3} (Engelen et al. 1996). It was obtained by the crystallization from aqueous solutions of ZnSeO$_3$ or Zn(HSeO$_3$)$_2$ (Engelen et al. 1996).

ZnSeO$_3$·2H$_2$O crystallizes in the monoclinic structure with the lattice parameters $a = 645$, $b = 880$, $c = 765 \pm 1$ pm, and $\beta = 82°0' \pm 10'$ and calculated and experimental densities 3.52 and 3.5 ± 0.1 g·cm^{-3} (Markovskiy and Sapozhnikov 1960, Gladkova and Kondrashev 1964).

Zn$_3$(SeO$_3$)$_3$·H$_2$O crystallizes in the triclinic structure with the lattice parameters $a = 804.32 \pm 0.03$, $b = 808.05 \pm 0.03$, $c = 888.36 \pm 0.03$ pm, $\alpha = 65.119° \pm 0.001°$, $\beta = 67.516° \pm 0.001°$, and $\gamma = 68.326° \pm 0.001°$ and calculated density 4.216 g·cm^{-3} (Harrison and Phillips 1999). To obtain this compound, a mixture of "H$_2$SeO$_3$" solution (dissolved SeO$_2$) (4 mL, 0.5 M), Zn(NO$_3$)$_2$ solution (2 mL, 1 M), and water

(6 mL) was sealed in a 23 mL Teflon-lined hydrothermal bomb, which was heated to 150°C for 2 days. The bomb was removed from the furnace and cooled to ambient temperature over a period of 2–3 h. Air stable perfectly faceted transparent single crystals were recovered by vacuum filtration, rinsing with acetone and drying in air.

$Zn(HSeO_3)_2 \cdot 2H_2O$ crystallizes in the monoclinic structure with the lattice parameters $a = 715.5$, $b = 843.4$, $c = 688.8$ pm, and $\beta = 61°35'$ (Kondrashev et al. 1979). Single crystals of this compound were obtained from aqueous solution of zinc selenite with H_2SeO_3 excess. Earlier investigations (Lieder and Gattow 1967) gave for $Zn(HSeO_3)_2 \cdot 2H_2O$ the formula $ZnSe_2O_5 \cdot 3H_2O$ (the monoclinic structure with the lattice parameters $a = 1267 \pm 1.5$, $b = 689 \pm 1$, $c = 838 \pm 1$ pm, and $\beta = 82.5° \pm 0.1°$ and calculation and experimental densities 3.271 and 3.212 $g \cdot cm^{-3}$, respectively).

$Zn(HSeO_3)_2 \cdot 4H_2O$ crystallizes in the monoclinic structure with the lattice parameters $a = 1468.3 \pm 0.2$, $b = 755.8 \pm 0.1$, $c = 1103.1 \pm 0.1$ pm, and $\beta = 126.79° \pm 0.01°$ and calculated and experimental densities 2.665 and 2.68 $\pm$ 0.01 $g \cdot cm^{-3}$, respectively (Engelen et al. 1995). Single crystals of this compound could be obtained from aqueous solution of $Zn(HSeO_3)_2$.

2.4   ZINC–POTASSIUM–GALLIUM–SELENIUM

ZnSe–K₂Se–Ga₂Se₃. The phase diagram is not constructed. The $KZn_4Ga_5Se_{12}$ quaternary compound is formed in this system, which crystallizes in the hexagonal structure with the lattice parameters $a = 1402.32 \pm 0.21$ and $c = 951.66 \pm 0.16$ pm ($a = 869.56 \pm 0.1$ and $\alpha = 107.480° \pm 0.006°$ in rhombohedral settings) and calculated density 4.908 $g \cdot cm^{-3}$ (Schwer et al. 1993). In order to produce $KZn_4Ga_5Se_{12}$, $ZnGa_2Se_4$ was solved in KBr in molar ratio 1:9 in a quartz-glass ampoule, which was sealed under vacuum and heated to 900°C. After 1 day, the furnace was switched off and the sample cooled down slowly. The reguli were treated with hot water to remove the unreacted halide.

2.5   ZINC–POTASSIUM–TIN–SELENIUM

ZnSe–K₂Se–SnSe₂. The phase diagram is not constructed. The $K_6Zn_4Sn_5Se_{17}$ quaternary compound is formed in this system (Kanatzidis et al. 1997a,b). The reaction of Sn, Zn, K_2Se, and Se at 400°C for 4 days upon a cooling rate of 4°C/h afforded yellowish white crystals of $K_6Zn_4Sn_5Se_{17}$. The product must be washed with degassed dimethylformamide and dried with acetone and ether.

2.6   ZINC–POTASSIUM–PHOSPHORUS–SELENIUM

$K_2ZnP_2Se_6$ quaternary compound exists in this system (Chondroudis and Kanatzidis 1998). To obtain this compound, a mixture of Zn, P_2Se_5, K_2Se, and Se was sealed under vacuum in a Pyrex tube and heated to 500°C for 4 days followed by cooling to 150°C at 2°C/h. $K_2ZnP_2Se_6$ is a wide-gap semiconductor and crystallizes in the monoclinic structure.

2.7   ZINC–RUBIDIUM–YTTERBIUM–SELENIUM

ZnSe–Rb$_2$Se–Yb$_2$Se$_3$. The phase diagram is not constructed. The RbZnYbSe$_3$ quaternary compound is formed in this system. It crystallizes in the orthorhombic structure with the lattice parameters a = 407.37 ± 0.04, b = 1506.6 ± 0.2, and c = 1078.8 ± 0.1 pm at 153 ± 2 K; energy gap 2.07 eV for (010) face; and calculated density 5.626 g·cm^{-3} (Mitchell et al. 2004). RbZnYbSe$_3$ was synthesized using the mixture of Rb$_2$Se$_3$, Yb, Zn, Se, and RbI. The sample was heated to 900°C in 48 h, kept at this temperature for 96 h, and cooled to room temperature in 148 h.

2.8   ZINC–RUBIDIUM–TIN–SELENIUM

ZnSe–Rb$_2$Se–SnSe$_2$. The phase diagram is not constructed. The Rb$_2$ZnSn$_2$Se$_6$ quaternary compound is formed in this system (Kanatzidis et al. 1997a,b). The reaction of Sn, Zn, Rb$_2$Se, and Se at 400°C for 4 days upon a cooling rate of 4°C/h afforded brownish orange crystals of Rb$_2$ZnSn$_2$Se$_6$. The product must be washed with degassed dimethylformamide and dried with acetone and ether.

2.9   ZINC–CESIUM–YTTRIUM–SELENIUM

ZnSe–Cs$_2$Se–Y$_2$Se$_3$. The phase diagram is not constructed. The CsZnYSe$_3$ quaternary compound is formed in this system. It crystallizes in the orthorhombic structure with the lattice parameters a = 414.09 ± 0.04, b = 1581.45 ± 0.15, and c = 1092.84 ± 0.11 pm at 153 ± 2 K; energy gap 2.41 eV for (010) crystal face and 2.29 eV for (001) crystal face; and calculated density 4.864 g·cm^{-3} (Mitchell et al. 2002). CsZnYSe$_3$ was obtained from the reaction of Cs$_2$Se$_3$, Y, Zn, and Se with the addition of CsI as flux. The sample was heated to 1000°C in 48 h, kept at this temperature for 50 h, and cooled at 4°C/h to 200°C and then the furnace was turned off.

2.10   ZINC–CESIUM–SAMARIUM–SELENIUM

ZnSe–Cs$_2$Se–Sm$_2$Se$_3$. The phase diagram is not constructed. The CsZnSmSe$_3$ quaternary compound is formed in this system. It crystallizes in the orthorhombic structure with the lattice parameters a = 419.35 ± 0.14, b = 1575.3 ± 0.5, and c = 1110.2 ± 0.4 pm at 153 ± 2 K; energy gap 2.63 eV for (010) crystal face and 2.43 eV for (001) crystal face; and calculated density 5.303 g·cm^{-3} (Mitchell et al. 2002). CsZnSmSe$_3$ was obtained from the reaction of Cs$_2$Se$_3$, Sm, Zn, and Se with the addition of CsI as flux. The sample was heated to 1000°C in 48 h, kept at this temperature for 50 h, and cooled at 4°C/h to 200°C and then the furnace was turned off.

2.11   ZINC–CESIUM–GADOLINIUM–SELENIUM

ZnSe–Cs$_2$Se–Gd$_2$Se$_3$. The phase diagram is not constructed. The CsZnGdSe$_3$ quaternary compound is formed in this system. It crystallizes in the orthorhombic structure with the lattice parameters a = 416.84 ± 0.07, b = 1576.5 ± 0.3, and c = 1100.89 ± 0.18 pm at 153 ± 2 K and calculated density 5.439 g·cm^{-3} (Huang et al. 2001). CsZnGdSe$_3$ was obtained from the reaction of Cs$_2$Se$_3$, Gd, Zn, and Se with the addition of CsI as flux. The sample was heated to 1000°C in 48 h, kept at this temperature for 50 h, and cooled at 4°C/h to 200°C and then the furnace was turned off.

2.12   ZINC–CESIUM–TERBIUM–SELENIUM

ZnSe–Cs$_2$Se–Tb$_2$Se$_3$. The phase diagram is not constructed. The CsZnTbSe$_3$ quaternary compound is formed in this system. It crystallizes in the orthorhombic structure with the lattice parameters $a = 415.14 \pm 0.06$, $b = 1578.5 \pm 0.2$, and $c = 1096.34 \pm 0.15$ pm at 153 ± 2 K and calculated density 5.493 g·cm^{-3} (Mitchell et al. 2002). CsZnTbSe$_3$ was obtained from the reaction of Cs$_2$Se$_3$, Tb, Zn, and Se with the addition of CsI as flux. The sample was heated to 1000°C in 48 h, kept at this temperature for 50 h, and cooled at 4°C/h to 200°C and then the furnace was turned off.

2.13   ZINC–CESIUM–DYSPROSIUM–SELENIUM

ZnSe–Cs$_2$Se–Dy$_2$Se$_3$. The phase diagram is not constructed. The CsZnDySe$_3$ quaternary compound is formed in this system. It crystallizes in the orthorhombic structure with the lattice parameters $a = 412.82 \pm 0.09$, $b = 1576.7 \pm 0.4$, and $c = 1091.0 \pm 0.3$ pm at 153 ± 2 K and calculated density 5.590 g·cm^{-3} (Mitchell et al. 2002). CsZnDySe$_3$ was obtained from the reaction of Cs$_2$Se$_3$, Dy, Zn, and Se with the addition of CsI as flux. The sample was heated to 1000°C in 48 h, kept at this temperature for 50 h, and cooled at 4°C/h to 200°C and then the furnace was turned off.

2.14   ZINC–CESIUM–HOLMIUM–SELENIUM

ZnSe–Cs$_2$Se–Ho$_2$Se$_3$. The phase diagram is not constructed. The CsZnHoSe$_3$ quaternary compound is formed in this system. It crystallizes in the orthorhombic structure with the lattice parameters $a = 412.75 \pm 0.05$, $b = 1581.4 \pm 0.2$, and $c = 1089.74 \pm 0.14$ pm at 153 ± 2 K and calculated density 5.604 g·cm^{-3} (Mitchell et al. 2002). CsZnHoSe$_3$ was obtained from the reaction of Cs$_2$Se$_3$, Ho, Zn, and Se with the addition of CsI as flux. The sample was heated to 1000°C in 48 h, kept at this temperature for 50 h and cooled at 4°C/h to 200°C and then the furnace was turned off.

2.15   ZINC–CESIUM–ERBIUM–SELENIUM

ZnSe–Cs$_2$Se–Er$_2$Se$_3$. The phase diagram is not constructed. The CsZnErSe$_3$ quaternary compound is formed in this system. It crystallizes in the orthorhombic structure with the lattice parameters $a = 411.66 \pm 0.06$, $b = 1581.6 \pm 0.2$, and $c = 1085.74 \pm 0.17$ pm at 153 ± 2 K; energy gap 2.63 eV for (010) crystal face and 2.56 eV for (001) crystal face; and calculated density 5.660 g·cm^{-3} (Mitchell et al. 2002). CsZnErSe$_3$ was obtained from the reaction of Cs$_2$Se$_3$, Er, Zn, and Se with the addition of CsI as flux. The sample was heated to 1000°C in 48 h, kept at this temperature for 50 h, and cooled at 4°C/h to 200°C and then the furnace was turned off.

2.16   ZINC–CESIUM–THULIUM–SELENIUM

ZnSe–Cs$_2$Se–Tm$_2$Se$_3$. The phase diagram is not constructed. The CsZnTmSe$_3$ quaternary compound is formed in this system. It crystallizes in the orthorhombic structure with the lattice parameters a = 410.4 ± 0.3, b = 1582.5 ± 1.3, and c = 1081.8 ± 0.9 pm at 153 ± 2 K and calculated density 5.711 g·cm^{-3} (Mitchell et al. 2002). CsZnTmSe$_3$ was obtained from the reaction of Cs$_2$Se$_3$, Tm, Zn, and Se with the addition of CsI as flux. The sample was heated to 1000°C in 48 h, kept at this temperature for 50 h, and cooled at 4°C/h to 200°C and then the furnace was turned off.

2.17   ZINC–CESIUM–YTTERBIUM–SELENIUM

ZnSe–Cs$_2$Se–Yb$_2$Se$_3$. The phase diagram is not constructed. The CsZnYbSe$_3$ quaternary compound is formed in this system. It crystallizes in the orthorhombic structure with the lattice parameters a = 407.852 ± 0.005, b = 1572.99 ± 0.02, and c = 1079.47 ± 0.01 pm at 4 K (Mitchell et al. 2004); a = 408.53 ± 0.04, b = 1578.64 ± 0.15, and c = 1080.68 ± 0.10 pm at 153 ± 2 K (Mitchell et al. 2002); a = 409.37 ± 0.01, b = 1586.17 ± 0.03, and c = 1081.75 ± 0.01 pm at 295 K (Mitchell et al. 2004); energy gap 1.93 [2.10 (Mitchell et al. 2003)] eV for (010) crystal face and 1.88 [1.97 (Mitchell et al. 2003)] eV for (001) crystal face; and calculated density 5.796 g·cm^{-3} (Mitchell et al. 2002).

CsZnYbSe$_3$ was obtained from the reaction of Cs$_2$Se$_3$, Yb, Zn, and Se with the addition of CsI as flux (Mitchell et al. 2002). The sample was heated to 1000°C in 48 h, kept at this temperature for 50 h, and cooled at 4°C/h to 200°C and then the furnace was turned off.

2.18   ZINC–CESIUM–GERMANIUM–SELENIUM

ZnSe–Cs$_2$Se–GeSe$_2$. The phase diagram is not constructed. The Cs$_2$ZnGe$_3$Se$_8$ quaternary compound is formed in this system. It crystallizes in the orthorhombic structure with the lattice parameters a = 757.56 ± 0.05, b = 1261.91 ± 0.10, and c = 1765.43 ± 0.11 pm; calculated density 4.647 g·cm^{-3}; and energy gap 2.31 eV (Morris et al. 2013). To obtain Cs$_2$ZnGe$_3$Se$_8$, a mixture of Cs$_2$Se$_2$, Zn, Ge, and Se in the appropriate ratios was added to a fused silica tube inside a nitrogen-filled glove box. The tube was evacuated, flame sealed, and used in a flame-melting reaction. After cooling in air, a phase pure ingot containing yellowish-orange platelike crystals was obtained.

2.19   ZINC–CESIUM–TIN–SELENIUM

ZnSe–Cs$_2$Se–SnSe$_2$. The phase diagram is not constructed. The Cs$_2$ZnSn$_3$Se$_8$ quaternary compound is formed in this system. It crystallizes in the orthorhombic structure with the lattice parameters a = 781.46 ± 0.03, b = 1265.34 ± 0.04, and c = 1837.47 ± 0.05 pm; calculated density 4.822 g·cm^{-3}; and energy gap 2.12 eV (Morris et al. 2013). To obtain Cs$_2$ZnSn$_3$Se$_8$, a mixture of Cs$_2$Se$_2$, Zn, Sn, and Se in the appropriate ratios was added to a fused silica tube inside a nitrogen-filled glovebox. The tube was evacuated, flame sealed, and used in a flame-melting reaction. After cooling in air, a phase pure ingot containing reddish-orange platelike crystals was obtained.

2.20 ZINC–COPPER–ALUMINUM–SELENIUM

ZnSe–CuAlSe$_2$. The phase diagram is not constructed. $(2ZnSe)_{1-x}(CuAlSe_2)_x$ solid solutions crystallize in the chalcopyrite-type structure at $x > 0.7$ and in the sphalerite-type structure at $x < 0.7$ (Gebicki et al. 1996, Bodnar' 2002). Order–disorder phase transition from chalcopyrite to disordered sphalerite structure was observed. This phase transition is not sharp and a compositional range of two-phase coexistence may be found (Gebicki et al. 1996).

Concentration dependence of the lattice parameters could be expressed by the next equations: $a = 560.5 + 0.04x$ and $c = 1097 + 0.005x$. Concentration dependence of the energy gap is characterized by the presence of a minimum (Bodnar' 2002).

Using XRD, it was determined that the lattice parameters of solid solutions increase at first with CuAlSe$_2$ contents increasing up to 1 mol.% and then decrease (Donohue and Bierstadt 1974).

According to the data of Parthé et al. (1969), the CuZn$_2$AlSe$_4$ quaternary compound is formed in this system. It crystallizes in the cubic structure of sphalerite type with the lattice parameter $a = 562.4 \pm 0.5$ pm, but this compound can be one of the solid solution compositions.

This system was investigated through XRD. The ingots were annealed at 1000°C (Donohue and Bierstadt 1974). $(2ZnSe)_{1-x}(CuAlSe_2)_x$ semiconductor materials over all compositional range were grown by the chemical transport method using I$_2$ as the transport agent (Gebicki et al. 1996).

2.21 ZINC–COPPER–GALLIUM–SELENIUM

ZnSe–CuGaSe$_2$. The phase relations in this system are shown in Figure 2.1 (Sysa et al. 2000). Solid solutions based on Ga$_2$Se$_3$ and ZnSe crystallize primarily from the liquid state in this system. Four-phase equilibrium takes place at 1029°C.

Solid solutions with chalcopyrite and sphalerite structures are formed within the interval of 0–60 mol.% ZnSe and 0–25 mol.% CuGaSe$_2$, respectively (Lambrecht 1973). The lattice parameters of forming solid solutions change linearly with composition. Order–disorder phase transition from chalcopyrite to disordered sphalerite structure was observed (Gebicki et al. 1996). It was indicated (Gallardo 1992) that it is possible to predict the value at which the chalcopyrite phase in this system disappears (x_c) and $T_c(x)$ behavior for the $Zn_{2x}(CuGa)_{1-x}Se_2$ alloys for which a chalcopyrite–sphalerite-like disordering transition with extensive terminal solid solution formation exists.

According to the data of Parthé et al. (1969) and Garbato and Manca (1974), the CuZn$_2$GaSe$_4$ quaternary compound is formed in this system. It crystallizes in the cubic structure of sphalerite type with the lattice parameter $a = 565.3 \pm 0.5$ pm (Parthé et al. 1969) [$a = 564 \pm 2$ pm (Garbato and Manca 1974)], but this compound can be one of the solid solution compositions.

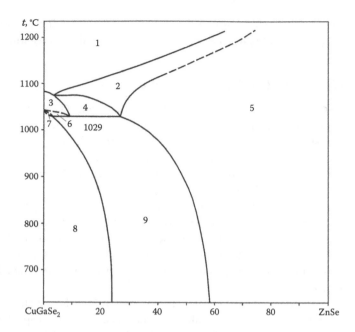

FIGURE 2.1 Phase relations in the system ZnSe–CuGaSe$_2$: 1, L; 2, L + (ZnSe); 3, L + (Ga$_2$Se$_3$); 4, L + (ZnSe) + (Ga$_2$Se$_3$); 5, (ZnSe); 6, L + (CuGaSe$_2$) + (Ga$_2$Se$_3$); 7, (CuGaSe$_2$) + (Ga$_2$Se$_3$); 8, (CuGaSe$_2$); and 9, (CuGaSe$_2$) + (ZnSe). (From Sysa, L.V. et al., *Fiz. i khim. tv. tila*, 1(2), 167, 2000.)

This system was investigated through differential thermal analysis (DTA), metallography, and XRD and the ingots were annealed at 600°C during 500 h (Garbato and Manca 1974, Sysa et al. 2000). (2ZnSe)$_x$(CuGaSe$_2$)$_{1-x}$ semiconductor materials over all compositional range were grown by the chemical transport method using I$_2$ as the transport agent (Gebicki et al. 1996).

2.22 ZINC–COPPER–INDIUM–SELENIUM

ZnSe–CuInSe$_2$. The phase diagram is shown in Figure 2.2 (Bodnar' and Chibusova 1998, Wagner et al. 2005). The two-phase region in the system 2(ZnSe)$_x$(CuInSe$_2$)$_{1-x}$ covers the composition range $0.10 \leq x \leq 0.36$, in which a tetragonal and a cubic phases are coexisting (Wagner et al. 2005). Within the composition range $0 \leq x \leq 0.10$, only tetragonal solid solutions exist. At concentration rates above 36 mol.% 2ZnSe, only cubic-structured solid solutions are found to be stable. However, in the range of 36 mol.% to about 60 mol.% 2ZnSe, tiny precipitates with stannite-like structure exist too. It was indicated (Gallardo 1992) that it is possible to predict the value, at which the chalcopyrite phase in this system disappears (x_c) and $T_c(x)$ behavior for the Zn$_{2x}$(CuIn)$_{1-x}$Se$_2$ alloys for which a chalcopyrite–sphalerite-like disordering transition with extensive terminal solid solution formation exists.

There are other data about two-phase region in this system: the chalcopyrite–sphalerite phase transition exists between $0.3 \leq x \leq 0.4$ (Bodnar' and Chibusova 1998,

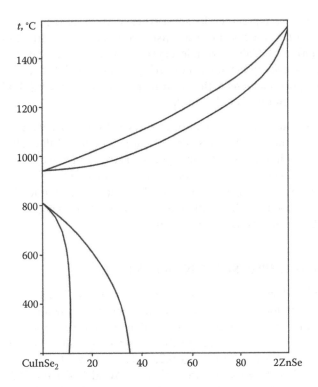

FIGURE 2.2 The $2ZnSe–CuInSe_2$ phase diagram. (From Bodnar', I.V. and Chibusova, L.V., *Zhurn. neorgan. khimii*, 43(11), 1913, 1998.)

Bodnar' and Gremenok 2003, Bodnar' et al. 2004). Solid solutions crystallize in the chalcopyrite structure from the $CuInSe_2$-rich side and in the sphalerite structure from the ZnSe-rich side (Lambrecht 1973, Bodnar' and Chibusova 1998, Bodnar' and Gremenok 2003). Within the single-phase region of the solid solution series, the lattice parameter follow Vegard's rule (Lambrecht 1973, Bodnar' and Chibusova 1998, Bodnar' et al. 2004, Schorr et al. 2005) [$a = 578.2 - 0.135x$ and $c = 1237 - 0.18x$ (Bodnar' and Gremenok 2003)]. The concentration dependence of the band-gap is nonlinear (Bodnar' and Gremenok 2003, Bodnar' et al. 2004).

According to the data of Parthé et al. (1969) and Garbato and Manca (1974), the $CuZn_2InSe_4$ quaternary compound is formed in this system. It crystallizes in the cubic structure of sphalerite type with the lattice parameter $a = 573.3 \pm 0.5$ pm (Parthé et al. 1969) [$a = 572 \pm 0$ pm (Garbato and Manca 1974)], but this compound can be one of the solid solution compositions.

This system was investigated through DTA, XRD, measuring of microhardness and density (Garbato and Manca 1974, Bodnar' and Chibusova 1998), and by TEM, neutron powder diffraction, diffraction using synchrotron radiation, and HRTEM with EDX (Schorr et al. 2005, Wagner et al. 2005). The ingots were annealed at 680°C–730°C for 500 h (Bodnar' et al. 2004) [at 800°C for 500 h (Bodnar' and Chibusova 1998)]. Powder sample $Zn_{2x}(CuIn)_{1-x}Se_2$ alloys with $0 \leq x \leq 0.1$ were prepared by solid-state reaction of the elements at 850°C, and samples with $0.4 < x < 1.0$ were also prepared by the directional freezing method (Schorr et al. 2005). Single crystals of the $(2ZnSe)_x(CuInSe_2)_{1-x}$ solid solutions were grown by the horizontal Bridgman method ($0 \leq x < 0.5$) and using chemical transport reactions with I_2 as transport agent ($0.5 < x \leq 1$) (Bodnar' and Gremenok 2003, Bodnar' et al. 2004).

$ZnSe-CuInSe_2-In_2Se_3$. The phase diagram is not constructed. According to the data of Chippaux and Deschanvres (1982), a large domain of thiogallate-type compounds was found in this quasiternary system. These compounds have a number of cation vacancies less or equal to that of $ZnIn_2Se_4$ compound.

This system was investigated through XRD and using atomic absorption. The ingots were obtained by the heating of mixtures of the chemical elements for 15–20 days (Chippaux and Deschanvres 1982).

2.23 ZINC–COPPER–SILICON–SELENIUM

$ZnSe-Cu_2SiSe_3$. The phase diagram is not constructed. The $Cu_2ZnSiSe_4$ quaternary compound is formed in this system. It melts incongruently at 973°C ± 5°C (Schäfer and Nitsche 1977) [according to the data of Yao et al. (1987), it decomposes at 470°C ± 10°C] and crystallizes in the orthorhombic structure with the lattice parameters $a = 782.6 \pm 0.3$, $b = 672.7 \pm 0.2$, and $c = 644.5 \pm 0.1$ pm (Yao et al. 1987) [$a = 783$, $b = 673$, and $c = 644$ pm (Schleich and Wold 1977); $a = 782.3$, $b = 672.0$, and $c = 644.0$ pm (Schäfer and Nitsche 1974, 1977); $a = 780$, $b = 670$, and $c = 646$ pm (Nitsche et al. 1967)]; calculation and experimental densities 5.25 and 5.22 g·cm^{-3}, respectively; and energy gap $E_g = 2.33$ eV [$E_g = 2.20$ eV (Yao et al. 1987)] (Schleich and Wold 1977, Nitsche et al. 1967).

Single crystals of $Cu_2ZnSiSe_4$ were obtained by the chemical transport reactions using I_2 as the transport agent (Nitsche et al. 1967, Schäfer and Nitsche 1974, Schleich and Wold 1977, Yao et al. 1987).

2.24 ZINC–COPPER–GERMANIUM–SELENIUM

$ZnSe-Cu_2GeSe_3$. The phase diagram is shown in Figure 2.3 (Parasyuk et al. 2001). The coordinates of the eutectic point are 775°C and 3 mol.% ZnSe. The $Cu_2ZnGeSe_4$ quaternary compound is formed in this system that melts incongruently at 890°C (Matsuhita et al. 2000, 2005, Parasyuk et al. 2001) [785°C ± 5°C (Schäfer and Nitsche 1977)] and has a polymorphous transformation at 798°C (Parasyuk et al. 2001). It crystallizes in the tetragonal structure with the lattice parameters $a = 561.043 \pm 0.008$ and $c = 1104.57 \pm 0.03$ pm (Parasyuk et al. 2001) [$a = 560.6$ and $c = 1104.2$ pm (Matsuhita et al. 2000, 2005); $a = 561.3$ and $c = 1104.8$ pm (Schäfer and Nitsche 1977); $a = 562.2$ and $c = 1106$ pm (Hahn and Schulze 1965, Schäfer and Nitsche 1974, Guen et al. 1979, Guen and Glaunsinger 1980); $a = 561$ and $c = 1105$ pm (Schleich and Wold 1977); $a = 561$ and $c = 1102$ pm (Nitsche et al. 1967)]; calculation and experimental densities 5.54 and 5.50 g·cm^{-3} [5.52 and 5.48 g·cm^{-3} (Hahn and Schulze 1965); 5.52 and 5.50 g·cm^{-3} (Guen et al. 1979,

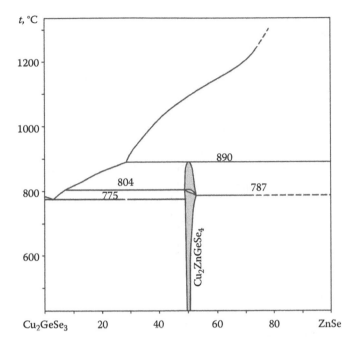

FIGURE 2.3 Phase diagram of the ZnSe–Cu$_2$GeSe$_3$ system. (From Parasyuk, O.V. et al., *J. Alloys Compd.*, 329(1–2), 202, 2001.)

Guen and Glaunsinger 1980)], respectively; and energy gap E_g = 1.29 eV (Schleich and Wold 1977) [E_g = 1.63 eV (Matsuhita et al. 2000, 2005)].

Because that energy between kesterite- and stannite-type structures is small, both structures can coexist in synthesized samples of Cu$_2$ZnGeSe$_4$. According to the first-principles calculations, kesterite-type structure of this compound is characterized by the tetragonal structure with the lattice parameters a = 560.2 and c = 1125.9 pm and energy gap 1.60 eV, and stannite-type structure is characterized also by the tetragonal structure with the lattice parameters a = 558.3 and c = 1132.5 pm and energy gap 1.32 eV (Chen and Ravindra 2013).

This system was investigated through DTA, XRD, metallography, and electron probe micro-analysis (EPMA) (Matsuhita et al. 2000, Parasyuk et al. 2001). The samples were annealed at 400°C during 500 h and after annealing, they were quenched in ice water (Parasyuk et al. 2001). Single crystals of Cu$_2$ZnGeSe$_4$ were obtained by the chemical transport reactions using I$_2$ as the transport agent (Nitsche et al. 1967, Schäfer and Nitsche 1974, Schleich and Wold 1977, Guen et al. 1979, Guen and Glaunsinger 1980, Yao et al. 1987, Parasyuk et al. 2001) or by the sintering of mixtures from binary compounds at 650°C–900°C (Hahn and Schulze 1965) or using the horizontal gradient method from respective melts (Matsuhita et al. 2000, 2005).

ZnSe–Cu$_8$GeSe$_6$. This system is not a quasibinary section of the Zn–Cu–Ge–Se quaternary system and crosses two three-phase and one two-phase region (Figure 2.4) (Romanyuk and Parasyuk 2003). There are two peritectic equilibria at 809°C

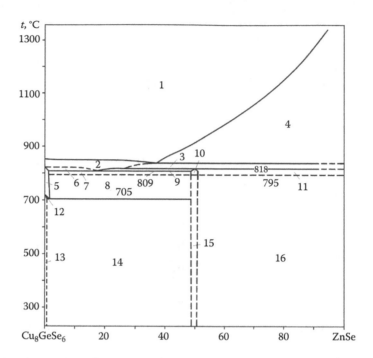

FIGURE 2.4　Phase relations in the ZnSe–Cu_8GeSe_6 system: 1, L; 2, L + (β-Cu_2Se); 3, L + (β-Cu_2Se) + ZnSe; 4, L + ZnSe; 5, (γ-Cu_8GeSe_6); 6, L + (β-Cu_2Se) + (γ-Cu_8GeSe_6); 7, (β-Cu_2Se) + (β-$Cu_2ZnGeSe_4$) + (γ-Cu_8GeSe_6); 8, (β-Cu_2Se) + (α-$Cu_2ZnGeSe_4$) + (γ-Cu_8GeSe_6); 9, L + (β-Cu_2Se) + (β-$Cu_2ZnGeSe_4$); 10, (β-Cu_2Se) + (β-$Cu_2ZnGeSe_4$); 11, (β-Cu_2Se) + (β-$Cu_2ZnGeSe_4$) + ZnSe; 12, (β-Cu_8GeSe_6) + (γ-Cu_8GeSe_6); 13, (β-Cu_8GeSe_6); 14, (β-Cu_2Se) + (α-$Cu_2ZnGeSe_4$) + (β-Cu_8GeSe_6); 15, (β-Cu_2Se) + (α-$Cu_2ZnGeSe_4$); and 16, (β-Cu_2Se) + (α-$Cu_2ZnGeSe_4$) + ZnSe. (From Romanyuk, Ya.E. and Parasyuk, O.V., *J. Alloys Compd.*, 348(1–2), 195, 2003.)

and 818°C in this section. The polymorphous transformation of $Cu_2ZnGeSe_4$ is depicted by a dotted line as its thermal effects were not observed in the thermograms. This system was investigated through DTA, XRD, and metallography. The samples were annealed at 400°C during 500 h followed by quenching in cold water.

ZnSe–Cu_2Se–$GeSe_2$.　The liquidus surface projection of this system is shown in Figure 2.5 (Romanyuk and Parasyuk 2003). It consists of seven fields of primary crystallization. Sixteen monovariant lines divide the various fields of primary crystallization. There are eight binary univariant points, five ternary invariant transition points, and two ternary eutectics in the ZnSe–Cu_2Se–$GeSe_2$ quasiternary system. Temperatures and compositions of the ternary invariant points in the ZnSe–Cu_2Se–$GeSe_2$ quasiternary system are given in Table 2.1. Isothermal section of the ZnSe–Cu_2Se–$GeSe_2$ quasiternary system at 400°C is given in Figure 2.6 (Romanyuk and Parasyuk 2003).

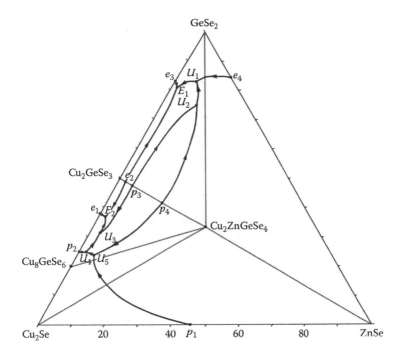

FIGURE 2.5 The liquidus surface projection of the ZnSe–Cu$_2$Se–GeSe$_2$ quasiternary system. (From Romanyuk, Ya.E. and Parasyuk, O.V., *J. Alloys Compd.*, 348(1–2), 195, 2003.)

TABLE 2.1

Temperatures and Compositions of the Ternary Invariant Points in the ZnSe–Cu$_2$Se–GeSe$_2$ Quasiternary System

Invariant Points	t, °C	Composition (mol.%)		
		Cu$_2$Se	ZnSe	GeSe$_2$
E_1	697	17.5	1.5	81
E_2	750	61	2	37
U_1	710	11	6	83
U_2	791	15	10	75
U_3	795	66	3	31
U_4	809	73	2	25
U_5	818	71	5	24

Source: Romanyuk, Ya.E. and Parasyuk, O.V., *J. Alloys Compd.*, 348(1–2), 195, 2003.

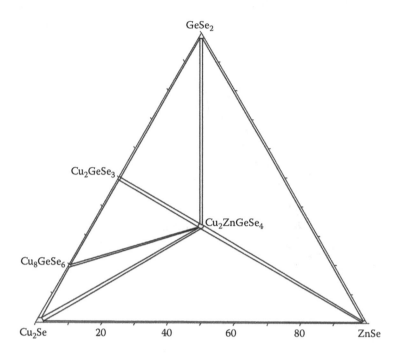

FIGURE 2.6   Isothermal section of the $ZnSe-Cu_2Se-GeSe_2$ quasiternary system at 400°C. (From Romanyuk, Ya.E. and Parasyuk, O.V., *J. Alloys Compd.*, 348(1–2), 195, 2003.)

2.25   ZINC–COPPER–TIN–SELENIUM

ZnSe–Cu$_2$SnSe$_3$. The phase diagram is shown in Figure 2.7 (Dudchak and Piskach 2001). The Cu$_2$ZnSnSe$_4$ quaternary compound is formed in this system according to the peritectic reaction at 790°C (Dudchak and Piskach 2001) [at 805°C (Matsuhita et al. 2000, 2005); at 796°C ± 5°C (Schäfer and Nitsche 1977)]. The composition of the peritectic point is 14 mol.% ZnSe. This compound has a homogeneity range and undergoes a polymorphous transformation in the temperature interval 616°C–619°C [Matsuhita et al. (2000) indicated that this compound has three polymorphous modifications]. A eutectic process occurs at 694°C with the eutectic point composition 2.5 mol.% ZnSe. Solubility of Cu$_2$SnSe$_3$ in ZnSe at the peritectic temperature reaches up to 4 mol.%.

Cu$_2$ZnSnSe$_4$ crystallizes in the tetragonal structure with the lattice parameters a = 568.82 ± 0.02 and c = 1133.78 ± 0.09 pm (Olekseyuk et al. 2002) [a = 585.5 ± 0.1 and c = 1137.9 ± 0.3 pm (Dudchak and Piskach 2001, Olekseyuk et al. 2001); a = 569.3 and c = 1133.3 pm (Matsuhita et al. 2000, 2005); a = 569.4 and c = 1134.7 pm (Schäfer and Nitsche 1977; a = 568.1 and c = 1134 pm (Hahn and Schulze 1965, Schäfer and Nitsche 1974, Guen et al. 1979, Guen and Glaunsinger 1980)]; calculation and experimental densities 5.69 and 5.62 g·cm^{-3} {5.69 and 5.68 g·cm^{-3} (Hahn and Schulze 1965); 5.68 [5.689 ± 0.03 (Olekseyuk et al. 2002)]; and 5.62 g·cm^{-3} (Guen et al. 1979)}, respectively (Guen and Glaunsinger 1980); and energy gap E_g = 1.44 eV (Matsuhita et al. 2000, 2005) [E_g = 0.96 eV (Chen et al. 2009); E_g ≈ 1.0 (Persson 2010)].

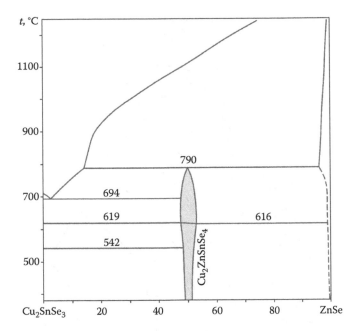

FIGURE 2.7 Phase diagram of the ZnSe–Cu₂SnSe₃ system. (From Dudchak, I. and Piskach, L., *Visnyk L'viv. un-tu. Ser. khim.*, (40), 73, 2001.)

Using the first-principles density functional method, it was shown that the low-energy crystal structure of $Cu_2ZnSnSe_4$ is a kesterite-type structure (Chen et al. 2009). However, the stannite or partially disordered kesterite structure can also exist in synthesized samples due to the small energy cost. According to the first-principles calculations, kesterite-type structure of this compound is characterized by the tetragonal structure with the lattice parameters $a = 564.2$ and $c = 1130.3$ pm [$a = 560.5$ and $c = 1120.0$ pm (Persson 2010)] and calculated density 5.7871 g·cm⁻³, and stannite-type structure is characterized also by the tetragonal structure with the lattice parameters $a = 565.0$ and $c = 1127.0$ pm [$a = 560.4$ and $c = 1120.8$ pm (Persson 2010)] and calculated density 5.7882 g·cm⁻³ (Bensalem et al. 2014).

This system was investigated through DTA, XRD, metallography, and electron probe micro-analysis (EPMA) (Matsuhita et al. 2000, 2005, Dudchak and Piskach 2001), and the ingots were annealed at 400°C during 250 h (Dudchak and Piskach 2001). Single crystals of $Cu_2ZnSnSe_4$ were obtained by the chemical transport reactions (Schäfer and Nitsche 1974, Guen et al. 1979, Guen and Glaunsinger 1980) or by the sintering of mixtures from binary compounds at 650°C–900°C (Hahn and Schulze 1965) or using the horizontal gradient method from respective melts (Matsuhita et al. 2000, 2005) or by crystallization from the solution in the melt (Olekseyuk et al. 2002).

ZnSe–Cu₂Se–SnSe₂. The liquidus surface of this system (Figure 2.8) consists of five fields of primary crystallization of Cu_2SnSe_3 and solid solutions based on Cu_2Se, ZnSe, SnSe₂, and $Cu_2ZnSnSe_4$, respectively (Olekseyuk et al. 2001, Dudchak and Piskach 2003). There are 10 monovariant lines and 10 invariant points (2 binary

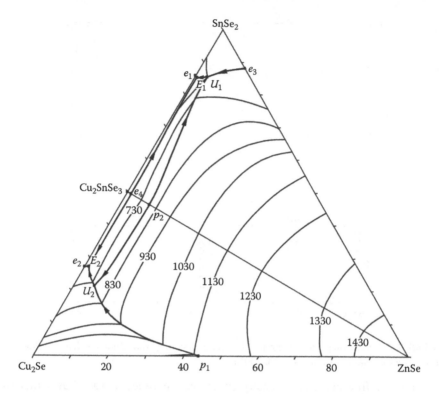

FIGURE 2.8 The liquidus surface of the ZnSe–Cu$_2$Se–SnSe$_2$ quasiternary system. (From Olekseyuk, I.D. et al., *Fiz. i khim. tv. tila*, 2(2), 195, 2001; Dudchak, I.V. and Piskach, L.V., *J. Alloys Compd.*, 351, 145, 2003.)

peritectics, 4 binary eutectics, 2 ternary transition points, and 2 ternary eutectics) in the liquidus surface. The field of primary crystallization of ZnSe occupies the largest area of the concentration triangle as the most refractory component. The field of primary crystallization of Cu$_2$ZnSnSe$_4$ is elongated along the boundary side Cu$_2$Se–SnSe$_2$ with the temperature maximum in the quasibinary section ZnSe–Cu$_2$SnSe$_3$. The polymorphous transformation of the solid solutions based on Cu$_2$ZnSnSe$_4$ compound occurs at 627°C in the quasiternary system ZnSe–Cu$_2$SnSe$_3$–Cu$_2$Se and at 583°C for the ZnSe–Cu$_2$SnSe$_3$–SnSe$_2$ quasiternary system.

The isothermal section of the ZnSe–Cu$_2$Se–SnSe$_2$ quasiternary system at 400°C is shown in Figure 2.9 (Olekseyuk et al. 2001, Dudchak and Piskach 2003). The solubility in Cu$_2$Se is lower than 2 mol.% and is elongated along the Cu$_2$Se–SnSe$_2$ boundary side; in ZnSe and SnSe$_2$, it is lower than 1 mol.%; and in Cu$_2$SnSe$_3$, it is lower than 0.5 mol.%. The miscibility gap of Cu$_2$ZnSnSe$_4$ was found to be 3 mol.% along the triangulated section ZnSe–Cu$_2$SnSe$_3$.

This system was investigated through DTA, XRD, and metallography, and the ingots were annealed at 400°C during 250 h (Olekseyuk et al. 2001, Dudchak and Piskach 2003).

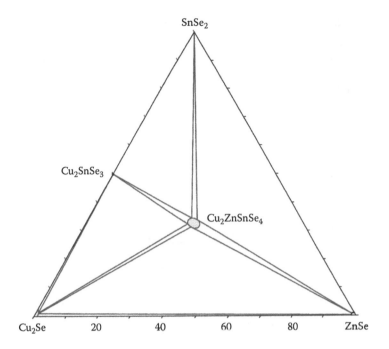

FIGURE 2.9 Isothermal section of the ZnSe–Cu$_2$Se–SnSe$_2$ quasiternary system at 400°C. (From Olekseyuk, I.D. et al., *Fiz. i khim. tv. tila*, 2(2), 195, 2001; Dudchak, I.V. and Piskach, L.V., *J. Alloys Compd.*, 351, 145, 2003.)

2.26 ZINC–SILVER–GALLIUM–SELENIUM

ZnSe–AgGaSe$_2$. The phase diagram of this system is given in Figure 2.10 (Kozer et al. 2008). The eutectic contains 7 mol.% ZnSe and crystallizes at 844°C. AgZn$_2$GaSe$_4$ quaternary compound is formed in the ZnSe–AgGaSe$_2$ system. Upon heating, it decomposes according to a solid-state reaction at 777°C. Below 696°C, AgZn$_2$GaSe$_4$ decomposes forming AgGaSe$_2$ and ZnSe. It crystallizes in the tetragonal structure with the lattice parameters $a = 572.43 \pm 0.03$ and $c = 1133.7 \pm 0.1$ pm. This system was investigated by DTA and XRD.

ZnSe–Ag$_9$GaSe$_6$. The phase diagram is of eutectic type (Figure 2.11) (Kozer et al. 2008). The coordinates of eutectic points are ~5 mol.% ZnSe and 752°C (this system is regarded as ZnSe–1/10 Ag$_9$GaSe$_6$). This system was investigated through DTA and XRD and the ingots were annealed at 600°C during 250 h with the next cooling in the cold water.

ZnSe–Ag$_2$Se–Ga$_2$Se$_3$. The isothermal section of the ZnSe–Cu$_2$Se–SnSe$_2$ quasiternary system at 600°C is shown in Figure 2.12 (Kozer et al. 2008). It consists of six one-phase, nine two-phase, and four three-phase regions. Quaternary compounds were not found at this temperature.

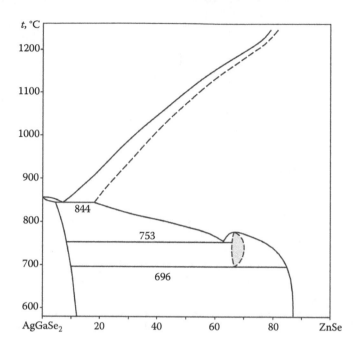

FIGURE 2.10   Phase diagram of the ZnSe–AgGaSe$_2$ system. (From Kozer, V.R. et al., *Nauk. visnyk Volyn. nats. un-tu. Khim. nauky*, (13), 20, 2008.)

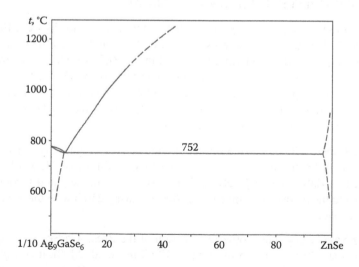

FIGURE 2.11   Phase diagram of the ZnSe–1/10 Ag$_9$GaSe$_6$ system. (From Kozer, V.R. et al., *Nauk. visnyk Volyn. nats. un-tu. Khim. nauky*, (13), 20, 2008.)

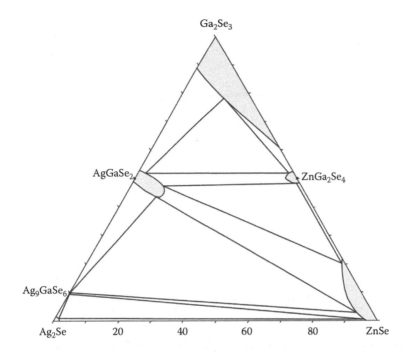

FIGURE 2.12 Isothermal section of the ZnSe–Ag$_2$Se–Ga$_2$Se$_3$ quasiternary system at 600°C. (From Kozer, V.R. et al., *Nauk. visnyk Volyn. nats. un-tu. Khim. nauky*, (13), 20, 2008.)

2.27 ZINC–SILVER–SILICON–SELENIUM

ZnSe–Ag$_8$SiSe$_6$. The phase diagram is a eutectic type (Figure 2.13) (Piskach et al. 2006). The eutectic composition and temperature are 62 ± 2 mol.% ZnSe and 926°C, respectively. Solubility of ZnSe in Ag$_8$SiSe$_6$ at the eutectic temperature is 57 mol.% and less than 2 mol.% at room temperature (Parasyuk et al. 2003, Piskach et al. 2006). This system was investigated by DTA, XRD, and metallography (Piskach et al. 2006).

ZnSe–Ag$_2$Se–SiSe$_2$. Isothermal section of this quasiternary system at room temperature is shown in Figure 2.14 and no quaternary phases were found (Parasyuk et al. 2003). The solid solubility range of Ag$_8$SiSe$_6$ is less than 2 mol.% ZnSe. The ingots were annealed at 400°C during 250 h and further cooling was done by turning off the furnace.

2.28 ZINC–SILVER–GERMANIUM–SELENIUM

ZnSe–AgGeSe$_2$. According to the data of Parthé et al. (1969), the AgZn$_2$GeSe$_4$ quaternary compound is formed in this system. It crystallizes in the hexagonal structure with the lattice parameters $a = 426.9 \pm 0.5$ and $c = 565.9 \pm 0.5$ pm but this compound can be one of the solid solution compositions. This compound was obtained by the heating of mixtures from chemical elements at 600°C–800°C.

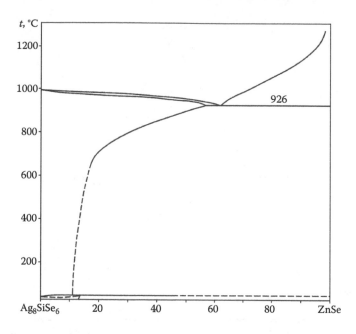

FIGURE 2.13 The ZnSe–Ag$_8$SiSe$_6$ phase diagram. (From Piskach, L.V. et al., *J. Alloys Compd.*, 421(1–2), 98, 2006.)

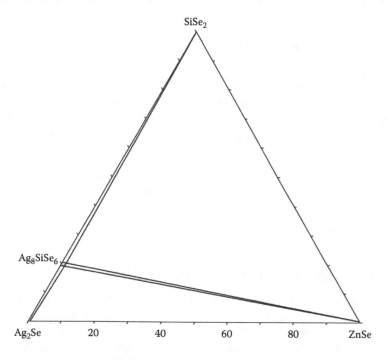

FIGURE 2.14 Isothermal section of ZnSe–Ag$_2$Se–SiSe$_2$ quasiternary system at room temperature. (From Parasyuk, O.V. et al., *J. Alloys Compd.*, 354(1–2), 138, 2003.)

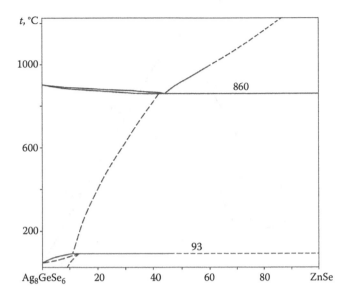

FIGURE 2.15 The ZnSe–Ag$_8$GeSe$_6$ phase diagram. (From Piskach, L.V. et al., *J. Alloys Compd.*, 421(1–2), 98, 2006.)

ZnSe–Ag$_8$GeSe$_6$. The phase diagram is a eutectic type (Figure 2.15) (Piskach et al. 2006). The eutectic composition and temperature are 44 ± 2 mol.% ZnSe and 860°C, respectively. Solubility of ZnSe in Ag$_8$GeSe$_6$ at the eutectic temperature is 42 mol.%.

ZnSe–Ag$_2$Se–GeSe$_2$. The glass-formation region in this quasiternary system is shown in Figure 2.16 (Olekseyuk et al. 2009). Such region in the Ag$_2$Se–GeSe$_2$ boundary system is limited to the range 53–56 mol.% GeSe$_2$. The maximum amount of ZnSe that can be introduced in the glass is 10 mol.%. The maximum GeSe$_2$ content is 63 mol.% at 4–6 mol.% ZnSe. Characteristic temperatures of the glassy alloys, namely, the glass transition temperature (T_g), the crystallization temperature (T_c), and the melting temperature (T_m) of the crystallized alloys, have been measured. It was determined that T_g of the ZnSe-containing glasses lies in a fairly narrow range (253°C ± 8°C), probably because the region of glass existence is rather small. This system was investigated by DTA and XRD.

2.29   ZINC–SILVER–TIN–SELENIUM

ZnSe–Ag$_8$SnSe$_6$. The phase diagram is a eutectic type (Figure 2.17) (Piskach et al. 2006). The eutectic composition and temperature are 37 ± 1 mol.% ZnSe and 715°C, respectively. Solubility of ZnSe in Ag$_8$SiSe$_6$ at the eutectic temperature is 33 mol.%. This system was investigated by DTA, XRD, and metallography.

2.30   ZINC–MAGNESIUM–TELLURIUM–SELENIUM

ZnSe + MgTe ⇔ ZnTe + MgSe. The phase diagram is not constructed. The correlation function method was used to estimate the energy gap of the Zn$_{1-x}$Mg$_x$Se$_y$Te$_{1-y}$ solid solutions over the entire composition range (Figure 2.18) (Shim et al. 2000).

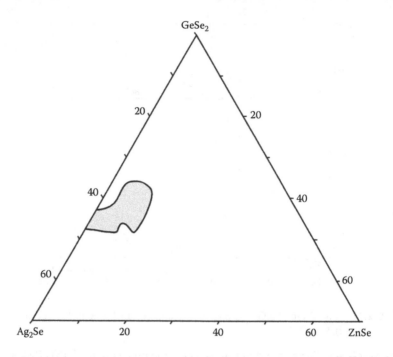

FIGURE 2.16   The glass-formation region in the ZnSe–Ag$_2$Se–GeSe$_2$ quasiternary system. (From Olekseyuk, I.D. et al., *Chem. Met. Alloys*, 2(3–4), 146, 2009.)

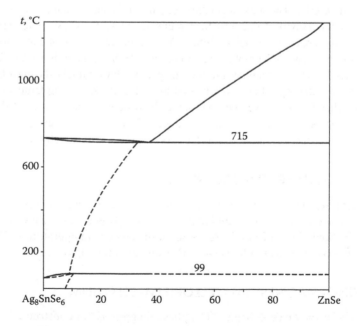

FIGURE 2.17   The ZnSe–Ag$_8$SnSe$_6$ phase diagram. (From Piskach, L.V. et al., *J. Alloys Compd.*, 421(1–2), 98, 2006.)

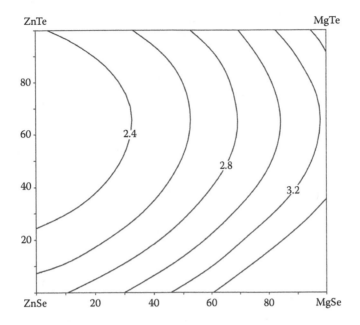

FIGURE 2.18 The bandgap variation (eV) over the entire range of alloy compositions for $Zn_{1-x}Mg_xSe_yTe_{1-y}$ solid solutions. (From Shim, K. et al., *J. Cryst. Growth*, 214–215, 350, 2000.)

Density functional theory (local density approximation) has been employed by Hassah El Haj et al. (2006) to study the structural and electronic properties of such quaternary alloys, but these authors used too small value of the ZnTe energy bandgap. The calculated formation energy showed that the least stable quaternary alloys are around 60 at.% for both Mg and Se atoms.

2.31 ZINC–STRONTIUM–OXYGEN–SELENIUM

$SrZn(SeO_3)_2$ ($SrZnSe_2O_6$) quaternary compound is formed in the Zn–Sr–O–Se quaternary system (Johnston and Harrison 2001). It crystallizes in the monoclinic structure with the lattice parameters $a = 448.36 \pm 0.03$, $b = 1475.76 \pm 0.10$, $c = 942.67 \pm 0.07$ pm, and $\beta = 95.273° \pm 0.002°$ and calculated density 4.35 g·cm^{-3}. This compound was synthesized by the hydrothermal synthesis using a mixture of $SrCO_3$, ZnO, SeO_2, and H_2O, which was heated at 150°C for 7 days.

2.32 ZINC–BARIUM–OXYGEN–SELENIUM

$BaZn(SeO_3)_2$ ($BaZnSe_2O_6$) quaternary compound is formed in the Zn–Ba–O–Se quaternary system (Jiang et al. 2006). It crystallizes in the monoclinic structure with the lattice parameters $a = 553.75 \pm 0.07$, $b = 1642.9 \pm 0.2$, $c = 716.83 \pm 0.09$ pm, and $\beta = 96.671° \pm 0.005°$ and optical bandgap $E_g = 3.8$ eV. This compound is stable up to 435°C. The weight loss occurred in the range of 435°C–721°C corresponding to the release on one SeO_2. Single crystals of $BaZn(SeO_3)_2$ were initially obtained by the solid-state reaction of $BaCO_3$, ZnO, and SeO_2 in an evacuated quartz tube

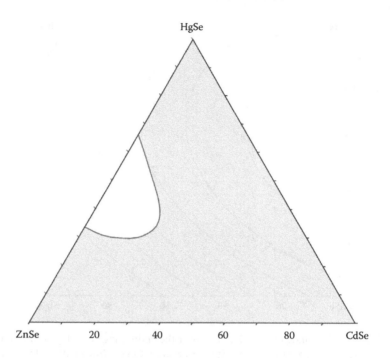

FIGURE 2.19 Calculated miscibility gap in the ZnSe–CdSe–HgSe quasiternary system. (From Ohtani, H. et al., *J. Alloys Compd.*, 182(1), 103, 1992.)

at 750°C for 6 days and then cooled to 350°C at 4°C/h before switching off the furnace. Subsequently, a pure powder sample of this compound was prepared quantitatively by reacted a mixture of BaO, ZnO and SeO$_2$ in molar ratio of 1:1:2 at 600°C.

2.33 ZINC–CADMIUM–MERCURY–SELENIUM

ZnSe–CdSe–HgSe. The position of the calculated miscibility gap in the ZnSe–CdSe–HgSe system at 300°C is shown in Figure 2.19 (Ohtani et al. 1992). It is seen that the surface of the miscibility gap extends from ZnSe–HgSe quasibinary system into the ternary.

2.34 ZINC–CADMIUM–ARSENIC–SELENIUM

CdSe–Zn$_3$As$_2$. The phase diagram is a eutectic type (Figure 2.20) (Golovey and Olekseyuk 1971). The eutectic composition and temperature are 55 mol.% 2CdSe and 770°C ± 5°C, respectively. An immiscibility region within the interval of 60–90 mol.% 2CdSe exists in this system. Solubility of 2CdSe in Zn$_3$As$_2$ at the eutectic temperature reaches 45 mol.% and decreases to 10 mol.% at room temperature. Solubility of Zn$_3$As$_2$ in 2CdSe at the same temperatures decreases from 5 mol.% to approximately 2.5 mol.%. Low-temperature Zn$_3$As$_2$ modification can dissolve up to 5 mol.% 2CdSe. This system was investigated through DTA, metallography, XRD, and measuring of microhardness.

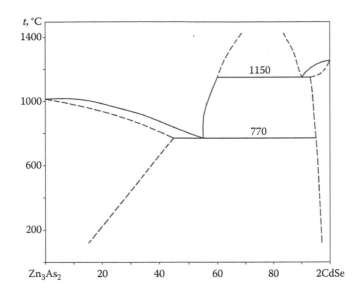

FIGURE 2.20 The CdSe–Zn₃As₂ phase diagram. (From Golovey, M.I. and Olekseyuk, I.D., *Izv. AN SSSR. Neorgan. materialy*, 7(6), 934, 1971.)

2.35 ZINC–CADMIUM–TELLURIUM–SELENIUM

ZnSe–CdTe. The phase diagram is shown in Figure 2.21 (Oleinik et al. 1977). Solid solutions over entire range of concentrations are formed in this system above 850°C. The azeotropic composition and temperature are 97 mol.% CdTe and 1070°C, respectively. The lattice parameters change linearly with composition (Yim et al. 1972, Oleinik et al. 1977).

The immiscibility gap exists in this system below 850°C with maximum at 68 mol.% ZnSe and 783°C (Tai and Hori 1977). At 500°C, 600°C, and 700°C, this region is situated within the interval of 20–85, 31–82, and 52–75 mol.% ZnSe, respectively. Calculated immiscibility gap in this system (Ohtani et al. 1992) practically coincides with the experimental data. The energy gap of $(ZnSe)_{0.91}(CdTe)_{0.09}$ solid solution is equal to 2.3 eV, which points to the nonlinear change of energy gap in this system (Yim et al. 1972).

This system was investigated through DTA, metallography, and XRD (Oleinik et al. 1977).

ZnTe–CdSe. This section is a nonquasibinary section of the Zn–Cd–Te–Se quaternary system (Vitrichovski 1977, Oleinik et al. 1978a). Solid solutions based on ZnTe and CdSe crystallize in the sphalerite and wurtzite structure, respectively. The width of two-phase region is approximately 7 mol.% (Figure 2.22) (Oleinik et al. 1978a). The immiscibility gap exists in this system.

The concentration dependence of energy gap passes through minimum (Vitrichovski 1977).

This system was investigated through DTA, metallography, and XRD (Vitrichovski 1977, Oleinik et al. 1978a).

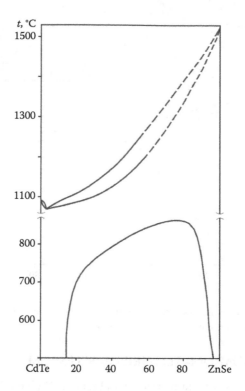

FIGURE 2.21   The ZnSe–CdTe phase diagram. (From Oleinik, G.S. et al., *Izv. AN SSSR. Neorgan. materialy*, 13(11), 1976, 1977.)

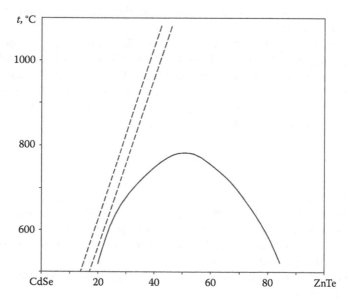

FIGURE 2.22   Phase relations in the ZnTe–CdSe system. (From Oleinik, G.S. et al., *Poluprovodn. tehnika i mikroelectron.*, (28), 56, 1978a.)

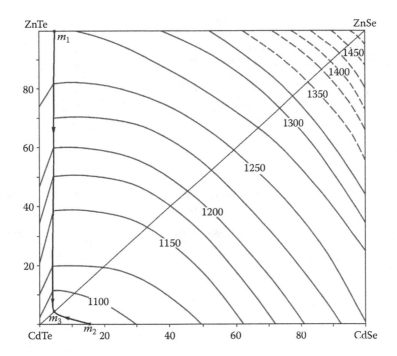

FIGURE 2.23 Liquidus surface of the ZnSe + CdTe ⇔ ZnTe + CdSe ternary mutual system. (From Oleinik, G.S. et al., *Izv. AN SSSR. Neorgan. materialy*, 14(3), 441, 1978b.)

ZnSe + CdTe ⇔ ZnTe + CdSe. Two fields of primary crystallization exist on the liquidus surface of this system (Figure 2.23) (Oleinik et al. 1978b). The field of primary crystallization of the solid solutions based on ZnSe occupies the most part of this liquidus surface. Minimum liquidus temperature is in the ZnSe–CdTe quasi-binary system and corresponds to 1070°C.

Isothermal sections of this ternary mutual system at 530°C, 630°C, 830°C, and 1020°C are shown in Figure 2.24 (Leute and Wulff 1992, 1993). There are structural and spinodal miscibility gaps in this system. The structural miscibility gap is situated from the CdSe-rich side and moves toward CdSe with temperature decreasing. The spinodal miscibility gap extends symmetrically to the diagonal ZnSe–CdTe, respectively, that this diagonal can be treated as a quasibinary section.

In order to determine the structural miscibility gap between wurtzite and sphalerite regions and the spinodal miscibility gap within the sphalerite region, samples with overall compositions corresponding to the regions of interest were prepared from the pure binary component. In the first step, all these samples were equilibrated at 830°C and thereupon the final equilibrium temperature in the region of 530°C–1020°C was adjusted. Depending on this final annealing temperature, the annealing times, required to establish equilibrium, ranged between 2 weeks and more than 10 months (Leute and Wulff 1992).

Calculated miscibility gap in the ZnSe + CdTe ⇔ ZnTe + CdSe ternary mutual system at 400°C–600°C is shown in Figure 2.25 (Ohtani et al. 1992). The position of the miscibility gap indicates that the phase separation occurs toward the

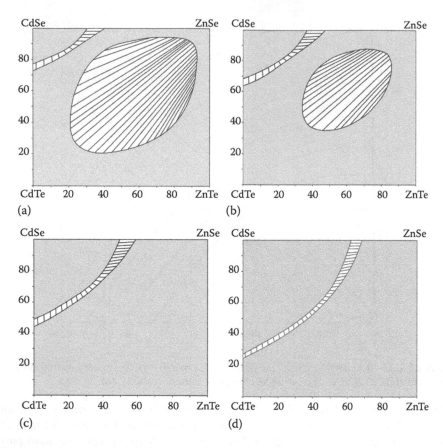

FIGURE 2.24 Isothermal sections of the ZnSe + CdTe ⇔ ZnTe + CdSe ternary mutual system at (a) 530°C, (b) 630°C, (c) 830°C, and (d) 1020°C. (From Leute, V. and Wulff, B., *Ber. Bunsenges. Phys. Chem.*, 96(2), 119, 1992.)

ZnSe and CdTe ends, which have the highest and lowest melting temperatures, respectively, among the four compounds.

2.36 ZINC–CADMIUM–CHLORINE–SELENIUM

CdSe–ZnCl₂–CdCl₂. The phase diagram is not constructed. Using oriented zone recrystallization and radioactive isotopes, equilibrium coefficient of $ZnCl_2$ distribution in the CdSe–CdCl₂ eutectic was determined: $k_0 = 1.55$ (Leonov and Chunarev 1977). Since $k_0 > 1$, evidently, the next chemical reaction $CdSe + ZnCl_2 = ZnSe + CdCl_2$ takes place in this system.

2.37 ZINC–CADMIUM–MANGANESE–SELENIUM

ZnSe–CdSe–MnSe. Isothermal section of this system at 960°C is shown in Figure 2.26 (Manhas et al. 1986). Polycrystalline samples were produced by the standard melt and anneal technique. The alloys were annealed at 960°C for 2 weeks.

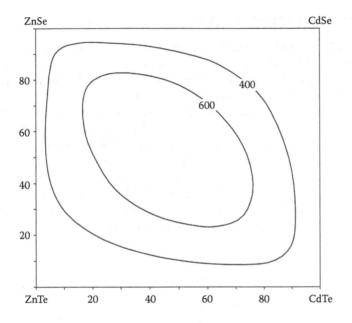

FIGURE 2.25 Calculated miscibility gap in the ZnSe + CdTe ⇔ ZnTe + CdSe quasiternary system. (From Ohtani, H. et al., *J. Alloys Compd.*, 182(1), 103, 1992.)

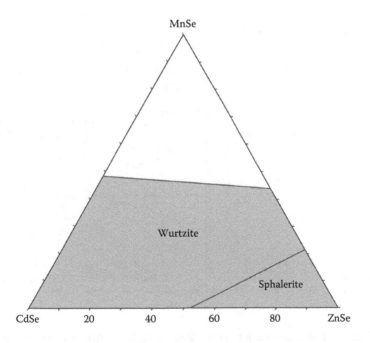

FIGURE 2.26 Isothermal section of the ZnSe–CdSe–MnSe system at 960°C. (From Manhas, S. et al., *Phys. Status Solidi (a)*, 94(1), 213, 1986.)

2.38 ZINC–MERCURY–TELLURIUM–SELENIUM

ZnSe + HgTe ⇔ ZnTe + HgSe. The phase diagram is not constructed. The isothermal sections of this ternary mutual system at 530°C, 650°C, 730°C, and 830°C are shown in Figure 2.27 (Leute and Plate 1989). The first two isothermal sections show that in spite of complete solid solubility along the edges of the phase squares there is a much extended ternary miscibility gap (Figure 2.27a, b). An addition of only 0.1 mol.% of Hg chalcogenides to ZnSe-rich solid solutions is sufficient to give rise to a decomposition into a HgTe-rich and a ZnSe-rich compound. More extended ternary solid solutions regions only occur near the HgSe and ZnTe corner.

At 730°C, a narrow region of liquid Hg chalcogenides is limited by two liquid solid gaps situated on the HgSe–HgTe and ZnTe–HgTe edges (Figure 2.27c). This section also exhibits two extended three-phase regions in which a decomposition into one liquid phase and two regions of solid sphalerite phase occur.

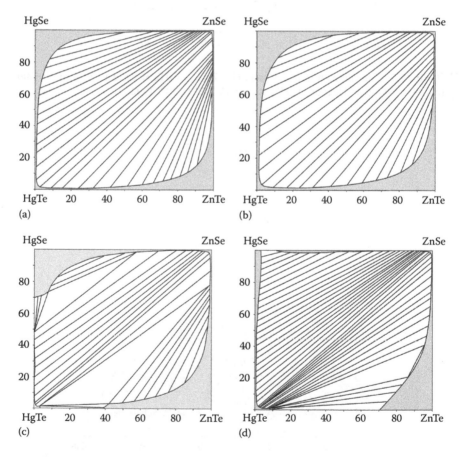

FIGURE 2.27 Isothermal sections of the ZnSe + HgTe ⇔ ZnTe + HgSe ternary mutual system at (a) 530°C, (b) 650°C, (c) 730°C, and (d) 830°C. (From Leute, V. and Plate H., *Ber. Bunsenges. Phys. Chem.*, 93(7), 757, 1989.)

At 830°C, the liquid range is more extended, the liquid solid gap on the HgSe–HgTe edge has disappeared, a small liquid solid gap on the ZnSe–HgSe edge has developed, and one of the two three-phase regions has disappeared (Figure 2.27d). At still higher temperatures, the second three-phase region will also disappear, and there remains only one extended gap between the liquid Hg chalcogenides and the solid Zn chalcogenides. Even at 1530°C, there is no complete solubility in the quasiternary liquid because the reaction enthalpy causes a liquid miscibility gap in spite of ideal solubility on the quasibinary edges (Leute and Plate 1989).

Calculated miscibility gap in the ZnSe + HgTe ⇔ ZnTe + HgSe ternary mutual system at 600°C–1200°C is shown in Figure 2.28 (Ohtani et al. 1992). The position of the miscibility gap indicates that the phase separation occurs toward the ZnSe and HgTe ends, which have the highest and lowest melting temperatures, respectively, among the four compounds.

A thermodynamic analysis of stability for the $Zn_xHg_{1-x}Te_ySe_{1-x}$ quaternary solid solutions has been developed by Kazakov et al. (2001). The locations of the unstable regions in the temperature region of 100°C–550°C were calculated using both the disordered solid solution approach and the short-range clustering approach but the obtained results are in disagreement with the results obtained in Leute and Plate (1989).

This system was investigated through DTA and XRD. The ingots were equilibrated at 830°C and thereupon the final equilibrium temperature in the region of 530°C–1020°C was adjusted. Depending on this final annealing temperature, the annealing times, required to establish equilibrium, ranged between 66 days and more than 270 days (Leute and Plate 1989).

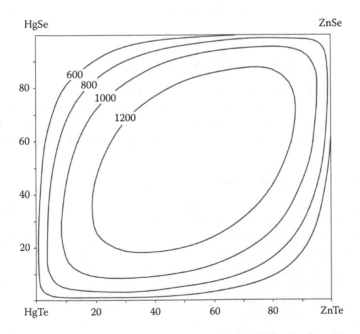

FIGURE 2.28   Calculated miscibility gap in the ZnSe + HgTe ⇔ ZnTe + HgSe quasiternary system. (From Ohtani, H. et al., *J. Alloys Compd.*, 182(1), 103, 1992.)

2.39   ZINC–MERCURY–BROMINE–SELENIUM

ZnSe–HgSe–HgBr$_2$. The phase diagram is not constructed. ZnHg$_3$Se$_3$Br$_2$ quaternary compound is formed in this system. It crystallizes in the hexagonal structure with the lattice parameters $a = 738.3 \pm 0.7$ and $c = 1092.4 \pm 0.9$ pm and calculated density 6.852 g·cm^{-3} (Chen et al. 2012). This compound was synthesized from a reaction of ZnBr$_2$, HgBr$_2$, and Se. The starting materials were mixed and loaded into a glass tube, which was flame sealed under a vacuum. The tube was heated to 500°C in 6 h from room temperature and kept for 10 days, followed by cooling to 100°C at a rate of 6°C/h to promote crystal growth, then cooled to 35°C in 5 h and power off.

2.40   ZINC–ALUMINUM–CHLORINE–SELENIUM

ZnSe–Al$_2$Se$_3$–AlCl$_3$. The phase diagram is not constructed. The Zn$_2$AlSe$_3$Cl quaternary compound is formed in this system (Range and Handrick 1988). It crystallizes in the cubic structure with the lattice parameter $a = 559.79 \pm 0.09$ pm. This compound was obtained at 600°C–1100°C by the interaction of ZnSe, ZnCl$_2$, and Al$_2$Se$_3$.

2.41   ZINC–ALUMINUM–BROMINE–SELENIUM

ZnSe–Al$_2$Se$_3$–AlBr$_3$. The phase diagram is not constructed. The Zn$_2$AlSe$_3$Br quaternary compound is formed in this system (Range and Handrick 1988). It crystallizes in the cubic structure with the lattice parameter $a = 560.24 \pm 0.02$ pm. This compound was obtained at 600°C–1100°C by the interaction of ZnSe, ZnBr$_2$, and Al$_2$Se$_3$.

2.42   ZINC–ALUMINUM–IODINE–SELENIUM

ZnSe–Al$_2$Se$_3$–AlI$_3$. The phase diagram is not constructed. The Zn$_2$AlSe$_3$I quaternary compound is formed in this system (Range and Handrick 1988). It crystallizes in the cubic structure with the lattice parameter $a = 560.64 \pm 0.03$ pm. This compound was obtained at 600°C–1100°C by the interaction of ZnSe, ZnI$_2$, and Al$_2$Se$_3$.

2.43   ZINC–GALLIUM–GERMANIUM–SELENIUM

ZnSe–Ga$_2$Se$_3$–GeSe$_2$. The liquidus surface of this quasiternary system (Figure 2.29) consists of four fields of primary crystallization, which belong to the solid solutions based on ZnSe, Ga$_2$Se$_3$ and GeSe$_2$, and the ZnGa$_2$Se$_4$ ternary compound (Olekseyuk et al. 2003). No quaternary compounds were found. The solid solution based on ZnSe occupies the largest area of the concentration triangle. The fields of primary crystallization are separated by five monovariant lines and six invariant points, four of which correspond to binary reactions and two to ternary ones. Ternary eutectic E (3 mol.% ZnSe, 13 mol.% Ga$_2$Se$_3$, and 84 mol.% GeSe$_2$) crystallizes at 651°C, and the temperature of the ternary transition point U (6 mol.% ZnSe, 9 mol.% Ga$_2$Se$_3$, and 85 mol.% GeSe$_2$) is equal to 693°C.

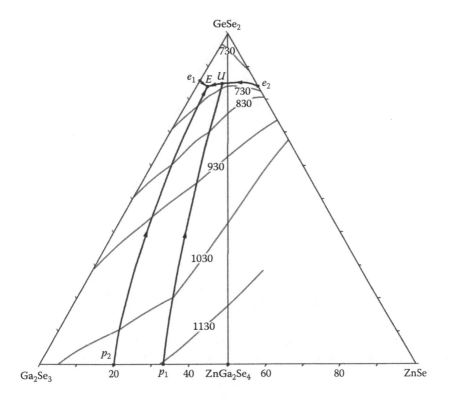

FIGURE 2.29 Liquidus surface of the ZnSe–Ga$_2$Se$_3$–GeSe$_2$ quasiternary system. (From Olekseyuk, I.D. et al., *J. Alloys Compd.*, 351(1–2), 171, 2003.)

Isothermal section of the ZnSe–Ga$_2$Se$_3$–GeSe$_2$ quasiternary system at 400°C is given in Figure 2.30 (Olekseyuk et al. 2003). The solid solution ranges of ZnSe and Ga$_2$Se$_3$ were found to be elongated along the ZnSe–Ga$_2$Se$_3$ quasibinary section. The GeSe$_2$ content in the solid solution based on ZnSe does not exceed 3 mol.%, while in the solid solution based on Ga$_2$Se$_3$, it remains below 4 mol.%.

The glass-formation region in this quasiternary system is shown in Figure 2.31 (Olekseyuk et al. 1997). It is elongated along the Ga$_2$Se$_3$–GeSe$_2$ quasibinary system. The maximum amount of Ga$_2$Se$_3$ that can be introduced in the glass is 28 mol.% and the ZnSe content in this glass does not exceed 3 mol.%.

This system was investigated through DTA, XRD, and metallography and all samples were annealed at 400°C for 500 h with subsequent quenching in cold water (Olekseyuk et al. 1997, 2003).

2.44 ZINC–GALLIUM–TIN–SELENIUM

ZnSe–Ga$_2$Se$_3$–SnSe$_2$. The liquidus surface of this quasiternary system is formed by four fields of primary crystallization (Figure 2.32) (Parasyuk et al. 2004). Three of them correspond to the solid solutions based on ZnSe, Ga$_2$Se$_3$, and SnSe$_2$.

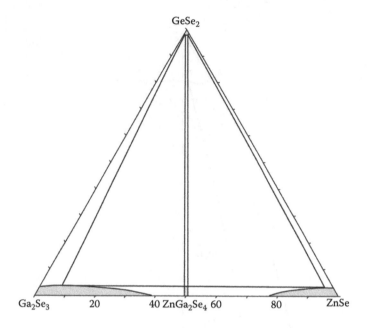

FIGURE 2.30 Isothermal section of the ZnSe–Ga$_2$Se$_3$–GeSe$_2$ quasiternary system at 400°C. (From Olekseyuk, I.D. et al., *J. Alloys Compd.*, 351(1–2), 171, 2003.)

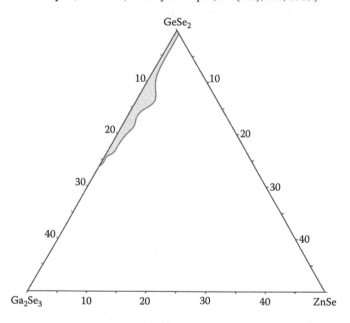

FIGURE 2.31 The glass-formation region in the ZnSe–Ga$_2$Se$_3$–GeSe$_2$ quasiternary system. (From Olekseyuk, I.D. et al., *Fizyka kondens. vysokomolek. system. Nauk. zap. Rinens'kogo pedinstytutu*, (3), 148, 1997.)

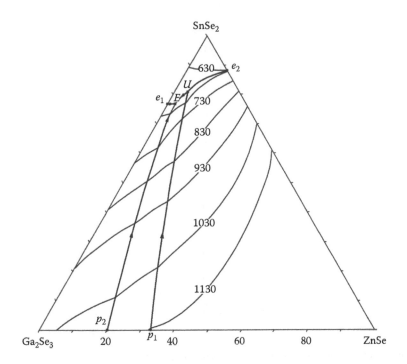

FIGURE 2.32 Liquidus surface of the ZnSe–Ga₂Se₃–SnSe₂ quasiternary system. (From Parasyuk, O.V. et al., *J. Alloys Compd.*, 379(1–2), 143, 2004.)

The other one belongs to the solid solution based on $ZnGa_2Se_4$ ternary compound. No quaternary compounds were found. The fields of primary crystallization are separated by five monovariant lines and six invariant points, four of which correspond to binary reactions and two to ternary ones. Ternary eutectic E (2 mol.% ZnSe, 21 mol.% Ga_2Se_3, and 77 mol.% $SnSe_2$) crystallizes at 578°C, and the temperature of the ternary transition point U (4 mol.% ZnSe, 15 mol.% Ga_2Se_3 and 81 mol.% $SnSe_2$) is equal to 591°C.

Isothermal section of the ZnSe–Ga₂Se₃–SnSe₂ quasiternary system at 400°C is shown in Figure 2.33 (Parasyuk et al. 2004). ZnSe and Ga_2Se_3 form solid solution ranges, which are elongated along the quasibinary section ZnSe–Ga₂Se₃ and the maximum $SnSe_2$ content in them equals 2 and 3 mol.%, respectively.

The glass-formation region in this quasiternary system is shown in Figure 2.34 (Olekseyuk et al. 1999). Glass transition temperature (T_g), the crystallization temperature (T_c), and the melting temperature (T_m) were determined for obtained glasses. The glasses have considerable tendency to crystallization and they can be obtained only in the case of the rigid hardening.

This system was investigated through DTA, XRD, and metallography, and all samples were annealed at 400°C for 500 h and then quenched in cold water (Olekseyuk et al. 1999, Parasyuk et al. 2004).

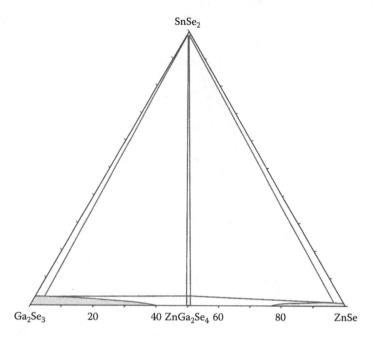

FIGURE 2.33 Isothermal section of the ZnSe–Ga$_2$Se$_3$–SnSe$_2$ quasiternary system at 400°C. (From Parasyuk, O.V. et al., *J. Alloys Compd.*, 379(1–2), 143, 2004.)

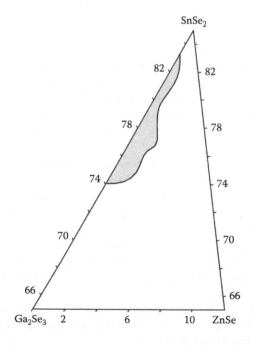

FIGURE 2.34 The glass-formation region in the ZnSe–Ga$_2$Se$_3$–SnSe$_2$ quasiternary system. (From Olekseyuk, I.D. et al., *Funct. Mater.*, 6(3), 474, 1999.)

2.45 ZINC–GALLIUM–PHOSPHORUS–SELENIUM

ZnSe–GaP. The phase diagram is shown in Figure 2.35 (Shumilin et al. 1977, Ufimtsev et al. 1980). Solid solutions over the entire range of concentration are formed in this system. Lattice parameters of forming solid solutions change linearly with composition (Yim 1969). Sonomura et al. (1973) supposes the presence of immiscibility gap within the interval of 40–90 mol.% ZnSe.

This system was investigated using high-pressure chamber (Shumilin et al. 1977). Liquidus temperatures were determined by sight, and solidus line was calculated using a model of regular solutions. Single crystals of solid solutions were obtained by the chemical transport reactions (Voitsehovski and Stetsenko 1976) or by the crystallization from the solutions in the Sn or Zn melts (Sonomura et al. 1973).

ZnSe–Ga–GaP. The liquidus surface of this quasiternary system near Ga corner (Figure 2.36) was constructed using DTA, local XRD, measuring of microhardness, and mathematical simulation of experiment (Shumilin et al. 1977).

2.46 ZINC–GALLIUM–ARSENIC–SELENIUM

ZnSe–GaAs. The phase diagram of this quasibinary system is shown in Figure 2.37 (Lakeenkov ct al. 1975, Vasiliev and Novikova 1977). Solid solutions over the entire range of concentrations are formed in this system (Goriunova and Fedorova 1959, Kirovskaya and Mulikova 1975). The composition and temperature of azeotropic

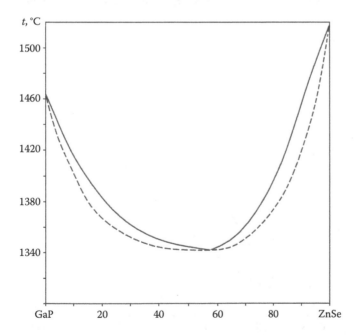

FIGURE 2.35 The ZnSe–GaP phase diagram. (From Shumilin, V.P. et al., *Izv. AN SSSR. Neorgan. materialy*, 13(9), 1560, 1977; Ufimtsev, V.B. et al., *Zavodsk. laboratoria*, 46(6), 525, 1980.)

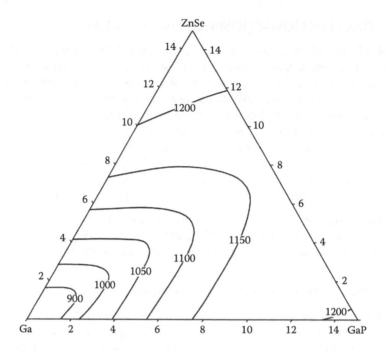

FIGURE 2.36   Liquidus surface of the ZnSe–Ga–GaP quasiternary system near Ga corner. (From Shumilin, V.P. et al., *Izv. AN SSSR. Neorgan. materialy*, 13(9), 1560, 1977.)

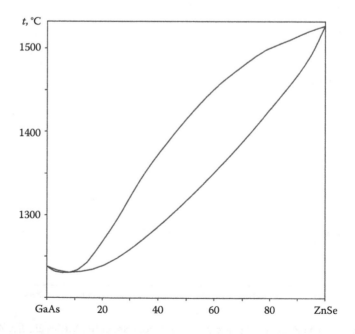

FIGURE 2.37   The ZnSe–GaAs phase diagram. (From Lakeenkov, V.M. et al., *Izv. AN SSSR. Neorgan. materialy*, 11(7), 1311, 1975.)

point are 8 mol.% ZnSe and 1230°C, respectively (Lakeenkov et al. 1975) [9 mol.% ZnSe and 1233°C ± 5°C (Vasiliev and Novikova 1977)].

Energy gap of the forming solid solutions at 27°C is within the interval of 1.42–1.96 eV (Kirovskaya and Mulikova 1975), increases gradually at the ZnSe contents increasing, and has negative deviation from linearity (Ku and Bodi 1968).

Discrepancies in the determination of concentration dependence of the lattice parameters [corresponding to Vegard's rule (Goriunova and Fedorova 1959, Lakeenkov et al. 1975); there is a deviation from Vegard's rule with maximum at 50–60 mol.% ZnSe (Yim 1969)] can be probably explained by the vicinity of the lattice parameters of ZnSe and GaAs [a_{ZnSe} = 566.87 pm and a_{GaAs} = 565.33 pm (Goriunova and Fedorova 1959)].

Solid solutions in this system were obtained by the melting of ZnSe and GaAs mixtures (Goriunova and Fedorova 1959, Ku and Bodi 1968, Yim 1969) or by the chemical transport reactions and recrystallization from the solutions in the gallium melt and floating-zone refining through the liquid gallium (Ku and Bodi 1968) or by the annealing of fine-dispersed powder of initial binary compounds at 1200°C (Kirovskaya and Mulikova 1973).

This system was investigated through DTA, metallography, XRD, and measuring of microhardness (Goriunova and Fedorova 1959, Kirovskaya and Mulikova 1975, Lakeenkov ct al. 1975, Vasiliev and Novikova 1977).

ZnSe–Ga–GaAs. A part of liquidus surface of this quasiternary system near the Ga corner (Figure 2.38) was constructed according to four vertical sections (Figure 2.39) (Novikova et al. 1974, 1977). Eutecic is degenerated from the Ga-rich side at 26°C in all vertical sections. There is a "valley" on the liquidus surface, which is elongated along the Ga–GaAs subsystem (Lakeenkov et al. 1974, Novikova et al. 1974, 1977). Using XRD, it was determined that solid materials extracted from the gallium are $(ZnSe)_{1-x}(GaAs)_x$ solid solutions with sphalerite structure. Thus, practically all investigated liquidus surface is a field of $(ZnSe)_{1-x}(GaAs)_x$ solid solution crystallization. This system was investigated through DTA and XRD.

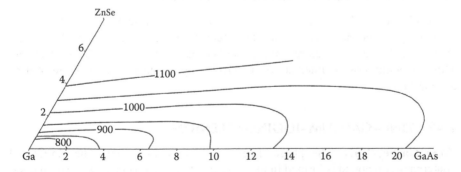

FIGURE 2.38 A part of liquidus surface of the ZnSe–Ga–GaAs quasiternary system near the Ga corner. (From Lakeenkov, V.M. et al., *Izv. AN SSSR. Neorgan. materialy*, 11(7), 1311, 1975.)

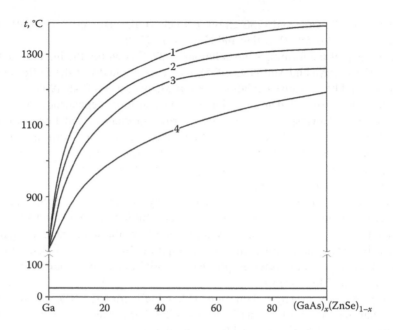

FIGURE 2.39 The $Ga–(ZnSe)_{1-x}(GaAs)_x$ vertical sections: $x = 0.2$ (1), 0.4 (2), 0.6 (3), and 0.8 (4). (From Novikova, E.M. et al., *Deposited in VINITI*, № 4439–77Dep, 1977.)

2.47 ZINC–GALLIUM–CHLORINE–SELENIUM

$ZnSe–Ga_2Se_3–GaCl_3$. The phase diagram is not constructed. The Zn_2GaSe_3Cl quaternary compound is formed in this system (Range and Handrick 1988). It crystallizes in the cubic structure with the lattice parameter $a = 559.45 \pm 0.03$ pm. This compound was obtained at 600°C–1100°C by the interaction of $ZnSe$, $ZnCl_2$, and Ga_2Se_3.

2.48 ZINC–GALLIUM–BROMINE–SELENIUM

$ZnSe–Ga_2Se_3–GaBr_3$. The phase diagram is not constructed. The Zn_2GaSe_3Br quaternary compound is formed in this system (Range and Handrick 1988). It crystallizes in the cubic structure with the lattice parameter $a = 559.53 \pm 0.04$ pm. This compound was obtained at 600°C–1100°C by the interaction of $ZnSe$, $ZnBr_2$, and Ga_2Se_3.

2.49 ZINC–GALLIUM–IODINE–SELENIUM

$ZnSe–Ga_2Se_3–GaI_3$. The phase diagram is not constructed. The Zn_2GaSe_3I quaternary compound is formed in this system (Range and Handrick 1988). It crystallizes in the cubic structure with the lattice parameter $a = 559.51 \pm 0.02$ pm. This compound was obtained at 600°C–1100°C by the interaction of $ZnSe$, ZnI_2, and Ga_2Se_3.

TABLE 2.2
Solubility of Zn + Se (cm^{-3}) in InAs

t, °C	Zn/Se = 3:1	Zn/Se = 1:1	Zn/Se = 1:3
700	$1.9 \cdot 10^{20}$	$5.4 \cdot 10^{20}$	$2.8 \cdot 10^{20}$
800	$2.4 \cdot 10^{20}$	$7.6 \cdot 10^{20}$	$3.7 \cdot 10^{20}$
850	$3.1 \cdot 10^{20}$	$14.5 \cdot 10^{20}$	$8.6 \cdot 10^{20}$
900	$14.7 \cdot 10^{20}$	$19 \cdot 10^{20}$	$11.2 \cdot 10^{20}$

Source: Glazov, V.M. et al., *Izv. AN SSSR. Neorgan. materialy*, 15(3), 390, 1979.

2.50 ZINC–INDIUM–ARSENIC–SELENIUM

ZnSe–InAs. The phase diagram is not constructed. The solubility of Zn + Se in InAs at Zn/Se = 3:1, 1:1, and 1:3 (Table 2.2) (Glazov et al. 1979) was investigated. Maximum solubility takes place at the equimolar ratio of doping elements. This system was investigated through metallography and measuring of microhardness. The ingots were annealed at 700°C, 800°C, 850°C, and 900°C for 1200, 1000, 800, and 500 h, respectively.

2.51 ZINC–INDIUM–ANTIMONY–SELENIUM

ZnSe–InSb. The phase diagram is not constructed. According to the data of XRD and bandgap measuring, solid solutions based on InSb contain up to 30 mol.% ZnSe (Kirovskaya et al. 2002).

2.52 ZINC–INDIUM–OXYGEN–SELENIUM

$ZnIn_2(SeO_3)_4$ quaternary compound is formed in this system. It is stable up to 450°C and crystallizes in the monoclinic structure with the lattice parameters $a = 843.31 \pm 0.07$, $b = 478.19 \pm 0.04$, $c = 1465.83 \pm 0.14$ pm, and $\beta = 101.684° \pm 0.006°$ and calculated density 4.606 g·cm^{-3} at 200 K (Lee et al. 2012). Crystals of this compound were obtained by standard solid-state reaction. ZnO, In_2O_3, and SeO_2 in stoichiometric ratio were thoroughly mixed and introduced into fused silica tube that was subsequently evacuated and sealed. The tube was gradually heated to 380°C for 5 h and then to 600°C for 48 h. The sample was cooled at a rate of 6°C/h to room temperature.

2.53 ZINC–INDIUM–CHROMIUM–SELENIUM

ZnSe–In$_2$Se$_3$–Cr$_2$Se$_3$. The phase diagram is not constructed. The sections $Zn_{1-x}In_{0.667x}Cr_2Se_4$ and $ZnCr_{2-y}In_ySe_4$ as well as some samples of compositions outside these joins of this quasiternary system were studied using XRD of quenched samples (Figure 2.40) (Okońska-Kozłowska et al. 1988). Whereas no detectable amounts of chromium can be incorporated into $ZnIn_2Se_4$ of the thiogallate structure

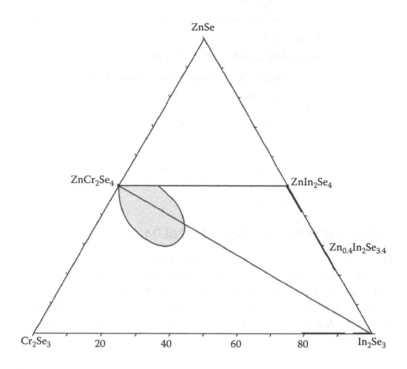

FIGURE 2.40   Isothermal section of the $ZnSe-In_2Se_3-Cr_2Se_3$ quasiternary system at 900°C. (From Okońska-Kozłowska, I. et al., *Z. anorg. und allg. Chem.*, 559(4), 201, 1988.)

in the case of the spinel $ZnCr_2Se_4$ (a = 1050.0 pm), up to 21 at.% Cr and up to 20 at.% of Zn can be substituted by In. However, spinel-type solid solutions with larger In content (up to a = 1076 pm) are formed by coincident substitution of both Zn and Cr corresponding to $Zn_{1-x}In_{0.667x}Cr_{2-y}In_ySe_4$ (0 < $x + y$ < 0.6) with indium in both tetrahedral and octahedral lattice sites.

2.54   ZINC–INDIUM–CHLORINE–SELENIUM

$ZnSe-In_2Se_3-InCl_3$. The phase diagram is not constructed. The Zn_2InSe_3Cl quaternary compound is formed in this system (Range and Handrick 1988). It crystallizes in the cubic structure with the lattice parameter a = 568.78 ± 0.02 pm. This compound was obtained at 600°C–1100°C by the interaction of ZnSe, $ZnCl_2$, and In_2Se_3.

2.55   ZINC–INDIUM–BROMINE–SELENIUM

$ZnSe-In_2Se_3-InBr_3$. The phase diagram is not constructed. The Zn_2InSe_3Br quaternary compound is formed in this system (Range and Handrick 1988). It crystallizes in the cubic structure with the lattice parameter a = 569.11 ± 0.04 pm. This compound was obtained at 600°C–1100°C by the interaction of ZnSe, $ZnBr_2$, and In_2Se_3.

2.56 ZINC–INDIUM–IODINE–SELENIUM

ZnSe–In$_2$Se$_3$–InI$_3$. The phase diagram is not constructed. The Zn$_2$InSe$_3$I quaternary compound is formed in this system (Range and Handrick 1988). It crystallizes in the cubic structure with the lattice parameter $a = 569.26 \pm 0.04$ pm. This compound was obtained at 600°C–1100°C by the interaction of ZnSe, ZnI$_2$, and In$_2$Se$_3$.

2.57 ZINC–LANTHANUM–OXYGEN–SELENIUM

ZnS–La$_2$O$_2$Se. The phase diagram is not constructed. According to the data of Baranov et al. (1996), this system is a quasibinary system of the Zn–La–O–Se quaternary system. New phases were not found in the ZnS–La$_2$O$_2$Se system. This system was investigated through XRD. The ingots were annealed at 730°C–830°C for 250 h.

2.58 ZINC–TIN–TELLURIUM–SELENIUM

ZnSe–SnTe. The phase diagram is a eutectic type (Figure 2.41) (Dubrovin et al. 1984). The eutectic composition and temperature are 6 mol.% ZnSe and 780°C, respectively. Solubility of ZnSe in SnTe at the eutectic temperature is equal to 5 mol.% and decreases to 4 mol.% at 400°C and solubility of SnTe in ZnSe is not higher than 1 mol.%. This system was investigated through DTA, metallography, XRD, and measuring of microhardness. The ingots were annealed at 700°C and 400°C for 120 and 250 h, respectively.

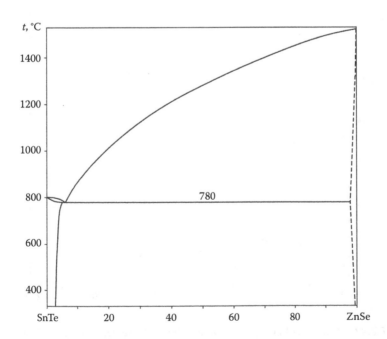

FIGURE 2.41 The ZnSe–SnTe phase diagram. (From Dubrovin, I.V. et al., *Izv. AN SSSR. Neorgan. materialy*, 20(4), 571, 1984.)

2.59   ZINC–LEAD–CHLORINE–SELENIUM

ZnSe–PbCl$_2$. The phase diagram is shown in Figure 2.42 (Triboulet et al. 1982). The eutectic composition and temperature are 10.5 mol.% ZnSe and 450°C, respectively. At 584°C, a peritectic transformation takes place in this system. The composition of peritectic point corresponds to 24 mol.% ZnSe (eutectic and peritectic compositions and temperatures are taken from Figure 2.42). This system was investigated through DTA.

2.60   ZINC–LEAD–BROMINE–SELENIUM

ZnSe–PbBr$_2$. The phase diagram is not constructed. According to the data of (Triboulet et al. 1982), it is of eutectic type with a eutectic that crystallizes at approximately 340°C and is near PbBr$_2$ composition. This system was investigated through DTA.

2.61   ZINC–NITROGEN–PHOSPHORUS–SELENIUM

ZnSe–Zn$_3$N$_2$–P$_3$N$_5$. The phase diagram is not constructed. The Zn$_8$[P$_{12}$N$_{24}$]Se$_2$ quaternary compound is formed in this system. It crystallizes in the cubic structure with the lattice parameter $a = 823.91 \pm 0.01$ pm and calculated density 4.12 g·cm^{-3} (Wester and Schnick 1996). This compound was obtained by the reaction of HPN$_2$ with ZnSe at 750°C.

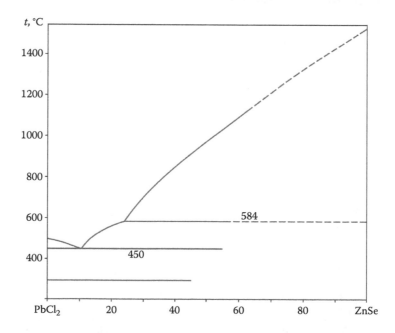

FIGURE 2.42   The ZnSe–PbCl$_2$ phase diagram. (From Triboulet, R. et al., *J. Cryst. Growth*, 59(1–2), 172, 1982.)

2.62 ZINC–BISMUTH–TELLURIUM–SELENIUM

ZnTe–Bi$_2$Se$_3$. This section is nonquasibinary section of the ZnSe + Bi$_2$Te$_3$ ⇔ ZnTe + Bi$_2$Se$_3$ ternary mutual system (Figure 2.43) (Odin 1996). According to the data of XRD, the equilibrium in this system is shifted to the ZnSe and Bi$_2$Te$_3$ formation. This system was investigated through DTA, metallography, and XRD. The ingots were annealed at the temperatures 20°C below the temperatures of nonvariant equilibria with liquidus presence for 1000 h.

3ZnSe + Bi$_2$Te$_3$ ⇔ 3ZnTe + Bi$_2$Se$_3$. There are two fields of primary crystallization on the liquidus surface of this system (Figure 2.44) (Odin 1996): ZnSe$_x$Te$_{1-x}$ and Bi$_2$Se$_{3(1-x)}$Te$_{3x}$ solid solutions (Bi$_2$Se$_{3(1-x)}$Te$_{3x}$ solid solutions contain some quantities of zinc chalcogenides). Isothermal section of the 3ZnSe + Bi$_2$Te$_3$ ⇔ 3ZnTe + Bi$_2$Se$_3$ ternary mutual system at 450°C is shown in Figure 2.45 (Odin and Marugin 1991). Solubility of ZnSe$_x$Te$_{1-x}$ in Bi$_2$Se$_{3(1-x)}$Te$_{3x}$ at this temperature reaches 3 mol.% and solubility of Bi$_2$Se$_{3(1-x)}$Te$_{3x}$ in ZnSe$_x$Te$_{1-x}$ at the same temperature is insignificant.

This system was investigated through DTA, metallography, and XRD. The ingots were annealed at the temperatures 20°C below the temperatures of nonvariant equilibria with liquidus presence for 1000 h (Odin 1996) [at 450°C for 920 h (Odin and Marugin 1991)].

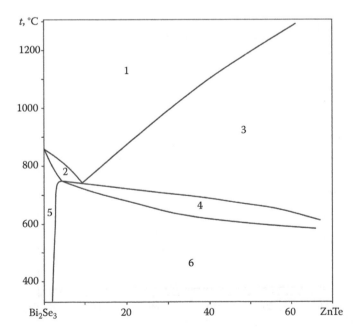

FIGURE 2.43 Phase relations in the ZnTe–Bi$_2$Se$_3$ system: 1, L; 2, L + Bi$_2$Se$_{3(1-x)}$Te$_{3x}$; 3, L + ZnTe$_{1-x}$Se$_x$; 4, L + ZnTe$_{1-x}$Se$_x$ + Bi$_2$Se$_{3(1-x)}$Te$_{3x}$; 5, Bi$_2$Se$_{3(1-x)}$Te$_{3x}$; 6, Bi$_2$Se$_{3(1-x)}$Te$_{3x}$ + ZnTe$_{1-x}$Se$_x$. (From Odin, I.N., *Zhurn. neorgan. khimii*, 41(6), 941, 1996.)

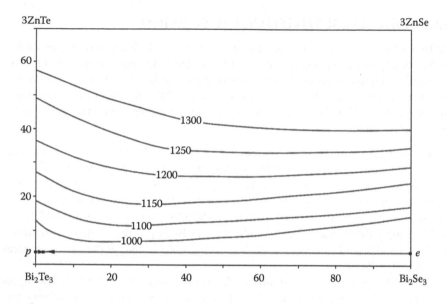

FIGURE 2.44 Liquidus surface of the 3ZnSe + Bi$_2$Te$_3$ ⇔ 3ZnTe + Bi$_2$Se$_3$ ternary mutual system. (From Odin, I.N., *Zhurn. neorgan. khimii*, 41(6), 941, 1996.)

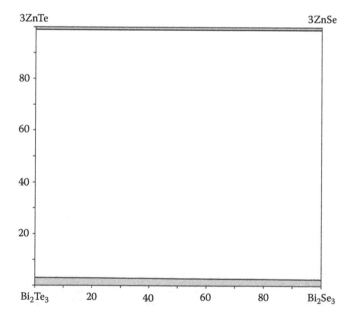

FIGURE 2.45 Isothermal section of the 3ZnSe + Bi$_2$Te$_3$ ⇔ 3ZnTe + Bi$_2$Se$_3$ ternary mutual system at 450°C. (From Odin, I.N. and Marugin, V.V., *Zhurn. neorgan. khimii*, 36(9), 2379, 1991.)

2.63 ZINC–VANADIUM–OXYGEN–SELENIUM

$ZnVSe_2O_7$ quaternary compound exists in this system. It crystallizes in the monoclinic structure with the lattice parameters $a = 744.2 \pm 0.3$, $b = 1253.0 \pm 0.4$, $c = 748.0 \pm 0.3$ pm, and $\beta = 117.066° \pm 0.004°$; calculated density 4.131 $g \cdot cm^{-3}$; and energy gap 0.80 eV (Jiang et al. 2008).

To obtain it, a mixture of $Zn(CH_3COO)_2 \cdot 2H_2O$ (0.84 mmol), V_2O_5 (0.28 mmol), SeO_2 (1.08 mmol), and H_2O (5 mL) was sealed in an autoclave equipped with a Teflon liner (23 mL) and heated at 200°C for 4 days followed by slow cooling to room temperature at a rate of 6°C/h (Jiang et al. 2008). Dark cyan prism-shaped crystals were recovered. To prepare its single-phase product, VO_2 was used directly as the starting materials. A mixture of $Zn(CH_3COO)_2 \cdot 2H_2O$ (0.74 mmol), VO_2 (0.37 mmol), and SeO_2 (1.85 mmol) in 5 mL H_2O was allowed to react under the same conditions previously described. This compound is stable under an air atmosphere up to 390°C.

2.64 ZINC–OXYGEN–MOLYBDENUM–SELENIUM

$Zn_2(MoO_4)SeO_3$ quaternary compound is formed in this system. It crystallizes in the monoclinic structure with the lattice parameters $a = 518.09 \pm 0.04$, $b = 832.38 \pm 0.07$, $c = 715.41 \pm 0.06$ pm, and $\beta = 98.413° \pm 0.001°$; calculated density 4.545 $g \cdot cm^{-3}$; and calculated energy gap 3.3 eV (Nguyen ct al. 2011). This compound releases SeO_2 at ~500°C and subsequently decomposes to ZnO, $ZnMoO_4$, and $Zn_3Mo_2O_9$.

Bulk crystalline and crystals of $Zn_2(MoO_4)SeO_3$ were prepared by combining 2 mmol of ZnO, 1 mmol of SeO_2, and 1 mmol of MoO_3 (Nguyen et al. 2011). The mixture was thoroughly ground and pressed into a pellet and placed in a fused silica tube that was evacuated and flame sealed. The sealed ampoule was heated to 380°C for 24 h, to 500°C for 24 h, and finally to 550°C for 48 h before being cooling to room temperature at 6°C/h. The product consists of colorless rod-shaped crystals and white powder.

2.65 ZINC–OXYGEN–CHLORINE–SELENIUM

$Zn_2(SeO_3)Cl_2$ quaternary compound is formed in the Zn–O–Fe–Se quaternary system. It crystallizes in two polymorph modifications. α-$Zn_2(SeO_3)Cl_2$ crystallizes in the orthorhombic structure with the lattice parameters $a = 1025.1 \pm 0.4$, $b = 1522.3 \pm 0.2$, and $c = 766.6 \pm 0.5$ pm and calculated density 3.64 $g \cdot cm^{-3}$ (Vergasova et al. 1989, Semenova et al. 1992). β-$Zn_2(SeO_3)Cl_2$ crystallizes in the monoclinic structure with the lattice parameters $a = 766.99 \pm 0.08$, $b = 1026.12 \pm 0.11$, $c = 765.71 \pm 0.08$ pm, and $\beta = 100.004° \pm 0.002°$ at 123 K (Johnsson and Törnroos 2007).

β-$Zn_2(SeO_3)Cl_2$ was synthesized by chemical transport reactions in sealed evacuated soda-glass tubes (Johnsson and Törnroos 2007). $ZnCl_2$, ZnO, and SeO_2 were used as starting materials. Equimolar amounts of these compounds were mixed in a mortar and placed in a glass tube, which was evacuated and heated at 427°C for 72 h. The product appeared as colorless transparent platelike single crystals and as powder. These crystals are hygroscopic.

2.66 ZINC–OXYGEN–MANGANESE–SELENIUM

ZnSe–MnSe–O. The phase diagram is not constructed. According to the data of thermodynamic calculations, an oxidation of the $Zn_xMn_{1-x}Se$ solid solutions begins from the oxidation of MnSe (Medvedev and Berchenko 1994).

2.67 ZINC–OXYGEN–IRON–SELENIUM

$ZnFe_2^{3+}(SeO_3)_4$ ($ZnFe_2Se_4O_{12}$) and $Zn_3Fe_2^{3+}(SeO_3)_6$ ($Zn_3Fe_2Se_6O_{18}$) quaternary compounds are formed in the Zn–O–Fe–Se quaternary system. $ZnFe_2^{3+}(SeO_3)_4$ and $Zn_3Fe_2^{3+}(SeO_3)_6$ crystallize in the monoclinic structure with the lattice parameters $a = 819.6 \pm 0.4$, $b = 799.7 \pm 0.4$, $c = 803.3 \pm 0.4$ pm, and $\beta = 92.27° \pm 0.03°$ (Giester 1996) and $a = 1249.1 \pm 0.8$, $b = 833.1 \pm 0.5$, $c = 769.6 \pm 0.5$ pm, and $\beta = 95.96° \pm 0.04°$ (Giester 1995), respectively. Calculated density of $Zn_3Fe_2^{3+}(SeO_3)_6$ is equal to 4.46 g·cm^{-3} (Giester 1995).

$ZnFe_2^{3+}(SeO_3)_4$ could be obtained at low-hydrothermal conditions at 220°C (Giester 1996). $Zn_3Fe_2^{3+}(SeO_3)_6$ was synthesized at 230°C using aqueous solutions of mixture of SeO_2, $FeC_2O_4 \cdot 2H_2O$, and $ZnSeO_3$ (Giester 1995).

REFERENCES

Baranov I.Yu., Dolgih V.A., Popovkin B.A. Investigation of phase correlations in the La_2O_2X–MX (M = Zn, Cd, Hg; X = S, Se) systems [in Russian], *Zhurn. neorgan. khimii*, **41**(11), 1916–1919 (1996).

Bensalem S., Chegaar M., Maouche D., Bouhemadou A. Theoretical study of structural, elastic and thermodynamic properties of CZTX (X = S and Se) alloys, *J. Alloys Compd.*, **589**, 137–142 (2014).

Bochmann M., Bwembya G., Webb K.J., Malik M.A., Walsh J.R., O'Brien P. Arene chalcogenolato complexes of zinc and cadmium, *Inorg. Synthesis*, **31**, 19–24 (1997).

Bochmann M, Coleman A.P., Powell A.K. Synthesis of some alkyl metal selenolato complexes of zinc, cadmium and mercury. X-ray crystal structure of Me, Hg, Se(2,4,6-Pr$_3^i$C$_6$H$_2$), *Polyhedron*, **11**(5), 507–512 (1992).

Bochmann M., Webb K. Novel precursors for the deposition of II-VI semiconductor films, *Mater. Res. Symp. Proc.*, **204**, 149–154 (1991).

Bochmann M., Webb K., Hursthouse M.B., Mazid M. Sterically hindered chalcogenolato complexes. Mono- and di-meric thiolates and selenolates of zinc and cadmium; structure of [{Cd(SeC$_6$H$_2$Bu$_3^t$-2,4,6)$_2$}$_2$], the first three-co-ordinate cadmium–selenium complex, *J. Chem. Soc., Dalton Trans.*, (9), 2317–2323 (1991).

Bodnar' I.V. Physico-chemical properties of the $(CuAlSe_2)_x(2ZnSe)_{1-x}$ solid solutions [in Russian], *Neorgan. materialy*, **38**(1), 12–16 (2002).

Bodnar' I.V., Chibusova L.V. Phase diagram of the $CuInSe_2$–2ZnSe system [in Russian], *Zhurn. neorgan. khimii*, **43**(11), 1913–1915 (1998).

Bodnar' I.V., Gremenok V.F. Growth and properties of the $(CuInSe_2)_x(2ZnSe)_{1-x}$ solid solutions crystals [in Russian], *Neorgan. materialy*, **39**(11), 1301–1305 (2003).

Bodnar I.V., Gremenok V.F., Schmitz W., Bente K., Doering Th. Preparation and investigation of $(CuInSe_2)_{1-x}(2ZnSe)_x$ and $(CuInTe_2)_{1-x}(2ZnTe)_x$ solid solutions single crystals, *Cryst. Res. Technol.*, **39**(4), 301–307 (2004).

Brennan J.G., Siegrist T., Carroll P.J., Stuczynski S.M., Reynders P., Brus L.E., Steigerwald M.L. Bulk and nanostructure group II-VI compounds from molecular organometallic precursors, *Chem. Mater.*, **2**(4), 403–409 (1990).

Chaus I.S., Kompanichenko N.M., Andreichenko V.G., Sheka I.A. Investigation of zinc sulfide hydrazine complex [in Russian], *Izv. AN SSSR. Neorgan. materialy*, **19**(5), 730–732 (1983).

Chen D., Ravindra N.M. Electronic and optical properties of Cu_2ZnGeX_4 (X = S, Se and Te) quaternary semiconductors, *J. Alloys Compd.*, **579**, 468–472 (2013).

Chen H.-L., Kuang H.-M., Chen W.-T. Synthesis and characterization of a novel quaternary metal chalcogenide—$Hg_3ZnSe_3Br_2$, *Rus. J. Inorg. Chem.*, **57**(8), 1064–1066 (2012); *translated from Zhurn. neorgan. khimii*, **57**(8), 1140–1142 (2012).

Chen S., Gong X.G., Walsh A., Wei S.-H. Crystal and electronic band structure of Cu_2ZnSnX_4 (X = S and Se) photovoltaic absorbers: First-principles insights, *Appl. Phys. Lett.*, **94**(4), 041903_1–041903_3 (2009).

Chippaux D., Deschanvres A. Étude du diagramme ternaire $CuInSe_2$–$ZnSe$–In_2Se_3, *J. Solid State Chem.*, **45**(2),200–211 (1982).

Chondroudis K., Kanatzidis M.G. Group-10 and 12 one-dimensional selenodiphosphates: $A_2MP_2Se_6$ (A = K, Rb, Cs; M = Pd, Zn, Cd, Hg), *J. Solid State Chem.*, **138**(2), 321–328 (1998).

Coates G.E., Ridley D. Alkoxy-, thio-, and amino-derivatives of methylzinc, *J. Chem. Soc.*, 1870–1877 (1965).

Donohue P.C., Bierstadt P. E. Cell dimensions and electrical properties of the solid solutions: $(1-x)ZnS/xCuAlS_2$ and $(1-x)ZnSe/xCuAlSe_2$ where x = 0 to 0,33, *J. Electrochem. Soc.*, **121**(3), 327–329 (1974).

Dubrovin I.V., Budionnaya L.D., Mizetskaya I.B., Sharkina E.V. The SnTe–ZnSe system [in Russian], *Izv. AN SSSR. Neorgan. materialy*, **20**(4), 571–573 (1984).

Dudchak I., Piskach L. Phase equilibria in the Cu_2SnSe_3–ZnSe system [in Ukrainian], *Visnyk L'viv. un-tu. Ser. khim.*, (40), 73–76 (2001).

Dudchak I.V., Piskach L.V. Phase equilibria in the Cu_2SnSe_3–$SnSe_2$–ZnSe system, *J. Alloys Compd.*, **351**, 145–150 (2003).

Engelen B., Bäumer U., Hermann B., Müller H., Unterderweide K. Zur Polymorphie und Pseudosymmetrie der Hydrate $MSeO_3 \cdot H_2O$ (M = Mn, Co, Ni, Zn, Cd), *Z. anorg. und allg. Chem.*, **622**(11), 1886–1892 (1996).

Engelen B., Boldt K., Unterderweide K., Bäumer U. Zur Kenntnis der Hydrate $M(HSeO_3)_2$ $4H_2O$ (M = Mg, Co, Ni, Zn). Röntgenstrukturanalytische, schwingungsspektros-kopische und thermoanalytische Untersuchungen, *Z. anorg. und allg. Chem.*, **621**(2), 331–339 (1995).

Gallardo P.G. Order-disorder phase transitions in $D_{2x}^{II}(A^IB^{III})_{1-x}C_2^{VI}$ alloys systems, *Phys. Status Solidi (a)*, **134**(1), 119–125 (1992).

Garbato L., Manca P. Synthesis and characterization of some chalcogenides of $A^IB_2^{II}C^{III}D_4^{VI}$, *Mater. Res. Bull.*, **9**(4), 511–517 (1974).

Gebicki W., Filipowicz J., Bacewicz R. Raman scattering in novel $(CuAlSe_2)_x(2ZnSe)_{1-x}$ and $(CuGaSe_2)_x(2ZnSe)_{1-x}$ mixed crystals, *J. Phys.: Condens. Matter*, **8**(44), 8695–8703 (1996).

Giester G. The crystal structures of the isotypic compounds $M_3^{2+}Fe_2^{3+}(SeO_3)_6$ (M = Cu, Zn), *Acta Chem. Scand.*, **49**, 824–828 (1995).

Giester G. The crystal structures of $ZnFe_2^{3+}(SeO_3)_4$, *Monatsh. Chem.*, **127**(4), 347–354 (1996).

Gladkova V.F., Kondrashev Yu.D. Crystal structure of $ZnSeO_3 \cdot 2H_2O$ [in Russian], *Kristallografiya*, **9**(2), 190–196 (1964).

Glazov V.M., Kiselev A.N., Shvedkov E.I. Solubility and donor-acceptor interaction in InAs doping by S, Se and Zn [in Russian], *Izv. AN SSSR. Neorgan. materialy*, **15**(3), 390–394 (1979).

Golovey M.I., Olekseyuk I.D. The $(Zn_3As_2)_{1-x}$–$(2CdSe)_x$ system [in Russian], *Izv. AN SSSR. Neorgan. materialy*, **7**(6), 934–938 (1971).

Goriunova N.A., Fedorova N.N. About solid solutions in the ZnSe–GaAs system [in Russian], *Fizika tverdogo tela*, **1**(2), 344–345 (1959).

Guen L., Glaunsinger W.S. Electrical, magnetic and EPR studies of the quaternary chalcogenides $Cu_2A^{II}B^{IV}X_4$ prepared by iodine transport, *J. Solid State Chem.*, **35**(1), 10–21 (1980).

Guen L., Glaunsinger W.S., Wold A. Physical properties of the quaternary chalcogenides $Cu_2^IB^{II}C^{IV}X_4$ (B^{II} = Zn, Mn, Fe, Co; C^{IV} = Si, Ge, Sn; X = S, Se), *Mater. Res. Bull.*, **14**(4), 463–467 (1979).

Hahn H., Schulze H. Über quaternäre Chalkogenide des Germanium und Zinns, *Naturwissenschaften*, **52**(14) 426 (1965).

Harrison W.T.A., Phillips M.L.F. An extreme case of ZnO_6 octahedral distortion in trizinc tris(selenite) hydrate, $Zn_3(SeO_3)_3 \cdot H_2O$, *Acta Crystallogr.*, **C55**(12), 1980–1983 (1999).

Hassah El Haj F., Hashemifar S.J., Akbarzadeh H. Density functional study of $Zn_{1-x}Mg_xSe_yTe_{1-y}$ quaternary semiconductor alloys, *Phys. Rev. B*, **73**(19), 195202_1–195202_6 (2006).

Huang F.Q., Mitchell K., Ibers J.A. New layered materials: Synthesis, structures and optical and magnetic properties of $CsGdZnSe_3$, $CsZrCuSe_3$, $CsUCuSe_3$ and $BaGdCuSe_3$, *Inorg. Chem.*, **40**(20), 5123–5126 (2001).

Jiang H.-L., Feng M.L., Mao J.-G. Synthesis, crystal structures and characterizations of $BaZn(SeO_3)_2$ and $BaZn(TeO_3)Cl_2$, *J. Solid State Chem.*, **179**(6), 1911–1917 (2006).

Jiang H.-L., Kong F., Fan Y., Mao J.-G. $ZnVSe_2O_7$ and $Cd_6V_2Se_5O_{21}$: New d_{10} transition-metal selenites with V(IV) or V(V) cations, *Inorg. Chem.*, 47(16), 7430–7437 (2008).

Johnsson M., Törnroos K.W. Zinc selenium oxochloride, β-$Zn_2(SeO_3)Cl_2$, a synthetic polymorph of the mineral sophiite, *Acta Crystallogr.*, **C63**(5), i34–i36 (2007).

Johnston M.G., Harrison W.T.A. $SrZn(SeO_3)_2$ containing novel ZnO_{4+2} bicapped tetrahedra, *Inorg. Chem.*, **40**(25), 6518–6520 (2001).

Kanatzidis M.G., Liao J.H., Marking G.A. Alkali metal quaternary chalcogenides and process for the preparation of thereof, USA Patent 5 614 128. Appl. № 606565. Filed 26.02.96; Data of Patent 25.03.1997 (1997a).

Kanatzidis M.G., Liao J.H., Marking G.A. Alkali metal quaternary chalcogenides and process for the preparation of thereof, USA Patent 5 618 471. Appl. № 606886. Filed 26.02.96; Data of Patent 08.04.1997 (1997b).

Kazakov A.I., Stepanov M.A., Shlikher V.A. Calculation of the immiscibility regions in the four-component $Cd_xHg_{1-x}Te_ySe_{1-y}$ and $Zn_xHg_{1-x}Te_ySe_{1-x}$ solid solutions [in Russian], *Trudy Odes. politekhn. un-ta*, (2), 145–150 (2001).

Kirovskaya I.A., Azarova O.P., Shubenkova E.G., Dubina O.N. Synthesis and optical absorption of the solid solutions of the InSb–(II-VI) systems [in Russian], *Neorgan. materialy*, **38**(2), 135–138 (2002).

Kirovskaya I.A., Mulikova G.M. About obtaining and identification of the substitution solid solutions based on GaAs and ZnSe [in Russian], *Tr. Tom. un-ta*, **240**(8), 155–166 (1973).

Kirovskaya I.A., Mulikova G.M. The GaAs–ZnSe system [in Russian], *Izv. AN SSSR. Neorgan. materialy*, **11**(6), 1131–1132 (1975).

Kondrashev Yu.D., Nozik Yu.Z., Fykin L.E., Shibanova T.A. Neutron diffraction studies of the crystal structure of acid zinc selenite dihydrate $Zn(HSeO_3)_2 \cdot 2H_2O$ [in Russian], *Kristallografiya*, **24**(3), 586–589 (1979).

Kozer V.R., Olekseyuk I.D., Parasyuk O.V. Phase equilibria in the Ag_2Se–Ga_2Se_3–ZnSe quasiternary system [in Ukrainian], *Nauk. visnyk Volyn. nats. un-tu. Khim. nauky*, (13), 20–24 (2008).

Ku S.M., Bodi L.J. Synthesis and some properties of ZnSe:GaAs solid solutions, *J. Phys. Chem. Sol.*, **29**(12), 2077–2082 (1968).

Lakeenkov V.M., Mil'vidski M.G., Pelevin O.V. Physico-chemical investigation of the Ga–GaAs–ZnSe quasiternary system [in Russian], *Deposited in VINITI*, № 520–74Dep (1974).

Lakeenkov V.M., Mil'vidski M.G., Pelevin O.V. Phase diagram of the GaAs–ZnSe system [in Russian], *Izv. AN SSSR. Neorgan. materialy*, **11**(7), 1311–1312 (1975).

Lambrecht, Jr. V.G. Preparation and crystal growth of materials in the pseudo-binary $CuInSe_2$–ZnSe and $CuGaSe_2$–ZnSe systems, *Mater. Res. Bull.*, **8**(12), 1383–1387 (1973).

Lee D.W., Bak D.-B., Kim S.B., Kim J., Ok K.M. Effect of the framework flexibility on the centricities in centrosymmetric $In_2Zn(SeO_3)_4$ and noncentrosymmetric $Ga_2Zn(TeO_3)_4$, *Inorg. Chem.*, **51**(14), 7844–7850 (2012).

Leonov V.V., Chunarev E.N. Distribution of potassium and zinc chlorides at the crystallization of eutectics of cadmium chalcogenides with its chloride [in Russian]. In: *Physico-khim. processy v geterogennyh systemah.* Krasnoyarsk, Russia: Krasnoyar. Un-t Publish., pp. 59–64 (1977).

Leute V., Plate H. The phase diagram of the semiconductor alloy $Zn_kHg_{(1-k)}Se_lTe_{(1-l)}$, *Ber. Bunsenges. Phys. Chem.*, **93**(7), 757–763 (1989).

Leute V., Wulff B. The phase diagram of the quasiternary system $(Zn_kCd_{1-k})(Se_lTe_{1-l})$, *Ber. Bunsenges. Phys. Chem.*, **96**(2), 119–128 (1992).

Leute V., Wulff B. Solid state reactions in the quasiternary system $(Zn_kCd_{1-k})(Se_lTe_{1-l})$, *Ber. Bunsenges. Phys. Chem.*, **97**(7), 923–928 (1993).

Lieder O.J., Gattow G. Über Diselenite vom Typ $Me^{II}Se_2O_5 \cdot 3H_2O$. *Naturwissenschaften*, 54(12), 318–319 (1967).

Manhas S., Manoogian A., Lamarche G., Woolley J.C. Lattice parameters and magnetic properties of the $Cd_xZn_yMn_zSe$ alloy system, *Phys. Status Solidi (a)*, **94**(1), 213–221 (1986).

Markovskiy L.Ya., Sapozhnikov Yu.P. About various forms and some properties of the average zinc selenite [in Russian], *Zhurn. strukt. khimii*, **1**(3), 346–352 (1960).

Matsuhita H., Ichikawa T., Katsui A. Structural, thermodynamic and optical properties of Cu_2–II–IV–VI_4 quaternary compounds, *J. Mater. Sci.*, **40**(8), 2003–2005 (2005).

Matsuhita H., Maeda T., Katsui A., Takizawa T. Thermal analysis and synthesis from the melts of Cu-based quaternary compounds Cu–III–IV–VI_4 and Cu_2–II–IV–VI_4 (II – Zn, Cd; III – Ga, In; IV – Ge, Sn; VI – Se), *J. Cryst. Growth*, **208**(1–4), 416–422 (2000).

Medvedev Yu.V., Berchenko N.N. Analysis of the phase compositions at the interface of $Mn_{1-x}A_x^{II}B^{VI}(A^{II} = Zn, Cd, Hg; B^{VI} = Te, Se)$ solid solution with own oxides [in Russian], *Zhurn. neorgan. khimii*, **39**(5), 846–848 (1994).

Mitchell K., Haynes Ch.L., McFarland A.D., Van Duyne R.P., Ibers J.A. Tuning of optical band gaps: Synthesis, structures, magnetic properties and optical properties of $CsLnZnSe_3$ Ln = Sm, Tb, Dy, Ho, Er, Tm, Yb and Y, *Inorg. Chem.*, **41**(5), 1199–1204 (2002).

Mitchell K., Huang F.Q., Caspi E.N., McFarland A.D., Haynes Ch.L., Somers R.C., Jorgensen J.D., Van Duyne R.P., Ibers J.A. Synthesis, structure and selected physical properties of $CsLnMnSe_3$ (Ln = Sm, Gd, Tb, Dy, Ho, Er, Tm, Yb, Y) and $AYbZnQ_3$ (A = Rb, Cs; Q = S, Se, Te), *Inorg. Chem.*, **43**(3), 1082–1089 (2004).

Mitchell K., Huang F.Q., McFarland A.D., Haynes Ch.L., Somers R.C., Van Duyne R.P., Ibers J.A. The $CsLnMSe_3$ semiconductors (Ln = rare-earth element, Y; M = Zn, Cd, Hg), *Inorg. Chem.*, **42**(13), 4109–4116 (2003).

Morris C.D., Li H., Jin H., Malliakas C.D., Peters J.A., Trikalitis P.N., Freeman A.J., Wessels B.W., Kanatzidis M.G. $Cs_2M^{II}M_3^{IV}Q_8$ (Q = S, Se, Te): An extensive family of layered semiconductors with diverse band gaps, *Chem. Mater.*, **25**(16), 3344–3356 (2013).

Nguyen S.D., Kim S.-H., Halasyamani P.S. Synthesis, characterization, and structure–property relationships in two new polar oxides: $Zn_2(MoO_4)(SeO_3)$ and $Zn_2(MoO_4)(TeO_3)$, *Inorg. Chem.*, **50**(11), 5215–5222 (2011).

Nitsche R., Sargent D.F., Wild P. Crystal growth of quaternary $1_2 4 6_4$ chalcogenides by iodine vapor transport, *J. Cryst. Growth*, **1**(1), 52–53 (1967).

Novikova E.M., Ahverdov O.S., Ershova S.A. Investigation of the Ga–GaAs–ZnSe quasiternary system in high temperature region [in Russian], *Deposited in VINITI*, № 4439–77Dep (1977).

144   Quaternary Alloys Based on II-VI Semiconductors

Novikova E.M., Vasil'ev M.G., Krapuhin V.V., Evseev V.A., Ershova S.A. Phase diagram of the Ga–GaAs–ZnSe quasiternary system from the Ga-rich side [in Russian], *Deposited in VINITI*, № 3038–74Dep (1974).

Odin I.N. Physico-chemical analysis of ternary and ternary mutual systems, containing cadmium, zinc, silicon, bismuth chalcogenides and properties of ingots in these systems [in Russian], *Zhurn. neorgan. khimii*, **41**(6), 941–953 (1996).

Odin I.N., Marugin V.V. Electrophysical properties of solid solutions with tetradymite structure in the $Bi_2Se_3+Zn_3Te_3 \Leftrightarrow Bi_2Te_3+Zn_3Se_3$ system [in Russian], *Zhurn. neorgan. khimii*, **36**(9), 2379–2382 (1991).

Ohtani H., Kojima K., Ishida K., Nishizawa T. Miscibility gap in II-VI semiconductor systems, *J. Alloys Compd.*, **182**(1), 103–114 (1992).

Okońska-Kozłowska I., Kuske P., Lutz H.D. Das System $ZnSe–Cr_2Se_3–In_2Se_3$, *Z. anorg. und allg. Chem.*, **559**(1), 201–207 (1988).

Oleinik G.S., Tomashik V.N., Mizetskaya I.B. Phase relations in the ZnTe–CdSe system [in Russian], *Poluprovodn. tehnika i mikroelectron.*, (28), 56–58 (1978a).

Oleinik G.S., Tomashik V.N., Mizetskaya I.B. CdTe + ZnSe $\Leftrightarrow$ CdSe + ZnTe ternary mutual system [in Russian], *Izv. AN SSSR. Neorgan. materialy*, **14**(3), 441–443 (1978b).

Oleinik G.S., Tomashik V.N., Mizetskaya I.B., Novitskaya G.N., Chalyi V.P. Phase diagram of the CdTe–ZnSe pseudobinary system [in Russian], *Izv. AN SSSR. Neorgan. materialy*, **13**(11), 1976–1979 (1977).

Olekseyuk I.D., Dudchak I.V., Piskach L.V. Phase equilibria in the $Cu_2Se–ZnSe–Cu_2SnSe_3$ quasiternary system" [in Ukrainian], *Fiz. i khim. tv. tila*, **2**(2), 195–200 (2001).

Olekseyuk I.D., Gulay L.D., Dudchak I.V., Piskach L.V., Parasyuk O.V., Marchuk O.V. Single crystal preparation and crystal structure of the $Cu_2Zn(Cd,Hg)SnSe_4$ compounds, *J. Alloys Compd.*, **340**(1–2), 141–145 (2002).

Olekseyuk I.D., Kogut Yu.M., Parasyuk O.V., Piskach L.V., Gorgut G.P., Kus'ko O.P., Pekhnyo V.I., Volkov S.V. Glass-formation in the $Ag_2Se–Zn(Cd, Hg)Se–GeSe_2$, *Chem. Met. Alloys*, **2**(3–4), 146–150 (2009).

Olekseyuk I.D., Mazurets I.I., Parasyuk O.V. Phase relations in the $ZnSe–Ga_2Se_3–GeSe_2$ system, *J. Alloys Compd.*, **351**(1–2), 171–175 (2003).

Olekseyuk I.D., Parasyuk O.V., Bozhko V.V., Galyan V.V., Petrus' I.I. Glassforming in the $Zn(Cd,Hg)Se–Ga_2Se_3–GeSe_2$ systems [in Ukrainian], *Fizyka kondens. vysokomolek. system. Nauk. zap. Rinens'kogo pedinstytutu*, (3), 148–152 (1997).

Olekseyuk I.D., Parasyuk O.V., Bozhko V.V., Petrus' I.I., Galyan V.V. Formation and properties of the quasiternary $Zn(Cd,Hg)Se–Ga_2Se_3–SnSe_2$ system glasses, *Funct. Mater.*, **6**(3), 474–477 (1999).

Parasyuk O.V., Gulay L.D., Romanyuk Ya.E., Piskach L.V. Phase diagram of the $Cu_2GeSe_3–ZnSe$ system and crystal structure of the $Cu_2ZnGeSe_4$ compound, *J. Alloys Compd.*, **329**(1–2), 202–207 (2001).

Parasyuk O.V., Olekseyuk I.D., Gulay L.D., Piskach L.V. Phase diagram of the $Ag_2Se–Zn(Cd)Se–SiSe_2$ systems and crystal structure of the Cd_4SiSe_6 compound, *J. Alloys Compd.*, **354**(1–2), 138–142 (2003).

Parasyuk O.V., Olekseyuk I.D., Mazurets I.I., Piskach L.V. Phase equilibria in the quasiternary $ZnSe–Ga_2Se_3–SnSe_2$ system, *J. Alloys Compd.*, **379**(1–2), 143–147 (2004).

Parthé E., Yvon K., Deitch R.H. The crystal structure of Cu_2CdGeS_4 and other quaternary normal tetrahedral structure compounds, *Acta Crystallogr. B*, **25**(6), 1164–1174 (1969).

Persson C. Electronic and optical properties of Cu_2ZnSnS_4 and $Cu_2ZnSnSe_4$, *J. Appl. Phys.*, **107**(5), 053710_1–053710_8 (2010).

Piskach L.V., Parasyuk O.V., Olekseyuk I.D., Romanyuk Y.E., Volkov S.V., Pekhnyo V.I. Interaction of argyrodite family compounds with the chalcogenides of II-b elements, *J. Alloys Compd.*, **421**(1–2), 98–104 (2006).

Range K.-J., Handrick K. Synthese und Hochdruckverhalten quaternärer Chalkogen-idhalogenide $M_2M'X_3Y$ (M = Zn, Cd; M' = Al, Ga, In; X = Se, Te; Y = Cl, Br, I), *Z. Naturforsch.*, **B43**(2), 153–158 (1988).

Romanyuk Ya.E., Parasyuk O.V. Phase equilibria in the quasi-ternary Cu_2Se–$ZnSe$–$GeSe_2$ system, *J. Alloys Compd.*, **348**(1–2), 195–202 (2003).

Schäfer W., Nitsche R. Tetrahedral quaternary chalcogenides of the type Cu_2–II–IV–$S_4(Se_4)$, *Mater. Res. Bull.*, **9**(5), 645–654 (1974).

Schäfer W., Nitsche R. Zur Systematik tetraedrischer Verbindungen vom Typ $Cu_2Me^{II}Me^{IV}$ Me_4^{VI}(Stannite und Wurtzstannite), *Z. Kristallogr.*, **145**(5–6), 356–370 (1977).

Schleich D.M., Wold A. Optical and electrical properties of quaternary chalcogenides, *Mater. Res. Bull.*, **12**(2), 111–114 (1977).

Schorr S., Tovar M., Sheptyakov D., Keller L., Geandier G. Crystal structure and cations distribution in the solid solution series 2(ZnX)–$CuInX_2$(X = S, Se, Te), *J. Phys. Chem. Solids*, **66**(11), 1961–1965 (2005).

Schwer H., Keller E., Krämer V. Crystal structure and twinning of $KCd_4Ga_5S_{12}$ and some isotypic $AB_4C_5X_{12}$ compounds, *Z. Kristallogr.*, **204**(Pt. 2), 203–213 (1993).

Semenova T.F., Rozhdestvenskaya I.V., Filatov S.K., Vergasova L.P. Crystal structure and physical properties of sophiite [sofiite], Sophiite $Zn_2(SeO_3)Cl_2$, a new mineral, *Mineral. Mag.*, **56**, 241–245 (1992).

Shim K., Rabitz H., Chang J.-H., Yao T. Energy band gap of the alloy $Zn_{1-x}Mg_xSe_yTe_{1-y}$ lattice matched to ZnTe, InAs and InP, *J. Cryst. Growth*, **214–215**, 350–354 (2000).

Shumilin V.P., Cherviakov A.I., Lobanov A.A. Interaction of Zn and Se at the growth of $(GaP)_x(ZnSe)_{1-x}$ solid solutions by the Czochralski method and method of liquid epitaxy [in Russian], *Izv. AN SSSR. Neorgan. materialy*, **13**(9), 1560–1564 (1977).

Sonomura H., Uragaki T., Miyauchi T. Synthesis and some properties of solid solutions in the GaP–ZnS and GaP–ZnSe pseudobinary systems, *Jap. J. Appl. Phys.*, **12**(7), 968–973 (1973).

Sysa L.V., Olekseyuk I.D., Galka V.O., Parasyuk O.V. Crystallochemical method of sphalerite and chalcopyrite boundary reciprocal solubility calculation in the $CuGaSe_2$–ZnSe system as example [in Ukrainian], *Fiz. i khim. tv. tila*, **1**(2), 167–176 (2000).

Tai H., Hori S.. Mutual solubility and bandgap energy of the $(CdTe)_{1-x}$–$(ZnSe)_x$ system [in Japanese], *J. Jap. Inst. Metals*, **41**(1), 33–37 (1977).

Triboulet R., Rabago F., Legros R., Lozykowski H., Didier G. Low-temperature growth of ZnSe crystals, *J. Cryst. Growth*, **59**(1–2), 172–177 (1982).

Ufimtsev V.B., Cherviakov A.I., Shumilin V.P. Thermal analysis method of dissociated semiconductor systems [in Russian], *Zavodsk. laboratoria*, **46**(6), 525–527 (1980).

Vasiliev M.G., Novikova E.M. Investigation of equilibria in the Me–ZnSe, Sn–GaAs, ZnSe–GaAs vertical sections of Me–GaAs–ZnSe ternary systems, where Me–Ga, Sn [in Russian], *Deposited in VINITI*, № 971–77Dep (1977).

Vergasova L.P., Filatov S.K., Semenova T.F., Filosofova T.M. Sophiite $Zn_2(SeO_3)Cl_2$—A new mineral from volcanic sublimates [in Russian], *Zap. Vses. mineralogy. obshch.*, **118**(1), 65–69 (1989).

Vitrichovski N.I. The $(ZnTe)_x(CdSe)_{1-x}$ solid solutions [in Russian], *Izv. AN SSSR. Neorgan. materialy*, **13**(3), 437–440 (1977).

Voitsehovski A.V., Stetsenko T.P. About obtaining of single crystals of $(GaP)_x(ZnSe)_{1-x}$ solid solutions by the chemical transport reactions [in Russian]. In: *Issled. po molekuliar. fizike i fizike tverdogo tela*. Kiev, Ukraine: Kiev. ped. in-t, pp. 38–40 (1976).

Wagner G., Lehmann S., Schorr S., Spemann D., Doering Th. The two-phase region in $2(ZnSe)_x(CuInSe_2)_{1-x}$ alloys and structural relation between the tetragonal and cubic phases, *J. Solid State Chem.*, **178**(2), 3631–3638 (2005).

Wester F., Schnick W. Nitrido-sodalithe: III. Synthese, Struktur und Eigenschaften von $Zn_8[P_{12}N_{14}]X_2$ mit X = O, S, Se, Te, *Z. anorg. und allg. Chem.*, **622**(8), 1281–1286 (1996).

Yao G.-Q., Shen H.-S., Honig E.D., Kershaw R., Dwight K., Wold A. Preparation and characterization of the quaternary chalcogenides $Cu_2B(II)C(IV)X_4$[B(II) = Zn, Cd; C(IV) = Si, Ge; X = S, Se], *Solid State Ionics*, **24**(3), 249–252 (1987).

Yim M.F. Solid solutions in the pseudobinary (III-V)–(II-VI) systems and their optical energy gap, *J. Appl. Phys.*, **40**(6), 2617–2623 (1969).

Yim W.M., Dismukes J.P., Stofko E.J., Ulmer R.J. Miscibility between ZnSe and CdTe, *Phys. Status Solidi (a)*, **13**(1), K57–K61 (1972).

3 Systems Based on ZnTe

3.1 ZINC–HYDROGEN–CARBON–TELLURIUM

The $Zn(TeC_6H_2Me_3-2,4,6)_2$ or $C_{18}H_{22}Te_2Zn$ [bis(2,4,6-trimethylbenzenetellurolato) zinc] quaternary compound is formed in this system. To obtain this compound, $Zn[N(SiMe_3)_2]_2$ was added via syringe to a cold [−30°C (Bochmann et al. 1997); −78°C (Bochmann et al. 1995)] solution of 2,4,6-trimethylbenzenetellurolato in light petroleum prepared in situ from $(2,4,6-Me_3C_6H_2Te)_2$ (Bochmann et al. 1995, 1997). A white to pale beige solid precipitates from the reddish solution. The mixture was stirred at −10°C for 4 h [at −78°C for 30 min (Bochmann et al. 1995)], allowed to warm to room temperature, and filtered. The colorless residue was washed with warm toluene and light petroleum (2 × 20 mL) [with light petroleum ether (3 × 30 mL) (Bochmann et al. 1995)] and dried in vacuo to give an off-white to beige microcrystalline solid, which melts with decomposition at 224°C–225°C (Bochmann et al. 1997). The reaction was carried out under inert gas using standard vacuum-line techniques. Heating a solution of $C_{18}H_{22}Te_2Zn$ in mesitylene leads to formation of ZnTe (Bochmann et al. 1995, 1997). The thermolysis of this compound in refluxing 4-ethylpiridine proceeded more slowly giving nanoscale particles of ZnTe (Bochmann et al. 1995).

3.2 ZINC–HYDROGEN–NITROGEN–TELLURIUM

The $ZnTe(N_2H_4)_2$ quaternary compound is formed in the Zn–H–N–Te quaternary system (Mitzi 2005). It has two polymorph modifications. Both modifications crystallize in the monoclinic structure with the lattice parameters $a = 721.57 \pm 0.04$, $b = 1154.39 \pm 0.06$, $c = 739.09 \pm 0.04$ pm, and $\beta = 101.296° \pm 0.001°$ for α-$ZnTe(N_2H_4)_2$ and $a = 813.01 \pm 0.05$, $b = 695.80 \pm 0.05$, $c = 1073.80 \pm 0.07$ pm, and $\beta = 91.703° \pm 0.001°$ for β-$ZnTe(N_2H_4)_2$.

3.3 ZINC–HYDROGEN–CHLORINE–TELLURIUM

ZnTe–HCl. Figure 3.1 shows the HCl concentration dependence of the gas partial pressures in this system according to the thermodynamic calculations (Nishio and Ogava 1986). The gas partial pressures of $ZnCl_2$, Te_2, HCl, and H_2 increase monotonically with the increase of the HCl concentration. On the other hand, the partial pressure of Zn is small and decreases with the increase of the HCl concentration (not shown in Figure 3.1). The gas pressures of Cl_2 and Cl are small enough to be neglected. Accordingly, the dominant species are $ZnCl_2$, Te_2, HCl, and H_2.

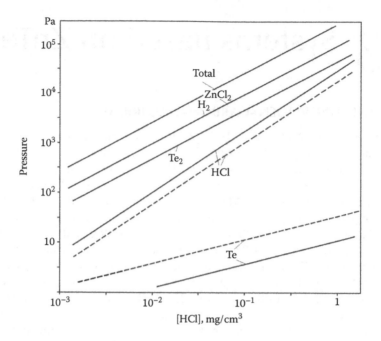

FIGURE 3.1 Partial pressures of gas species versus HCl concentration in the ZnTe–HCl system. (From Nishio, M. and Ogava, H., *J. Cryst. Growth*, 78(2), 218, 1986.)

3.4 ZINC–LITHIUM–OXYGEN–TELLURIUM

The Li_4ZnTeO_6 quaternary compound is formed in this system. It crystallizes in the monoclinic structure with the lattice parameters $a = 521.14 \pm 0.03$, $b = 892.88 \pm 0.04$, $c = 517.68 \pm 0.03$ pm, and $\beta = 110.873° \pm 0.003°$ (Nalbandyan et al. 2013) [$a = 522.16 \pm 0.03$, $b = 892.79 \pm 0.06$, $c = 516.83 \pm 0.03$ pm, and $\beta = 110.827° \pm 0.005°$ (Kumar et al. 2012)].

This compound has been prepared by solid-state reactions of the mixture of Li_2CO_3, ZnO, and TeO_2, which was calcined at 850°C two times for 2 h with intermediate grinding and pressing (Nalbandyan et al. 2013). Bulk polycrystalline samples of Li_4ZnTeO_6 were also synthesized from the mixture of Li_2CO_3, ZnO, TeO_2, and H_6TeO_6, which was heated at 650°C for 12 h followed by heating at 800°C and/or 900°C for the next 12 h (Kumar et al. 2012).

3.5 ZINC–SODIUM–OXYGEN–TELLURIUM

The $Na_2Zn_2TeO_6$ quaternary compound is formed in this system. It crystallizes in the hexagonal structure with the lattice parameters $a = 527.96 \pm 0.02$ and $c = 1129.41 \pm 0.04$ (Evstigneeva et al. 2011). This compound has also several extreme reflections, which only can be indexed on a primitive orthorhombic supercell with $a = 527.01 \pm 0.05$, $b = 914.39 \pm 0.08$, and $c = 1128.96 \pm 0.09$ pm, although no line splitting was observed. Almost all these reflections vanish at 300°C but appear again on cooling. $Na_2Zn_2TeO_6$ has no phase transition in the temperature range from 25°C to 300°C.

$Na_2Zn_2TeO_6$ was prepared by conventional solid-state reactions (Evstigneeva et al. 2011). Starting materials were Na_2CO_3, $NaNO_3$, TeO_2, and basic carbonate of Zn. The first three substances were dried at 200°C. The reagents were mixed in desired ratios with mortar and pestle, pressed, and calcined: first at 600°C–650°C for 2–10 h and then at 800°C–820°C for 2–4 h.

3.6 ZINC–POTASSIUM–COPPER–TELLURIUM

$ZnTe-K_2Te-Cu_2Te$. The phase diagram is not constructed. The $KCuZnTe_2$ quaternary compound is formed in this system. It crystallizes in the tetragonal structure with the lattice parameters $a = 426.9 \pm 0.1$ and $c = 1491.2 \pm 0.2$ and calculated density 5.172 g·cm^{-3} (Heulings et al. 1998).

This compound could be obtained by the direct reactions of stoichiometric ratio of K_2Te, Cu_2Te, and ZnTe. Single crystals of $KCuZnTe_2$ were grown from a K_2Te/Te flux at 450°C. A sample containing K_2Te, Cu, Zn, and Te was mixed in an Ar-filled glove box and was transferred to a Pyrex tube, which was then sealed under vacuum. The container was placed in a furnace, and the temperature was brought up slowly to 450°C and kept there for 4 days. The tube was then allowed to cool at a rate of 4°C/h until it reaches 150°C. Once the sample reached 150°C, it was cooled naturally to room temperature (Heulings et al. 1998).

3.7 ZINC–RUBIDIUM–COPPER–TELLURIUM

$ZnTe-Rb_2Te-Cu_2Te$. The phase diagram is not constructed. The $RbCuZnTe_2$ quaternary compound is formed in this system. It could be obtained by the direct reactions of stoichiometric ratio of Rb_2Te, Cu_2Te, and ZnTe (Heulings et al. 1998).

3.8 ZINC–RUBIDIUM–YTTERBIUM–TELLURIUM

$ZnTe-Rb_2Te-Yb_2Te_3$. The phase diagram is not constructed. The $RbZnYbTe_3$ quaternary compound is formed in this system. It crystallizes in the orthorhombic structure with the lattice parameters $a = 433.77 \pm 0.06$, $b = 1670.4 \pm 0.2$, and $c = 1150.9 \pm 0.2$ pm at 153 ± 2 K and calculated density 6.007 g·cm^{-3} (Mitchell et al. 2004). $RbZnYbTe_3$ was synthesized using the mixture of Rb_2Te_3, Zn, Yb, Te, and RbI. The sample was heated to 900°C in 24 h, kept at this temperature for 96 h, cooled to 200°C, and then rapidly cooled to room temperature.

3.9 ZINC–CESIUM–YTTRIUM–TELLURIUM

$ZnTe-Cs_2Te-Y_2Te_3$. The phase diagram is not constructed. The $CsZnYTe_3$ quaternary compound is formed in this system. It crystallizes in the orthorhombic structure with the lattice parameters $a = 438.64 \pm 0.03$, $b = 1671.15 \pm 0.12$, and $c = 1161.02 \pm 0.08$ pm at 153 ± 2 K, energy gap ~2.12 eV, and calculated density 5.229 g·cm^{-3} (Yao et al. 2004). $CsZnYTe_3$ was obtained from the reaction of Zn, Y, and Te with the addition of CsCl as flux. The sample was heated to 850°C in 30 h, kept at this temperature for 96 h, cooled at 4°C/h to 300°C, and then cooled to room temperature. This compound is air sensitive, decomposing after a few hours of exposure.

3.10   ZINC–CESIUM–LANTHANUM–TELLURIUM

$ZnTe-Cs_2Te-La_2Te_3$. The phase diagram is not constructed. The $CsZnLaTe_3$ quaternary compound is formed in this system. It crystallizes in the orthorhombic structure with the lattice parameters $a = 453.64 \pm 0.04$, $b = 1662.03 \pm 0.15$, and $c = 1208.54 \pm 0.11$ pm at 153 ± 2 K, energy gap ~2.12 eV, and calculated density 5.248 g·cm^{-3} (Yao et al. 2004). $CsZnLaTe_3$ was obtained from the reaction of Zn, La, and Te with the addition of CsCl as flux. The sample was heated to 850°C in 30 h, kept at this temperature for 96 h, cooled at 4°C/h to 300°C, and then cooled to room temperature. This compound is air sensitive, decomposing after a few hours of exposure.

3.11   ZINC–CESIUM–CERIUM–TELLURIUM

$ZnTe-Cs_2Te-Ce_2Te_3$. The phase diagram is not constructed. The efforts to synthesize the $CsZnCeTe_3$ quaternary compound were unsuccessful (Yao et al. 2004).

3.12   ZINC–CESIUM–PRASEODYMIUM–TELLURIUM

$ZnTe-Cs_2Te-Pr_2Te_3$. The phase diagram is not constructed. The $CsZnPrTe_3$ quaternary compound is formed in this system. It crystallizes in the orthorhombic structure with the lattice parameters $a = 448.96 \pm 0.05$, $b = 1664.7 \pm 0.2$, and $c = 1192.94 \pm 0.15$ pm at 153 ± 2 K, energy gap ~2.12 eV, and calculated density 5.379 g·cm^{-3} (Yao et al. 2004). $CsZnPrTe_3$ was obtained from the reaction of Zn, Pr, and Te with the addition of CsCl as flux. The sample was heated to 850°C in 30 h, kept at this temperature for 96 h, cooled at 4°C/h to 300°C, and then cooled to room temperature. This compound is air sensitive, decomposing after a few hours of exposure.

3.13   ZINC–CESIUM–NEODYMIUM–TELLURIUM

$ZnTe-Cs_2Te-Nd_2Te_3$. The phase diagram is not constructed. The $CsZnNdTe_3$ quaternary compound is formed in this system. It crystallizes in the orthorhombic structure with the lattice parameters $a = 447.02 \pm 0.05$, $b = 1666.48 \pm 0.17$, and $c = 1186.56 \pm 0.12$ pm at 153 ± 2 K, energy gap ~2.12 eV, and calculated density 5.450 g·cm^{-3} (Yao et al. 2004). $CsZnNdTe_3$ was obtained from the reaction of Zn, Nd, and Te with the addition of CsCl as flux. The sample was heated to 850°C in 30 h, kept at this temperature for 96 h, cooled at 4°C/h to 300°C, and then cooled to room temperature. This compound is air sensitive, decomposing after a few hours of exposure.

3.14   ZINC–CESIUM–SAMARIUM–TELLURIUM

$ZnTe-Cs_2Te-Sm_2Te_3$. The phase diagram is not constructed. The $CsZnSmTe_3$ quaternary compound is formed in this system. It crystallizes in the orthorhombic structure with the lattice parameters $a = 443.68 \pm 0.05$, $b = 1668.40 \pm 0.17$, and

c = 1177.24 ± 0.12 pm at 153 ± 2 K, energy gap ~2.12 eV, and calculated density 5.575 g·cm^{-3} (Yao et al. 2004). CsZnSmTe$_3$ was obtained from the reaction of Zn, Sm, and Te with the addition of CsCl as flux. The sample was heated to 850°C in 30 h, kept at this temperature for 96 h, cooled at 4°C/h to 300°C, and then cooled to room temperature. This compound is air sensitive, decomposing after a few hours of exposure.

3.15 ZINC–CESIUM–GADOLINIUM–TELLURIUM

ZnTe–Cs$_2$Te–Gd$_2$Te$_3$. The phase diagram is not constructed. The CsZnGdTe$_3$ quaternary compound is formed in this system. It crystallizes in the orthorhombic structure with the lattice parameters a = 441.61 ± 0.04, b = 1671.48 ± 0.15, and c = 1169.89 ± 0.11 pm at 153 ± 2 K, energy gap ~2.12 eV, and calculated density 5.679 g·cm^{-3} (Yao et al. 2004). CsZnGdTe$_3$ was obtained from the reaction of Zn, Gd, and Te with the addition of CsCl as flux. The sample was heated to 850°C in 30 h, kept at this temperature for 96 h, cooled at 4°C/h to 300°C, and then cooled to room temperature. This compound is air sensitive, decomposing after a few hours of exposure.

3.16 ZINC–CESIUM–TERBIUM–TELLURIUM

ZnTe–Cs$_2$Te–Tb$_2$Te$_3$. The phase diagram is not constructed. The CsZnTbTe$_3$ quaternary compound is formed in this system. It crystallizes in the orthorhombic structure with the lattice parameters a = 439.92 ± 0.04, b = 1670.44 ± 0.14, and c = 1166.34 ± 0.09 pm at 153 ± 2 K, energy gap ~2.12 eV, and calculated density 5.735 g·cm^{-3} (Yao et al. 2004). CsZnTbTe$_3$ was obtained from the reaction of Zn, Tb, and Te with the addition of CsCl as flux. The sample was heated to 850°C in 30 h, kept at this temperature for 96 h, cooled at 4°C/h to 300°C, and then cooled to room temperature. This compound is air sensitive, decomposing after a few hours of exposure.

3.17 ZINC–CESIUM–DYSPROSIUM–TELLURIUM

ZnTe–Cs$_2$Te–Dy$_2$Te$_3$. The phase diagram is not constructed. The CsZnDyTe$_3$ quaternary compound is formed in this system. It crystallizes in the orthorhombic structure with the lattice parameters a = 438.38 ± 0.04, b = 1670.90 ± 0.14, and c = 1161.94 ± 0.10 pm at 153 ± 2 K, energy gap ~2.12 eV, and calculated density 5.803 g·cm^{-3} (Yao et al. 2004). CsZnDyTe$_3$ was obtained from the reaction of Zn, Dy, and Te with the addition of CsCl as flux. The sample was heated to 850°C in 30 h, kept at this temperature for 96 h, cooled at 4°C/h to 300°C, and then cooled to room temperature. This compound is air sensitive, decomposing after a few hours of exposure.

3.18 ZINC–CESIUM–HOLMIUM–TELLURIUM

ZnTe–Cs$_2$Te–Ho$_2$Te$_3$. The phase diagram is not constructed. The CsZnHoTe$_3$ quaternary compound is formed in this system. It crystallizes in the orthorhombic structure with the lattice parameters a = 437.61 ± 0.14, b = 1672.1 ± 0.5, and c = 1158.2 ± 0.4 pm at 153 ± 2 K, energy gap ~2.12 eV, and calculated density 5.847 g·cm^{-3}

(Yao et al. 2004). $CsZnHoTe_3$ was obtained from the reaction of Zn, Ho, and Te with the addition of CsCl as flux. The sample was heated to 850°C in 30 h, kept at this temperature for 96 h, cooled at 4°C/h to 300°C, and then cooled to room temperature. This compound is air sensitive, decomposing after a few hours of exposure.

3.19 ZINC–CESIUM–ERBIUM–TELLURIUM

$ZnTe-Cs_2Te-Er_2Te_3$. The phase diagram is not constructed. The $CsZnErTe_3$ quaternary compound is formed in this system. It crystallizes in the orthorhombic structure with the lattice parameters $a = 436.04 \pm 0.08$, $b = 1670.6 \pm 0.3$, and $c = 1153.4 \pm 0.2$ pm at 153 ± 2 K, energy gap ~2.12 eV, and calculated density 5.916 g·cm⁻³ (Yao et al. 2004). $CsZnErTe_3$ was obtained from the reaction of Zn, Er, and Te with the addition of CsCl as flux. The sample was heated to 850°C in 30 h, kept at this temperature for 96 h, cooled at 4°C/h to 300°C, and then cooled to room temperature. This compound is air sensitive, decomposing after a few hours of exposure.

3.20 ZINC–CESIUM–THULIUM–TELLURIUM

$ZnTe-Cs_2Te-Tm_2Te_3$. The phase diagram is not constructed. The $CsZnTmTe_3$ quaternary compound is formed in this system. It crystallizes in the orthorhombic structure with the lattice parameters $a = 434.71 \pm 0.07$, $b = 1671.2 \pm 0.3$, and $c = 1151.57 \pm 0.19$ pm at 153 ± 2 K, energy gap ~2.12 eV, and calculated density 5.955 g·cm⁻³ (Yao et al. 2004). $CsZnTmTe_3$ was obtained from the reaction of Zn, Tm, and Te with the addition of CsCl as flux. The sample was heated to 850°C in 30 h, kept at this temperature for 96 h, cooled at 4°C/h to 300°C, and then cooled to room temperature. This compound is air sensitive, decomposing after a few hours of exposure.

3.21 ZINC–CESIUM–YTTERBIUM–TELLURIUM

$ZnTe-Cs_2Te-Yb_2Te_3$. The phase diagram is not constructed. The $CsZnYbTe_3$ quaternary compound is formed in this system. It crystallizes in the orthorhombic structure with the lattice parameters $a = 432.28 \pm 0.05$, $b = 1597.0 \pm 0.2$, and $c = 1148.0 \pm 0.1$ pm at 153 ± 2 K and calculated density 5.923 g·cm⁻³ (Mitchell et al. 2004). $CsZnYbTe_3$ was synthesized using the mixture of Cs_2Te_3, Zn, Yb, Te, and CsI. The sample was heated to 900°C in 24 h, kept at this temperature for 96 h, cooled to 200°C, and then rapidly cooled to room temperature.

3.22 ZINC–CESIUM–GERMANIUM–TELLURIUM

$ZnTe-Cs_2Te-GeTe_2$. The phase diagram is not constructed. The $Cs_2ZnGe_3Te_8$ quaternary compound is formed in this system. It crystallizes in the orthorhombic structure with the lattice parameters $a = 813.45 \pm 0.06$, $b = 1311.45 \pm 0.13$, and $c = 1882.63 \pm 0.19$ pm, calculated density 5.192 g·cm⁻³, and energy gap 1.07 eV (Morris et al. 2013). To obtain $Cs_2ZnGe_3Te_8$, a mixture of Cs_2Te_3, Zn, Ge, and Te in the appropriate ratios was added to a fused silica tube inside a nitrogen-filled glove box. The tube was evacuated, flame sealed, and used in a flame-melting reaction. After cooling in air, a phase-pure ingot containing black platelike crystals was obtained.

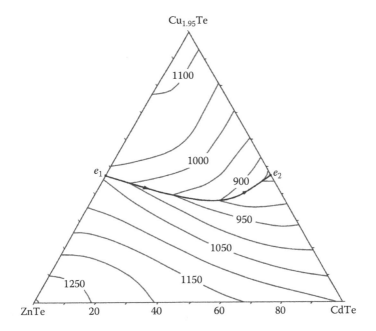

FIGURE 3.2 Liquidus surface of the $ZnTe–Cu_{1.95}Te–CdTe$ quasiternary system. (From Trishchuk, L.I. et al., *Chem. Met. Alloys*, 1(1), 58, 2008.)

3.23 ZINC–COPPER–CADMIUM–TELLURIUM

$ZnTe–Cu_{1.95}Te–CdTe$. The liquidus surface of this quasiternary system (Figure 3.2) contains two fields of primary crystallization: the fields of primary crystallization of the $Cu_{2-x}Te$ and $Zn_xCd_{1-x}Te$ solid solutions (Trishchuk et al. 2008). These fields are separated by the line of secondary crystallization e_1e_2, which connects the eutectics in the quasibinary systems based on copper telluride. The eutectic of the quasibinary system $Cu_{1.95}Te–CdTe$ is the lowest melting point in this quasiternary system.

3.24 ZINC–COPPER–ALUMINUM–TELLURIUM

$ZnTe–CuAlTe_2$. The phase diagram of this system is shown in Figure 3.3 (Bodnar' 2001). The complete series of the solid solutions is formed in the $2ZnTe–CuAlTe_2$ system. Solid solutions containing more than 70 mol.% $CuAlTe_2$ crystallize in the chalcopyrite-type structure, and the solid solutions containing up to 70 mol.% $CuAlTe_2$ crystallize in the sphalerite-type structure. Lattice parameters of the obtained solid solutions change linearly with the changing $CuAlTe_2$ content, and the bandgap has a minimum at 50 mol.% $CuAlTe_2$.

According to the data by Parthé et al. (1969), the $CuZn_2AlTe_4$ quaternary compound is formed in this system. It crystallizes in the cubic structure of sphalerite type with the lattice parameter $a = 604.3 \pm 0.5$ pm, but this compound can be one of the solid solution compositions.

This system was investigated using differential thermal analysis (DTA) and x-ray diffraction (XRD) (Bodnar' 2001).

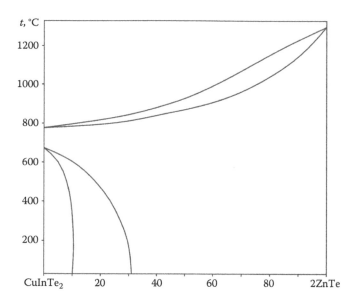

FIGURE 3.5 The 2ZnTe–CuInTe$_2$ phase diagram. (From Bodnar', I.V. et al., *Zhurn. neorgan. khimii*, 45(9), 1538, 2000; Roussak, L. et al., *J. Solid State Chem.*, 178(11), 3476, 2005.)

The variation of bandgap with respect to crystal composition of the obtained solid solutions can be described by a quadratic expression (Bodnar' et al. 2000a).

It cannot be excluded that more than two phases, possibly including other metastable and unknown phases, exist in the two-phase region (Roussak et al. 2005). In the low-temperature region, for samples annealed at 300°C over 60 days and then quenched, a third phase could be detected. This randomly localized phase represents In$_x$Te$_y$ with x/y ratios ranging from 1 to 3.5, but the majority of this phase was found to be In rich.

According to the data of Parthé et al. (1969) and Garbato and Manca (1974), the CuZn$_2$InTe$_4$ quaternary compound is formed in this system. It crystallizes in the cubic structure of sphalerite type with the lattice parameter $a = 615.3 \pm 0.5$ pm (Parthé et al. 1969) [$a = 613 \pm 0$ pm (Garbato and Manca 1974)], but this compound can be one of the solid solution compositions.

This system was investigated through DTA, XRD, metallography, and measuring of microhardness (Garbato and Ledda 1979, Bodnar' et al. 2000a,b). The ingots were annealed at 700°C–800°C during 700 h (Bodnar' et al. 2000b) [at 410°C–980°C during 720 h (Bodnar' et al. 2000a)]. Subsolidus phase relationships were investigated by TEM experiments combined with EDX analysis, selected area diffraction method, and HRTEM (Roussak et al. 2005, Schorr et al. 2005). The samples were prepared by the solid-state reaction of the elements during long annealing times, followed either by quenching in ice-cold water or by controlled cooling at different rates. Powder samples of the $Zn_{2x}(CuIn)_{1-x}Te_2$ alloys with $0 \leq x \leq 0.1$ were prepared by solid-state reaction of the elements at 700°C, and samples with $0.4 < x < 1.0$ were also prepared by the directional freezing method (Schorr et al. 2005). Single crystals of the solid solutions in this system have been obtained by the directional freezing method (Bodnar' et al. 2000a,b) or by the horizontal Bridgman method ($0 \leq x < 0.5$) and using chemical transport reactions with I_2 as transport agent ($0.5 < x \leq 1$) (Bodnar' et al. 2004).

3.27 ZINC–COPPER–SILICON–TELLURIUM

ZnTe–Cu$_2$SiTe$_3$. The phase diagram is not constructed. The $Cu_2ZnSiTe_4$ quaternary compound is formed in this system. It melts at 700°C and crystallizes in the tetragonal structure with the lattice parameters $a = 597.2 \pm 0.1$ and $c = 1179.7 \pm 0.4$ pm (Haeuseler et al. 1991) [$a = 598$ and $c = 1178$ pm and energy gap 1.47 eV (Matsuhita et al. 2005)].

Single crystals of $Cu_2ZnGeSe_4$ were obtained by the chemical transport reactions using I_2 as the transport agent or using the horizontal gradient freezing method (Matsuhita et al. 2005).

3.28 ZINC–COPPER–GERMANIUM–TELLURIUM

ZnTe–Cu$_2$GeTe$_3$. The phase relations in this system are shown in Figure 3.6 (Parasyuk et al. 2005). Since Cu_2GeTe_3 forms incongruently, there are phase fields related to phases that form upon decomposition of the ternary compound above the solidus. The liquidus of the system is represented by three fields of primary crystallization. The largest one belongs to the primary crystallization of ZnTe (~12–100 mol.% ZnTe). The other two fields correspond to the primary crystallization of β_1-$Cu_{2-x}Te$ and α-$Cu_2ZnGeTe_4$; they are small and located in the concentration range 0–12 mol.% ZnTe. α-$Cu_2ZnGeTe_4$ melts incongruently at 550°C and crystallizes in the tetragonal structure with the lattice parameters $a = 595.40 \pm 0.04$ and $c = 1184.8 \pm 0.1$ pm (Parasyuk et al. 2005) [$a = 599.9 \pm 0.2$ and $c = 1191.8 \pm 0.5$ pm (Haeuseler et al. 1991)]. The peritectic point is at ~12 mol.% ZnTe. According to the data of Matsuhita et al. (2005), $Cu_2ZnGeTe_4$ has the mixed phases of ZnTe and Cu_2GeTe_3.

Because the energy between kesterite- and stannite-type structures is small, both structures can coexist in synthesized samples of $Cu_2ZnGeTe_4$. According to the first-principles calculations, kesterite-type structure of this compound is characterized by the tetragonal structure with the lattice parameters $a = 610.2$ and $c = 1212.6$ pm and energy gap 0.81 eV, and stannite-type structure is characterized also by the tetragonal structure with the lattice parameters $a = 609.4$ and $c = 1222.0$ pm and energy gap 0.55 eV (Chen and Ravindra 2013).

This system was investigated by DTA and XRD and the ingots were annealed at 400°C for 700 h (Parasyuk et al. 2005).

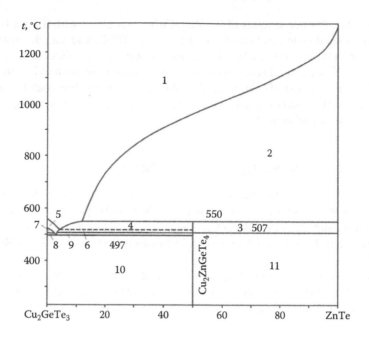

FIGURE 3.6 Phase relations in the 2ZnTe–Cu$_2$GeTe$_3$ system: 1, L; 2, L + ZnTe; 3, α-Cu$_2$ZnGeTe$_4$ + ZnTe; 4, L + α-Cu$_2$ZnGeTe$_4$; 5, L + β$_1$-Cu$_{2-x}$Te; 6, L + β$_1$-Cu$_{2-x}$Te + α-Cu$_2$ZnGeTe$_4$; 7, L + β$_1$-Cu$_{2-x}$Te + β-GeTe; 8, L + β$_1$-Cu$_{2-x}$Te + Cu$_2$GeTe$_3$,; 9, L + β$_1$-Cu$_{2-x}$Te + β-Cu$_2$ZnGeTe$_4$; 10, Cu$_2$GeTe$_3$ + β-Cu$_2$ZnGeTe$_4$; and 11, β-Cu$_2$ZnGeTe$_4$ + ZnTe. (From Parasyuk, O.V. et al., *J. Alloys Compd.*, 397(1–2), 169, 2005.)

3.29 ZINC–COPPER–TIN–TELLURIUM

ZnTe–Cu$_2$SnTe$_3$. The phase diagram is not constructed. The Cu$_2$ZnSnTe$_4$ quaternary compound is formed in this system (Haeuseler et al. 1991). It crystallizes in the tetragonal structure with the lattice parameters $a = 608.8 \pm 0.1$ and $c = 1218.0 \pm 0.4$ pm (Haeuseler et al. 1991) [$a = 607.5$ and $c = 1201$ pm and calculation and experimental densities 6.15 and 6.01 g·cm^{-3}, respectively (Hahn and Schulze 1965)]. According to the data of Matsuhita et al. (2005), Cu$_2$ZnSnTe$_4$ has the mixed phases of ZnTe, SnTe, and Cu$_2$GeTe$_3$.

3.30 ZINC–COPPER–OXYGEN–TELLURIUM

Cu$_{1.5}$Zn$_{1.5}$TeO$_6$ and Cu$_5$Zn$_4$Te$_3$O$_{18}$ quaternary compounds are formed in this system. Cu$_{1.5}$Zn$_{1.5}$TeO$_6$ crystallizes in the cubic structure with the lattice parameter $a = 955.7 \pm 0.1$, and Cu$_5$Zn$_4$Te$_3$O$_{18}$ crystallizes in the monoclinic structure with the lattice parameters $a = 1483.4 \pm 0.2$, $b = 880.1 \pm 0.1$, $c = 1037.5 \pm 0.2$ pm, and β = 93.27° ± 0.02° (Wulff and Müller-Buschbaum 1998). These compounds were obtained by the interaction of TeO$_2$, CuCO$_3$·Cu(OH)$_2$, and ZnCO$_3$ at 750°C.

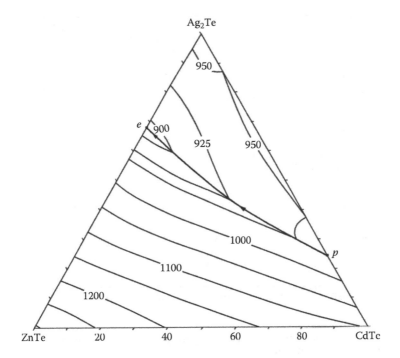

FIGURE 3.7 Liquidus surface of the ZnTe–Ag$_2$Te–CdTe quasiternary system. (From Trishchuk, L.I. et al., *Chem. Met. Alloys*, 1(1), 58, 2008; Trishchuk, L.I., *Optoelektronika i poluprovodn. tekhnika*, (44), 99, 2009.)

3.31 ZINC–SILVER–CADMIUM–TELLURIUM

ZnTe–Ag$_2$Te–CdTe. The liquidus surface of this quasiternary system (Figure 3.7) contains two fields of primary crystallization: the fields of primary crystallization of the solid solution based on Ag$_2$Te and the field of Zn$_x$Cd$_{1-x}$Te solid solution (Trishchuk et al. 2008, Trishchuk 2009). These fields are separated by the line of secondary crystallization *ep*, which connects the eutectic in the quasibinary system Ag$_2$Te–ZnTe and the peritectic in the quasibinary system Ag$_2$Te–CdTe. The liquidus surface ascends sharply from the line of secondary crystallization in the direction of the high-temperature melting component (ZnTe). The eutectic of the quasibinary system Ag$_2$Te–ZnTe is the lowest melting point in the Ag$_2$Te–ZnTe–CdTe quasiternary system.

3.32 ZINC–SILVER–INDIUM–TELLURIUM

ZnTe–AgInTe$_2$. This section is nonquasibinary section of the Zn–Ag–In–Te quaternary system (Figure 3.8) (Tovar et al. 1989). Two-solid-phase fields, the chalcopyrite α-phase corresponding to the solid solutions based on AgInTe$_2$ and the sphalerite β-phase corresponding to the solid solutions based on ZnTe, plus a two-solid-phase (α + β) field occur in the composition diagram.

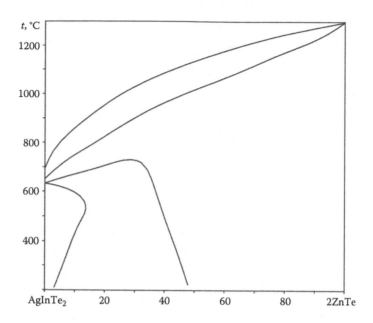

FIGURE 3.8   Phase relations in the ZnTe–AgInTe$_2$ system. (From Tovar, R. et al., *Phys. Status Solidi (a)*, 111(2), 405, 1989.)

It is seen from Figure 3.8 that the α-phase appears for $x = 0$ at $t \leq 635°C$ and that β-phase extends up to 660°C, and above this temperature, it splits up into two phases (L + β$_2$) that occur up to the liquidus point. The β- and β$_2$-phases are both sphalerite but of different compositions. The field of α-phase achieves a maximum width of approximately $x = 0.14$ at 550°C [Zn$_{2x}$(AgIn)$_{1-x}$Te$_2$].

The shape of the boundary between the β and (α + β) fields indicates that the tie lines of the (α + β) field cannot lie in the plane of the diagram; that is, the section is not quasibinary.

For samples with $0 \leq x \leq 0.4$, the DTA data showed peaks at temperatures of 500°C–530°C that are due to segregation of another phase. This additional phase may be associated with segregation of tellurium from the main phase (Tovar et al. 1989).

It was indicated (Gallardo 1992) that it is possible to predict the value at which the chalcopyrite phase in this system disappears (x_c) and $T_c(x)$ behavior for the Zn$_{2x}$(CuIn)$_{1-x}$Te$_2$ alloys for which a chalcopyrite–sphalerite-like disordering transition with extensive terminal solid solution formation exists.

According to the data of Parthé et al. (1969), the AgZn$_2$InTe$_4$ quaternary compound is formed in this system. It crystallizes in the cubic structure of sphalerite type with the lattice parameter $a = 625.3 \pm 0.5$ pm, but this compound can be one of the solid solution compositions.

This system was investigated through DTA and XRD and the ingots were annealed at 600°C for 30–40 days (Tovar et al. 1989).

3.33   ZINC–CALCIUM–OXYGEN–TELLURIUM

The $Ca_3Zn_3Te_2O_{12}$ quaternary compound is formed in this system. It crystallizes in the cubic structure with the lattice parameter $a = 1258$ pm (Kasper 1968).

3.34   ZINC–CADMIUM–MERCURY–TELLURIUM

Some features of the Zn–Cd–Hg–Te quaternary system have been calculated by Yu and Brebrick (1992, 1993).

ZnTe–CdTe–HgTe. The location of the miscibility gap in this system has been predicted using the functions based on the Kohler and Redlich–Kister–Muggianu procedures (Voronin and Pentin 2005, Pentin et al. 2006). There are differences between the data obtained by the Kohler method (Figure 3.9) and those found using the Redlich–Kister–Muggianu procedure (Figure 3.10).

Isothermal section of the ZnTe–CdTe–HgTe system at 112°C calculated by using the Redlich–Kister–Muggianu method is shown in Figure 3.11 (Voronin and Pentin 2005). Figure 3.12 shows the position of the calculated miscibility gap in this system at 100°C according to the data of Ohtani et al. (1992). These results should be considered only as a first prediction of the miscibility gap location in the ZnTe–CdTe–HgTe system. But clearly, additional experimental data are required for more precise calculations.

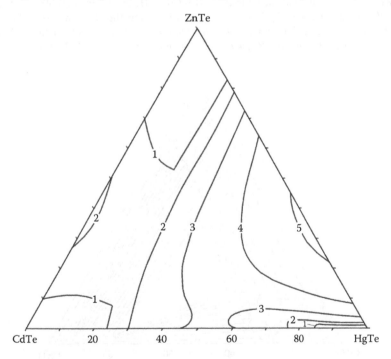

FIGURE 3.9   The binodal curves in the ZnTe–CdTe–HgTe system at (1) 80°C, (2) 130°C, (3) 180°C, (4) 230°C, and (5) 380°C, calculated by using the Kohler method. (From Pentin, I.V. et al., *CALPHAD: Comput. Coupling Phase Diagr. Thermochem.*, 30(2), 191, 2006.)

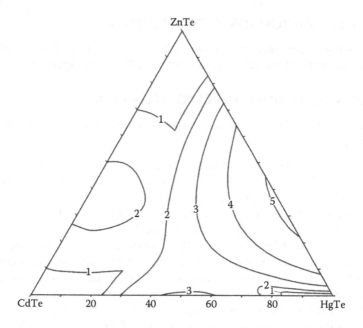

FIGURE 3.10   The binodal curves in the ZnTe–CdTe–HgTe system at (1) 80°C, (2) 130°C, (3) 180°C, (4) 230°C, and (5) 380°C, calculated by using the Redlich–Kister–Muggianu method. (From Voronin, G.F. et al., *Zhurn. fiz. khimii*, 79(10), 1771, 2005; Pentin, I.V. et al., *CALPHAD: Comput. Coupling Phase Diagr. Thermochem.*, 30(2), 191, 2006.)

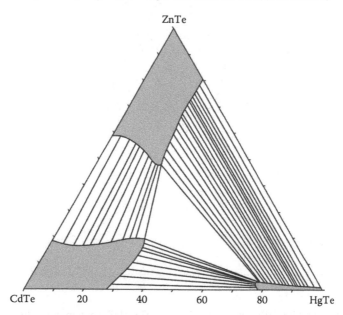

FIGURE 3.11   Isothermal section of the ZnTe–CdTe–HgTe system at 112°C calculated by using the Redlich–Kister–Muggianu method. (From Voronin, G.F. et al., *Zhurn. fiz. khimii*, 79(10), 1771, 2005.)

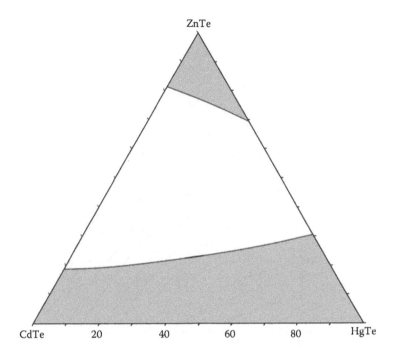

ZnTe

CdTe   20   40   60   80   HgTe

FIGURE 3.12 Calculated miscibility gap in the ZnTe–CdTe–HgTe quasiternary system at 100°C. (From Ohtani, H. et al., *J. Alloys Compd.*, 182(1), 103, 1992.)

3.35   ZINC–CADMIUM–GALLIUM–TELLURIUM

ZnTe–CdTe–Ga. Liquidus surface of this quasiternary system (Figure 3.13) was constructed using mathematical simulation of experiments (Radautsan and Maksimova 1976). This system was investigated through DTA, metallography, and XRD.

3.36   ZINC–CADMIUM–PHOSPHORUS–TELLURIUM

CdTe–Zn$_3$P$_2$. The phase diagram is not constructed. Complex interaction of CdTe and Zn$_3$P$_2$ exists in this system (Goriunova et al. 1970). Probably at the interaction, the next reaction takes place: $(4 - x)Zn + 2P + xCd + Te \rightarrow Zn_{1-x}Cd_xTe + Zn_3P_2$. This system was investigated through DTA, XRD, and chemical analysis.

3.37   ZINC–CADMIUM–ARSENIC–TELLURIUM

ZnTe–Cd$_3$As$_2$. This section is nonquasibinary section of the Zn–Cd–As–Te quaternary system (Figure 3.14) (Lakiza 1980, Lakiza and Olekseyuk 1981). Solid solutions based on β-Cd$_3$As$_2$ (β-phase) and based on ZnTe (γ-phase) primarily crystallize in this system. The solidus of this section includes three lines: the lines of the β-phase and the γ-phase crystallization and the line of L + γ ⇔ β three-phase peritectic reaction. The solubility of 2ZnTe in Cd$_3$As$_2$ reaches 20 mol.% (Serginov et al. 1968, 1970) and decreases to 15 mol.% at 250°C (Lakiza 1980, Lakiza and Olekseyuk 1981).

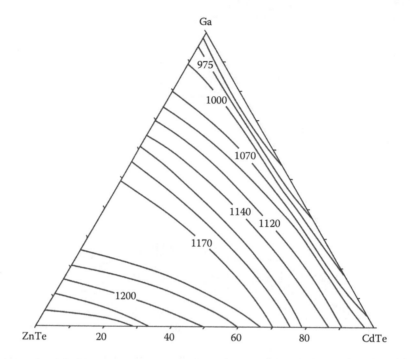

FIGURE 3.13 Liquidus surface of the ZnTe–CdTe–Ga quasiternary system. (From Radautsan, S.I. and Maksimova, O.G., Phase interaction in the Te–ZnTe–CdTe and Ga–ZnTe–CdTe systems and growth of the $Zn_xCd_{1-x}Te$ crystals from solution in melt [in Russian], in: *Poluprovodn. materialy i ih primenienie*, Kishinev, Republic of Moldova, 1976, p. 3.)

The solubility of Cd_3As_2 in 2ZnTe is not higher than 1 mol.% (Lakiza 1980, Lakiza and Olekseyuk 1981). Solid solutions based on Cd_3As_2 have $\beta \Leftrightarrow \gamma$ polymorphous transformation. The $Zn_xCd_{1-x}Te$ solid solutions are formed in this system at the content more than 20 mol.% 2ZnTe.

This system was investigated through DTA, metallography, XRD, and measuring of microhardness (Lakiza 1980, Lakiza and Olekseyuk 1981). Single crystals of solid solutions were grown by the chemical transport reactions (Serginov et al. 1968, 1970).

CdTe–Zn₃As₂. This section is nonquasibinary section of the Zn–Cd–As–Te quaternary system (Figure 3.15) (Lakiza and Olekseyuk 1978). Solid solutions based on β-Zn_3As_2 (β-phase) and CdTe (γ-phase) primarily crystallize in this system. The composition of peritectic point is 25 mol.% 2CdTe. The solidus of this section includes three lines: the lines of the β- and γ-solid solution crystallization and the line of the three-phase peritectic reaction $L + \gamma \Leftrightarrow \beta$.

Large regions of solid solutions based on Zn_3As_2 exist in this system (Serginov et al. 1968, 1970, Kradinova et al. 1969, Lakiza and Olekseyuk 1978). The solubility of 2CdTe in Zn_3As_2 at 870°C reaches 40 mol.%, and the solubility of Zn_3As_2 in 2CdTe at room temperature is not higher than 1 mol.% (Lakiza and Olekseyuk 1978).

The temperature of β-Zn_3As_2 → α-Zn_3As_2 polymorphous transformation decreases with CdTe contents increasing. The region of solid solutions with

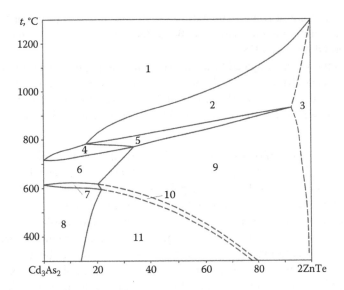

FIGURE 3.14   Phase relations in the 2ZnTe–Cd$_3$As$_2$ system: 1, L; 2, L + γ; 3, γ; 4, L + β; 5, L + β + γ; 6, β; 7, α + β; 8, α; 9, β + γ; 10, α + β + γ; and 11, α + γ. (From Lakiza, S.N., Investigation of the interaction and the Zn$_3$As$_2$ + 3CdTe ⇔ Cd$_3$As$_2$ + 3ZnTe ternary mutual system [in Russian], in: *Razrab. i issled. novykh materialov i kompozitsiy na ikh osnove*, Kiev, 1980, p. 46; Lakiza, S.N. and Olekseyuk, I.D., *Zhurn. neorgan. khimii*, 26(4), 1118, 1981.)

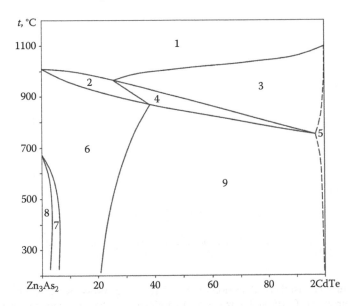

FIGURE 3.15   Phase relations in the 2CdTe–Zn$_3$As$_2$ system: 1, L; 2, L + β; 3, L + γ; 4, L + β + γ; 5, γ; 6, β; 7, α + β; 8, α; and 9, β + γ. (From Lakiza, S.N., Investigation of the interaction and the Zn$_3$As$_2$ + 3CdTe ⇔ Cd$_3$As$_2$ + 3ZnTe ternary mutual system [in Russian], in: *Razrab. i issled. novykh materialov i kompozitsiy na ikh osnove*, Kiev, 1980, p. 46; Lakiza, S.N. and Olekseyuk, I.D., *Zhurn. neorgan. khimii*, 26(4), 1118, 1981.)

tetragonal structure at room temperature exists within the interval of 0–3 mol.% 2CdTe, and solid solutions based on β-Zn_3As_2 contain up to 20 mol.% 2CdTe (Lakiza and Olekseyuk 1978).

This system was investigated through DTA, metallography, XRD, and measuring of microhardness (Kradinova et al. 1969, Lakiza and Olekseyuk 1978). The ingots were annealed at 500°C for 720 h (Lakiza and Olekseyuk 1978). Single crystals of the solid solutions were grown by the chemical transport reactions (Serginov et al. 1968, 1970).

3ZnTe + Cd_3As_2 ⇔ 3CdTe + Zn_3As_2. The isothermal section of this ternary mutual system at 250°C is shown in Figure 3.16 (Lakiza 1980, Lakiza and Olekseyuk 1981). Liquidus surface includes the fields of primary crystallization of $Zn_xCd_{3-x}As_2$ and $Zn_xCd_{1-x}Te$ solid solutions. No invariant points exist in this system.

The solidus surface consists of two surfaces of γ- and β-solid solutions crystallization and surface of the L + γ ⇔ β peritectic reaction, which proceeds with excess of γ-solid solutions. Two intermediate surfaces of the beginning and the end of L + γ ⇔ β peritectic reaction exist between the liquidus and solidus surfaces. There is a wide region of solid solutions in this system, which adjoins to the Zn_3As_2–Cd_3As_2 quasibinary system and includes some areas. The β-solid solution region has a wedge form and is based on the region of the β-solid solutions in

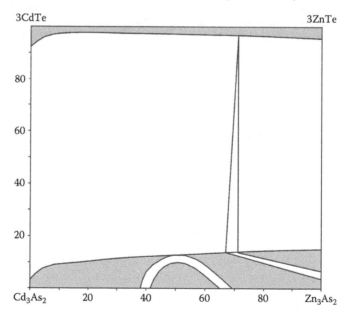

FIGURE 3.16 The isothermal section of the 3ZnTe + Cd_3As_2 ⇔ 3CdTe + Zn_3As_2 ternary mutual system at 250°C. (From Lakiza, S.N., Investigation of the interaction and the Zn_3As_2 + 3CdTe ⇔ Cd_3As_2 + 3ZnTe ternary mutual system [in Russian], in: *Razrab. i issled. novykh materialov i kompozitsiy na ikh osnove*, Kiev, 1980, p. 46; Lakiza, S.N. and Olekseyuk, I.D., *Zhurn. neorgan. khimii*, 26(4), 1118, 1981.)

the $3ZnTe-Zn_3As_2$ quasibinary system. Cubic structure of the β-solid solutions becomes destabilized and transforms into tetragonal structure of α-solid solutions at the increase of Cd_3As_2 contents. The regions of α- and β-solid solutions are divided by the narrow two-phase region. The region of α-solid solutions is disrupted by the δ-superstructure that penetrates inside the concentration square. The region of solid solutions based on Zn(Cd)Te is not higher than 1 mol.%. Continuous transition from peritectoid transformation in the $3CdTe-Cd_3As_2$ system to eutectoid transformation in the $3ZnTe-Zn_3As_2$ system takes place in this ternary mutual system. This system was investigated through DTA, metallography, XRD, using infrared spectrum, and measuring of microhardness (Lakiza 1980, Lakiza and Olekseyuk 1981).

3.38 ZINC–CADMIUM–OXYGEN–TELLURIUM

The $Zn_3Cd_3Te_2O_{12}$ quaternary compound is formed in this system. It crystallizes in the cubic structure with the lattice parameter $a = 1253$ pm (Kasper 1968).

3.39 ZINC–CADMIUM–CHLORINE–TELLURIUM

$CdTe-ZnCl_2-CdCl_2$. The phase diagram is not constructed. Using oriented zone-recrystallization and radioactive isotopes, there was a determined equilibrium coefficient of $ZnCl_2$ distribution in the $CdTe-CdCl_2$ eutectic: $k_0 = 1.39$ (Leonov and Chunarev 1977). As $k_0 > 0$, it is obvious the next exchange reaction $CdTe + ZnCl_2 = ZnTe + CdCl_2$ takes place in this system.

3.40 ZINC–CADMIUM–MANGANESE–TELLURIUM

$ZnTe-CdTe-MnTe$. Polycrystalline samples of the $Zn_xCd_yMn_{1-x-y}Te$ solid solutions were produced by a melt and anneal technique (annealing at 800°C for 2 weeks) (Brun del Re et al. 1983), and its single crystals were grown by the Bridgman technique (Beckett et al. 1988). The comparison of the initial growth material composition with the initial melt composition indicates that the liquidus and solidus sheets have a temperature separation of less than 5°C. The phase diagram for the ZnTe–CdTe–MnTe quasiternary system with the lines of constant lattice parameter and constant energy gap is given in Figure 3.17 (Brun del Re et al. 1983).

3.41 ZINC–MERCURY–MANGANESE–TELLURIUM

$ZnTe-HgTe-MnTe$. Single crystals of the $Zn_xHg_{1-x}Mn_yTe$ ($x \le 0.10$; $y \le 0.15$) solid solutions were grown using the Bridgman method (Frasunyak 2000). Dependence of the energy gap versus temperature and composition could be expressed by the next equation: E_g (eV) = $- 0.302 + 0.023x^{1/2} + 2.731x + 4.24y + (5.1 - 2.839x^{1/2} - 6.837x - 23.03y) \cdot 10^{-4}T - 1.258x^2 + 2.132x^3 - 4.42y^2 - 0.272y^3$.

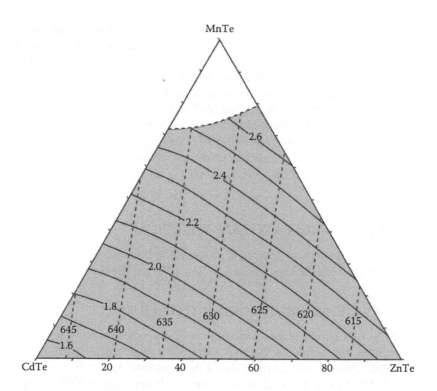

FIGURE 3.17 Phase diagram for the ZnTe–CdTe–MnTe quasiternary system with the lines of constant lattice parameter (in pm) and constant magnetic transition temperature (in K). (From Brun, del Re R. et al., *Nuovo cim.*, D2(6), 1911, 1983.)

3.42 ZINC–ALUMINUM–IODINE–TELLURIUM

ZnTe–Al$_2$Te$_3$–AlI$_3$. The phase diagram is not constructed. The Zn$_2$AlTe$_3$I quaternary compound is formed in this system (Range and Handrick 1988). It crystallizes in the cubic structure with the lattice parameter a = 604.75 ± 0.04 pm. This compound was obtained at 600°C–1100°C by the interaction of ZnTe, ZnI$_2$, and Al$_2$Te$_3$.

3.43 ZINC–GALLIUM–ARSENIC–TELLURIUM

ZnTe–GaAs. The phase diagram is a eutectic type (Figure 3.18) (Ufimtseva et al. 1973). The eutectic composition and temperature are 53 mol.% GaAs and 1142°C, respectively. The solubility of ZnTe in GaAs is equal to 4.0 ± 0.4 mol.% at 1180°C and decreases to 1.5 mol.% at room temperature [according to the data of Glazov et al. (1975d), the solubility of ZnTe in GaAs reaches 15 mol.%]. The solubility of GaAs in ZnTe reaches 12 ± 1.1 mol.% at 1180°C and 2.0 ± 0.2 mol.% at room temperature (Ufimtseva et al. 1973) [is not higher than 5 mol.% (Glazov et al. 1975d)]. The lattice parameters in the homogeneity regions change linearly with composition. The liquidus and solidus temperatures from the GaAs-rich side are given in Table 3.1 (Anishchenko et al. 1980).

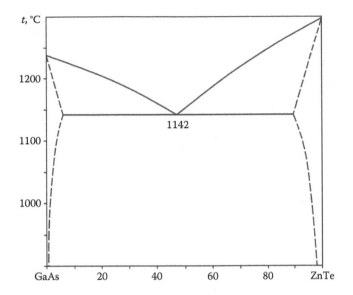

FIGURE 3.18 The ZnTe–GaAs phase diagram. (From Ufimtseva, E.V. et al., *Izv. AN SSSR. Neorgan. materialy*, 9(4), 587, 1973.)

TABLE 3.1
Liquidus and Solidus Temperatures of the
ZnTe–GaAs System from the GaAs-Rich Side

x_{GaAs}, mol.%	t_S, °C	t_L, °C
100	1238	1238
95	1215	1225
90	1210	1220
85	1190	1215

Source: Anishchenko, V.A. et al., *Izv. AN SSSR. Neorgan. materialy*, 16(2), 354, 1980.

This system was investigated through DTA, metallography, local XRD, and measuring of microhardness (Ufimtseva et al. 1973, Glazov et al. 1975d). The ingots were annealed at 1235°C for 100 h. Single crystals of the solid solutions were grown by the oriented crystallization, floating-zone refining, and chemical transport reactions (Anishchenko et al. 1980).

3.44 ZINC–GALLIUM–ANTIMONY–TELLURIUM

ZnTe–GaSb. The phase diagram is a eutectic type (Figure 3.19) (Glazov et al. 1975b,c). The eutectic crystallizes at 690°C and contains approximately 10 mol.% ZnTe, but according to the data of Burdiyan and Korolevski (1966), Burdiyan (1970, 1971), and Kirovskaya et al. (2007a), the solubility of ZnTe in GaSb is equal 10–20 mol.%.

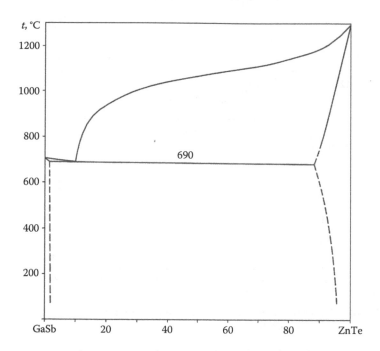

FIGURE 3.19   The ZnTe–GaSb phase diagram. (From Glazov, V.M. et al., *Izv. AN SSSR. Neorgan. materialy*, 11(3), 418, 1975c.)

The energy gap of forming solid solutions changes almost linearly with the composition (Burdiyan 1970, 1971). The solubility of GaSb in ZnTe reaches 15 mol.% (Kirovskaya et al. 2007a) [is not higher than 10 mol.% (Glazov et al. 1975b,c)]. This system was investigated through DTA and metallography (Burdiyan and Korolevski 1966, Glazov et al. 1975b,c).

3.45   ZINC–GALLIUM–OXYGEN–TELLURIUM

The $ZnGa_2(TeO_3)_4$ quaternary compound is formed in this system. It is stable up to melting point at 800°C and crystallizes in the cubic structure with the lattice parameters $a = 1057.94 \pm 0.08$ at 200 K and calculated density 5.089 g·cm^{-3} (Lee et al. 2012). Crystals of this compound were obtained by standard solid-state reaction. ZnO, In_2O_3, and TeO_2 in stoichiometric ratio were thoroughly mixed and introduced into fused silica tube that was subsequently evacuated and sealed. The tube was gradually heated to 380°C for 5 h and then to 700°C for 48 h. The sample was cooled at a rate of 6°C/h to room temperature.

3.46   ZINC–GALLIUM–CHLORINE–TELLURIUM

ZnTe–Ga₂Te₃–GaCl₃. The phase diagram is not constructed. The Zn_2GaTe_3Cl quaternary compound is formed in this system (Range and Handrick 1988). It crystallizes in the cubic structure with the lattice parameter $a = 603.16 \pm 0.03$ pm. This compound was obtained at 600°C–1100°C by the interaction of ZnTe, $ZnCl_2$, and Ga_2Te_3.

3.47   ZINC–GALLIUM–BROMINE–TELLURIUM

ZnTe–Ga$_2$Te$_3$–GaBr$_3$. The phase diagram is not constructed. The Zn$_2$GaTe$_3$Br quaternary compound is formed in this system (Range and Handrick 1988). It crystallizes in the cubic structure with the lattice parameter $a = 601.99 \pm 0.05$ pm. This compound was obtained at 600°C–1100°C by the interaction of ZnTe, ZnBr$_2$, and Ga$_2$Te$_3$.

3.48   ZINC–GALLIUM–IODINE–TELLURIUM

ZnTe–Ga$_2$Te$_3$–GaI$_3$. The phase diagram is not constructed. The Zn$_2$GaTe$_3$I quaternary compound is formed in this system (Range and Handrick 1988). It crystallizes in the cubic structure with the lattice parameter $a = 604.02 \pm 0.04$ pm. This compound was obtained at 600°C–1100°C by the interaction of ZnTe, ZnI$_2$, and Ga$_2$Te$_3$.

3.49   ZINC–INDIUM–PHOSPHORUS–TELLURIUM

ZnTe–InP. The phase diagram is a eutectic type (Figure 3.20) (Glazov et al. 1972, 1973). The eutectic composition and temperature are 48 mol.% ZnTe and 420°C ± 10°C, respectively. The solubility of ZnTe in InP at the eutectic temperature is equal to 3 mol.%, and the solubility of InP in ZnTe at the same temperature reaches 25 mol.%. This system was investigated through DTA and metallography.

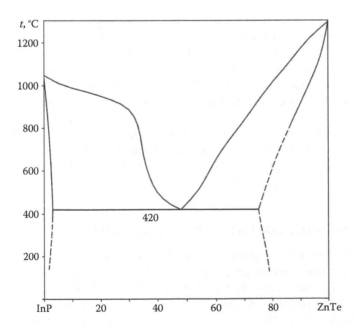

FIGURE 3.20   The ZnTe–InP phase diagram. (From Glazov, V.M. et al., *Elektron. tehnika. Ser. 6: Materialy*, (4), 127, 1972; Glazov, V.M. et al., *Izv. AN SSSR. Neorgan. materialy*, 9(11), 1883, 1973.)

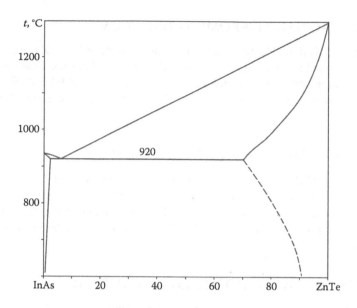

FIGURE 3.21 The ZnTe–InAs phase diagram. (From Shumilin, V.P. et al., *Izv. AN SSSR. Neorgan. materialy*, 10(8), 1414, 1974.)

3.50 ZINC–INDIUM–ARSENIC–TELLURIUM

ZnTe–InAs. The phase diagram is a eutectic type (Figure 3.21) (Shumilin et al. 1974). The eutectic composition and temperature are 94 mol.% InAs and 920°C, respectively. The solubility of ZnTe in InAs is not higher than 1 mol.% [2 mol.% (Glazov et al. 1975a)], and the solubility of InAs in ZnTe reaches 30 mol.% (Shumilin et al. 1974). The regions of the solid solutions in the Zn–Te–InAs quasiternary system based on InAs at the different temperatures are shown in Figure 3.22 (Glazov et al. 1975a).

This system was investigated through DTA, metallography, local XRD, and measuring of microhardness. The ingots were annealed at 750°C, 800°C, and 850°C for 350 h (Shumilin et al. 1974) [at 700°C, 800°C, 850°C, and 900°C for 1400, 1050, 700, and 350 h (Glazov et al. 1975a)].

3.51 ZINC–INDIUM–ANTIMONY–TELLURIUM

ZnTe–InSb. The phase diagram is a eutectic type (Figure 3.23) (Puris et al. 1970). The eutectic composition and temperature are 2.4 mol.% (2 mass.%) ZnTe and 520°C, respectively. The maximum solubility of ZnTe in InSb at 450°C is equal to 2.0 mol.% (1.6 mass.%). The solubility of InSb in ZnTe is not determined. According to the data of Kirovskaya et al. (2002), solid solution based on InSb with sphalerite structure contains up to 10 mol.% ZnTe [up to 20 mol.% ZnTe (Kirovskaya et al.) 2007b]. This system was investigated through DTA, metallography, and XRD (Puris et al. 1970, Kirovskaya et al. 2002).

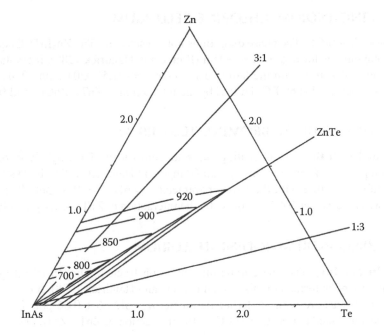

FIGURE 3.22   Isotherms of solubility in the Zn–Te–InAs quasiternary system from the InAs-rich side. (From Glazov, V.M. et al., *Izv. AN SSSR. Neorgan. materialy*, 11(7), 1181, 1975a.)

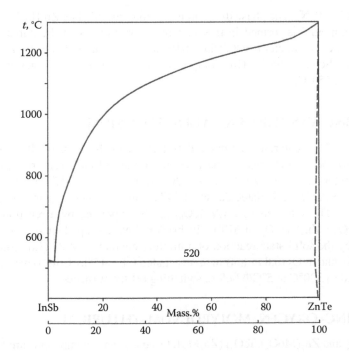

FIGURE 3.23   The ZnTe–InSb phase diagram. (From Puris, T.E. et al., *Izv. AN SSSR. Neorgan. materialy*, 6(10), 1811, 1970.)

3.52 ZINC–INDIUM–CHLORINE–TELLURIUM

ZnTe–In$_2$Te$_3$–InCl$_3$. The phase diagram is not constructed. The Zn$_2$InTe$_3$Cl quaternary compound is formed in this system (Range and Handrick 1988). It crystallizes in the cubic structure with the lattice parameter a = 611.51 ± 0.02 pm. This compound was obtained at 600°C–1100°C by the interaction of ZnTe, ZnCl$_2$, and In$_2$Te$_3$.

3.53 ZINC–INDIUM–BROMINE–TELLURIUM

ZnTe–In$_2$Te$_3$–InBr$_3$. The phase diagram is not constructed. The Zn$_2$InTe$_3$Br quaternary compound is formed in this system (Range and Handrick 1988). It crystallizes in the cubic structure with the lattice parameter a = 611.05 ± 0.04 pm. This compound was obtained at 600°C–1100°C by the interaction of ZnTe, ZnBr$_2$, and In$_2$Te$_3$.

3.54 ZINC–INDIUM–IODINE–TELLURIUM

ZnTe–In$_2$Te$_3$–InI$_3$. The phase diagram is not constructed. The Zn$_2$InTe$_3$I quaternary compound is formed in this system (Range and Handrick 1988). It crystallizes in the cubic structure with the lattice parameter a = 612.06 ± 0.06 pm. This compound was obtained at 600°C–1100°C by the interaction of ZnTe, ZnI$_2$, and In$_2$Te$_3$.

3.55 ZINC–NITROGEN–PHOSPHORUS–TELLURIUM

ZnTe–Zn$_3$N$_2$–P$_3$N$_5$. The phase diagram is not constructed. The Zn$_8$[P$_{12}$N$_{24}$]Te$_2$ quaternary compound is formed in this system. It crystallizes in the cubic structure with the lattice parameter a = 829.63 ± 0.01 pm and calculated density 4.32 g·cm^{-3} (Wester and Schnick 1996). This compound was obtained by the reaction of HPN$_2$ with ZnTe at 750°C.

3.56 ZINC–VANADIUM–OXYGEN–TELLURIUM

The Zn$_3$V$_2$TeO$_{10}$ quaternary compound is formed in this system. It is thermally stable up to about 840°C and crystallizes in the monoclinic structure with the lattice parameters a = 526.29 ± 0.05, b = 3053.4 ± 0.3, c = 550.54 ± 0.05 pm, and β = 98.653° ± 0.003°; calculated density 4.447 g·cm^{-3}; and energy gap 2.96 eV (Jiang et al. 2008). The synthesis of Zn$_3$V$_2$TeO$_{10}$ can be expressed by a reaction 3ZnO + V$_2$O$_5$ + TeO$_2$ = Zn$_3$V$_2$TeO$_{10}$ at 670°C. Its brick-red plate-shaped single crystals were prepared by the solid-state reaction of a mixture containing ZnO (1.6 mmol), V$_2$O$_5$ (0.8 mmol), and TeO$_2$ (0.8 mmol) in an evacuated tube at 700°C for 6 days and were then cooled to 300°C at 5°C/h before switching off the furnace.

3.57 ZINC–OXYGEN–MOLYBDENUM–TELLURIUM

ZnMoTeO$_6$ and Zn$_2$(MoO$_4$)(TeO$_3$) (Zn$_2$MoTeO$_7$) quaternary compounds are formed in this system. ZnMoTeO$_6$ crystallizes in the orthorhombic structure with the lattice parameters a = 526.30 ± 0.02, b = 504.26 ± 0.03, and c = 891.05 ± 0.04 pm (Doi et al. 2009)

[$a = 525.5 \pm 0.1$, $b = 504.4 \pm 0.1$, and $c = 890.9 \pm 0.2$ pm and calculated and experimental densities 5.412 and 5.30 g·cm^{-3}, respectively (Tieghi and Forzatti 1978)].

$Zn_2(MoO_4)(TeO_3)$ melts congruently at approximately 831°C and crystallizes in the monoclinic structure with the lattice parameters $a = 517.8 \pm 0.4$, $b = 840.9 \pm 0.6$, $c = 724.1 \pm 0.5$ pm, and $\beta = 99.351° \pm 0.008°$; calculated density 4.978 g·cm^{-3}; and calculated energy gap 3.1 eV (Nguyen et al. 2011).

To obtain $ZnMoTeO_6$, ZnO, TeO_2, and MoO_3 were weighted in appropriate ratios and ground intimately (Doi et al. 2009). The mixtures were pressed into pellets, and then the pellets were sealed in an evacuated silica tube to prevent the loss of reagents by the volatilization. These ampoules were heated to 550°C 12 times each for 3 h with intermediate grinding and pelletizing. This compound could be also obtained by the solid-state reaction of $ZnMoO_4$ with orthorhombic TeO_2 for 2–3 h at 500°C (Forzatti 1978, Tieghi and Forzatti 1978).

Crystals of $Zn_2(MoO_4)(TeO_3)$ were grown using hydrothermal techniques (Nguyen et al. 2011). Two millimoles of ZnO, 1 mmol of TeO_2, 1 mmol of MoO_3, and 3 mL of NH_4Cl/NH_3 and 1 M buffer solution were placed in a 23 mL Teflon-lined autoclave that was subsequently closed. The autoclave was heated to 230°C for 2 days and then cooled slowly to room temperature at a rate of 6°C/h. The product consisted of colorless rod-shaped crystals and white polycrystalline powder. Polycrystalline $Zn_2(MoO_4)(TeO_3)$ was prepared by combining 2 mmol of ZnO, 1 mmol of TeO_2, and 1 mmol of MoO_3. This mixture was ground and pressed into a pellet, which was heated to 550°C in air for 2 days and cooled to room temperature at a rate of 3°C/h. The product was then reground and the earlier process was repeated three times in order to obtain a single phase.

3.58 ZINC–OXYGEN–CHLORINE–TELLURIUM

Single crystals of dizinc tellurium dichloride trioxide, $Zn_2(TeO_3)Cl_2$, were synthesized in the Zn–O–Cl–Te quaternary system via a transport reaction in sealed, evacuated glass tubes (Johnsson and Törnroos 2003). This compound has a layered structure.

3.59 ZINC–OXYGEN–BROMINE–TELLURIUM

Single crystals of dizinc tellurium dibromide trioxide, $Zn_2(TeO_3)Br_2$, were synthesized in the Zn–O–Br–Te quaternary system via a transport reaction in sealed, evacuated silica tubes (Zhang and Johnsson 2008). It crystallizes in the orthorhombic structure with the lattice parameters $a = 1054.46 \pm 0.02$, $b = 1609.28 \pm 0.02$, and $c = 772.42 \pm 0.01$ pm and calculated density 4.725 g·cm^{-3}. The preparation of this compound was made from a nonstoichiometric mixture of $ZnO–ZnBr_2–TeO_2 = 1:5:4$, which after mixing in a mortar was put into an evacuated silica tube. The tube was heated for 120 h at 560°C.

3.60 ZINC–OXYGEN–MANGANESE–TELLURIUM

ZnTe–MnTe–O. The phase diagram is not constructed. According to the data of thermodynamic calculations, an oxidation of the $Zn_xMn_{1-x}Te$ solid solutions begins from the oxidation of MnTe (Medvedev and Berchenko 1994).

3.61 ZINC–OXYGEN–COBALT–TELLURIUM

The Zn_4CoTeO_8 quaternary compound is formed in this system. It crystallizes in the cubic structure with the lattice parameter $a = 856.0 \pm 0.1$ pm (Kasper 1967). This compound was obtained from the mixture of ZnO, basic Co carbonate, and $Te(OH)_6$ or TeO_2. This mixture was thoroughly grinded, heated up to 1040°C, and calcinated at this temperature for 0.5 h.

3.62 ZINC–OXYGEN–NICKEL–TELLURIUM

The Zn_4NiTeO_8 quaternary compound is formed in this system. It crystallizes in the cubic structure with the lattice parameter $a = 853.6 \pm 0.1$ pm (Kasper 1967). This compound was obtained from the mixture of ZnO, basic Ni carbonate, and $Te(OH)_6$ or TeO_2. This mixture was thoroughly grinded, heated up to 1040°C, and calcinated at this temperature for 0.5 h.

REFERENCES

Anishchenko V.A., Voitsehovski A.V., Pashun A.D. Some physico-chemical properties of the ingots of the GaAs–ZnTe system [in Russian], *Izv. AN SSSR. Neorgan. materialy*, **16**(2), 354–355 (1980).

Beckett D.J.S., Bernard P., Lamarch G., Woolley J.C. Growth of bulk single crystals of $Cd_xZn_yMn_zTe$, *J. Solid State Chem.*, **73**(2), 585–587 (1988).

Bochmann M., Bwembya G.C., Powell A.K., Song X. Zinc(II) arene tellurolato complexes as precursors to zinc telluride. The crystal and molecular structure of $[Zn(TeC_6H_2Me_3-2,4,6)_2(pyridine)_2]$, *Polyhedron*, **14**(23–24), 3495–3500 (1995).

Bochmann M., Bwembya G., Webb K.J., Malik M.A., Walsh J.R., O'Brien P. Arene chalcogenolato complexes of zinc and cadmium, *Inorg. Synthes.*, **31**, 19–24 (1997).

Bodnar' I.V. $(CuAlTe_2)_x(2ZnTe)_{1-x}$ solid solutions [in Russian], *Zhurn. neorgan. khimii*, **46**(4), 655–658 (2001).

Bodnar I.V., Eifler A., Doering Th., Schmitz W., Bente K., Gremenok V.F., Victorov I.V., Riede V. Growth and characterization of $(CuInTe_2)_{1-x}(2ZnTe)_x$ solid solutions single crystals, *Cryst. Res. Technol.*, **35**(10), 1135–1140 (2000a).

Bodnar I.V., Gremenok V.F., Schmitz W., Bente K., Doering Th. Preparation and investigation of $(CuInSe_2)_{1-x}(2ZnSe)_x$ and $(CuInTe_2)_{1-x}(2ZnTe)_x$ solid solutions single crystals, *Cryst. Res. Technol.*, **39**(4), 301–307 (2004).

Bodnar' I.V., Korzun B.V., Chibusova L.V. Investigation of physico-chemical properties of the $(CuInTe_2)_{1-x}(2ZnTe)_x$ solid solutions [in Russian], *Zhurn. neorgan. khimii*, **45**(9), 1538–1541 (2000b).

Brun del Re R., Donofrio T., Avon J., Majid J., Wollex J.C. Lattice parameters and optical-energy gap values for $Cd_xZn_yMn_zTe$ alloys, *Nuovo cim.*, **D2**(6), 1911–1918 (1983).

Burdiyan I.I. The solid solutions of gallium antimonide with cadmium, zinc and mercury tellurides [in Russian]. In: *Nekot. voprosy khimii i fiz. poluprovodn. slozhn. sostava*. Uzhgorod: Uzhgorod. Un-t Publish., pp. 190–195 (1970).

Burdiyan I.I. Investigation of solid solutions of GaSb with cadmium, zinc and mercury tellurides [in Russian], *Izv. AN SSSR. Neorgan. materialy*, **7**(3), 414–416 (1971).

Burdiyan I.I., Korolevski B.P. About possibility of solid solution formation in the GaSb–ZnTe system [in Russian], *Uchen. zap. Tirasp. ped. in-ta*, **16**, 127–128 (1966).

Chen D., Ravindra N.M. Electronic and optical properties of Cu_2ZnGeX_4 (X = S, Se and Te) quaternary semiconductors, *J. Alloys Compd.*, **579**, 468–472 (2013).

Doi Y., Suzuki R., Hinatsu Y., Ohoyama K. Magnetic and neutron diffraction study on quaternary oxides $MTeMoO_6$ (M = Mn and Zn), *J. Phys.: Condens. Matter*, **21**(4), 04606 (6pp.) (2009).

Evstigneeva M.A., Nalbandyan V.B., Petrenko A.A., Medvedev B.S., Kataev A.A. A new family of fast sodium ion conductors: $Na_2M_2TeO_6$ (M = Ni, Co, Zn, Mg), *Chem. Mater.*, **23**(5), 1174–1181 (2011).

Forzatti P. $ZnTeMoO_6$: A new metal tellurium molybdate, *Gazz. Chim. Ital.*, **108**(1–2), 73–75 (1978).

Frasunyak V.M. Electrical and optical properties of the $Hg_{1-x-y}Zn_xMn_yTe$ crystal [in Russian], *Neorgan. materialy*, **36**(5), 631–634 (2000).

Gallardo P.G. Order-disorder phase transitions in $D_{2x}^{II}(A^IB^{III})_{1-x}C_2^{VI}$ alloys systems, *Phys. Status Solidi (a)*, **134**(1), 119–125 (1992).

Garbato L., Ledda F. Phase diagram of the $(CuInTe_2)_{1-x}–(2ZnTe)_x$ system and crystal growth of $CuZn_2InTe_4$ from zinc chloride flux, *J. Solid State Chem.*, **30**(2), 189–195 (1979).

Garbato L., Ledda F., Manca P. Flux growth and characterization of multinary phases in the $CuGaTe_2–ZnTe$ system, *Jap. J. Appl. Phys.*, **19**(Suppl.) 19–3, 67–70 (1980).

Garbato L., Manca P. Synthesis and characterization of some chalcogenides of $A^IB_2^{II}C^{III}D_4^{VI}$, *Mater. Res. Bull.*, **9**(4), 511–517 (1974).

Glazov V.M., Krestovnikov A.N., Nagiev V.A., Rzaev F.R. Investigation of phase equilibria in the InP–ZnTe and InP–CdTe quasibinary system [in Russian], *Elektron. tehnika. Ser. 6: Materialy*, (4), 127–129 (1972).

Glazov V.M., Krestovnikov A.N., Nagiev V.A., Rzaev F.R Phase equilibria in the InP–ZnTe and InP–CdTe quasibinary system [in Russian], *Izv. AN SSSR. Neorgan. materialy*, **9**(11), 1883–1889 (1973).

Glazov V.M., Nagiev V.A., Glagoleva N.N. Separated and combined solubility of Zn, Cd and Te in InAs [in Russian], *Izv. AN SSSR. Neorgan. materialy*, **11**(7), 1181–1183 (1975a).

Glazov V.M., Pavlova L.M., Griazeva N.L. Investigation of phase equilibria and analysis of intermolecular interaction in the GaSb–Zn(Cd)Te quasibinary systems [in Russian]. In: *Termodinamicheskiye svoistva metallicheskih splavov*, Baku: Elm, pp. 368–371 (1975b).

Glazov V.M., Pavlova L.M., Griazeva N.L. Phase equilibria and analysis of intermolecular interaction in the GaSb–Zn(Cd)Te quasibinary systems [in Russian], *Izv. AN SSSR. Neorgan. materialy*, **11**(3), 418–423 (1975c).

Glazov V.M., Pavlova L.M., Lebedeva L.V. Thermodynamic analysis of interaction of gallium arsenide with zinc and cadmium tellurides [in Russian]. In: *Termodinamicheskiye svoistva metallicheskih splavov*, Baku: Elm, pp. 372–375 (1975d).

Goriunova N.A., Golovey M.I., Olekseyuk I.D., Rigan M.Yu., Shved V.V. Investigation of the $(Cd_3P_2)_{1-x}–(CdTe)_{2x}$ and $(Zn_3P_2)_{1-x}–(CdTe)_{2x}$ systems [in Russian], *Izv. AN SSSR. Neorgan. materialy*, **6**(7), 1272–1275 (1970).

Haeuseler H., Ohrendori F., Himmrich M. Zur Kenntnis quaternärer Telluride $Cu_2MM'Te_4$ mit Tetrahederstrukturen, *Z. Naturforsch.*, **B46**(8), 1049–1052 (1991).

Hahn H., Schulze H. Über quaternäre Chalkogenide des Germanium und Zinns, *Naturwissenschaften*, **52**(14), 426 (1965).

Heulings H.R. IV, Li J., Proserpio D.M. A quaternary $ThCr_2Si_2$ type structure: Crystal growth of $KCuZnTe_2$ from molten salt, *Main Group Met. Chem.*, **21**(4), 225–229 (1998).

Jiang H.-L., Huang S.-P., Fan Y., Mao J.-G., Cheng W.-D. Explorations of new types of second-order nonlinear optical materials in $Cd(Zn)-V^V-Te^{IV}-O$ systems, *Chem. Eur. J.*, **14**(6), 1972–1981 (2008).

Johnsson M., Törnroos K.W. A synthetic zinc tellurium oxochloride, $Zn_2(TeO_3)Cl_2$, *Acta Crystallogr. C*, **59**(6), i53–i54 (2003).

Kasper H. Die Lichtabsorption von Ni^{2+} und Co^{2+} in 2,6-Spinellen, *Z. anorg. und allg. Chem.*, **354**(1–2), 78–84 (1967).

Kasper H.M. Preparation of $\{A_3^{2+}\}[Te_2](B_3^{2+})$, *Mater. Res. Bull.*, **3**(9), 765–766 (1968).

Kirovskaya I.A., Azarova O.P., Shubenkova E.G., Dubina O.N. Synthesis and optical absorption of the solid solutions of the InSb–(II-VI) systems [in Russian], *Neorgan. materialy*, **38**(2), 135–138 (2002).

Kirovskaya I.A., Novgorodtseva L.V., Vasina M.V., Baranovskaya M.V., Chkalova A.L., Kuznetsova I.Yu. New materials based on the GaSb–$A^{II}B^{VI}$ systems [in Russian], *Sovrem. naukoemk. tekhnol.*, (6), 96–97 (2007a).

Kirovskaya I.A., Shubenkova E.G., Mironova E.V., Rud'ko T.A., Bykova E.I. Obtaining and properties of the solid solutions of the InSb–$A^{II}B^{VI}$ system [in Russian], *Sovremen. naukoemk. tekhnol.*, (8), 31–33 (2007b).

Kradinova L.V., Vaipolin A.A., Goriunova N.A. Solid solutions in the $A_3^{II}B_2^{V}$–$A^{II}b^{VI}$ systems [in Russian]. In: *Khim. sviaz' v kristallah*. Minsk: Nauka i tehnika Publish., pp. 417–422 (1969).

Kumar V., Bhardwaj N., Tomar N., Thakral V., Uma S. Novel lithium-containing honeycomb structures, *Inorg. Chem.*, **51**(20), 10471–10473 (2012).

Lakiza S.N. Investigation of the interaction and the Zn_3As_2 + 3CdTe ⇔ Cd_3As_2 + 3ZnTe ternary mutual system [in Russian]. In: *Razrab. i issled. novykh materialov i kompozitsiy na ikh osnove*, Kiev Nauk. Dumka Publish., pp. 46–50 (1980).

Lakiza S.N., Olekseyuk I.D. The Zn_3As_2–2CdTe system [in Russian], *Zhurn. neorgan. khimii*, **23**(8), 2190–2194 (1978).

Lakiza S.N., Olekseyuk I.D. Phase diagram of the Zn_3As_2 + 3CdTe ⇔ Cd_3As_2 + 3ZnTe ternary mutual system [in Russian], *Zhurn. neorgan. khimii*, **26**(4), 1118–1124 (1981).

Lee D.W., Bak D.-B., Kim S.B., Kim J., Ok K.M. Effect of the framework flexibility on the centricities in centrosymmetric $In_2Zn(SeO_3)_4$ and noncentrosymmetric $Ga_2Zn(TeO_3)_4$, *Inorg. Chem.*, **51**(14), 7844–7850 (2012).

Leonov V.V., Chunarev E.N. Distribution of potassium and zinc chlorides at the crystallization of eutectics of cadmium chalcogenides with its chloride [in Russian]. In: *Physico-khim. processy v geterogennyh systemah*. Krasnoyarsk: Krasnoyar. Un-t Publish., pp. 59–64 (1977).

Manca P., Garbato L. Phase relationships, crystal growth and stoichiometry defects in $A^I C^{III} D_2^{YI}$/$B^{II}D^{VI}$ heterojunction-forming system, *Sol. Cells*, **16**, 101–121 (1986).

Matsuhita H., Ichikawa T., Katsui A. Structural, thermodynamic and optical properties of Cu_2–II–IV–VI_4 quaternary compounds, *J. Mater. Sci.*, **40**(8), 2003–2005 (2005).

Medvedev Yu.V., Berchenko N.N. Analysis of the phase compositions at the interface of $Mn_{1-x}A_x^{II}B^{VI}(A^{II}$ = Zn, Cd, Hg; B^{II} = Te, Se) solid solution with own oxides [in Russian], *Zhurn. neorgan. khimii*, **39**(5), 846–848 (1994).

Mitchell K., Huang F.Q., Caspi E.N., McFarland A.D., Haynes Ch.L., Somers R.C., Jorgensen J.D., Van Duyne R.P., Ibers J.A. Synthesis, structure and selected physical properties of $CsLnMnSe_3$ (Ln = Sm, Gd, Tb, Dy, Ho, Er, Tm, Yb, Y) and $AYbZnQ_3$ (A = Rb, Cs; Q = S, Se, Te), *Inorg. Chem.*, **43**(3), 1082–1089 (2004).

Mitzi D.B. Polymorphic one-dimensional $(N_2H_4)_2ZnTe$: Soluble precursors for the formation of hexagonal or cubic zinc telluride, *Inorg. Chem.*, **44**(20), 7078–7086 (2005).

Morris C.D., Li H., Jin H., Malliakas C.D., Peters J.A., Trikalitis P.N., Freeman A.J., Wessels B.W., Kanatzidis M.G. $Cs_2M^{II}M_3^{IV}Q_8$ (Q = S, Se, Te): An extensive family of layered semiconductors with diverse band gaps, *Chem. Mater.*, **25**(16), 3344–3356 (2013).

Nalbandyan V.B., Avdeev M., Evstigneeva M.A. Crystal structure of Li_4ZnTeO_6 and revision of $Li_3Cu_2SbO_6$, *J. Solid State Chem.*, **199**, 62–65 (2013).

Nguyen S.D., Kim S.-H., Halasyamani P.S. Synthesis, characterization, and structure—property relationships in two new polar oxides: $Zn_2(MoO_4)(SeO_3)$ and $Zn_2(MoO_4)$ (TeO_3), *Inorg. Chem.*, **50**(11), 5215–5222 (2011).

Nishio M., Ogava H. Chemical vapor transport in the ZnTe–HCl closed-tube system and its thermodynamic analysis, *J. Cryst. Growth*, **78**(2), 218–226 (1986).

Ohtani H., Kojima K., Ishida K., Nishizawa T. Miscibility gap in II-VI semiconductor systems, *J. Alloys Compd.*, **182**(1), 103–114 (1992).

Parasyuk O.V., Olekseyuk I.D., Piskach L.V. X-ray powder diffraction refinement of $Cu_2ZnGeTe_4$ structure and phase diagram of the Cu_2GeTe_3–ZnTe system, *J. Alloys Compd.*, **397**(1–2), 169–172 (2005).

Parthé E., Yvon K., Deitch R.H. The crystal structure of Cu_2CdGeS_4 and other quaternary normal tetrahedral structure compounds, *Acta Crystallogr. B*, **25**(6), 1164–1174 (1969).

Pentin I.V., Grosheva A.A., Kozhemyakina N.V. The miscibility gap in cadmium, mercury and zinc telluride systems: Theoretical description, *CALPHAD: Comput. Coupling Phase Diagr. Thermochem.*, **30**(2), 191–195 (2006).

Puris T.E., Belaya A.D., Zemskov V.S., Shvarts N.N. Phase equilibria in the In–Sb–Zn–Te system [in Russian], *Izv. AN SSSR. Neorgan. materialy*, **6**(10), 1811–1815 (1970).

Radautsan S.I., Maksimova O.G. Phase interaction in the Te–ZnTe–CdTe and Ga–ZnTe–CdTe systems and growth of the $Zn_xCd_{1-x}Te$ crystals from solution in melt [in Russian]. In: *Poluprovodn. materialy i ih primenienie*, Kishinev, Republic of Moldova, Shtiintsa Publish., pp. 3–12 (1976).

Range K.-J., Handrick K. Synthese und Hochdruckverhalten quaternärer Chalkogenidhalogenide $M_2M'X_3Y$ (M = Zn, Cd; M' = Al, Ga, In; X = Se, Te; Y = Cl, Br, I), *Z. Naturforsch.*, **B43**(2), 153–158 (1988).

Roussak L., Wagner G., Schorr S., Bente K. Phase relationships in the pseudo-binary 2(ZnTe)–$CuInTe_2$ system, *J. Solid State Chem.*, **178**(11), 3476–3484 (2005).

Schorr S., Tovar M., Sheptyakov D., Keller L., Geandier G. Crystal structure and cation distribution in the solid solution series 2(ZnX)–$CuInX_2$(X = S, Se, Te), *J. Phys. Chem. Solids*, **66**(11), 1961–1965 (2005).

Serginov M., Goriunova N.A., Kradinova L.V., Vaipolin A.A., Prochuhan V.D. Solid solutions in the Zn–Cd–As–Te system [in Russian], *Tr. Kishin. polytehn. in-ta*, (12), 15–19 (1968).

Serginov M., Mamaev S., Prochuhan V.D. Investigation of some properties of Zn_3As_2–(2CdTe) and Cd_3As_2–(2ZnTe) solid solutions [in Russian], *Izv. AN TurkmSSR. Ser. fiz.-tehn., khim. i geol. nauk*, (4), 97–99 (1970).

Shumilin V.P., Uglichina G.N., Ufimtsev V.B., Gimel'farb F.A. Phase equilibria in the InAs–ZnTe system [in Russian], *Izv. AN SSSR. Neorgan. materialy*, **10**(8), 1414–1417 (1974).

Tieghi G., Forzatti P. Crystal data for $MnTeMoO_6$, $CoTeMoO_6$ and $ZnTeMoO_6$, *J. Appl. Cryst.*, **11**(4), 291–292 (1978).

Tovar R., Quintero M., Grima P., Woolley J.C. Phase diagram, lattice parameter, and optical energy gap values for the $Zn_{2x}(AgIn)_{1-x}Te_2$, *Phys. Status Solidi (a)*, **111**(2), 405–410 (1989).

Trishchuk L.I. Growth of the zinc and cadmium tellurides single crystals by the crystallization from the solution-melt [in Ukrainian], *Optoelektronika i poluprovodn. tekhnika*, (44), 99–106 (2009).

Trishchuk L.I., Oliynyk G.S., Tomashyk V.M. Phase diagrams of the $Cu_{1.95}(Ag_2)Te$–ZnTe(CdTe) quasibinary systems and liquidus surfaces of the $Cu_{1.95}(Ag_2)Te$–ZnTe–CdTe quasiternary systems, *Chem. Met. Alloys*, **1**(1), 58–61 (2008).

Ufimtseva E.V., Vigdorovich V.N., Pelevin O.V. Phase equilibria in the GaAs–ZnTe system [in Russian], *Izv. AN SSSR. Neorgan. materialy*, **9**(4), 587–591 (1973).

Voronin G.F., Pentin I.V. Decomposition of the solid solutions based on cadmium, mercury and zinc tellurides [in Russian], *Zhurn. fiz. khimii*, **79**(10), 1771–1778 (2005).

Wester F., Schnick W. Nitrido-sodalithe: III. Synthese, Struktur und Eigenschaften von $Zn_8[P_{12}N_{14}]X_2$ mit X = O, S, Se, Te, *Z. anorg. und allg. Chem.*, **622**(8), 1281–1286 (1996).

Wulff L., Müller-Buschbaum Hk. Zur Kristallchemie der Kupfer(II)-Zinktellurate $Cu_5Zn_4Te_3O_{18}$ und $Cu_{1.5}Zn_{1.5}TeO_6$, mit einer Notiz über $Cu_{1.5}Co_{1.5}TeO_6$, *Z. Naturforsch.*, **B53**(1), 53–57 (1998).

Yao J., Deng B., Sherry L.J., McFarland A.D., Ellis D.E., Van Duyne R.P. Syntheses, structure, some band gaps and electronic structures of $CsLnZnTe_3$ (Ln = La, Pr, Nd, Sm, Gd, Tb, Dy, Ho, Er, Tm, Y), *Inorg. Chem.*, **43**(24), 7735–7740 (2004).

Yu T.C., Brebrick R.F. The Hg–Cd–Zn–Te phase diagram, *J. Phase Equilib.*, **13**(5), 476–496 (1992).

Yu T.C., Brebrick R.F. Supplement: The Hg–Cd–Zn–Te phase diagram, *J. Phase Equilib.*, **14**(3), 271–272 (1993).

Zhang D., Johnsson M. $Zn_2(TeO_3)Br_2$, *Acta Crystallogr. E*, **64**(5), i26 (2008).

4 Systems Based on CdS

4.1 CADMIUM–HYDROGEN–CARBON–SULFUR

Some compounds are formed in this quaternary system.

CdMe$_2$S or C$_2$H$_6$SCd, (methylcadmium methyl sulfide). To obtain this compound, methanethiol (124 mL) was condensed on CdMe$_2$ (0.87 g) and toluene (20 mL) at −196°C (Coates and Lauder 1966). As the mixture warms up to ca. −78°C, CH$_4$ evolution was seen, and CdMe$_2$S was precipitated. The sulfide was washed with hexane and with toluene. It turned brown above 150°C and was hydrolyzed moderately rapidly by H$_2$O and vigorously by dilute H$_2$SO$_4$.

[Cd(MeS)$_2$]$_n$ or C$_2$H$_6$S$_2$Cd, (methanthiolato)cadmium. This compound could be prepared by a reaction of CdCl$_2$ with NaMeS: CdCl$_2$ (27 mmol) was dissolved in H$_2$O (100 mL) (Osakada and Yamamoto 1987, 1991). A small amount of undissolved solid was removed by filtration. Aqueous NaMeS (15%) (27 mL, ca. 4.1 g NaMeS) was added dropwise to the solution at room temperature to cause immediate precipitation of a white solid. The solid was separated from the solution by centrifugation followed by careful decantation of the solution from the mixture. Addition of H$_2$O (100 mL) to the solid with stirring followed by the centrifugation and decantation procedure was repeated until no Cl$^-$ was detected in the washings. Drying the resulting white solid under vacuum gave C$_2$H$_6$S$_2$Cd as a white powder. Thermolysis of C$_2$H$_6$S$_2$Cd gives CdS as a mixture of α- and β-form (Osakada and Yamamoto 1987, 1991).

(MeCdPhS)$_x$ or C$_7$H$_8$SCd, (methylcadmium phenyl sulfide). To obtain this compound, benzenethiol was condensed on CdMe$_2$ in hexane at −78°C (Coates and Lauder 1966). After precipitation, it was washed with benzene. It became yellow when heated above 60°C.

Cd(Bu^{n}S)$_2$ or C$_8$H$_{18}$S$_2$Cd, bis(n-butylthiolato)cadmium. This compound could be prepared electrochemically using a simple cell: Pt(−)/CH$_3$CN + n-C$_4$H$_9$SH/Cd(+) (Said and Tuck 1982). The anode was a piece of Cd, approximately 2 cm^2 area, suspended on a Pt wire. When the electrolysis began, a white solid formed at the anode almost immediately. The resulted solid was collected, washed several times with acetonitrile and petroleum ether, and dried. C$_8$H$_{18}$S$_2$Cd is air stable.

Cd(Bu^{t}S)$_2$ or C$_8$H$_{18}$S$_2$Cd, bis(t-butylthiolato) cadmium. This compound could be prepared by the same procedure as for bis(n-butylthiolato)cadmium using t-C$_4$H$_9$SH instead of n-C$_4$H$_9$SH (Said and Tuck 1982). This compound is air stable.

Cd(Bu^{n}S)$_2$·CS$_2$ or C$_9$H$_{18}$S$_4$Cd. Cd(Bu^{n}S)$_2$ is readily soluble in CS$_2$ at room temperature, and a clear yellow solution resulted when 1.03 mmol of the solid was stirred with 30 mL of CS$_2$ at room temperature (Black et al. 1986). Slow removal of the solvent yielded Cd(Bu^{n}S)$_2$·CS$_2$ as a yellow solid.

Cd(PhS)$_2$ or C$_{12}$H$_{10}$S$_2$Cd, bis(phenylthiolato)cadmium. This compound melts at 339°C–344°C, crystallizes in the orthorhombic structure with the lattice parameters a = 1549.0 ± 0.2, b = 1562.6 ± 0.2, and c = 2080.3 ± 0.3 pm, calculation and experimental density 1.77 and 1.74 g·cm^{-3}, respectively (Craig et al. 1986a,b, Dance et al. 1987a), and could be prepared by the same procedure as for bis(n-butylthiolato)cadmium using C$_6$H$_5$SH instead of n-C$_4$H$_9$SH (Said and Tuck 1982). It was also obtained by standard methods from benzenethiol, triethylamine, and Cd(NO$_3$)$_2$ in ethanol and crystallized from dimethylformamide by addition of alcohols, acetone, or acetonitrile (Craig et al. 1986a,b, Dance et al. 1987a). The same crystalline form of Cd(PhS)$_2$ was obtained by other standard method, including reaction of CdCO$_3$ or Cd(CH$_3$COO)$_2$ with benzenethiol in MeOH at high and low temperatures and reaction of Cd(NO$_3$)$_2$ with benzenethiol plus triethylamine in MeOH (Dance et al. 1987a). Its crystals were grown by layering EtOH over a solution of solid compound in dimethylformamide. Acetone or acetonitrile can also be used as precipitant. This compound is air stable.

Cd(PhS)$_2$·CS$_2$ or C$_{13}$H$_{10}$S$_4$Cd. Cd(PhS)$_2$ is insoluble in CS$_2$ at room temperature, and no evidence of reaction after stirring for 10 h was found, but when the suspension (1.03 mmol) was boiled for 8–10 h, solid Cd(PhS)$_2$·CS$_2$ was recovered (Black et al. 1986). This compound could be also obtained electrochemically. Electrochemical oxidation of a Cd anode in a solution phase of acetonitrile, 5 mL of PhSH (48.6 mmol), 20 mL of CS$_2$ (0.43 mmol), and 20 mg of tetraethylammonium perchlorate for 5 h at 20 V and a current of 20 mA led to the dissolution of 220 mg of Cd. The reaction mixture was filtered to remove any precipitated particles of metal, and the filtrate was slowly evaporated. The yellow crystals, which deposited on cooling, were washed several times with acetonitrile and then petroleum ether and dried in vacuo.

Cd(SC$_6$H$_4$Me-4)$_2$ or C$_{14}$H$_{14}$S$_2$Cd, bis(methylphenylthiolato)cadmium. This compound melts at 285°C with decomposition; crystallizes in the monoclinic structure with the lattice parameters a = 2719.6 ± 0.9, b = 1572.2 ± 0.5, c = 3704.6 ± 1.3 pm, and β = 132.03° ± 0.01° and calculation and experimental densities 1.62 and 1.61 g·cm^{-3}, respectively; and could be prepared by the same procedure as for Cd(PhS)$_2$ using methylbenzenethiol instead of benzenethiol (Dance et al. 1987a,b). Its crystals were grown from a solution in dimethylformamide.

Cd(SC$_6$H$_2$Me$_3$-2,4,6)$_2$ or C$_{18}$H$_{22}$S$_2$Cd. This compound was obtained by a pyrolysis of Cd[N(SiMe$_3$)$_2$]$_2$ with 2,4,6-Me$_3$C$_6$H$_2$SH as polymeric solids, which dissolved only in coordinating solvents (Bochmann et al. 1991). It melts at the temperature higher than 295°C (Bochmann and Webb 1991). The experiment was carried out under Ar atmosphere using standard vacuum-line techniques. Crystals suitable for x-ray diffraction (DTA) could not be grown (Bochmann et al. 1991).

(MeCdSBu^t)$_4$ or C$_{20}$H$_{48}$S$_4$Cd$_4$, (methylcadmium t-butyl sulfide). To obtain this compound, t-butylthiol in hexane was slowly added to CdMe$_2$ in hexane at ca. −78°C (Coates and Lauder 1966). Methane is rapidly evolved at this temperature and (MeCdSBu^t)$_4$ crystallized. On competition of gas evolution, more hexane was added and the solution heated until all solid had dissolved. On cooling, the product crystallized as colorless prism. It turned brown when heated above 100°C and was hydrolyzed moderately rapidly by cold H$_2$O and vigorously by 2N H$_2$SO$_4$.

(MeCdSPr^i)$_6$ or C$_{24}$H$_{60}$S$_6$Cd$_6$, (methylcadmium isopropyl sulfide). To obtain this compound, propane-2-thiol (2.15 mL, 1 M) in hexane (20 mL) was slowly added to CdMe$_2$ (3.25 g) in hexane (20 mL) at *ca.* −78°C (Coates and Lauder 1966). Methane is rapidly evolved at this temperature and (MeCdSPr^i)$_6$ crystallized. On competition of gas evolution, more hexane was added and the solution heated until all solid had dissolved. On cooling, the product crystallized as colorless prism. When heated in a sealed tube, it turned increasingly brown from 30°C to 300°C and did not melt. It was very slowly hydrolyzed by cold H$_2$O and only slowly with 2 N H$_2$SO$_4$.

Cd(SC$_6$H$_2$Pr$^i_3$-2,4,6)$_2$ or C$_{30}$H$_{46}$S$_2$Cd. The reaction of Cd[N(SiMe$_3$)$_2$]$_2$ with 2,4,6-Pr$^i_3$C$_6$H$_2$SH gives C$_{30}$H$_{46}$S$_2$Cd as a fine crystalline white solid, which is poorly soluble in light petroleum but dissolves in warm toluene (Bochmann et al. 1991). This compound sublimes at 200°C (1.3 Pa), and the melt decomposes at 300°C to give polycrystalline CdS (Bochmann et al. 1991, Bochmann and Webb 1991). The experiment was carried out under Ar atmosphere using standard vacuum-line techniques (Bochmann et al. 1991).

Cd(SC$_6$H$_2$Bu$^t_3$-2,4,6)$_2$ or C$_{36}$H$_{58}$S$_2$Cd. This compound melts at 320°C, sublimes at 300°C (0.13 Pa), and deposits CdS film from the vapor phase at 480°C (Bochmann et al. 1990a,b, 1997, Bochmann and Webb 1991). It crystallizes in the monoclinic structure with the lattice parameters $a = 1006.8 \pm 0.2$, $b = 1999.1 \pm 0.2$, $c = 1787.4 \pm 0.2$ pm, and $\beta = 91.76° \pm 0.01°$ and calculated density 1.23 g·cm^{-3} (Bochmann et al. 1990a,b).

To obtain this compound, into a 100 mL, three-necked flask, equipped with a magnetic stirring bar and connected to the inert gas supply of a vacuum line via a stopcock adaptor, are injected tri-*t*-butylbenzenethiol (3.39 mmol) and light petroleum (20 mL) (Bochmann et al. 1990a,b, Bochmann and Webb 1991, 1997). Cd[N(SiMe$_3$)$_2$]$_2$ (2.5 mmol) is added to the stirred solution via syringe. After a few seconds, a white solid precipitates. The reaction is warmed on a warm water bath (*ca.* 40°C) for 1 h. The supernatant liquid is filtered off and the residue recrystallized from toluene at 10°C to give colorless crystals. The second fraction is obtained from the mother liquor.

Cd$_{10}$S$_{16}$Ph$_{12}$ or C$_{72}$H$_{60}$S$_{16}$Cd$_{10}$. Bulk samples of this compound were prepared by heating (Me$_4$N)$_2$[S$_4$Cd$_{10}$(SPh)$_{16}$] in quartz tubes attached to a high vacuum line (Farneth et al. 1992). Samples were heated at 10°C/min to 300°C and held there for 15 min, all in dynamic vacuum. The originally white/colorless crystals become pale yellow.

4.2 CADMIUM–LITHIUM–SODIUM–SULFUR

CdS–Li$_2$S–Na$_2$S. The phase diagram is not constructed. The LiNaCdS$_2$ quaternary compound is formed in this system, which crystallizes in the trigonal structure with the lattice parameters $a = 413.20 \pm 0.03$ and $c = 686.66 \pm 0.11$ pm, energy gap 2.37 eV, and calculated density 3.377 g·cm^{-3} (Deng et al. 2007). LiNaCdS$_2$ has been synthesized by the reaction of Cd and Li$_2$S/S/Na$_2$S flux at 500°C.

4.3   CADMIUM–LITHIUM–GERMANIUM–SULFUR

CdS–Li$_2$S–GeS$_2$. The phase diagram is not constructed. The Li$_2$CdGeS$_4$ quaternary compound is formed in this system, which crystallizes in the orthorhombic structure with the lattice parameters a = 773.74 ± 0.01, b = 684.98 ± 0.01, and c = 636.88 ± 0.01 pm and energy gap 3.10 eV (Lekse et al. 2009). According to the first-principles calculation within the local density approximation (LDA) and generalized gradient approximation (GGA), the lattice parameters of this compound are a = 780.1, b = 682.7, and c = 641.6 pm and bulk modulus B_0 = 57.7 GPa and pressure derivative bulk modulus B_0' = 4.31 (LDA) and a = 750.3, b = 651.2, and c = 619.7 pm and bulk modulus B_0 = 57.4 GPa and pressure derivative bulk modulus B_0' = 4.53 (GGA) (Li et al. 2013). Li$_2$CdGeS$_4$ is not mechanically stable above 8.6 GPa.

Polycrystalline powder of this compound was synthesized by heating stoichiometric amounts of Li$_2$S, Cd, Ge, and S at 525°C for 163.5 h and then quenching in ice water. Its single crystals were grown from a lithium polysulfide flux at 650°C (Lekse et al. 2009).

4.4   CADMIUM–LITHIUM–TIN–SULFUR

CdS–Li$_2$S–SnS$_2$. The phase diagram is not constructed. The Li$_2$CdSnS$_4$ quaternary compound is formed in this system, which crystallizes in the orthorhombic structure with the lattice parameters a = 795.55 ± 0.03, b = 696.84 ± 0.03, and c = 648.86 ± 0.03 pm and energy gap 3.26 eV (Lekse et al. 2009) [a = 796.5 ± 0.2, b = 649.2 ± 0.2, and c = 696.85 ± 0.12 pm (Devi and Vidyasagar 2002)].

Li$_2$CdSnS$_4$ has been synthesized by the reaction of CdS, SnS, Li$_2$S, and S in an evacuated, sealed quartz tube (Devi and Vidyasagar 2002). The mixture was heated at 560°C for 96 h and then cooled to room temperature over a period of 96 h. This compound could not be prepared by conventional high-temperature solid-state reaction from a stoichiometric mixture of reactants. Both single crystals and polycrystalline samples of Li$_2$CdSnS$_4$ were synthesized using polychalcogenide flux at 750°C and 650°C, respectively (Lekse et al. 2009).

4.5   CADMIUM–SODIUM–TIN–SULFUR

CdS–Na$_2$S–SnS$_2$. The phase diagram is not constructed. The Na$_2$CdSnS$_4$ and Na$_6$CdSn$_4$S$_{12}$ quaternary compounds are formed in this system. Na$_2$CdSnS$_4$ crystallizes in the monoclinic structure with the lattice parameters a = 928.2 ± 0.1, b = 942.1 ± 0.3, c = 659.3 ± 0.9 pm, and β = 134.83° ± 0.09° and energy gap 1.52 eV. Na$_6$CdSn$_4$S$_{12}$ also crystallizes in the monoclinic structure with the lattice parameters a = 662.2 ± 0.4, b = 1148.9 ± 0.8, c = 699.9 ± 0.2 pm, and β = 108.56° ± 0.04° (Devi and Vidyasagar 2002).

A mixture of CdS, SnS, and Na$_2$S$_5$ heated at 780°C gives a biphasic mixture of Na$_2$CdSnS$_4$ and Na$_6$CdSn$_4$S$_{12}$ (Devi and Vidyasagar 2002). These compounds could not be prepared by conventional high-temperature solid-state reaction from a stoichiometric mixture of reactants. The attempts to prepare compound Na$_2$CdSnS$_4$ or Na$_6$CdSn$_4$S$_{12}$ as a single phase have always resulted in a mixture of both items.

4.6 CADMIUM–SODIUM–ARSENIC–SULFUR

CdS–Na$_2$S–As$_2$S$_3$. The phase diagram is not constructed. The NaCdAsS$_3$ quaternary compound is formed in this system. It crystallizes in the monoclinic structure with the lattice parameters $a = 565.61 \pm 0.08$, $b = 1654.87 \pm 0.15$, $c = 569.54 \pm 0.08$ pm, and $\beta = 90.315° \pm 0.011°$ and calculated density 3.819 g·cm^{-3} (Wu and Bensch 2012).

NaCdAsS$_3$ was prepared by a reaction mixture of Na$_2$S, Cd, As$_2$S$_3$, and additional S powder in 2:1:1:4 molar ratios. The mixture was heated in evacuated ampoule to 500°C at a rate of 0.5°C/min and kept at this temperature for 4 days before the sample was cooled down to 100°C at a rate of 3°C/h, and then the furnace was switched off to cool to room temperature (Wu and Bensch 2012). This compound is plagued with crystal twinning and acceptable crystal structure refinement could only be obtained by identifying the type of the twinning laws.

4.7 CADMIUM–SODIUM–ANTIMONY–SULFUR

CdS–Na$_2$S–Sb$_2$S$_3$. The phase diagram is not constructed. The NaCdSbS$_3$ quaternary compound is formed in this system. It crystallizes in the monoclinic structure with the lattice parameters $a = 813.29 \pm 0.06$, $b = 812.96 \pm 0.06$, $c = 1726.00 \pm 0.13$ pm, and $\beta = 103.499° \pm 0.006°$ and calculated density 4.230 g·cm^{-3} (Wu and Bensch 2012).

NaCdSbS$_3$ was prepared by a reaction mixture of Na$_2$S, Cd, Sb, and additional S powder in 1:1:1:8 molar ratios. The mixture was heated in evacuated ampoule to 500°C at a rate of 0.5°C/min and kept at this temperature for 4 days before the sample was cooled down to 100°C at a rate of 3°C/h, and then the furnace was switched off to cool to room temperature (Wu and Bensch 2012). This compound is plagued with crystal twinning and acceptable crystal structure refinement could only be obtained by identifying the type of the twinning laws.

4.8 CADMIUM–SODIUM–CHLORINE–SULFUR

CdS–NaCl. The phase diagram is not constructed. Cadmium sulfide and NaCl do not interact in the inert atmosphere (Pivneva et al. 1977). The next reaction CdS + 2NaCl = CdCl$_2$ + Na$_2$S is thermodynamically unlikely. This fact is confirmed by the chemical analysis.

4.9 CADMIUM–POTASSIUM–GOLD–SULFUR

CdS–K$_2$S–Au$_2$S. The phase diagram is not constructed. The K$_2$Au$_4$CdS$_4$ quaternary compound is formed in this system. It crystallizes in the orthorhombic structure with the lattice parameters $a = 1056.8 \pm 0.3$, $b = 697.3 \pm 0.2$, and $c = 1469.0 \pm 0.5$ pm and energy gap 2.79 eV (Axtell and Kanatzidis 1998). This compound has been synthesized by the reaction of K$_2$S, Au, Cd, and S in reactive K$_2$S$_x$ flux at 550°C.

4.10 CADMIUM–POTASSIUM–GALLIUM–SULFUR

$CdS-K_2S-Ga_2S_3$. The phase diagram is not constructed. The $KCd_4Ga_5S_{12}$ quaternary compound is formed in this system, which crystallizes in the hexagonal structure with the lattice parameters $a = 1378.23 \pm 0.30$ and $c = 932.95 \pm 0.27$ pm ($a = 854.89 \pm 0.07$ and $\alpha = 107.545° \pm 0.004°$ in rhombohedral settings) and calculated density 3.961 g·cm^{-3} (Schwer et al. 1993).

In order to produce $KCd_4Ga_5S_{12}$, $CdGa_2S_4$ was solved in KBr in molar ratio 1:9 in quartz-glass ampoule, which was sealed under vacuum and heated to 900°C. After 1 day, the furnace was switched off and the sample cooled down slowly. The reguli were treated with hot water to remove the unreacted halide (Schwer et al. 1993). The dried filter sludges had a light yellow color. Solving $CdGa_2S_4$ in KI yielded $KCd_4Ga_5S_{12}$, too.

4.11 CADMIUM–POTASSIUM–GERMANIUM–SULFUR

$CdS-K_2S-GeS_2$. The phase diagram is not constructed. The glass-forming region in this system is shown in Figure 4.1 (Imaoka and Jamadzaki 1967).

4.12 CADMIUM–POTASSIUM–CHLORINE–SULFUR

$CdS-KCl-CdCl_2$. The phase diagram is not constructed. Using oriented zone recrystallization and radioactive isotopes, there was determined equilibrium coefficient of KCl distribution in the $CdS-CdCl_2$ eutectic: $k_0 = 0.39$ (Leonov and Chunarev 1977).

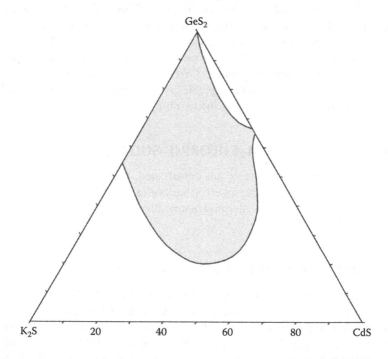

FIGURE 4.1 Glass-forming region in the $CdS-K_2S-GeS_2$ quasiternary system. (From Imaoka, M. and Jamadzaki, T., *Monthly J. Inst. Industr. Sci. Univ. Tokyo*, 19(9), 261, 1967.)

4.13 CADMIUM–RUBIDIUM–GALLIUM–SULFUR

CdS–Rb$_2$S–Ga$_2$S$_3$. The phase diagram is not constructed. The RbCd$_4$Ga$_5$S$_{12}$ quaternary compound is formed in this system, which crystallizes in the hexagonal structure with the lattice parameters a = 1382.62 ± 0.54 and c = 934.89 ± 0.23 pm (a = 856.98 ± 0.24 and α = 107.567° ± 0.003° in rhombohedral settings) and calculated density 4.085 g·cm^{-3} (Schwer et al. 1993). This compound was obtained if CdGa$_2$S$_4$ was kept in RbBr at the temperature above 150°C of its melting point for 1 day and then cooled down slowly.

4.14 CADMIUM–RUBIDIUM–GOLD–SULFUR

CdS–Rb$_2$S–Au$_2$S. The phase diagram is not constructed. The Rb$_2$Au$_2$Cd$_2$S$_4$ quaternary compound is formed in this system. It crystallizes in the orthorhombic structure with the lattice parameters a = 1393.0 ± 0.8, b = 575.7 ± 0.3, and c = 674.2 ± 0.1 pm and energy gap 2.81 eV (Axtell and Kanatzidis 1998). Rb$_2$Au$_2$Cd$_2$S$_4$ has been synthesized by the reaction of Rb$_2$S, Au, Cd, and S in reactive Rb$_2$S$_x$ flux at 550°C.

4.15 CADMIUM–RUBIDIUM–ARSENIC–SULFUR

Rb$_4$CdAs$_2$S$_9$ or Rb$_8$[Cd$_2$(AsS$_4$)$_2$(AsS$_5$)$_2$] quaternary compound is formed in the Cd–Rb–As–S quaternary system (Iyer and Kanatzidis 2004). It crystallizes in the triclinic structure with the lattice parameters a = 912.2 ± 0.2, b = 928.5 ± 0.2, and c = 1240.0 ± 0.3 pm; α = 11.700° ± 0.006°, β = 108.744° ± 0.005°, and γ = 90.163° ± 0.5°; and calculated density 3.239 g·cm^{-3} at 173 K.

This compound is the product of the reaction of Rb$_2$S, Cd, As$_2$S$_3$, and S in the molar ratio 2:1:1:10 (Iyer and Kanatzidis 2004). The reactants were loaded in fused silica tube and sealed under vacuum. The tube was then heated to 500°C in 10 h. It was isothermed at this temperature for 60 h before cooling down to 250°C at the rate of 5°C/h, followed by rapid quenching to room temperature. Rb$_4$CdAs$_2$S$_9$ is soluble in water.

4.16 CADMIUM–CESIUM–GOLD–SULFUR

CdS–Cs$_2$S–Au$_2$S. The phase diagram is not constructed. The Cs$_2$Au$_2$Cd$_2$S$_4$ quaternary compound is formed in this system. It crystallizes in the orthorhombic structure with the lattice parameters a = 658.33 ± 0.09, b = 1405.5 ± 0.3, and c = 603.69 ± 0.08 pm and energy gap 3.06 eV (Axtell and Kanatzidis 1998). This compound has been synthesized by the reaction of Cs$_2$S, Au, Cd, and S in reactive Cs$_2$S$_x$ flux at 550°C.

4.17 CADMIUM–CESIUM–GERMANIUM–SULFUR

CdS–Cs$_2$S–GeS$_2$. The phase diagram is not constructed. The Cs$_2$CdGe$_3$S$_8$ quaternary compound is formed in this system. It crystallizes in the orthorhombic structure with the lattice parameters a = 741.92 ± 0.02, b = 1250.99 ± 0.02, and

$c = 1708.43 \pm 0.06$ pm, calculated density 3.571 g·cm^{-3}, and energy gap 3.38 eV (Morris et al. 2013). This compound was obtained by the interaction of Cs_2S, Cd, Ge, and S in stoichiometric ratio using a flame-melting reaction.

4.18 CADMIUM–CESIUM–TIN–SULFUR

$CdS–Cs_2S–SnS_2$. The phase diagram is not constructed. The $Cs_{10}Cd_4Sn_4S_{17}$ quaternary compound is formed in this system. It crystallizes in the tetragonal structure with the lattice parameters $a = 1504.6 \pm 0.3$ and $c = 1053.4 \pm 0.3$ pm, calculated density 3.897 g·cm^{-3}, and energy gap 3.34 eV (Palchik et al. 2004). This compound decomposes at 714°C, and the main phase after decomposition is CdS.

$Cs_{10}Cd_4Sn_4S_{17}$ was synthesized from a mixture of 0.3 mmol of Cd, 0.3 mmol of Sn, 1.5 mmol of Cs_2S, and 3.6 mmol of S. The reagent was mixed, sealed in an evacuated silica tube, and heated at 500°C for 4 days. This was followed by cooling to room temperature at 5°C/h. The excess flux was removed with MeOH to reveal yellowish-white transparent rectangular crystals, which appear to be air stable (Palchik et al. 2004).

4.19 CADMIUM–COPPER–MERCURY–SULFUR

$CdS–CuS–HgS$. The phase diagram is not constructed. The lines that do not belong to the starting sulfides were determined on the roentgenograms of the heated ingots, obtained by the coprecipitation of CdS with CuS and HgS using hydrogen sulfide (Kislinskaya 1974). This indicates that complex processes take place at the interaction of binary components in this system.

4.20 CADMIUM–COPPER–ALUMINUM–SULFUR

$CdS–CuAlS_2$. The phase equilibria in this system up to 1000°C obtained using XRD are shown in Figure 4.2 (Robbins and Lambrecht 1973).

4.21 CADMIUM–COPPER–GALLIUM–SULFUR

$CdS–CuGaS_2$. The phase equilibria in this system up to 1000°C obtained using XRD are shown in Figure 4.3 (Robbins and Lambrecht 1973).

4.22 CADMIUM–COPPER–INDIUM–SULFUR

$CdS–CuInS_2$. The phase diagram of this system is shown in Figure 4.4. A continuous range of solid solutions is formed between γ-$CuInS_2$, which has a wurtzite structure and CdS with the same structure (Olekseyuk et al. 2000). At 600°C, the extent of the γ-solid solution range is 0–11 mol.% CdS, β-solid solution region does not exceed 7 mol.% (37–44 mol.% CdS), and the α-solid solution range of CdS exists in the interval 56–100 mol.% CdS.

This system was also partially investigated by Robbins and Lambrecht (1973) where it was found that at the temperatures up to 1000°C, the solid solutions are formed in the concentration ranges 0–15 and 50–100 mol.% CdS.

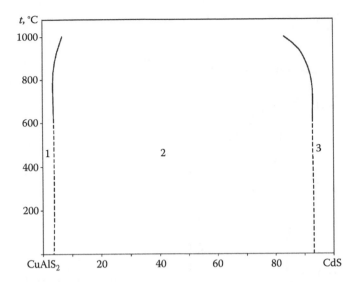

FIGURE 4.2 Phase equilibria in the CdS–CuAlS$_2$ system up to 1000°C: 1, chalcopyrite; 2, chalcopyrite + wurtzite; 3, wurtzite. (From Robbins, M. and Lambrecht, V.G., *J. Solid State Chem.*, 6(3), 402, 1973.)

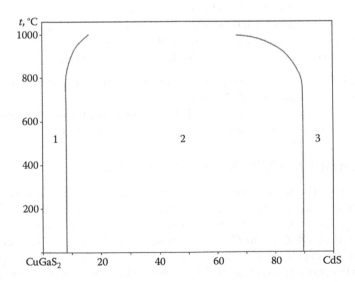

FIGURE 4.3 Phase equilibria in the CdS–CuGaS$_2$ system up to 1000°C: 1, chalcopyrite; 2, chalcopyrite + wurtzite; 3, wurtzite. (From Robbins, M. and Lambrecht, V.G., *J. Solid State Chem.*, 6(3), 402, 1973.)

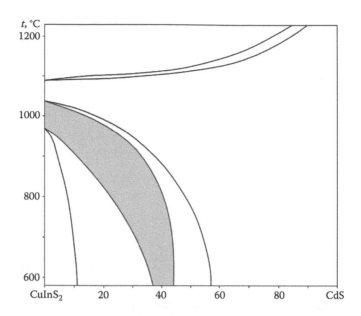

FIGURE 4.4 The CdS–CuInS$_2$ phase diagram. (From Olekseyuk, I.D. et al., *J. Alloys Compd.*, 309(1–2), 39, 2000.)

According to the data of Parthé et al. (1969), the CuCd$_2$InS$_4$ quaternary compound is formed in this system. It crystallizes in the hexagonal structure of the wurtzite type with the lattice parameters $a = 404.7 \pm 0.5$ and $c = 661.7 \pm 0.5$ pm, but this compound can be one of the solid solution compositions.

This system was investigated through differential thermal analysis (DTA), XRD, and metallography (Robbins and Lambrecht 1973, Olekseyuk et al. 2000). The ingots were annealed at 600°C for 250 h (Olekseyuk et al. 2000).

CdS–Cu$_2$S–In$_2$S$_3$. The isothermal section of this quasiternary system at 600°C is given in Figure 4.5 (Kozer et al. 2009b).

This system was investigated through DTA and XRD and the ingots were annealed at 600°C for 250 h and then cooled in cold water (Kozer et al. 2009b).

4.23 CADMIUM–COPPER–SILICON–SULFUR

CdS–Cu$_2$SiS$_3$. The phase diagram is shown in Figure 4.6 (Piskach et al. 1997, 1999). The eutectic composition and temperature are 6 mol.% CdS and 895°C, respectively. The Cu$_2$CdSiS$_4$ quaternary compound is formed in this system. It melts incongruently at 978°C [at 1016°C $\pm$ 5°C (Schäfer and Nitsche 1977); according to the data of Yao et al. (1987), it decomposes at 510°C $\pm$ 10°C] and has two polymorphous modifications (Piskach et al. 1997, 1999). The temperature of this transformation is 937°C from the CdS-rich side and 845°C from the Cu$_2$SiSe$_3$ side. This indicates on the existence of homogeneity region for the Cu$_2$CdSiS$_4$ compound (45–50 mol.% CdS at the eutectic temperature). Low-temperature modification crystallizes in the orthorhombic structure with the lattice parameters $a = 759.7 \pm 0.3$,

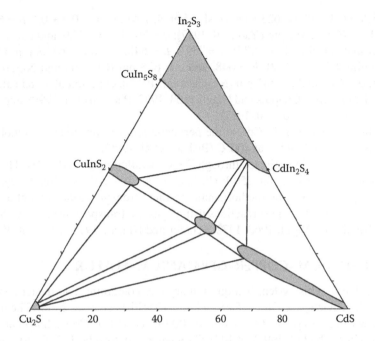

FIGURE 4.5 Isothermal section of the CdS–Cu$_2$S–In$_2$S$_3$ quasiternary system at 400°C. (From Kozer, V.R. and Parasyuk, O.V., *Chem. Met. Alloys*, 2(1–2), 102, 2009b.)

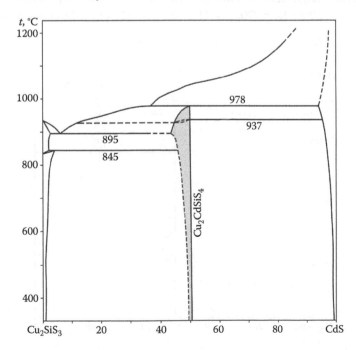

FIGURE 4.6 The CdS–Cu$_2$SiS$_3$ phase diagram. (From Piskach, L.V. et al., *Zhurn. neorgan. khimii*, 44(5), 823, 1999.)

$b = 648.5 \pm 0.3$, and $c = 623.1$ pm (Piskach et al. 1999) [$a = 760.9 \pm 0.1$, $b = 648.5 \pm 0.1$, and $c = 625.1 \pm 0.1$ pm (Yao et al. 1987); $a = 761.4$, $b = 648.9$, and $c = 625.4$ pm (Schäfer and Nitsche 1974, 1977); $a = 759.8$, $b = 648.6$, and $c = 625.8$ pm (Chapuis and Niggli 1972); $a = 760$, $b = 648$, and $c = 625$ pm (Chapuis and Niggli 1968); $a = 758$, $b = 644$, and $c = 617$ pm (Nitsche et al. 1967)], experimental and calculated density 4.27 g·cm⁻³ (Chapuis and Niggli 1968, 1972, Piskach et al. 1999), and energy gap 2.45 ± 0.05 eV (Yao et al. 1987).

Solubility of Cu_2SiS_3 in CdS at the peritectic temperature reaches 6 mol.% and decreases down to 2 mol.% at 400°C (Piskach et al. 1999).

This system was investigated through DTA, metallography, and XRD. The ingots were annealed at 400°C for 250 h (Piskach et al. 1997, 1999). Single crystals of Cu_2CdSiS_4 were grown by the horizontal gradient freezing (Matsuhita et al. 2005) or by the chemical transport reactions using I_2 as the transport agent (Nitsche et al. 1967, Chapuis and Niggli 1968, 1972, Schäfer and Nitsche 1974, Yao et al. 1987).

4.24 CADMIUM–COPPER–GERMANIUM–SULFUR

CdS–Cu₂GeS₃. This system is a quasibinary one containing two quaternary compounds: Cu_2CdGeS_4 and $Cu_2Cd_3GeS_6$ (Figure 4.7) (Piskach et al. 2000). The first one forms according to a peritectic reaction at 1009°C [at 1020°C (Matsuhita et al. 2005); at 957°C (Piskach et al. 1997); at 1021°C (Schäfer and Nitsche 1977)] and possesses a narrow homogeneity region. Cu_2CdGeS_4 crystallizes in the orthorhombic structure with the lattice parameters $a = 770.3 \pm 0.1$, $b = 655.49 \pm 0.09$, and $c = 631.2 \pm 0.1$ pm (Piskach et al. 2000) [$a = 770$, $b = 655$, and $c = 628$ pm (Matsuhita et al. 2005);

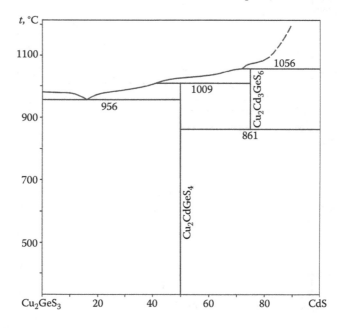

FIGURE 4.7 Phase diagram of the CdS–Cu₂GeS₃ system. (From Piskach, L.V. et al., *J. Alloys Compd.*, 299(1–2), 227, 2000.)

$a = 770.5 \pm 0.8$, $b = 655.8 \pm 0.6$, and $c = 630.8 \pm 0.3$ pm (Filonenko et al. 1991); $a = 769.0$, $b = 655.5$, and $c = 629.3$ pm (Schäfer and Nitsche 1977); $a = 769.2 \pm 0.2$, $b = 655.5 \pm 0.2$, and $c = 629.9 \pm 0.2$ pm (Parthé et al. 1969, Schäfer and Nitsche 1974); $a = 772$, $b = 657$, and $c = 629$ pm (Nitsche et al. 1967)]. A eutectic exists between Cu_2GeS_3 and Cu_2CdGeS_4 at 16 mol.% CdS and 956°C (Piskach et al. 2000) [14 mol.% CdS and 767°C (Piskach et al. 1997)]. Calculated density and energy gap of this compound are 4.6 g·cm^{-3} (Parthé et al. 1969) and 2.05 eV (Davydyuk et al. 2002, 2005) [1.9–2.0 eV (Filonenko et al. 1991)], respectively.

According to the preliminary data of Piskach et al. (1997), the CdS–Cu_2GeS_3 system was considered as nonquasibinary due to the incongruent melting of Cu_2GeS_3 [later it was shown (Piskach et al. 2000) that this ternary compound melts congruently].

The composition of the second quaternary phase was estimated from the concentration-dependent size of the endothermal effects and corresponds to 75 mol.% CdS ($Cu_2Cd_3GeS_6$). This phase is formed at 1056°C according to the peritectic reaction and is unstable and decomposes at 861°C. The solid solubility ranges of the components of this section do not exceed 2 mol.% (Piskach et al. 2000).

The investigation of this system was carried out using DTA, XRD, and metallography. All alloys were annealed at 400°C for 250 h and then they were quenched in cold water (Piskach et al. 1997, 2000). Single crystals of the quaternary compound Cu_2CdGeS_4 were grown using chemical vapor transport reactions using I_2 as the transport agent and gradient freezing method (Nitsche et al. 1967, Parthé et al. 1969, Schäfer and Nitsche 1974, Yao et al. 1987, Filonenko et al. 1991, Davydyuk et al. 2002, 2005, Matsuhita et al. 2005).

CdS–Cu$_2$S–GeS$_2$. The isothermal section of this system at 400°C is shown in Figure 4.8 (Parasyuk et al. 2005c). Two quaternary compounds Cu_2CdGeS_4 and $Cu_6Cd_{0.83}Ge_{1.17}S_{6.57}$ are formed in this system.

This system was investigated through DTA, XRD, and scanning electron microscopy (SEM) with energy-dispersive x-ray spectroscopy (EDAX) and the alloys were annealed at 400°C during 500 h.

4.25   CADMIUM–COPPER–TIN–SULFUR

CdS–Cu$_2$SnS$_3$. The phase diagram is shown in Figure 4.9 (Olekseyuk and Piskach 1997, Piskach et al. 1997). The eutectic composition and temperature are 15 mol.% CdS and 853°C, respectively. The Cu_2CdSnS_4 quaternary compound is formed in this system with peritectic composition and temperature 27 mol.% CdS and 905°C [926°C (Matsuhita et al. 2005); 920°C (Schäfer and Nitsche 1977)], respectively. This compound crystallizes in the tetragonal structure with the lattice parameters $a = 558.09 \pm 0.01$ and $c = 1082.6 \pm 0.2$ pm (Olekseyuk and Piskach 1997) [$a = 559$ and $c = 1084$ pm (Matsuhita et al. 2005); $a = 558.6$ and $c = 1083.4$ pm (Schäfer and Nitsche 1974, 1977); $a = 558$ and $c = 1082$ pm (Nitsche et al. 1967); $a = 558.2$ and $c = 10.86$ pm (Hahn and Schulze 1965)] and calculation and experimental densities 4.77 and 4.75 g·cm^{-3}, respectively (Hahn and Schulze 1965, Olekseyuk and Piskach 1997). The energy gap of this compound is equal to 1.4 eV (Davydyuk et al. 2005).

Cu_2CdSnS_4 as mineral černýite crystallized in the tetragonal structure with the lattice parameters $a = 548.7 \pm 0.2$ and $c = 1084.8 \pm 0.3$ pm (Szymański 1978)

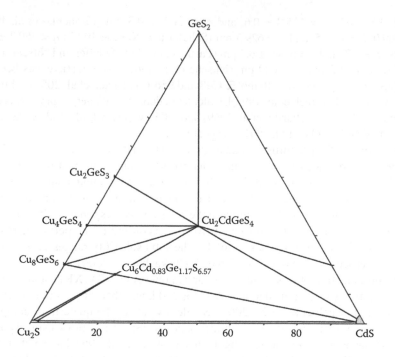

FIGURE 4.8 Isothermal section of the $CdS–Cu_2S–GeS_2$ system at 400°C. (From Parasyuk, O.V. et al., *J. Alloys Compd.*, 397(1–2), 85, 2005c.)

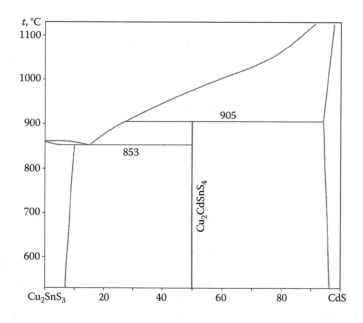

FIGURE 4.9 The $CdS–Cu_2SnS_3$ phase diagram. (From Olekseyuk, I.D. et al., *Zhurn. neorgan. khimii*, 42(2), 331, 1997.)

[a = 553.30 and c = 1082.66 pm and a = 548.71 and c = 1084.54 pm for the mineral from two localities (Kissin et al. 1978, Fleischer et al. 1983)] and calculated density 4.62 g·cm⁻³ (Szymański 1978) [4.776 and 4.618 g·cm⁻³ (Kissin et al. 1978, Fleischer et al. 1979)].

Solubility of CdS in Cu_2SnS_3 is equal 10 mol.% at the eutectic temperature and decreases to 8 mol.% at 550°C, and solubility of Cu_2SnS_3 in CdS at the peritectic temperature reaches 6 mol.% and decreases with temperature decreasing (Olekseyuk and Piskach 1997).

This system was investigated through DTA, metallography, XRD, and measuring of microhardness and density (Olekseyuk and Piskach 1997). The Cu_2CdSnS_4 compound was obtained by the horizontal gradient freezing (Matsuhita et al. 2005), or by the chemical transport reactions (Nitsche et al. 1967, Schäfer and Nitsche 1974, Davydyuk et al. 2005), or by the heating of binary compounds at 650°C–900°C (Hahn and Schulze 1965).

CdS–Cu₄SnS₄. This section is a nonquasibinary one and crosses the fields of primary crystallization of the solid solutions based on Cu_2S, Cu_2SnS_3, and CdS (Figure 4.10) (Piskach et al. 1998). In the subsolidus part, this section crosses one-phase fields of Cu_4SnS_4 and solid solution based on CdS, two-phase field (Cu_2S) + (CdS), and two ternary regions Cu_4SnS_4 + (Cu_2S) + Cu_2CdSnS_4 and (CdS) + (Cu_2S) + Cu_2CdSnS_4. The two ternary regions are separated by the secondary quasibinary system $Cu_2S–Cu_2CdSnS_4$.

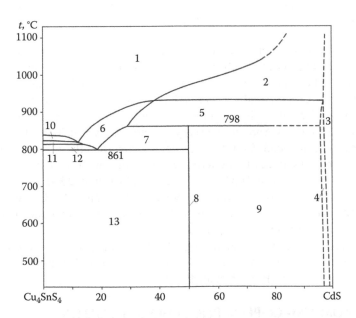

FIGURE 4.10 Phase relations in the system CdS–Cu₄SnS₄: 1, L; 2, L + (CdS); 3, (CdS); 4, (γ-Cu₂S) + (CdS); 5, L + (γ-Cu₂S) + (CdS); 6, L + (γ-Cu₂S); 7, L + (γ-Cu₂S) + Cu₂CdSnS₄; 8, (γ-Cu₂S) + Cu₂CdSnS₄; 9, (γ-Cu₂S) + Cu₂CdSnS₄ + (CdS); 10, L + (Cu₂SnS₃); 11, L + (Cu₂SnS₃) + (γ-Cu₂S); 12, L +(γ-Cu₂S) + Cu₄SnS₄; and 13, Cu₄SnS₄ + (γ-Cu₂S) + Cu₂CdSnS₄. (From Piskach, L.V. et al., *J. Alloys Compd.*, 279(2), 142, 1998.)

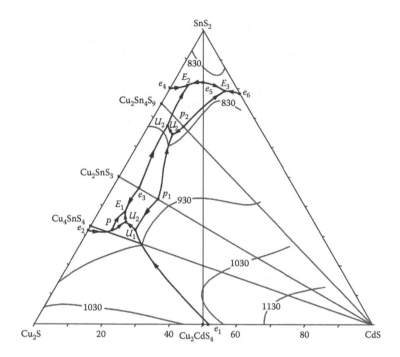

FIGURE 4.11 The liquidus surface projection of the $CdS-Cu_2S-SnS_2$ quasiternary system. (From Piskach, L.V. et al., *J. Alloys Compd.*, 279(2), 142, 1998.)

CdS–Cu₂S–SnS₂. Liquidus surface of this quasiternary system is shown in Figure 4.11 (Piskach et al. 1998). It consists of seven fields of primary crystallization of the solid solutions based on Cu_2S, CdS, SnS_2, and Cu_2SnS_3 and phases $Cu_2CdSn_3S_8$, Cu_2CdSnS_4, and Cu_4SnS_4. They are separated by 16 invariant points and 18 monovariant lines. Temperature and composition of the ternary invariant points in the $CdS-Cu_2S-SnS_2$ quasiternary system are given in Table 4.1.

Quaternary compound $Cu_2CdSn_3S_8$ melts incongruently at 818°C (Piskach et al. 1998) and crystallizes in the tetragonal structure with the lattice parameters $a = 738.27 \pm 0.04$ and $c = 1041.52 \pm 0.09$ pm (Chykhriy et al. 2000b).

The isothermal sections of this system at 500°C and 700°C are shown in Figure 4.12a and b (Piskach et al. 1998). Two quaternary compounds Cu_2CdSnS_4 and $Cu_2CdSn_3S_8$ exist at these temperatures.

This system was investigated through DTA, metallography, XRD, and SEM with EDAX, and the alloys were annealed at 500°C during 500 h with subsequent quenching in cold water (Piskach et al. 1998).

4.26 CADMIUM–COPPER–PHOSPHORUS–SULFUR

CdS–Cu₂S–P₂S₅. The phase diagram is not constructed. The $Cu_3Cd_2PS_6$ and $CuCd_3PS_6$ quaternary compounds are formed in this system (Odin et al. 2005a). $Cu_3Cd_2PS_6$ melts incongruently at 1103°C and is characterized by a polymorphous transformation. At 590°C, solid solutions within the interval of 59.0–66.5 mol.%

TABLE 4.1
Temperature and Composition of the Ternary Invariant Points in the CdS–Cu$_2$S–SnS$_2$ Quasiternary System

Invariant Points	Reaction	Composition, mol.%			t, °C
		Cu$_2$S	CdS	SnS$_2$	
E_1	L $\Leftrightarrow$ Cu$_4$SnS$_4$ + Cu$_2$CdSnS$_4$ + Cu$_2$SnS$_3$	54	8	38	770
E_2	L $\Leftrightarrow$ Cu$_2$CdSn$_3$S$_8$ + Cu$_2$SnS$_3$ + SnS$_2$	14	5	81	782
E_3	L $\Leftrightarrow$ SnS$_2$ + CdS + Cu$_2$CdSn$_3$S$_8$	4	17	79	762
P	L + Cu$_2$S + Cu$_2$SnS$_3$ $\Leftrightarrow$ Cu$_4$SnS$_4$	61	7	32	810
U_1	L + CdS $\Leftrightarrow$ Cu$_2$CdSnS$_4$ + Cu$_2$S	54	14	32	761
U_2	L + Cu$_2$S $\Leftrightarrow$ Cu$_4$SnS$_4$ + Cu$_2$CdSnS$_4$	55.5	9.5	35	798
U_3	L + CdS $\Leftrightarrow$ Cu$_2$CdSnS$_4$ + Cu$_2$CdSn$_3$S$_8$	27	9	64	805
U_4	L + Cu$_2$CdSnS$_4$ $\Leftrightarrow$ Cu$_2$CdSn$_3$S$_8$ + Cu$_2$SnS$_3$	27	6	67	795

Source: Piskach, L.V. et al., *J. Alloys Compd.*, 279(2), 142, 1998b.

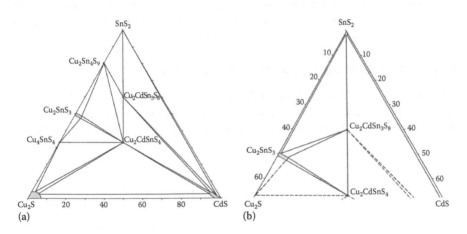

FIGURE 4.12 Isothermal sections of the CdS–Cu$_2$S–SnS$_2$ system at (a) 500°C and (b) 700°C. (From Piskach, L.V. et al., *J. Alloys Compd.*, 279(2), 142, 1998.)

Cu$_7$PS$_6$ along the section Cd$_7$P$_2$S$_{12}$–Cu$_7$PS$_6$ are formed on the base of the high-temperature modification of this compound. CuCd$_3$PS$_6$ is formed at 586°C according to the solid-state reaction and crystallizes in the monoclinic structure with the lattice parameters a = 1229 ± 1, b = 703.5 ± 0.2, c = 1228 ± 1 pm, and β = 110.45° ± 0.08° and calculation and experimental density 3.92 and 3.91 ± 0.01 g·cm^{-3}.

4.27 CADMIUM–COPPER–FLUORINE–SULFUR

CdS–CuF$_2$–CdF$_2$. The phase diagram is not constructed. The Cu$_2$Cd$_2$F$_6$S quaternary compound is formed in this system (Pannetier et al. 1972). It crystallizes in the rhombohedral structure with the lattice parameters a = 1071.3 pm and α = 90°39′ or in the hexagonal structure with the lattice parameters $a = 1523.8 \pm 0.7$ and $c = 1834 \pm 0.7$ pm. This compound was obtained by the annealing of mixtures from binary compounds at 700°C for 4 days or by their quick heating up to 1000°C during 0.5 h.

4.28 CADMIUM–COPPER–IODINE–SULFUR

CdS–CuI–CdI$_2$. The phase diagram is not constructed. The $Cu_2Cd_3S_2I_4$ quaternary compound is formed in this system (Blachnik and Dreisbach 1986). It crystallizes in the hexagonal structure with the lattice parameters $a = 424.7 \pm 0.1$ and $c = 695.2 \pm 1$ pm. This compound was obtained by the substitution of Hg by Cd in the CuHgSI quaternary compound.

4.29 CADMIUM–SILVER–ALUMINUM–SULFUR

CdS–AgAlS$_2$. The phase equilibria in this system up to 1000°C, obtained using XRD, are shown in Figure 4.13 (Robbins and Lambrecht 1973).

According to the data of Parthé et al. (1969), the $AgCd_2AlS_4$ quaternary compound is formed in this system. It crystallizes in the hexagonal structure with the lattice parameters $a = 413.4 \pm 0.5$ and $c = 672.3 \pm 0.5$ pm, but this compound can be one of the solid solution compositions. This compound was obtained by the heating of the mixture from chemical elements at 600°C–800°C.

4.30 CADMIUM–SILVER–GALLIUM–SULFUR

CdS–AgGaS$_2$. The phase diagram of this system is shown in Figure 4.14 (Chykhriy et al. 2000a). The coordinates of the eutectic point are 902°C and 37 mol.%. $AgCd_2GaS_4$ quaternary compound that melts incongruently at 1011°C is formed in

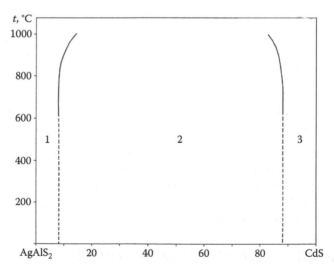

FIGURE 4.13 Phase equilibria in the CdS–AgAlS$_2$ system up to 1000°C: 1, chalcopyrite; 2, chalcopyrite + wurtzite; 3, wurtzite. (From Robbins, M. and Lambrecht, V.G., *J. Solid State Chem.*, 6(3), 402, 1973.)

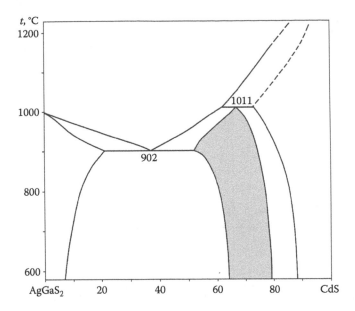

FIGURE 4.14 The CdS–AgGaS$_2$ phase diagram. (From Chykhrij, S.I. et al., *J. Alloys Compd.*, 312(1–2), 189, 2000a.)

this system. The peritectic point is located at 62 mol.% CdS. The homogeneity region of AgCd$_2$GaS$_4$ is within the interval 64–79 mol.% CdS at 600°C. The cell parameters increase with increasing CdS content with small negative deviations from Vegard's rule. AgCd$_2$GaS$_4$ crystallizes in the orthorhombic structure with the lattice parameters $a = 813.95 \pm 0.09$, $b = 693.94 \pm 0.08$, and $c = 660.14 \pm 0.07$ pm (Atuchin et al. 2005, Pervukhina et al. 2005) [$a = 814.60 \pm 0.04$, $b = 689.89 \pm 0.04$, $c = 659.32 \pm 0.04$ pm (Chykhriy et al. 2000a)] and calculated density 4.7557 ± 0.0008 g·cm^{-3} (Chykhriy et al. 2000a) [4.726 g·cm^{-3} (Atuchin et al. 2005, Pervukhina et al. 2005)]. The optical bandgap increases upon decreasing temperature from 2.15 eV at 27°C to 2.22 eV at −263°C (Olekseyuk et al. 2005, Atuchin et al. 2006). This compound is an *n*-type semiconductor.

The solid solubility range of CdS is 87–100 mol.% CdS at 600°C (Chykhriy et al. 2000a), which is approximately equal to the result reported by Robbins and Lambrecht (1973). The extent of the solid solutions based on AgGaS$_2$ is 0–7 mol.% CdS at 600°C (Chykhriy et al. 2000a). This concentration increases with increasing temperature and reaches a maximum (~21 mol.% CdS) at 902°C. The last value is also close to that obtained by Robbins and Lambrecht (1973).

This system was investigated through DTA, XRD, and metallography and the alloys were annealed at 600°C during 500 h (Chykhriy et al. 2000a, Pervukhina et al. 2005). Single crystals of the AgCd$_2$GaS$_4$ quaternary compound were grown by melting technique using AgGaS$_2$ as a solvent (Olekseyuk et al. 2005, Atuchin et al. 2006).

CdS–Ag$_9$GaS$_6$. The phase diagram of this system is a eutectic type (Figure 4.15) (Olekseyuk et al. 2001). The coordinates of the eutectic point are 724°C and

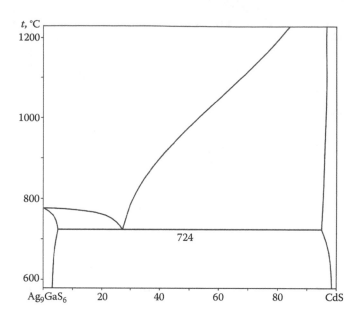

FIGURE 4.15   The CdS–Ag$_9$GaS$_6$ phase diagram. (From Olekseyuk, I.D. et al., *J. Alloys Compd.*, 325(1–2), 167, 2001.)

27 mol.% CdS. The solid solubility ranges of the system components amount to 5 mol.% at the eutectic temperature and decrease with decreasing temperature.

This system was investigated through DTA, XRD, and metallography and the alloys were annealed at 600°C during 500 h and then quenched in cold water (Olekseyuk et al. 2001).

CdS–Ag$_2$S–Ga$_2$S$_3$. The liquidus surface of this quasiternary system is shown in Figure 4.16 (Olekseyuk et al. 2001). It consists of 10 fields of primary crystallization of the phases that belong to the Ag$_2$S, CdS, Ag$_9$GaS$_6$, AgCd$_2$GaS$_4$, AgGaS$_2$, Cd$_5$Ga$_2$S$_8$, Cd$_2$GaS$_4$, Ag$_2$Ga$_{20}$S$_{31}$, and γ- and β-Ga$_2$S$_3$. The fields of primary crystallization are separated by 24 monovariant lines and by 24 invariant points, 15 of which are binary (10 binary eutectics and 5 binary peritectics) and 9 ternary ones (6 ternary eutectics and 3 transition points). The compositions and temperatures of the ternary invariant points are represented in Table 4.2.

The isothermal section of the CdS–Ag$_2$S–Ga$_2$S$_3$ quasiternary system at 600°C is given in Figure 4.17 (Olekseyuk et al. 2001).

4.31   CADMIUM–SILVER–INDIUM–SULFUR

CdS–AgInS$_2$. The phase equilibria in this system up to 1000°C obtained using XRD are shown in Figure 4.18 (Robbins and Lambrecht 1973).

According to the data of Kozer et al. (2009a), the solid solubility based on low-temperature modification of AgInS$_2$ at 600°C is less than 2 mol.% CdS. The solid solubility range of CdS at this temperature is 59–100 mol.% CdS.

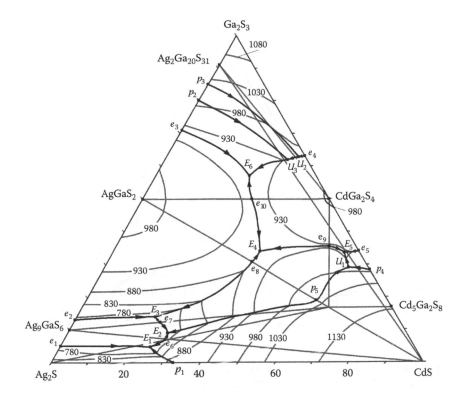

FIGURE 4.16 The liquidus surface projection of the CdS–Ag$_2$S–Ga$_2$S$_3$ quasiternary system. (From Olekseyuk, I.D. et al., *J. Alloys Compd.*, 325(1–2), 167, 2001.)

TABLE 4.2
Temperature and Composition of the Ternary Invariant Points in the CdS–Ag$_2$S–Ga$_2$S$_3$ Quasiternary System

Invariant Points	Reaction	Composition, mol.%			t, °C
		Ag$_2$S	CdS	Ga$_2$S$_3$	
E_1	L $\Leftrightarrow$ Ag$_2$S + CdS + Ag$_9$GaS$_6$	71	24	5	700
E_2	L $\Leftrightarrow$ CdS + AgCd$_2$GaS$_4$ + Ag$_9$GaS$_6$	64	27	9	714
E_3	L $\Leftrightarrow$ AgCd$_2$GaS$_4$ + AgGaS$_2$ + Ag$_9$GaS$_6$	65	21	14	741
E_4	L $\Leftrightarrow$ AgCd$_2$GaS$_4$ + AgGaS$_2$ + Cd$_2$GaS$_4$	27	34	39	889
E_5	L $\Leftrightarrow$ AgCd$_2$GaS$_4$ + Cd$_5$Ga$_2$S$_8$ + Cd$_2$GaS$_4$	4	62	34	900
E_6	L $\Leftrightarrow$ AgGaS$_2$ + Cd$_2$GaS$_4$ + Ag$_2$Ga$_{20}$S$_{31}$	18	25	57	895
U_1	L + CdS $\Leftrightarrow$ AgCd$_2$GaS$_4$ + Cd$_5$Ga$_2$S$_8$	5	66	29	950
U_2	L + β-Ga$_2$S$_3$ $\Leftrightarrow$ α-Ga$_2$S$_3$ + Cd$_2$GaS$_4$	1.5	35.5	63	951
U_3	L + α-Ga$_2$S$_3$ $\Leftrightarrow$ Ag$_2$Ga$_{20}$S$_{31}$ + Cd$_2$GaS$_4$	5	33	62	941

Source: Olekseyuk, I.D. et al., *J. Alloys Compd.*, 325(1–2), 167, 2001.

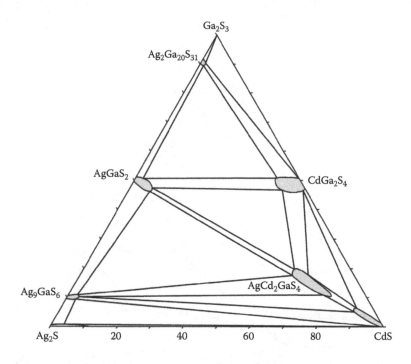

FIGURE 4.17 Isothermal section of the $CdS–Ag_2S–Ga_2S_3$ system at 600°C. (From Olekseyuk, I.D. et al., *J. Alloys Compd.*, 325(1–2), 167, 2001.)

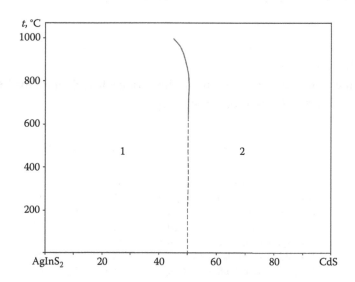

FIGURE 4.18 Phase equilibria in the $CdS–AgInS_2$ system up to 1000°C: 1, chalcopyrite + wurtzite; 2, wurtzite. (From Robbins, M. and Lambrecht, V.G., *J. Solid State Chem.*, 6(3), 402, 1973.)

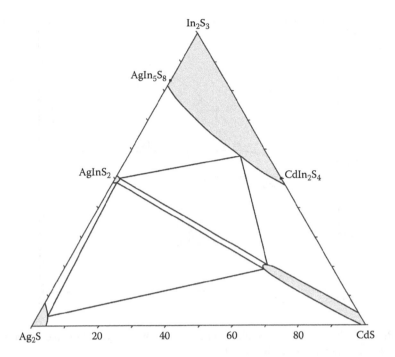

FIGURE 4.19 Isothermal section of the CdS–Ag$_2$S–In$_2$S$_3$ system at 600°C. (From Kozer, V.R. et al., *J. Alloys Compd.*, 480(2), 360, 2009a.)

Parthé et al. (1969) indicated that the AgCd$_2$InS$_4$ quaternary compound is formed in this system. It crystallizes in the hexagonal structure with the lattice parameters a = 411.2 ± 0.5 and c = 670.9 ± 0.5 pm, but this compound can be one of the solid solution compositions. This compound was obtained by the heating of the mixture from chemical elements at 600°C–800°C.

CdS–Ag$_2$S–In$_2$S$_3$. The isothermal section of this quasiternary system at 600°C is given in Figure 4.19 (Kozer et al. 2009a). No quaternary intermediate phase was found and a continuous solid solution series between In$_2$S$_3$, AgIn$_5$S$_8$, and CdIn$_2$S$_4$ was determined. A limited solid solution based on CdS is localized along the section CdS–AgInS$_2$.

4.32 CADMIUM–SILVER–SILICON–SULFUR

CdS–Ag$_2$SiS$_3$. The phase diagram is a eutectic type (Figure 4.20) (Parasyuk and Piskach 1999). The eutectic crystallizes at 668°C and contains 8 mol.% CdS. The quaternary compound Ag$_2$CdSiS$_4$ is formed in this system. It melts incongruently at 745°C and crystallizes in the orthorhombic structure with the lattice parameters a = 793.9 ± 0.5, b = 679.1 ± 0.4, and c = 652.7 ± 0.5 pm. The solubility of CdS in Ag$_2$SiS$_3$ is negligible, and the solid solution based on CdS does not exceed 2 mol.% Ag$_2$SiS$_3$. The homogeneity region of Ag$_2$CdSiS$_4$ is situated only from the Ag$_2$SiS$_3$ side and is less than 5 mol.% Ag$_2$SiS$_3$ at 400°C.

This system was investigated through DTA, XRD, and metallography and the alloys were annealed at 400°C during 250 h (Parasyuk and Piskach 1999).

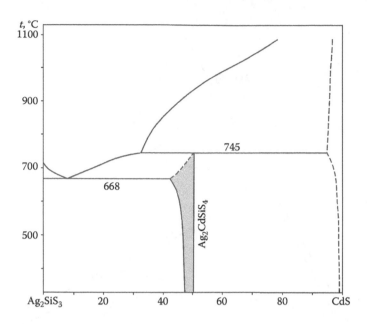

FIGURE 4.20 The CdS–Ag$_2$SiS$_3$ phase diagram. (From Parasyuk, O.V. and Piskach, L.V., *Zhurn. neorgan. khimii*, 44(6), 1032, 1999.)

CdS–Ag$_8$SiS$_6$. The phase diagram is a peritectic type (Figure 4.21) (Piskach et al. 2006). The peritectic composition and temperature are 70 ± 3 mol.% CdS and 1025°C, respectively. Solubility of CdS in Ag$_8$SiS$_6$ at the peritectic temperature is 93 mol.%. The crystallization of the solid solutions occurs in a narrow temperature range that does not exceed 20°C. These solid solutions are characterized by the polymorph transformation at 216°C. This system was investigated by DTA, XRD, and metallography.

4.33 CADMIUM–SILVER–GERMANIUM–SULFUR

CdS–Ag$_2$GeS$_3$. The phase diagram is a eutectic type (Figure 4.22) (Piskach and Parasyuk 1998). For the Ag$_2$GeS$_3$ compound, the congruent melting at 638°C has been confirmed. The eutectic crystallizes at 598°C and contains 4 mol.% CdS. The quaternary compounds Ag$_2$CdGeS$_4$ and Ag$_4$CdGe$_2$S$_7$ are formed in this system. Ag$_2$CdGeS$_4$ compound melts incongruently at 768°C and exhibits polymorphic transformation with a temperature decrease from 632°C to 612°C with an increase in CdS content. It crystallizes in the orthorhombic structure (space group *Pmn2$_1$*) with the lattice parameters for stoichiometric composition $a = 803.38 \pm 0.03$, $b = 686.80 \pm 0.02$, and $c = 658.66 \pm 0.03$ pm [$a = 801.6 \pm 0.5$, $b = 683.1 \pm 0.4$, and $c = 654.8 \pm 0.5$ pm (Piskach and Parasyuk 1998); $a = 804.4 \pm 0.8$, $b = 684.9 \pm 0.5$, and $c = 659.3 \pm 0.5$ pm (Parthé et al. 1969)] and calculated density 4.8334 ± 0.0006 g·cm^{-3} (Parasyuk et al. 2005b). According to the data of Brunetta et al. (2012), this compound crystallizes in the space group *Pna2$_1$* of the orthorhombic structure with the lattice parameters $a = 1374.15 \pm 0.08$, $b = 803.67 \pm 0.05$, and $c = 659.07 \pm 0.04$ pm, calculated density

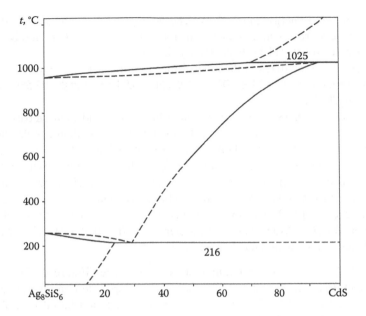

FIGURE 4.21 The CdS–Ag$_8$SiS$_6$ phase diagram. (From Piskach, L.V. et al., *J. Alloys Compd.*, 421(1–2), 98, 2006.)

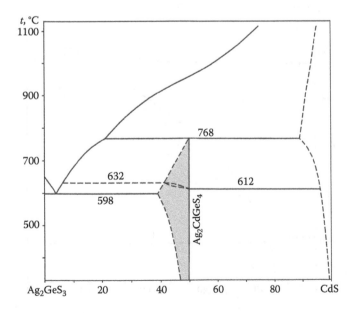

FIGURE 4.22 The CdS–Ag$_2$GeS$_3$ phase diagram. (From Piskach, L.V. et al., *Pol. J. Chem.*, 72(6), 1112, 1998.)

4.827 g·cm⁻³, and energy gap 2.32 eV. The homogeneity region of Ag_2CdGeS_4 at eutectic temperature is within the interval from 39 to 50 mol.% CdS and is less than 5 mol.% CdS at 400°C (Piskach and Parasyuk 1998).

$Ag_4CdGe_2S_7$ crystallizes in the monoclinic structure with the lattice parameters $a = 1743.64 \pm 0.08$, $b = 683.34 \pm 0.03$, $c = 1053.50 \pm 0.04$ pm, and $\beta = 93.589° \pm 0.003°$ (Gulay et al. 2002).

The solubility of CdS in Ag_2GeS_3 is insignificant and solid solution based on CdS contains 11 mol.% Ag_2GeS_3 at peritectic temperature and less than 2 mol.% at 400°C (Piskach and Parasyuk 1998). This system was investigated through DTA, XRD, and metallography and the alloys were annealed at 400°C during 250 h. Ag_2CdGeS_4 was synthesized via high-temperature solid-state synthesis (Brunetta et al. 2012). Its single crystals were produced using stoichiometric quantities of Ag, Cd, and Ge with some excess of S. The samples were heated to 800°C over 12 h and held at this temperature for 96 h. After a slow cooling step of 5°C/h (60 h) to 500°C, the samples were allowed to cool to ambient temperature.

CdS–Ag₈GeS₆. The phase diagram is a peritectic type (Figure 4.23) (Parasyuk et al. 2005b, Piskach et al. 2006). The peritectic composition and temperature are ~63 mol.% CdS and 975°C, respectively. The solid solution based on β-Ag₈GeS₆ contains up to 81 mol.% CdS at the peritectic temperature. The region of its existence substantially narrows with temperature decreasing, and at 195°C, it decomposes according to a eutectoid reaction.

This system was investigated by DTA, XRD, and metallography, and the alloys were annealed at 400°C during 250 h (Parasyuk et al. 2005b, Piskach et al. 2006).

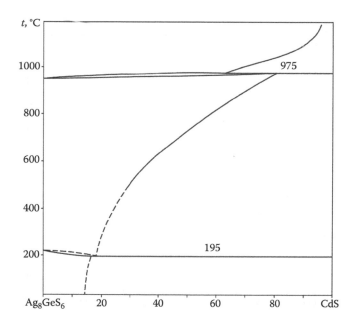

FIGURE 4.23 The CdS–Ag₈GeS₆ phase diagram. (From Parasyuk, O.V. et al., *J. Alloys Compd.*, 397(1–2), 95, 2005b; Piskach L.V. et al., *J. Alloys Compd.*, 421(1–2), 98, 2006.)

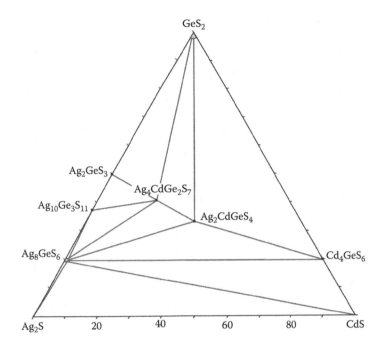

FIGURE 4.24 Isothermal section of the CdS–Ag$_2$S–GeS$_2$ system at 400°C. (From Parasyuk, O.V. et al., *J. Alloys Compd.*, 397(1–2), 95, 2005b.)

CdS–Ag$_2$S–GeS$_2$. The isothermal section of this quasiternary system at 400°C is presented in Figure 4.24 (Parasyuk et al. 2005b). The Ag$_2$CdGeS$_4$ and Ag$_4$CdGe$_2$S$_7$ quaternary compounds with narrow homogeneity regions exist in this system.

4.34 CADMIUM–SILVER–TIN–SULFUR

CdS–Ag$_2$SnS$_3$. The phase diagram is a eutectic type (Figure 4.25) (Parasyuk and Piskach 1998). The eutectic crystallizes at 646°C and contains 8 mol.% CdS. The quaternary compound Ag$_2$CdSnS$_4$ is formed in this system. It melts incongruently at 795°C and has two polymorphic modifications. The homogeneity region of Ag$_2$CdSnS$_4$ at 400°C is less than 5 mol.% CdS. It crystallizes in the orthorhombic structure with the lattice parameters for stoichiometric composition $a = 410.15 \pm 0.03$, $b = 702.24 \pm 0.04$, and $c = 669.46 \pm 0.04$ pm [$a = 411.1 \pm 0.5$, $b = 703.8 \pm 0.5$, and $c = 668.5 \pm 0.5$ pm (Parthé et al. 1969)] and calculated density 4.9521 ± 0.0009 g · cm^{-3} (Parasyuk et al. 2005a). According to the data of Parasyuk and Piskach (1998), the diffractogram of low-temperature Ag$_2$CdSnS$_4$ modification (the alloy of stoichiometric composition) confirms the existence of the orthorhombic structure with two possible variants of space groups. The lattice parameters for $Cmc2_1$ space group are $a = 1215.8 \pm 0.1$, $b = 704.8 \pm 0.1$, and $c = 665.9 \pm 0.1$ pm and for $Pnm2_1$, $a = 818.2 \pm 0.5$, $b = 701.5 \pm 0.3$, and $c = 666.6 \pm 0.5$ pm.

The solubility Ag$_2$SnS$_3$ in CdS and of CdS in Ag$_2$SnS$_3$ is less than 2 mol.% at 400°C (Parasyuk and Piskach 1998). This system was investigated through DTA, XRD, and metallography and the alloys were annealed at 400°C during 250 h.

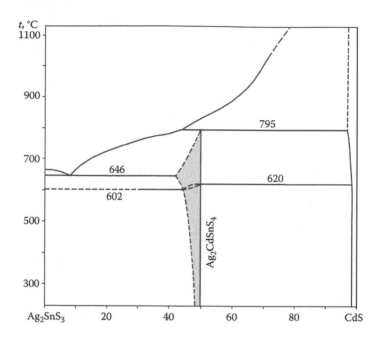

FIGURE 4.25 The CdS–Ag₂SnS₃ phase diagram. (From Parasyuk, O.V. et al., *J. Alloys Compd.*, 399(1–2), 173, 2005a.)

CdS–Ag₈SnS₆. The phase diagram is a eutectic type (Figure 4.26) (Parasyuk et al. 2005a, Piskach et al. 2006). The eutectic point coordinates are ~47 mol.% CdS and 809°C. The solubility of CdS in β-Ag₈SnS₆ exceeds 40 mol.% CdS at the eutectic temperature. As the temperature decreases, the homogeneity region of the solid solution based on Ag₈SnS₆ narrows and at 160°C, it undergoes eutectoid decomposition. At room temperature, the solubility of CdS in α-Ag₈SnS₆ determined only by the results of metallography lies between 10 and 20 mol.%.

This system was investigated by DTA, XRD, and metallography (Parasyuk et al. 2005a, Piskach et al. 2006).

CdS–Ag₂S–SnS₂. The isothermal section of this quasiternary system at room temperature is presented in Figure 4.27 (Parasyuk et al. 2005a). At this temperature, the existence of the Ag₂CdSnS₄ and Ag₂CdSn₃S₈ quaternary compounds is established. Ag₂CdSn₃S₈ crystallizes in the cubic structure of the chalcospinel type ($a = 1076.35 \pm 0.02$ nm and calculated density 5.0102 ± 0.0003 g·cm⁻³) or in the tetragonal structure of the rhodostannine type ($a = 761.63 \pm 0.06$ and $c = 1077.1 \pm 0.2$ pm and calculated density 5.000 ± 0.002 g·cm⁻³).

4.35 CADMIUM–SILVER–TELLURIUM–SULFUR

CdS–Ag₂Te. The phase diagram is a eutectic type (Figure 4.28) (Trishchuk et al. 1989). The eutectic composition and temperature are 13 mol.% CdS and 895°C, respectively. Solubility of CdS in γ-Ag₂Te reaches 11 mol.% and decreases as the temperature

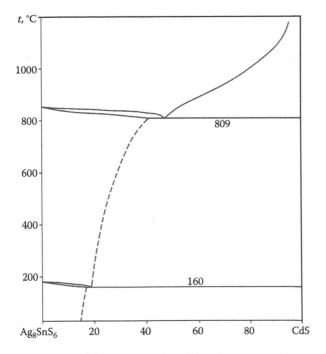

FIGURE 4.26 The CdS–Ag$_8$SnS$_6$ phase diagram. (From Parasyuk, O.V. et al., *J. Alloys Compd.*, 399(1–2), 173, 2005a; Piskach L.V. et al., *J. Alloys Compd.*, 421(1–2), 98, 2006.)

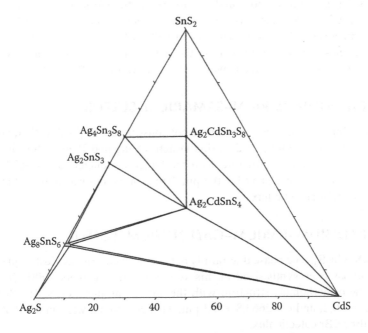

FIGURE 4.27 Isothermal section of the CdS–Ag$_2$S–SnS$_2$ system at room temperature. (From Parasyuk, O.V. et al., *J. Alloys Compd.*, 399(1–2), 173, 2005a.)

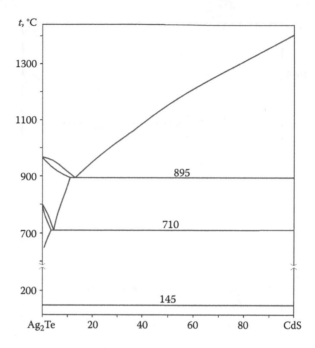

FIGURE 4.28 The CdS–Ag$_2$Te phase diagram. (From Trishchuk, L.I. et al., *Izv. AN SSSR. Neorgan. materialy*, 25(7), 1081, 1989.)

decreases. These solid solutions undergo the phase transformation according to the eutectoid reaction. The eutectoid composition and temperature are 4.5 mol.% CdS and 710°C, respectively. Maximum solubility of CdS in β-Ag$_2$Te is equal to 3.5 mol.%. Thermal effects at 145°C correspond to the polymorphous transformation of Ag$_2$Te (β-Ag$_2$Te → α-Ag$_2$Te). This system was investigated through DTA and metallography.

4.36 CADMIUM–BARIUM–SAMARIUM–SULFUR

CdS–BaS–Sm$_2$S$_3$. The phase diagram is not constructed. Ba$_4$Cd$_3$Sm$_2$S$_{10}$ quaternary compound has been synthesized in this system (Yang and Ibers 2000). It crystallizes in the orthorhombic structure with the lattice parameters a = 419.07 ± 0.08, b = 1793.1 ± 0.4, and c = 2659.4 ± 0.5 pm. This compound was obtained at 900°C using BaBr$_2$/KBr eutectic flux.

4.37 CADMIUM–BARIUM–GADOLINIUM–SULFUR

CdS–BaS–Gd$_2$S$_3$. The phase diagram is not constructed. Ba$_4$Cd$_3$Gd$_2$S$_{10}$ quaternary compound has been synthesized in this system (Yang and Ibers 2000). It crystallizes in the orthorhombic structure with the lattice parameters a = 417.26 ± 0.04, b = 1792.8 ± 0.2, and c = 2663.8 ± 0.3 pm. This compound was obtained at 900°C using BaBr$_2$/KBr eutectic flux.

4.38 CADMIUM–BARIUM–TERBIUM–SULFUR

$CdS–BaS–Tb_2S_3$. The phase diagram is not constructed. $Ba_4Cd_3Tb_2S_{10}$ quaternary compound has been synthesized in this system (Yang and Ibers 2000). It crystallizes in the orthorhombic structure with the lattice parameters $a = 416.43 \pm 0.08$, $b = 1791.6 \pm 0.4$, and $c = 2658.1 \pm 0.5$ pm. This compound was obtained at 900°C using $BaBr_2$/KBr eutectic flux.

4.39 CADMIUM–BARIUM–GERMANIUM–SULFUR

$CdS–BaS–GeS_2$. The phase diagram is not constructed. The $BaCdGeS_4$ quaternary compound is formed in this system, which crystallizes in the orthorhombic structure with the lattice parameters $a = 2169$, $b = 2125$, and $c = 1278$ pm and calculation and experimental density 4.08 and 3.96 $g \cdot cm^{-3}$, respectively (Teske 1980b).

4.40 CADMIUM–BARIUM–TIN–SULFUR

$CdS–BaS–SnS_2$. The phase diagram is not constructed. The $BaCdSnS_4$ and $Ba_3CdSn_2S_8$ quaternary compounds are formed in this system (Teske 1980a, 1985). $BaCdSnS_4$ crystallizes in the orthorhombic structure with the lattice parameters $a = 2186 \pm 2$, $b = 2169 \pm 1$, and $c = 1318 \pm 0.5$ pm and calculation and experimental density 4.23 and 4.11 $g \cdot cm^{-3}$, respectively (Teske 1980a). $Ba_3CdSn_2S_8$ crystallizes in the cubic structure with the lattice parameter $a = 1471.3$ pm and calculation and experimental density 4.25 and 4.21 ± 0.08 $g \cdot cm^{-3}$, respectively (Teske 1985).

4.41 CADMIUM–MERCURY–GERMANIUM–SULFUR

$CdS–HgS–GeS_2$. The phase diagram is not constructed. Using coprecipitation of CdS with HgS and GeS_2 by the hydrogen sulfide and then heating up to 300°C, any quaternary compounds were obtained (Kislinskaya 1974).

4.42 CADMIUM–MERCURY–TIN–SULFUR

$CdS–HgS–SnS_2$. The phase diagram is not constructed. Using coprecipitation of CdS with HgS and SnS_2 by the hydrogen sulfide and then heating up to 300°C–400°C, any quaternary compounds were obtained (Kislinskaya 1974).

4.43 CADMIUM–MERCURY–SELENIUM–SULFUR

$CdS + HgSe \Leftrightarrow CdSe + HgS$. The position of the calculated miscibility gap in this system at −100°C (173 K) is shown in Figure 4.29 (Ohtani et al. 1992).

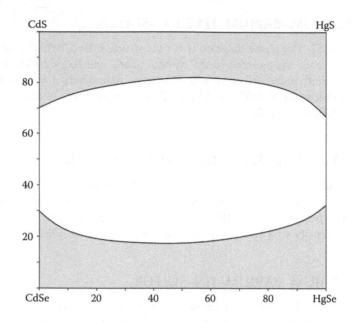

FIGURE 4.29 Calculated miscibility gap in the CdS + HgSe ⇔ CdSe + HgS ternary mutual system at 173 K. (From Ohtani H. et al., *J. Alloys Compd.*, 182(1), 103, 1992.)

4.44   CADMIUM–MERCURY–TELLURIUM–SULFUR

CdS + HgTe ⇔ CdTe + HgS. The calculated miscibility gaps in the CdS + HgTe ⇔ CdTe + HgS ternary mutual system at 600°C–1000°C are shown in Figure 4.30 (Ohtani et al. 1992). The position of the miscibility gaps indicate that the phase separation occurs toward the CdS and HgTe ends, which have the highest and lowest melting temperatures among the four compounds.

4.45   CADMIUM–MERCURY–BROMINE–SULFUR

CdS–Hg₂Br₂. The phase diagram is not constructed. The $Cd_2Hg_2Br_2S_2$ quaternary compound is formed in this system. It crystallizes in the orthorhombic structure with the lattice parameters $a = 905.33 \pm 0.15$, $b = 1846.7 \pm 0.3$, and $c = 463.54 \pm 0.10$ pm, calculated density 7.284 g·cm⁻³, and energy gap 2.41 eV (Zou et al. 2011). Its single crystals were prepared by the solid-state reaction of a mixture of 0.75 mmol Hg_2Br_2, 1 mmol Cd, and 1 mmol S at 350°C.

4.46   CADMIUM–BORON–OXYGEN–SULFUR

CdS–CdO–B₂O₃. The phase diagram is not constructed. The $Cd_3B_7O_{12.65}S_{0.85}$ quaternary compound (boracite) is formed in this system (Fouassier et al. 1970). It melts incongruently at 990°C and crystallizes in the cubic structure with the lattice parameter $a = 1248.4 \pm 0.4$ pm. This compound was obtained by the heating of mixtures from the binary compounds up to 920°C. Its single crystals were grown by the chemical transport reactions.

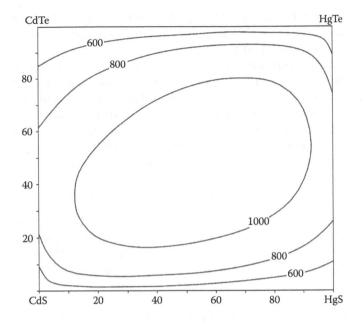

FIGURE 4.30 Calculated miscibility gaps in the CdS + HgTe ⇔ CdTe + HgS ternary mutual system. (From Ohtani H. et al., *J. Alloys Compd.*, 182(1), 103, 1992.)

4.47 CADMIUM–ALUMINUM–INDIUM–SULFUR

CdS–Al₂S₃–In₂S₃. The phase diagram is not constructed. According to the data of Moldovyan and Markus (1986), $CdAlInS_4$ quaternary compound is formed in this system. It crystallizes in the hexagonal structure with the lattice parameters $a = 385$ and $c = 1230$ pm and energy gap 3.2 eV. Abdullayev and Kyazumov (1989) investigated the structure of $Cd_xIn_yAl_zS_4$ ($0.9 \leq x \leq 1.0$; $1 \leq y$ 1.3; $0.7 \leq z \leq 1.0$) crystals and thin films and determined that they crystallize in the rhombohedral structure with the lattice parameters $a = 386.7$ and $c = 3697.8$ pm. Superstructure reflections were revealed with enlarging of the parameter in $\sqrt{3}$ times.

Single crystals of the $CdAlInS_4$ compound have been grown by the chemical transport reactions using iodine as a transport agent (Moldovyan and Markus 1986).

4.48 CADMIUM–GALLIUM–INDIUM–SULFUR

CdS–Ga₂S₃–In₂S₃. The phase diagram is not constructed. The $CdGaInS_4$ and $Cd_{0.5}Ga_2InS_5$ quaternary compounds are formed in this system (Shand 1970, Mehtiev et al. 1977, Kambas 1983, Kyazimov and Amiraslanov 1983, Razzetti et al. 1984, Haeuseler et al. 1988). $CdGaInS_4$ melts at 950°C ± 20°C and crystallizes in the trigonal structure with the lattice parameters $a = 386.9 \pm 0.3$ and $c = 2466.8 \pm 1.4$ pm and calculated density 4.455 g·cm⁻³ (Kyazimov and Amiraslanov 1983) [in the rhombohedral structure (Shand 1970) or in the hexagonal structure with the lattice parameters $a = 391.5$ and $c = 1272.5$ pm and calculation and experimental density 4.19 and 4.23 g·cm⁻³, respectively (Mehtiev et al. 1977)].

$Cd_{0.5}Ga_2InS_5$ crystallizes in the hexagonal structure with the lattice parameters $a = 380.9$ and $c = 3064$ pm (Haeuseler et al. 1988). This compound was synthesized from the binary sulfides at 600°C–800°C.

The $CdGaInS_4$ compound was obtained by sintering the chemical elements (Kambas 1983). Its single crystals were grown using chemical transport reactions with iodine as a transport agent (Kambas 1983, Kyazimov and Amiraslanov 1983, Razzetti et al. 1984). A superstructure at room temperature, which is produced by an ordering of Ga and In atoms within the layers, was found (it disappears at a rather high temperature) (Kambas 1983). One-, two-, and three-package polytypes are formed at the chemical transport reactions (Kyazimov and Amiraslanov 1983).

4.49   CADMIUM–GALLIUM–LANTHANUM–SULFUR

$CdS–Ga_2S_3–La_2S_3$. The phase diagram is not constructed. $CdGa_3LaS_7$ quaternary compound is formed in this system (Agaev et al. 1996). It melts at 880°C and crystallizes in the tetragonal structure with the lattice parameters $a = 942$ and $c = 618$ pm and energy gap 1.60 eV. Single crystals of this compound were grown by the chemical transport reactions.

4.50   CADMIUM–GALLIUM–PRASEODYMIUM–SULFUR

$CdS–Ga_2S_3–Pr_2S_3$. The phase diagram is not constructed. $CdGa_3PrS_7$ quaternary compound is formed in this system (Agaev et al. 1996). It melts at 925°C and crystallizes in the tetragonal structure with the lattice parameters $a = 937$ and $c = 617$ pm. Single crystals of this compound were grown by the zone recrystallization.

4.51   CADMIUM–GALLIUM–NEODYMIUM–SULFUR

$CdS–Ga_2S_3–Nd_2S_3$. The phase diagram is not constructed. $CdGa_3NdS_7$ quaternary compound is formed in this system (Agaev et al. 1996). It melts at 1005°C and crystallizes in the tetragonal structure with the lattice parameters $a = 947$ and $c = 603$ pm.

4.52   CADMIUM–GALLIUM–SAMARIUM–SULFUR

$CdS–Ga_2S_3–Sm_2S_3$. The phase diagram is not constructed. $CdGa_3SmS_7$ quaternary compound is formed in this system (Agaev et al. 1996). It melts at 987°C and crystallizes in the tetragonal structure with the lattice parameters $a = 953$ and $c = 614$ pm.

4.53   CADMIUM–GALLIUM–GERMANIUM–SULFUR

$CdS–Ga_2S_3–GeS_2$. The liquidus surface of this quasiternary system was depicted for the first time by Barnier et al. (1990). According to these data, the ternary system is triangulated by the quasibinary section $Cd_5Ga_2S_8–GeS_2$, though this section is quasibinary only in part of the subliquidus region because the $Cd_5Ga_2S_8$ phase is metastable.

Olekseyuk et al. (2006) noted that the liquidus surface of the $CdS–Ga_2S_3–GeS_2$ quasiternary system (Figure 4.31) consists of 8 fields of the primary crystallization, which are separated by 14 monovariant lines and by 6 nonvariant points of which 4 are the transition points and 2 are the ternary eutectics (Table 4.3).

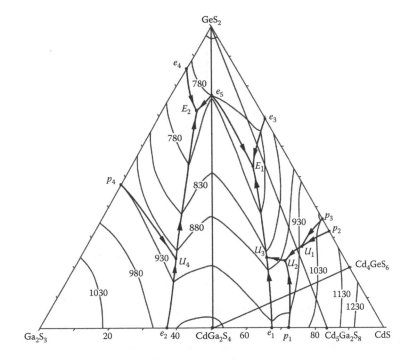

FIGURE 4.31 The liquidus surface projection of the $CdS-Ga_2S_3-GeS_2$ quasiternary system. (From Olekseyuk, I.D. et al., *J. Alloys Compd.*, 421(1–2), 91, 2006.)

TABLE 4.3

Temperature and Composition of the Ternary Invariant Points in the $CdS-Ga_2S_3-GeS_2$ Quasiternary System

Invariant Points	Reaction	Composition, mol.% CdS	Ga_2S_3	GeS_2	t, °C
U_1	$L + \beta\text{-}Cd_4GeS_6 \Leftrightarrow CdS + \alpha\text{-}Cd_4GeS_6$	63	11	26	1012
U_2	$L + CdS \Leftrightarrow \alpha\text{-}Cd_4GeS_6 + Cd_5Ga_2S_8$	70	18	12	900
U_3	$L + Cd_5Ga_2S_8 \Leftrightarrow CdGa_2S_4 + \alpha\text{-}Cd_4GeS_6$	54	23	23	860
U_4	$L + \beta\text{-}Ga_2S_3 \Leftrightarrow \alpha\text{-}Ga_2S_3 + Cd_2GaS_4$	28	49	23	744
E_1	$L \Leftrightarrow GeS_2 + CdGa_2S_4 + \alpha\text{-}Cd_4GeS_6$	35	12	53	735
E_2	$L \Leftrightarrow \alpha\text{-}Ga_2S_3 + GeS_2 + CdGa_2S_4$	10	18	72	726

Source: Olekseyuk, I.D. et al., *J. Alloys Compd.*, 421(1–2), 91, 2006.

The isothermal section of the $CdS-Ga_2S_3-GeS_2$ system at 400°C has been constructed by Olekseyuk et al. (2006) (Figure 4.32). Minor solid solutions exist for every system component. In the middle of the concentration triangle, the largest solid solubility is based on $CdGa_2S_4$ that stretches along the $CdGa_2S_4-GeS_2$ section to 8 mol.% GeS_2. The solid solution range of GeS_2 incorporates up to 3 mol.% $CdGa_2S_4$.

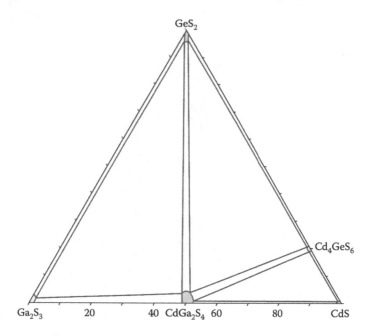

FIGURE 4.32 Isothermal section of the $CdS–Ga_2S_3–GeS_2$ system at 400°C. (From Olekseyuk, I.D. et al., *J. Alloys Compd.*, 421(1–2), 91, 2006.)

On the $CdGa_2S_4–Cd_4GeS_6$ section, the region of the solid solution based on $CdGa_2S_4$ is 3 mol.% and the solid solution range of Cd_4GeS_6 does not exceed 2 mol.%.

The glasses exist in a wide region in the $CdS–Ga_2S_3–GeS_2$ system (Figure 4.33) (Barnier et al. 1990, Wang et al. 2004). The glassy domain for the alloys quenched from 1000°C is widely extended toward Ga_2S_3 (Barnier et al. 1990). The formed glasses are yellow amber colored, and this color becomes redder with the simultaneous increase in concentration of Ga_2S_3 and CdS. Minimum and maximum vitreous temperatures appear at 270°C and 400°C, respectively. According to the data of Wang et al. (2004), the glass-forming region for the samples heating at 970°C is manly situated in the GeS_2-rich domain, and the amount of dissolved CdS is up to over 30 mol.% (Figure 4.33). The obtained glasses have relatively high glass transition temperatures (T_g = 375°C–436°C) and good thermal stability. The experimental results indicate (Gu et al. 2005) that the GeS_2 acts in the $CdS–Ga_2S_3–GeS_2$ glasses as the network former, the Ga_2S_3 as the net intermediate, and the CdS as the net modifier.

This system was investigated through DTA, XRD, and metallography (Barnier et al. 1990, Olekseyuk et al. 2006). The $CdS–Ga_2S_3–GeS_2$ glasses were prepared by conventional melt-quenching techniques (Gu et al. 2005).

4.54 CADMIUM–GALLIUM–ARSENIC–SULFUR

CdS–GaAs. The phase diagram is not constructed. According to the data of Kirovskaya and Zemtsov (2007) and Kirovskaya et al. (2007), continuous solid solutions are formed in this quasibinary system. Solid solutions containing up to

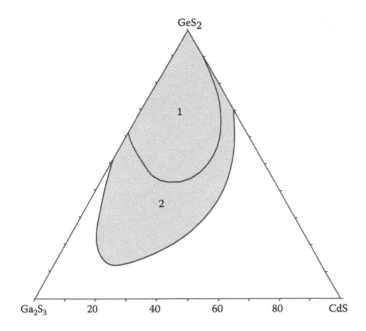

FIGURE 4.33 Glass region in the CdS–Ga$_2$S$_3$–GeS$_2$ system for the alloys quenched from 1000°C. (1—From Wang, X. et al., *Mater. Chem. Phys.*, 83(2–3), 284, 2004; 2—From Barnier, S. et al., *Mater. Sci. Eng.*, B7(3), 209, 1990.)

50 mol.% GaAs crystallize in the hexagonal structure of the wurtzite type, and solid solutions based on GaAs (containing >50 mol.% GaAs) crystallize in the cubic structure of the sphalerite type. Voitsehovski et al. (1970) indicated that the solubility of CdS in GaAs is equal to 10 mol.%.

The ingots were obtained by the melting of mixtures from chemical elements at 1250°C without any annealing (Voitsehovski et al. 1970).

4.55 CADMIUM–GALLIUM–CHROMIUM–SULFUR

CdS–Ga$_2$S$_3$–Cr$_2$S$_3$. The phase diagram is not constructed. CdGa$_3$CrS$_7$ quaternary compound is formed in this system (Agaev et al. 1996). It melts at 937°C and crystallizes in the tetragonal structure with the lattice parameters $a = 930$ and $c = 600$ pm.

4.56 CADMIUM–GALLIUM–TELLURIUM–SULFUR

CdS + GaTe ⇔ CdTe + GaS. The isothermal section of this ternary mutual system at 630°C is shown in Figure 4.34 (Odin et al. 2005c). Solid solutions based on the initial components of this system are participated in the phase equilibria. The ingots were annealing at 630°C during 1000 h.

3CdS + Ga$_2$Te$_3$ ⇔ 3CdTe + Ga$_2$S$_3$. The isothermal section of this ternary mutual system at 485°C is shown in Figure 4.35 (Odin and Rubina 2003). Solid solutions based on the initial components of this system and CdGa$_2$S$_4$, CdGa$_2$Te$_4$, and Ga$_2$S$_2$Te are participated in the phase equilibria.

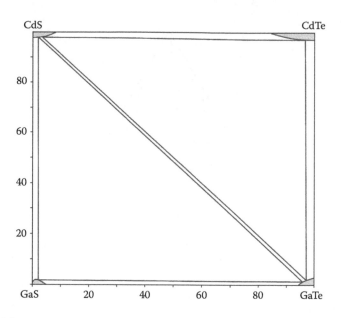

FIGURE 4.34 Isothermal section of the CdS + GaTe ⇔ CdTe + GaS ternary mutual system at 630°C. (From Odin, I.N. et al., *Zhurn. neorgan. khimii*, 50(4), 714, 2005c.)

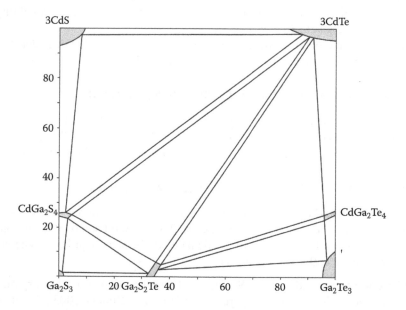

FIGURE 4.35 Isothermal section of the 3CdS + Ga$_2$Te$_3$ ⇔ 3CdTe + Ga$_2$S$_3$ ternary mutual system at 485°C. (From Odin, I.N. and Rubina, M.E., *Zhurn. neorgan. khimii*, 48(7), 1206, 2003.)

The metastable phases $Cd_2Ga_2S_2Te_3$ ($2CdS \cdot Ga_2Te_3$), $Cd_2Ga_6S_2Te_9$ ($2CdS \cdot 3Ga_2Te_3$), $Cd_3Ga_2S_2Te_4$ ($2CdS \cdot CdGa_2Te_4$), $Cd_5Ga_2S_2Te_6$ ($2CdS \cdot 3CdTe \cdot Ga_2Te_3$), and $CdGa_2S_2Te_2$ ($CdS \cdot 2/3Ga_2Te_3 \cdot 1/3Ga_2S_3$) have been obtained in this system within the temperature interval from 23°C to 485°C (Odin and Rubina 2003, Odin et al. 2005b). $CdGa_2S_2Te_2$ crystallizes in the tetragonal structure with the lattice parameters $a = 705.8 \pm 0.5$ and $c = 1016.4 \pm 0.7$ pm, $Cd_3Ga_2S_2Te_4$ is characterized by a face-centered cubic lattice with $a = 643.9 \pm 0.5$ pm (it may be a sublattice), $Cd_2Ga_2S_2Te_3$ is characterized by a rhombohedral distortion of the face-centered cubic lattice with $a = 639.5 \pm 0.5$ pm and $\alpha = 98.35° \pm 0.7°$, and most probably, the phase $Cd_2Ga_6S_2Te_9$ has a face-centered cubic sublattice with $a = 596.9 \pm 0.5$ pm (Odin and Rubina 2003).

$Cd_5Ga_2S_2Te_6$ has a homogeneity region within the interval of $Cd_3Ga_2S_2Te_4$–$Cd_5Ga_2S_2Te_6$ and crystallizes in the cubic structure of the sphalerite type with the lattice parameter $a = 641.5 \pm 0.5$ pm; calculation and experimental density 4.81 and 4.79 ± 0.01 g·cm^{-3}, respectively; and energy gap 1.29 eV (Odin et al. 2005b).

The metastable compounds have been obtained using the mixture of CdS, CdTe, and Ga_2Te_3 by the rapid quenching from the temperature of 1050°C (Odin et al. 2005b). The ingots were annealing at 485°C during 1000 h (Odin and Rubina 2003).

4.57 CADMIUM–GALLIUM–MANGANESE–SULFUR

$CdGa_3MnS_7$ quaternary compound is formed in the Cd–Ga–Mn–S quaternary system (Agaev et al. 1996). It melts at 1007°C and crystallizes in the tetragonal structure with the lattice parameters $a = 930$ and $c = 600$ pm. Single crystals of this compound were grown by the zone recrystallization.

4.58 CADMIUM–INDIUM–PHOSPHORUS–SULFUR

CdS–InP. The phase diagram is not constructed. Solubility of CdS in InP reaches 7 mol.% (Kirovskaya and Timoshenko 2007).

4.59 CADMIUM–INDIUM–ARSENIC–SULFUR

CdS–InAs. The phase diagram of this system belongs to the eutectic type (Figure 4.36) (Drobyazko and Kuznetsova 1983). The eutectic temperature and composition are 912°C and 19 mol.% CdS, respectively.

Using XRD, it was determined that the ingots containing up to 25 mol.% CdS crystallize in the sphalerite structure, and at the more CdS contents, two-phase region (sphalerite + wurtzite) exists in this system (Voitsehivski and Drobiazko 1967, Voitsehovski et al. 1968). According to the data of metallography, solubility of CdS in InAs reaches 18 mol.% (Drobyazko and Kuznetsova 1983) [20 mol.% (Voitsehivski and Drobiazko 1967, Voitsehovski et al. 1968)]. Lattice parameters of forming solid solutions change nearly linearly with composition.

This system was investigated through DTA, XRD, metallography, and measuring of microhardness (Voitsehivski and Drobiazko 1967, Voitsehovski et al. 1968, Drobyazko and Kuznetsova 1983). The ingots were annealing at 830°C, 780°C, and 730°C during 120, 240, and 480 h, respectively (Drobyazko and Kuznetsova 1983).

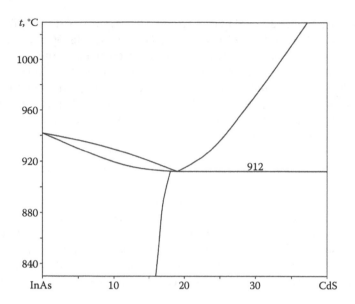

FIGURE 4.36 The CdS–InAs phase diagram. (From Drobyazko, V.P. and Kuznetsova, S.T., *Zhurn. neorgan. khimii*, 28(11), 2929, 1983.)

4.60 CADMIUM–INDIUM–ANTIMONY–SULFUR

CdS–InSb. The phase diagram is not constructed. Solubility of CdS in InSb is equal to 3–4 mol.% (Kirovskaya and Filatova 2007).

CdS–In$_2$S$_3$–Sb$_2$S$_3$. The phase diagram is not constructed. The CdInSbS$_4$ quaternary semiconductor compound is formed in this system (Azizov et al. 1986). It crystallizes in the cubic structure with the lattice parameter $a = 1080 \pm 3$ pm. This compound was obtained by the melting of mixtures from chemical elements. Its single crystals were grown by the chemical transport reactions.

4.61 CADMIUM–INDIUM–SELENIUM–SULFUR

CdS–In$_2$S$_3$–In$_2$Se$_3$. The phase diagram is not constructed. The CdIn$_2$Se$_2$S$_2$ quaternary compound is formed in this system (Paracchini et al. 1986). The energy gap of this compound at room temperature is $E_g = 1.95$ eV. This compound was obtained by the chemical precipitation from the gas phase.

4.62 CADMIUM–INDIUM–TELLURIUM–SULFUR

CdS–InTe. The phase diagram is a eutectic type (Figure 4.37) (Odin et al. 2005c). The eutectic crystallizes at 647°C ± 8°C and contains 23 ± 2 mol.% CdS. The solubility of CdS in InTe is equal to 0.5 mol.%, and the solubility of InTe in CdS is not higher than 0.8 mol.%. This system was investigated through DTA, XRD, and metallography and the ingots were annealed at 630°C during 1000 h.

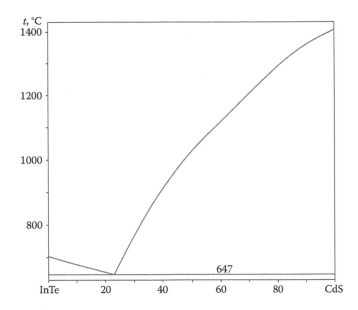

FIGURE 4.37 The CdS–InTe phase diagram. (From Odin, I.N. et al., *Zhurn. neorgan. khimii*, 50(4), 714, 2005c.)

CdS–CdTe–InTe. The isothermal section of this quasiternary system at 630°C is shown in Figure 4.38 (Odin et al. 2005c). Solid solutions based on the binary components are participated in the phase equilibria.

CdTe–CdIn₂S₄. The phase diagram is a eutectic type (Figure 4.39) (Odin et al. 2005d). The eutectic crystallizes at 1021°C ± 10°C and contains 55 ± 3 mol.% CdTe. This system was investigated through DTA, XRD, and metallography and the ingots were annealed at 740°C during 720 h.

3CdS + In₂Te₃ ⇔ 3CdTe + In₂S₃. The liquidus surface of the CdS–CdTe–CdIn₂S₄ quasiternary system is shown in Figure 4.40 (Odin et al. 2005d). Ternary eutectic *E* in this system crystallizes at 982°C ± 10°C.

The isothermal section of this ternary mutual system at 485°C is shown in Figure 4.41 (Odin et al. 2004). Odin et al. (2005d) constructed the isothermal section of the CdS–CdTe–CdIn₂S₄ quasiternary system at 740°C, which practically coincides with the results given in Figure 4.41. Solid solutions based on the binary components and CdIn₂S₄, CdIn₂Te₄, and CdIn₈Te₁₃ are participated in the phase equilibria.

The metastable phases Cd₃In₂S₂Te₄ (2CdS·CdIn₂Te₄) and Cd₂In₆S₂Te₉ (2CdS·3In₂Te₃) have been obtained in this system within the temperature interval from 23°C to 485°C (Odin et al. 2004, 2005b). Cd₃In₂S₂Te₄ has a homogeneity region within the interval of Cd₃In₂S₂Te₄–Cd₅In₂S₂Te₆ and crystallizes in the hexagonal structure with the lattice parameters *a* = 434.4 ± 0.4 and *c* = 710.3 ± 0.6 pm [*a* = 433.9 ± 0.4 and *c* = 708.1 ± 0.6 pm (Odin et al. 2004)]; calculation and experimental density 5.44 and 5.41 ± 0.01 g·cm⁻³, respectively; and energy gap 1.23 eV (Odin et al. 2005b).

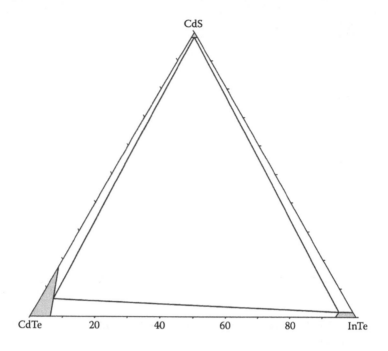

FIGURE 4.38 Isothermal section of the CdS–CdTe–InTe quasiternary system at 630°C. (From Odin, I.N. et al., *Zhurn. neorgan. khimii*, 50(4), 714, 2005c.)

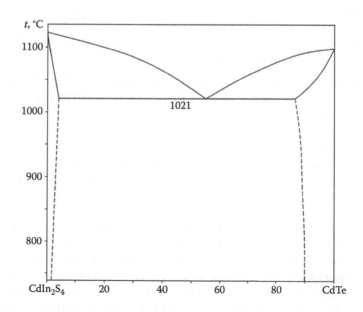

FIGURE 4.39 The CdTe–CdIn$_2$S$_4$ phase diagram. (From Odin, I.N. et al., *Zhurn. neorgan. khimii*, 50(6), 1018, 2005d.)

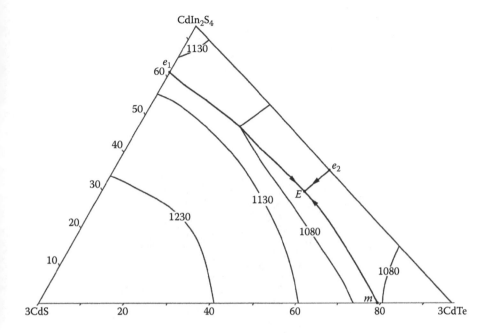

FIGURE 4.40 Liquidus surface of the 3CdS–3CdTe–CdIn$_2$S$_4$ quasiternary system. (From Odin, I.N. et al., *Zhurn. neorgan. khimii*, 50(6), 1018, 2005d.)

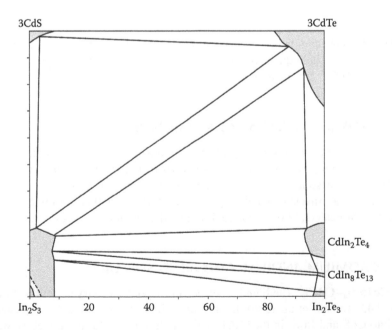

FIGURE 4.41 Isothermal section of the 3CdS + In$_2$Te$_3$ ⇔ 3CdTe + In$_2$S$_3$ ternary mutual system at 485°C. (From Odin, I.N. et al., *Zhurn. neorgan. khimii*, 49(5), 848, 2004.)

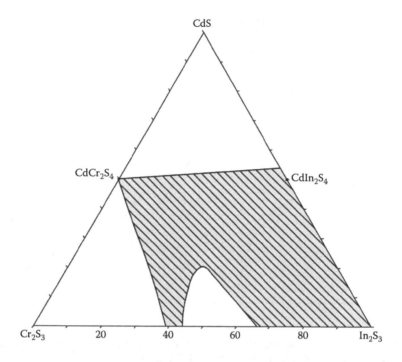

FIGURE 4.42 Isothermal section of the $CdS-In_2S_3-Cr_2S_3$ quasiternary system at 800°C. (From Lutz, H.D. et al., *J. Solid State Chem.*, 46(1), 56, 1983.)

The metastable compounds have been obtained using the mixture of CdS, CdTe, and In_2Te_3 by the rapid quenching of the melts (Odin et al. 2004). The ingots were annealing at 485°C during 1000 h.

4.63   CADMIUM–INDIUM–CHROMIUM–SULFUR

$CdS-In_2S_3-Cr_2S_3$. The phase diagram is not constructed. The isothermal section of this quasiternary system at 800°C is shown in Figure 4.42 (Lutz et al. 1983). A wide region of solid solutions with spinel structure exists at this temperature. This system was investigated through DTA, XRD, and thermogravimetry. The ingots were annealed at 600°C for 10 days and then at 800°C for 4 days.

4.64   CADMIUM–INDIUM–CHLORINE–SULFUR

$CdS-CdIn_2S_4-CdCl_2$. The liquidus surface of this quasiternary system is shown in Figure 4.43 (Buzhor et al. 1986). It was determined that $CdIn_2S_4$ decomposes partially into CdS and In_2S_3 in the $CdCl_2$ melt. Addition of CdS to the melt slows this decomposition.

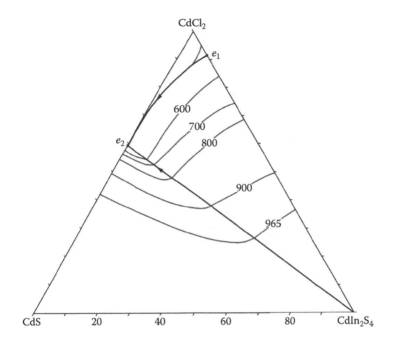

FIGURE 4.43   The liquidus surface of the $CdS–CdIn_2S_4–CdCl_2$ quasiternary system. (From Buzhor, V.P. et al., *Izv. AN SSSR. Neorgan. materialy*, 22(3), 497, 1986.)

4.65   CADMIUM–LANTHANUM–SILICON–SULFUR

$CdS–La_2S_3–SiS_2$. The isothermal section of this quasiternary system at 1050°C is shown in Figure 4.44 (Perez et al. 1970). The region of $La_6Cd_ySi_{2.5-y/2}\square_{1.5-y/2}S_{14}$ solid solutions with rhombohedral structure is strongly limited, and the formula of these solutions is closed to $La_6Cd\square Si_2S_{14}$ (the lattice parameters are $a = 1038.0 \pm 0.4$ and $c = 575.0 \pm 0.2$ pm).

4.66   CADMIUM–LANTHANUM–GERMANIUM–SULFUR

$CdS–La_2S_3–GeS_2$. The isothermal section of this quasiternary system at 1050°C is shown in Figure 4.45 (Perez et al. 1970). The region of $La_6Cd_yGe_{2.5-y/2}\square_{1.5-y/2}S_{14}$ solid solutions with rhombohedral structure is strongly limited, and the formula of these solutions is closed to $La_6Cd\square Ge_2S_{14}$ (the lattice parameters are $a = 1037.8 \pm 0.4$ and $c = 580.8 \pm 0.2$ pm).

4.67   CADMIUM–LANTHANUM–OXYGEN–SULFUR

$CdS–La_2O_2S$. The phase diagram is not constructed. According to the data of Baranov et al. (1996), this system is a quasibinary system of the Cd–La–O–S quaternary system. New phases were not found. This system was investigated through XRD. The ingots were annealed at 730°C–830°C for 250 h.

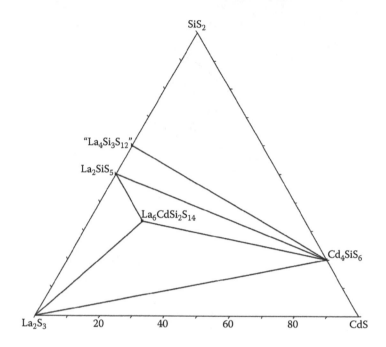

FIGURE 4.44 Isothermal section of the $CdS–La_2S_3–SiS_2$ quasiternary system at 1050°C. (From Perez, G. et al., *J. Solid State Chem.*, 2(1), 42, 1970.)

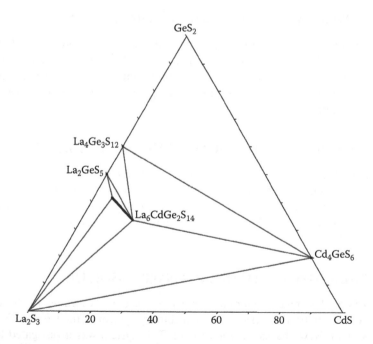

FIGURE 4.45 Isothermal section of the $CdS–La_2S_3–GeS_2$ quasiternary system at 1050°C. (From Perez, G. et al., *J. Solid State Chem.*, 2(1), 42, 1970.)

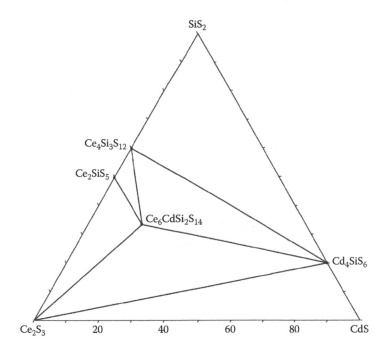

FIGURE 4.46 Isothermal section of the CdS–Ce$_2$S$_3$–SiS$_2$ quasiternary system at 1050°C. (From Perez, G. et al., *J. Solid State Chem.*, 2(1), 42, 1970.)

4.68 CADMIUM–CERIUM–SILICON–SULFUR

CdS–Ce$_2$S$_3$–SiS$_2$. The isothermal section of this quasiternary system at 1050°C is shown in Figure 4.46 (Perez et al. 1970). The region of Ce$_6$Cd$_y$Si$_{2.5-y/2}$□$_{1.5-y/2}$S$_{14}$ solid solutions with rhombohedral structure is strongly limited, and the formula of these solutions is closed to Ce$_6$Cd□Si$_2$S$_{14}$ (the lattice parameters are $a = 1022.5 \pm 0.4$ and $c = 570.5 \pm 0.2$ pm).

4.69 CADMIUM–CERIUM–GERMANIUM–SULFUR

CdS–Ce$_2$S$_3$–GeS$_2$. The isothermal section of this quasiternary system at 1050°C is shown in Figure 4.47 (Perez et al. 1970). The region of Ce$_6$Cd$_y$Ge$_{2.5-y/2}$□$_{1.5-y/2}$S$_{14}$ solid solutions with rhombohedral structure exists between the Ce$_6$Ge$_{2.5}$S$_{14}$ and Ce$_6$CdGe$_2$S$_{14}$; the lattice parameters of the quaternary compound are $a = 1022.5 \pm 0.4$ and $c = 577.0 \pm 0.2$ pm.

4.70 CADMIUM–PRASEODYMIUM–SILICON–SULFUR

CdS–Pr$_2$S$_3$–SiS$_2$. The isothermal section of this quasiternary system at 1050°C is shown in Figure 4.48 (Perez et al. 1970). The region of Pr$_6$Cd$_y$Si$_{2.5-y/2}$□$_{1.5-y/2}$S$_{14}$ solid solutions with rhombohedral structure is strongly limited, and the formula of these solutions is closed to Pr$_6$Cd□Si$_2$S$_{14}$ (the lattice parameters are $a = 1020.0 \pm 0.4$ and $c = 570.0 \pm 0.2$ pm).

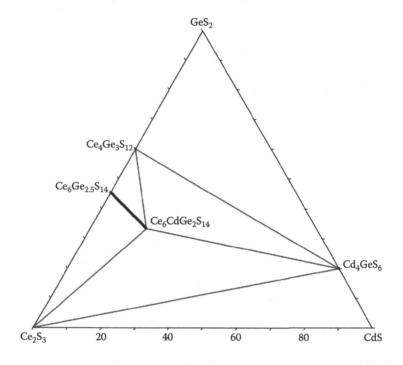

FIGURE 4.47   Isothermal section of the $CdS–Ce_2S_3–GeS_2$ quasiternary system at 1050°C. (From Perez, G. et al., *J. Solid State Chem.*, 2(1), 42, 1970.)

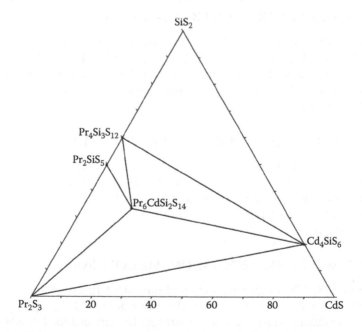

FIGURE 4.48   Isothermal section of the $CdS–Pr_2S_3–SiS_2$ quasiternary system at 1050°C. (From Perez, G. et al., *J. Solid State Chem.*, 2(1), 42, 1970.)

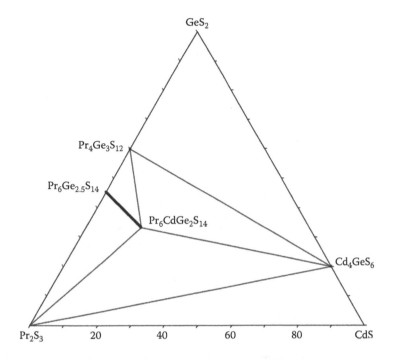

FIGURE 4.49 Isothermal section of the CdS–Pr_2S_3–GeS_2 quasiternary system at 1050°C. (From Perez, G. et al., *J. Solid State Chem.*, 2(1), 42, 1970.)

4.71 CADMIUM–PRASEODYMIUM–GERMANIUM–SULFUR

CdS–Pr_2S_3–GeS_2. The isothermal section of this quasiternary system at 1050°C is shown in Figure 4.49 (Perez et al. 1970). The region of $Pr_6Cd_yGe_{2.5-y/2}\square_{1.5-y/2}S_{14}$ solid solutions with rhombohedral structure exists between the $Pr_6Ge_{2.5}S_{14}$ and $Pr_6CdGe_2S_{14}$; the lattice parameters of the quaternary compound are $a = 1021.0 \pm 0.4$ and $c = 577.0 \pm 0.2$ pm.

4.72 CADMIUM–NEODYMIUM–SILICON–SULFUR

CdS–Nd_2S_3–SiS_2. The isothermal section of this quasiternary system at 1050°C is shown in Figure 4.50 (Perez et al. 1970). The region of $Nd_6Cd_ySi_{2.5-y/2}\square_{1.5-y/2}S_{14}$ solid solutions with rhombohedral structure is strongly limited, and the formula of these solutions is closed to $Nd_6Cd\square Si_2S_{14}$ (the lattice parameters are $a = 1016.0 \pm 0.4$ and $c = 568.5 \pm 0.2$ pm).

4.73 CADMIUM–NEODYMIUM–GERMANIUM–SULFUR

CdS–Nd_2S_3–GeS_2. The isothermal section of this quasiternary system at 1050°C is shown in Figure 4.51 (Perez et al. 1970). The region of $Nd_6Cd_yGe_{2.5-y/2}\square_{1.5-y/2}S_{14}$ solid solutions with rhombohedral structure exists between the $Nd_6Ge_{2.5}S_{14}$ and $Nd_6CdGe_2S_{14}$; the lattice parameters of the quaternary compound are $a = 1015.4 \pm 0.4$ and $c = 575.7 \pm 0.2$ pm.

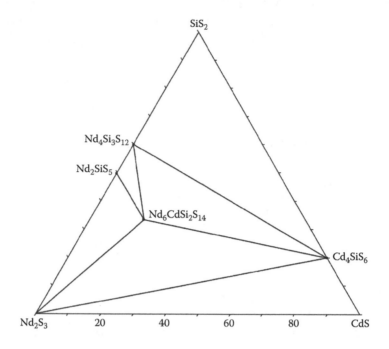

FIGURE 4.50 Isothermal section of the $CdS-Nd_2S_3-SiS_2$ quasiternary system at 1050°C. (From Perez, G. et al., *J. Solid State Chem.*, 2(1), 42, 1970.)

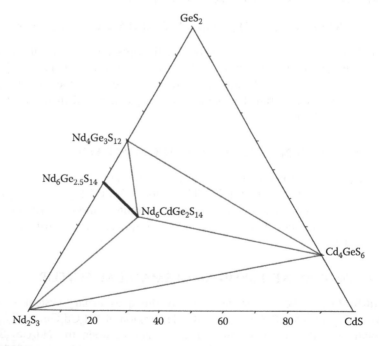

FIGURE 4.51 Isothermal section of the $CdS-Nd_2S_3-GeS_2$ quasiternary system at 1050°C. (From Perez, G. et al., *J. Solid State Chem.*, 2(1), 42, 1970.)

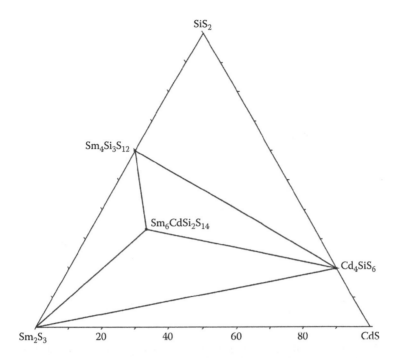

FIGURE 4.52 Isothermal section of the CdS–Sm_2S_3–SiS_2 quasiternary system at 1050°C. (From Perez, G. et al., *J. Solid State Chem.*, 2(1), 42, 1970.)

4.74 CADMIUM–SAMARIUM–SILICON–SULFUR

CdS–Sm_2S_3–SiS_2. The isothermal section of this quasiternary system at 1050°C is shown in Figure 4.52 (Perez et al. 1970). The region of $Sm_6Cd_ySi_{2.5-y/2}\square_{1.5-y/2}S_{14}$ solid solutions with rhombohedral structure is strongly limited, and the formula of these solutions is closed to $Sm_6Cd\square Si_2S_{14}$ (the lattice parameters are $a = 1006.0 \pm 0.4$ and $c = 566.0 \pm 0.2$ pm).

4.75 CADMIUM–SAMARIUM–GERMANIUM–SULFUR

CdS–Sm_2S_3–GeS_2. The isothermal section of this quasiternary system at 1050°C is shown in Figure 4.53 (Perez et al. 1970). The region of $Sm_6Cd_yGe_{2.5-y/2}\square_{1.5-y/2}S_{14}$ solid solutions with rhombohedral structure exists between the $Sm_6Ge_{2.5}S_{14}$ and $Sm_6CdGe_2S_{14}$; the lattice parameters of the quaternary compound are $a = 1005.2 \pm 0.4$ and $c = 573.3 \pm 0.2$ pm.

4.76 CADMIUM–GADOLINIUM–SILICON–SULFUR

CdS–Gd_2S_3–SiS_2. The isothermal section of this quasiternary system at 1050°C is shown in Figure 4.54 (Perez et al. 1970). The region of $Gd_6Cd_ySi_{2.5-y/2}\square_{1.5-y/2}S_{14}$ solid solutions with rhombohedral structure exists between the $Gd_6Si_{2.5}S_{14}$ and $Gd_6CdSi_2S_{14}$; the lattice parameters of the quaternary compound are $a = 998.5 \pm 0.4$ and $c = 563.5 \pm 0.2$ pm.

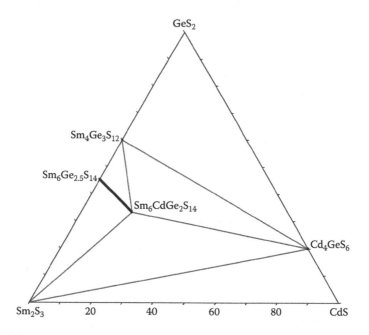

FIGURE 4.53   Isothermal section of the CdS–Sm_2S_3–GeS_2 quasiternary system at 1050°C.
(From Perez, G. et al., *J. Solid State Chem.*, 2(1), 42, 1970.)

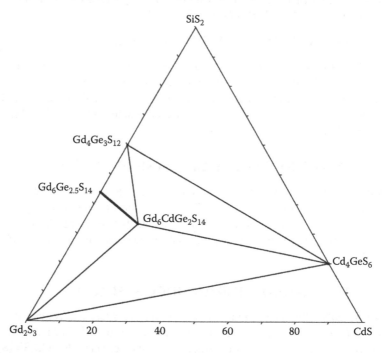

FIGURE 4.54   Isothermal section of the CdS–Gd_2S_3–SiS_2 quasiternary system at 1050°C.
(From Perez, G. et al., *J. Solid State Chem.*, 2(1), 42, 1970.)

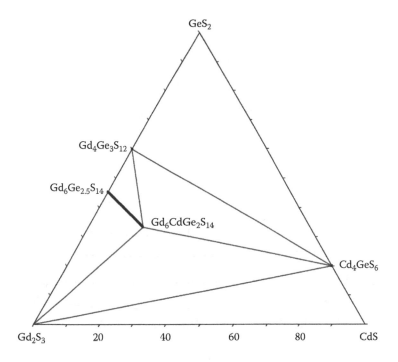

FIGURE 4.55 Isothermal section of the $CdS-Gd_2S_3-GeS_2$ quasiternary system at 1050°C. (From Perez, G. et al., *J. Solid State Chem.*, 2(1), 42, 1970.)

4.77 CADMIUM–GADOLINIUM–GERMANIUM–SULFUR

$CdS-Gd_2S_3-GeS_2$. The isothermal section of this quasiternary system at 1050°C is shown in Figure 4.55 (Perez et al. 1970). The region of $Gd_6Cd_yGe_{2.5-y/2}\square_{1.5-y/2}S_{14}$ solid solutions with rhombohedral structure exists between the $Gd_6Ge_{2.5}S_{14}$ and $Gd_6CdGe_2S_{14}$; the lattice parameters of the quaternary compound are $a = 997.0 \pm 0.4$ and $c = 572.5 \pm 0.2$ pm.

4.78 CADMIUM–CARBON–FLUORINE–SULFUR

$Cd(SC_6F_5)_2$, or $C_{12}F_{10}S_2Cd$, quaternary compound is formed in this system. To obtain this compound, a slight excess, maximum of 5%, of C_6F_5SH was added to approximately 3 mmol of Cd^{2+} in 25 mL H_2O or dilute acid (Peach 1968). The mixture was stirred magnetically until a solid formed and no more C_6F_5SH appeared to be present. The product was filtered off, washed, and dried. $Cd(SC_6F_5)_2$ melts at a temperature higher than 350°C.

4.79 CADMIUM–CARBON–CHLORINE–SULFUR

$Cd(SC_6Cl_5)_2$, or $C_{12}Cl_{10}S_2Cd$, quaternary compound is formed in this system. This compound was obtained when a suspension of C_6Cl_5SH and a solution of $Cd(NO_3)_2 \cdot 4H_2O$ in 95% ethanol were stirred at room temperature (Lucas and Peach 1969, 1970). It melts at a temperature higher than 300°C.

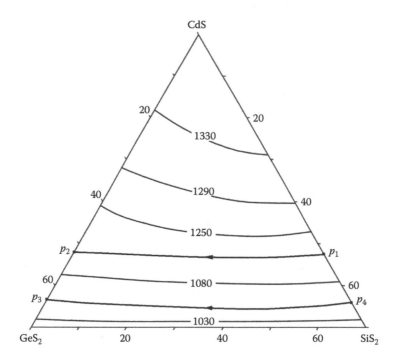

FIGURE 4.56 A part of the liquidus surface of the $CdS-GeS_2-SiS_2$ quasiternary system at the 70–100 mol.% CdS. (From Odin, I.N. and Chukichev, M.V., *Zhurn. neorgan. khimii*, 45(2), 255, 2000.)

4.80 CADMIUM–SILICON–GERMANIUM–SULFUR

$CdS-GeS_2-SiS_2$. The liquidus surface of this quasiternary system at the 70–100 mol.% CdS is shown in Figure 4.56 (Odin and Chukichev 2000). It consists of three fields of the primary crystallization of the solid solutions based on CdS and α- and β-$Cd_4Si_xGe_{1-x}S_6$ solid solutions. The ingots were annealed at 20°C lower than the temperature of the nonvariant equilibrium with liquid participation for 1000 h. The system was investigated using DTA, XRD, and metallography.

4.81 CADMIUM–SILICON–SELENIUM–SULFUR

$CdS-SiSe_2$. This section is a nonquasibinary section of the Cd–Si–Se–S quaternary system and is similar to the $CdSe-SiS_2$ section (Odin 1996). The $CdS-SiSe_2$ section was investigated through DTA, metallography, and XRD. The ingots were annealed at 20°C below the temperatures of nonvariant equilibria with liquid participation for 1000 h.

$CdSe-SiS_2$. This section is a nonquasibinary section of the Cd–Si–Se–S quaternary system (Figure 4.57) (Odin 1996). Solid solutions $CdSe_{1-x}S_x$ and $SiS_{2-x}Se_x$ crystallize primarily in this system. Phases based on β- and α-$Cd_4SiSe_{6-x}S_x$ also crystallize from the CdSe-rich side. The next peritectic reaction $L + CdSe_{1-x}S_x \Leftrightarrow β-Cd_4SiSe_{6-x}S_x$ takes place in this system at 80 mol.% CdSe.

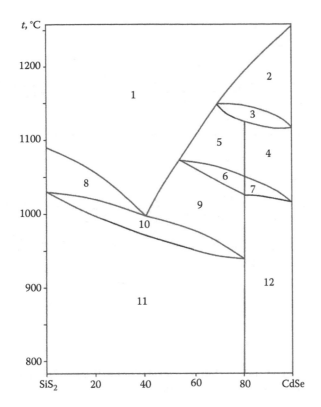

FIGURE 4.57 Phase relations in the CdSe–SiS$_2$ system: 1, L; 2, L + CdSe$_{1-x}$S$_x$; 3, L + CdSe$_{1-x}$S$_x$ + β-Cd$_4$SiSe$_{6-x}$S$_x$; 4, CdSe$_{1-x}$S$_x$ + β-Cd$_4$SiSe$_{6-x}$S$_x$; 5, L + β-Cd$_4$SiSe$_{6-x}$S$_x$; 6, L + α-Cd$_4$SiSe$_{6-x}$S$_x$ + β-Cd$_4$SiSe$_{6-x}$S$_x$; 7, CdSe$_{1-x}$S$_x$ + α-Cd$_4$SiSe$_{6-x}$S$_x$ + β-Cd$_4$SiSe$_{6-x}$S$_x$; 8, L + SiS$_{2(1-x)}$Se$_{2x}$; 9, L + α-Cd$_4$SiSe$_{6-x}$S$_x$; 10, L + SiS$_{2(1-x)}$Se$_{2x}$ + α-Cd$_4$SiSe$_{6-x}$S$_x$; 11, SiS$_{2(1-x)}$Se$_{2x}$ + α-Cd$_4$SiSe$_{6-x}$S$_x$; and 12, α-Cd$_4$SiSe$_{6-x}$S$_x$ + CdSe$_{1-x}$S$_x$. (From Odin, I.N., *Zhurn. neorgan. khimii*, 41(6), 941, 1996.)

This system was investigated through DTA, metallography, and XRD. The ingots were annealed at 20°C below the temperatures of nonvariant equilibria with liquid participation for 1000 h (Odin 1996).

2CdS + SiSe$_2$ ⇔ 2CdSe + SiS$_2$. The fields of primary crystallizations of CdSe$_{1-x}$S$_x$, SiS$_{2-x}$Se$_x$, β-Cd$_4$SiSe$_{6-x}$S$_x$, and α-Cd$_4$SiSe$_{6-x}$S$_x$ solid solutions exist on the liquidus surface of this ternary mutual system (Figure 4.58) (Odin 1996). The p_1p_3-, p_2p_4-, and e_1e_2-lines correspond to the peritectic formation of the β-Cd$_4$SiSe$_{6-x}$S$_x$ solid solutions, to the polymorphous transformation α-Cd$_4$SiSe$_{6-x}$S$_x$ → β-Cd$_4$SiSe$_{6-x}$S$_x$ with liquid participation, and to the crystallization of SiS$_{2-x}$Se$_x$ + α-Cd$_4$SiSe$_{6-x}$S$_x$ mixtures from the liquid, respectively. These lines are monovariant and no monovariant points exist in this system. This system was investigated through DTA, metallography, and XRD. The ingots were annealed at 20°C below the temperatures of nonvariant equilibria with liquid participation for 1000 h.

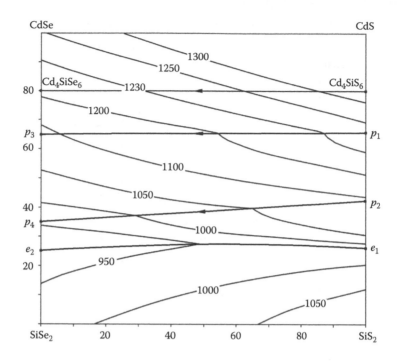

FIGURE 4.58 Liquidus surface of the $2CdS + SiSe_2 \Leftrightarrow 2CdSe + SiS_2$ ternary mutual system. (From Odin, I.N., *Zhurn. neorgan. khimii*, 41(6), 941, 1996.)

4.82 CADMIUM–GERMANIUM–SELENIUM–SULFUR

CdS–GeSe₂. The phase diagram is a eutectic type (Figure 4.59) (Mirzoyev et al. 2006). The eutectic composition and temperature are 30 mol.% CdS and 667°C, respectively. $Cd_4GeSe_2S_4$ quaternary compound is formed in this system. It melts incongruently at 877°C (peritectic point contains 50 mol.% CdS) and crystallizes in the monoclinic structure with the lattice parameters $a = 1261.8$, $b = 724.2$, $c = 1255.3$ pm, and $\beta = 110°80'$ and calculation and experimental densities 5.01 and 5.12 g·cm⁻³, respectively. The limited solid solutions based on CdS and GeSe₂ are formed in this system (Asadov 2006, Mirzoyev et al. 2006).

This system was investigated through DTA, XRD, metallography, and measuring of microhardness and emf of concentration chains (Asadov 2006, Mirzoyev et al. 2006).

4.83 CADMIUM–TIN–TELLURIUM–SULFUR

CdS–SnTe. This section is a nonquasibinary section of the Cd–Sn–Te–S quaternary system (Figure 4.60) (Dubrovin et al. 1985). Solubility of CdS in SnTe at 725°C is equal 12 mol.% and decreases to 2 mol.% at 400°C. Solubility of SnTe in CdS is not higher than 1 mol.%. Thermal effects at 605°C correspond to the polymorphous transformation of the solid solutions based on SnS (γ) forming at the exchange interaction of CdS and SnTe. This system was investigated through DTA, metallography, and XRD. The ingots were annealed at 400°C, 500°C, and 670°C for 250, 200, and 100 h, respectively.

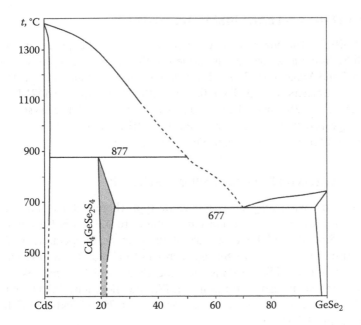

FIGURE 4.59 The CdS–GeSe$_2$ phase diagram. (From Mirzoyev, A.J. et al., *Zhurn. khim. problem*, (2), 322, 2006.)

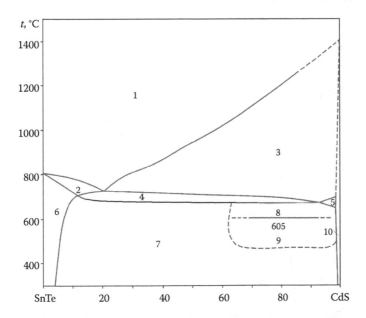

FIGURE 4.60 Phase relations in the CdS–SnTe system: 1, L; 2, L + (SnTe); 3, L + (CdS); 4, L + (CdS) + (SnTe); 5, L + (CdS) + γ; 6, (SnTe); 7, (CdS) + (SnTe); 8, (CdS) + (SnTe) + γ′; 9, (CdS) + (SnTe) + γ; and 10, (CdS). (From Dubrovin, I.V. et al., *Izv. AN SSSR. Neorgan. materialy*, 21(11), 1873, 1985.)

4.84 CADMIUM–LEAD–BISMUTH–SULFUR

CdS–PbS–Bi$_2$S$_3$. The phase diagram is not constructed. The quaternary compound (Cd,Pb)Bi$_2$S$_4$ (mineral kudriavite) is formed in this system (Chaplygin et al. 2005, Balić-Žunić and Makovicky 2007). It crystallizes in the monoclinic structure with the lattice parameters a = 1309.5 ± 0.1, b = 400.32 ± 0.03, c = 1471.1 ± 0.1 pm, and β = 115.59° ± 0.01° and calculated density 6.578 g·cm^{-3}. The crystal–chemical analysis suggests that the observed ratio Pb–Cd = 1:1 in natural kudriavite represents the upper limit for the substitution of Pb in the structure.

4.85 CADMIUM–LEAD–SELENIUM–SULFUR

CdS–PbSe. The phase diagram is a eutectic type (Figure 4.61) (Tomashik et al. 1981). The eutectic composition and temperature are 42 mol.% CdS and 1000°C, respectively. Solubility of PbSe in CdS at the eutectic temperature is 33 mol.% [30 mol.% (Leute and Böttner 1978)] and decreases to 11 mol.% at 800°C and 9 mol.% at 750°C. Solubility of PbSe in CdS is not higher than 1 mol.% (Tomashik et al. 1981).

This system was investigated through DTA, metallography, XRD, and measuring of microhardness. The ingots were annealed at 750°C and 800°C for 100 h (Tomashik et al. 1981).

CdSe–PbS. This section is a nonquasibinary section of the Cd–Pb–Se–S quaternary system since at the interaction of CdSe and PbS, lead selenide and CdS were determined (Leute and Böttner 1978).

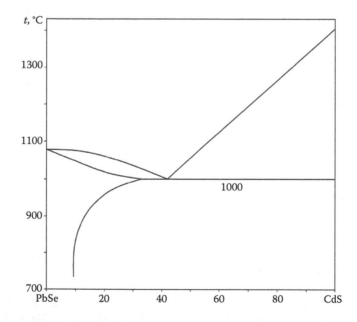

FIGURE 4.61 The CdS–PbSe phase diagram. (From Tomashik, Z.F. et al., *Izv. AN SSSR. Neorgan. materialy*, 17(12), 2155, 1981.)

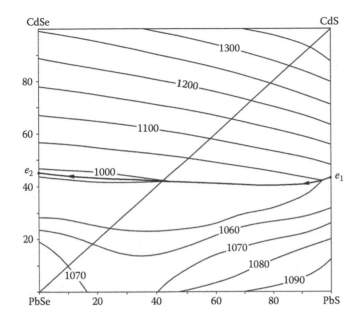

FIGURE 4.62 Liquidus surface of the CdS + PbSe ⇔ CdSe + PbS ternary mutual system. (From Tomashik, Z.F. and Tomashik, V.N. et al., *Izv. AN SSSR. Neorgan. materialy*, 20(4), 568, 1984.)

CdS + PbSe ⇔ CdSe + PbS. The fields of primary crystallization of CdS_xSe_{1-x} and PbS_xSe_{1-x} solid solutions exist on the liquidus surface of this ternary mutual system (Figure 4.62) (Tomashik and Tomashik 1984). This system was investigated through DTA and using mathematical planning of experiment.

4.86 CADMIUM–LEAD–TELLURIUM–SULFUR

CdS–PbTe. The phase diagram is a eutectic type (Figure 4.63) (Tomashik and Tomashik 1987). The eutectic composition and temperature are 20 mol.% CdS and 874°C, respectively. The solubility of CdS in PbTe at the eutectic temperature is 13 mol.% and the solubility of PbTe in CdS is not higher than 1 mol.%. This system was investigated through DTA and metallography.

CdS + PbTe ⇔ CdTe + PbS. The fields of primary crystallization of CdS_xTe_{1-x} and PbS_xTe_{1-x} solid solutions exist on the liquidus surface of this ternary mutual system (Figure 4.64) (Tomashik and Tomashik 1987). This system was investigated through DTA, metallography, and using mathematical simulation of experiment.

4.87 CADMIUM–LEAD–IODINE–SULFUR

CdS–PbI₂. The phase diagram is a eutectic type (Figure 4.65) (Odin 2001). The eutectic composition and temperature are 23 ± 1 mol.% CdS and 366°C, respectively. Mutual solubility of CdS and PbI_2 is negligible: the solubility of PbI_2

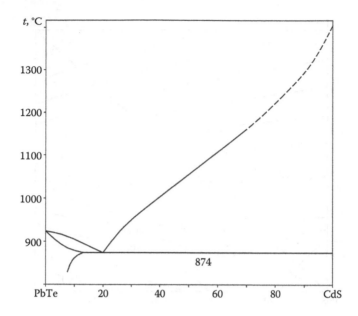

FIGURE 4.63 The CdS–PbTe phase diagram. (From Tomashik, Z.F. and Tomashik, V.N., *Izv. AN SSSR. Neorgan. materialy*, 23(12), 1981, 1987.)

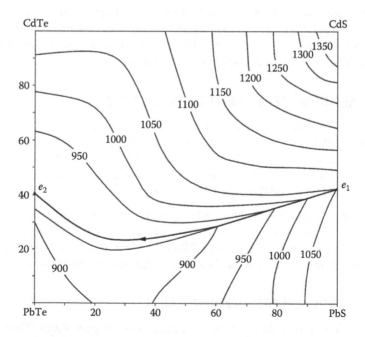

FIGURE 4.64 Liquidus surface of the CdS + PbTe ⇔ CdTe + PbS ternary mutual system. (From Tomashik, Z.F. and Tomashik, V.N., *Izv. AN SSSR. Neorgan. materialy*, 23(12), 1981, 1987.)

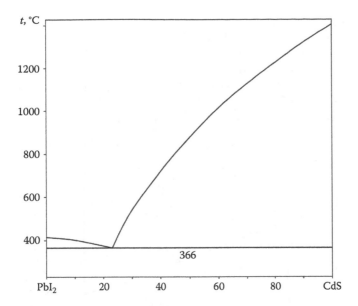

FIGURE 4.65 The CdS–PbI$_2$ phase diagram. (From Odin I.N., *Zhurn. neorgan. khimii*, 46(10), 1733, 2001.)

in CdS is 0.15 mol.%. The crystallization of the melts from the PbI$_2$ side leads to the formation of two polytypic forms of this compound: 6R-PbI$_2$ and 2H-PbI$_2$ (Odin 2001, Odin and Chukichev 2001). This system was investigated through DTA, XRD, and metallography, and ingots were annealed during 1000 h (Odin 2001).

CdS + PbI$_2$ ⇔ CdI$_2$ + PbS. The fields of primary crystallization of solid solutions based on CdS, PbS, and Pb$_5$S$_2$I$_6$ and Cd$_x$Pb$_{1-x}$I$_2$ solid solutions exist on the liquidus surface of this ternary mutual system (Figure 4.66) (Odin 2001). The reaction L + PbS ⇔ Pb$_5$S$_2$I$_6$ + CdS takes place in the transition point U at 409°C. The ternary eutectic E at 357°C is formed by the solid solutions based on CdS and Pb$_5$S$_2$I$_6$ and Cd$_x$Pb$_{1-x}$I$_2$ solid solutions. A minimum m exists on the e_2e_4 line of the secondary crystallization.

The isothermal section of this ternary mutual system at 350°C is shown in Figure 4.67 (Odin 2001).

Metastable phases can form at the crystallization of the melts (Odin and Chukichev 2001).

4.88   CADMIUM–NITROGEN–PHOSPHORUS–SULFUR

CdS–Cd$_3$N$_2$–P$_3$N$_5$. The phase diagram is not constructed. The Cd$_4$P$_6$N$_{12}$S quaternary compound is formed in this system (Ronis et al. 1988a,b,c, 1989). It crystallizes in the cubic structure with the lattice parameter a = 847.3 ± 0.1 pm. This compound was obtained by the interaction of CdS and P$_2$S$_5$ in ammonia at 900°C–970°C according to the reaction 4CdS + 3P$_2$S$_5$ + 12NH$_3$ = Cd$_4$P$_6$N$_{12}$S.

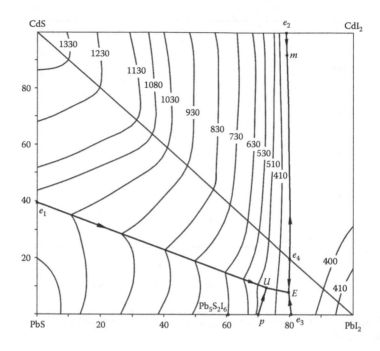

FIGURE 4.66 Liquidus surface of the CdS + PbI$_2$ ⇔ CdI$_2$ + PbS ternary mutual system. (From Odin I.N., *Zhurn. neorgan. khimii*, 46(10), 1733, 2001.)

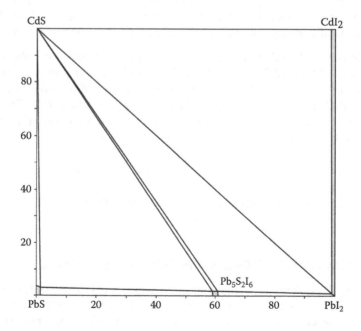

FIGURE 4.67 Isothermal section of the CdS + PbI$_2$ ⇔ CdI$_2$ + PbS ternary mutual system at 350°C. (From Odin I.N., *Zhurn. neorgan. khimii*, 46(10), 1733, 2001.)

4.89   CADMIUM–PHOSPHORUS–IODINE–SULFUR

CdS–CdI$_2$–P$_2$S$_5$. The phase diagram is not constructed. Cd$_{14-x}$P$_4$S$_{24-2x}$I$_{2x}$ crystal series with $0 \leq x \leq 1$ were synthesized from the elements in this system (Grieshaber et al. 1976). Cd$_{13}$P$_4$S$_{22}$I$_2$ has a reversible solid–solid phase transition at 380°C. Low-temperature modification of this compound is monoclinic with the lattice parameters $a = 1223.6 \pm 0.5$, $b = 701.6 \pm 0.4$, $c = 1228.9 \pm 0.6$ pm, and $\beta = 110.33° \pm 0.05°$ and calculation and experimental densities 4.269 and 4.10 ± 0.01 g·cm^{-3}, respectively. High-temperature modification crystallizes in the cubic structure with the lattice parameter $a = 996.9$ pm and experimental density 3.98 g·cm^{-3} (Bubenzer et al. 1976, Grieshaber et al. 1976). Cd$_{13}$P$_4$S$_{22}$I$_2$ quaternary compound begins to oxidize in air at 450°C, and in flowing nitrogen, it is stable up to 480°C, when this compound starts to decompose quantitatively into CdS, I$_2$, and volatile phosphorus sulfides (Grieshaber et al. 1976).

This system was investigated through DTA, XRD, and chemical analysis. Cd$_{14-x}$P$_4$S$_{24-2x}$I$_{2x}$ single crystals were prepared by I$_2$ vapor transport reactions (Bubenzer et al. 1976, Grieshaber et al. 1976).

4.90   CADMIUM–PHOSPHORUS–IRON–SULFUR

CdS–P$_2$S$_5$–Fe. The phase diagram is not constructed. The CdFeP$_2$S$_6$ quaternary compound is formed in this system (Lee et al. 1988). It crystallizes in the mono-clinic structure with the lattice parameters $a = 608.0$, $b = 1047.8$, $c = 680.1$ pm, and $\beta = 107.26°$. This compound belongs to the series of compounds with formula Cd$_x$Fe$_{1-x}$PS$_3$ where $0.2 \leq x \leq 0.8$, which were obtained by the interaction of chemical elements at 600°C.

4.91   CADMIUM–ARSENIC–SELENIUM–SULFUR

CdS–As$_2$Se$_3$. This section is a nonquasibinary section of the Cd–As–Se–S quaternary system (Figure 4.68) (Rustamov and Aliev 1986). Solid solutions based on As$_2$Se$_3$, CdS, and CdSe primarily crystallize in this system. Four-phase peritectic reaction L + CdSe ⇔ (As$_2$Se$_3$) + Cd$_3$As$_2$S$_3$Se$_3$ takes place at 250°C from the As$_2$Se$_3$-rich side. From the CdS-rich side, CdSe interacts with liquid at 280°C according to the next peritectic reaction: L + CdSe ⇔ Cd$_3$As$_2$S$_3$Se$_3$ + CdS. At slow cooling, the glass-forming region based on As$_2$Se$_3$ reaches 13 mol.% CdS and at the quenching, this region enlarges up to 18 mol.% CdS. Glass + crystal region exists within the interval of 13–40 mol.% CdS. This system was investigated through DTA, metallography, XRD, and measuring of microhardness and density.

4.92   CADMIUM–ARSENIC–TELLURIUM–SULFUR

CdS–As$_2$Te$_3$. This section is a nonquasibinary section of the Cd–As–Te–S quaternary system (Figure 4.69) (Aliev and Aliev 1991). Solid solutions based on As$_2$Te$_3$, As$_2$S$_3$, CdS, and CdTe primarily crystallize in this system. Primary crystallization

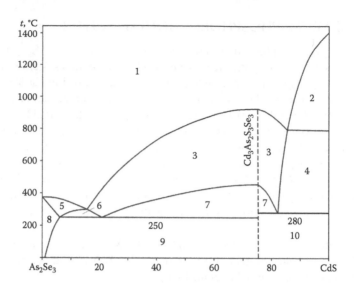

FIGURE 4.68 Phase relations in the CdS–As₂Se₃ system: 1, L; 2, L + CdS; 3, L + CdSe; 4, L + CdSe + CdS; 5, L + (As₂Se₃); 6, L + (As₂Se₃) + CdSe; 7, L + CdSe + Cd₃As₂S₃Se₃; 8, (As₂Se₃); 9, (As₂Se₃) + Cd₃As₂S₃Se₃; and 10, CdS + Cd₃As₂S₃Se₃. (From Rustamov, P.G. and Aliev, I.I., *Zhurn. neorgan. khimii*, 31(3), 771, 1986.)

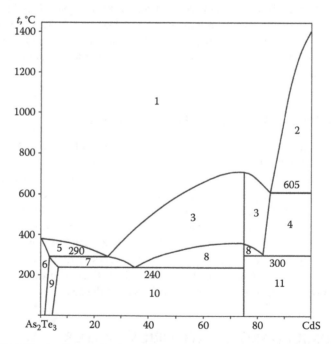

FIGURE 4.69 Phase relations in the CdS–As₂Te₃ system: 1, L; 2, L + CdS; 3, L + CdTe; 4, L + CdS + CdTe; 5, L + (As₂Te₃); 6, (As₂Te₃); 7, L + (As₂Te₃) + CdTe; 8, L + As₂S₃ + CdTe; 9, (As₂Te₃) + CdTe; 10, (As₂Te₃) + As₂S₃ + CdTe; 11, As₂S₃ + CdS + CdTe. (From Aliev, I.I. and Aliev, O.M., *Zhurn. neorgan. khimii*, 36(9), 2383, 1991.)

of (As_2Te_3) and CdTe takes place within the intervals of 0–25 and 25–75 mol.% CdS, respectively. Simultaneous crystallization of (As_2Te_3) and CdS ends in the binary eutectic, and simultaneous crystallization of (As_2Te_3), CdTe, and As_2S_3 takes place in the ternary eutectic. The composition and temperature of this eutectic are 35 mol.% CdS and 240°C, respectively. The composition and temperature of the eutectic from the CdS-rich side are 85 mol.% CdS and 605°C, respectively. Solubility of CdS in As_2Te_3 is not higher than 1.5 mol.%. This system was investigated through DTA, metallography, XRD, and measuring of microhardness and density. The ingots were annealed at 230°C (0–75 mol.% CdS) and 290°C (75–100 mol.% CdS) for 650 h.

4.93 CADMIUM–ANTIMONY–SELENIUM–SULFUR

CdS–Sb_2Se_3. The phase diagram is shown in Figure 4.70 (Safarov and Abilov 1987, Safarov et al. 1989a, 1990). The $CdSb_2Se_3S$ quaternary compound is formed in this system, which melts incongruently at 557°C. Solubility of CdS in Sb_2Se_3 and Sb_2Se_3 in CdS at room temperature reaches 6 and 18 mol.%, respectively. It is necessary to note that CdS has no phase transformations in the investigated temperature region; therefore, this phase diagram raises some doubts.

This system was investigated through DTA, metallography, XRD, and measuring of microhardness and density (Safarov and Abilov 1987, Safarov et al. 1989a, 1990). The ingots were annealed at 330°C [at 300°C (Safarov and Abilov 1987)] for 200 h (Safarov et al. 1989a, 1990).

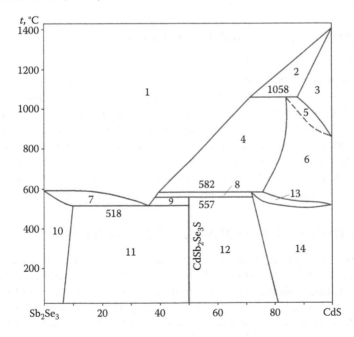

FIGURE 4.70 The CdS–Sb_2Se_3 phase diagram: 1, L; 2, L + γ-CdS; 3, γ-CdS; 4, L + β-CdS; 5, β-CdS + γ-CdS; 6, β-CdS; 7, L + (Sb_2Se_3); 8, L + α-CdS; 9, L + $CdSb_2Se_3S$; 10, (Sb_2Se_3); 11, $CdSb_2Se_3S$ + (Sb_2Se_3); 12, α-CdS + $CdSb_2Se_3S$; 13, α-CdS + β-CdS; and 14, α-CdS. (From Safarov, M.G. et al., *Zhurn. neorgan. khimii*, 34(6), 1583, 1989a.)

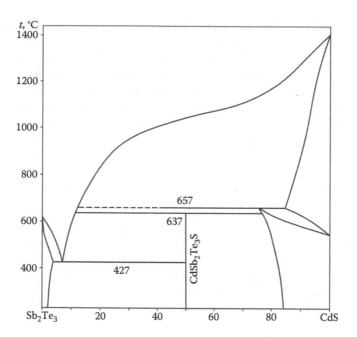

FIGURE 4.71 The CdS–Sb$_2$Te$_3$ phase diagram. (From Safarov, M.G. et al., *Zhurn. neorgan. khimii*, 34(7), 1906, 1989b.)

4.94 CADMIUM–ANTIMONY–TELLURIUM–SULFUR

CdS–Sb$_2$Te$_3$. The phase diagram is shown in Figure 4.71 (Safarov et al. 1989b). The eutectic composition and temperature are 7 mol.% CdS and 427°C, respectively. The CdSb$_2$Te$_3$S quaternary compound is formed in this system, which melts incongruently at 637°C. Solubility of CdS in Sb$_2$Te$_3$ and Sb$_2$Te$_3$ in CdS at room temperature reaches 3 and 12 mol.%, respectively (Safarov et al. 1989b, 2000). It is necessary to note that CdS has no phase transformations in the investigated temperature region; therefore, this phase diagram raises some doubts. This system was investigated through DTA, metallography, XRD, and measuring of microhardness and density. The ingots were annealed at 300°C for 100 h (Safarov et al. 1989b).

4.95 CADMIUM–ANTIMONY–CHLORINE–SULFUR

CdS–SbSCl. The phase diagram is not constructed. The CdSbS$_2$Cl quaternary compound is formed in this system (Wang et al. 2006). It crystallizes in the orthorhombic structure with the lattice parameters $a = 958.5 \pm 0.2$, $b = 399.10 \pm 0.08$, and $c = 1244.3 \pm 0.3$ pm. The stoichiometric synthesis of this compound was successfully carried out as follows: $3CdS + 2Sb + SbCl_3 + 3S = 3CdSbS_2Cl$. The mixture was first slowly (0.5°C/min) heated to 250°C and isothermed for 10 h and then heated at 1°C/min to 400°C–430°C, isothermed for 4 days, and slowly cooled (0.1°C/min) to room temperature. A stoichiometric yield of polycrystalline product was acquired by employing different starting materials according to equation $CdS + 2Sb + CdCl_2 + 3S = 2CdSbS_2Cl$ at a temperature as low as 450°C. When the same reaction mixture was used, larger single crystals were grown by first heating the reactants to 900°C followed by slow cooling to room temperature. Single crystals of this compound can also be grown by a solid-state reaction.

4.96   CADMIUM–ANTIMONY–BROMINE–SULFUR

CdS–SbSBr. The phase diagram is not constructed. The $CdSbS_2Br$ quaternary compound is formed in this system (Wang et al. 2006). It crystallizes in the monoclinic structure with the lattice parameters $a = 1293.8 \pm 0.3$, $b = 393.10 \pm 0.08$, $c = 966.10 \pm 0.19$ pm, and $\beta = 91.11° \pm 0.03°$. The stoichiometric synthesis of this compound was successfully carried out as follows: $3CdS + 2Sb + SbBr_3 + 3S = 3CdSbS_2Br$. The mixture was first slowly (0.5°C/min) heated to 250°C and isothermed for 10 h and then heated at 1°C/min to 400°C–430°C, isothermed for 4 days, and slowly cooled (0.1°C/min) to room temperature. A stoichiometric yield of polycrystalline product was acquired by employing different starting materials according to equation $CdS + 2Sb + CdBr_2 + 3S = 2CdSbS_2Br$ at a temperature as low as 450°C. When the same reaction mixture was used, larger single crystals were grown by first heating the reactants to 900°C followed by slow cooling to room temperature. Single crystals of this compound can also be grown by a solid-state reaction.

4.97   CADMIUM–ANTIMONY–IODINE–SULFUR

CdS–Sb$_2$S$_3$–SbSI. The phase diagram is not constructed. The $CdSb_6S_8I_4$ quaternary compound is formed in this system (Sirota et al. 1976). It crystallizes in the triclinic structure with the lattice parameters $a = 886.2 \pm 0.3$, $b = 891.3 \pm 0.2$, and $c = 948.7 \pm 0.2$ pm; $\alpha = 133.61° \pm 0.02°$, $\beta = 86.28° \pm 0.02°$, and $\gamma = 89.11° \pm 0.02°$; and calculated density 4.86 g·cm^{-3}.

4.98   CADMIUM–BISMUTH–TELLURIUM–SULFUR

CdS–Bi$_2$Te$_3$. The phase diagram is a eutectic type (Figure 4.72) (Datsenko et al. 1981). The eutectic is degenerated from the Bi_2Te_3-rich side and crystallizes at 578°C ± 3°C. Mutual solubility of CdS and Bi_2Te_3 is insignificant. This system was investigated through DTA and XRD.

4.99   CADMIUM–BISMUTH–CHLORINE–SULFUR

CdS–BiSCl. The phase diagram is not constructed. The $CdBiS_2Cl$ quaternary compound is formed in this system (Wang et al. 2006). It crystallizes in the orthorhombic structure with the lattice parameters $a = 954.1 \pm 0.2$, $b = 397.00 \pm 0.08$, and $c = 1254.5 \pm 0.3$ pm. Crystals of this compound were initially obtained from quaternary Bi/K/Ti/S reaction. Bi, K, Ti, and S were mixed in a 2:4:3:10 molar ratio. The reaction mixture was loaded in a carbon-coated tube with $CdCl_2$ and KCl eutectic flux (3:1) with a charge-to-flux ratio of 1:4 by weight. When a slow heating process was used to avoid an explosion due to the volatile S, the reaction mixture was heated

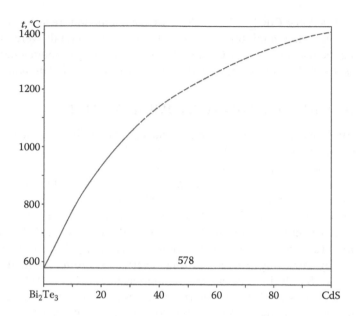

FIGURE 4.72   The CdS–Bi$_2$Te$_3$ phase diagram. (From Datsenko, A.M. et al., *Tr. Mosk. khim. tehnol. in-t*, (120), 32, 1981.)

to 990°C in three steps: first to 754°C (0.5°C/min), held for 1 day, and then to 990°C (0.75°C/min). The reaction was isothermed for 3 days and followed by slow cooling at a rate of 3°C/h to room temperature. Reddish transparent, needle-shaped crystals were isolated from the flux. To make large single crystals, a vapor transport method with I$_2$ was employed using the obtained polycrystalline product. Single crystals of this compound can also be grown by quenching the reaction melts at 500°C or directly grown by a solid-state reaction.

The stoichiometric synthesis of CdBiS$_2$Cl was successfully carried out as follows: 3CdS + 2Bi + BiCl$_3$ + 3S = 3CdBiS$_2$Cl (Wang et al. 2006). The mixture was first slowly (0.5°C/min) heated to 250°C and isothermed for 10 h and then heated at 1°C/min to 400°C–430°C, isothermed for 4 days, and slowly cooled (0.1°C/min) to room temperature. A stoichiometric yield of polycrystalline product was acquired by employing different starting materials according to equation CdS + 2Bi + CdCl$_2$ + 3S = 2CdBiS$_2$Cl at a temperature as low as 450°C. When the same reaction mixture was used, larger single crystals were grown by first heating the reactants to 900°C followed by slow cooling to room temperature.

4.100   CADMIUM–BISMUTH–BROMINE–SULFUR

CdS–BiSBr. The phase diagram is not constructed. The CdBiS$_2$Br quaternary compound is formed in this system (Wang et al. 2006). It crystallizes in the monoclinic structure with the lattice parameters $a = 1297.7 \pm 0.3$, $b = 401.20 \pm 0.08$, $c = 958.4 \pm 0.2$ pm, and $\beta = 91.07° \pm 0.03°$. Crystals of CdBiS$_2$Br could be obtained from quaternary Bi/K/Ti/S reaction as is described for the crystals of CdBiS$_2$Cl using

$CdBr_2$ + KBr eutectic flux instead of $CdCl_2$ and KCl eutectic flux. To make large single crystals, a vapor transport method with I_2 was employed using the obtained polycrystalline product. Single crystals of this compound can also be grown by a solid-state reaction.

The stoichiometric synthesis of this compound was successfully carried out as follows: $3CdS + 2Bi + BiBr_3 + 3S = 3CdBiS_2Br$ (Wang et al. 2006). The mixture was first slowly (0.5°C/min) heated to 250°C and isothermed for 10 h and then heated at 1°C/min to 400°C–430°C, isothermed for 4 days, and slowly cooled (0.1°C/min) to room temperature. A stoichiometric yield of polycrystalline product was acquired by employing different starting materials according to equation $CdS + 2Bi + CdBr_2 + 3S = 2CdBiS_2Br$ at a temperature as low as 450°C. When the same reaction mixture was used, larger single crystals were grown by first heating the reactants to 900°C followed by slow cooling to room temperature.

4.101   CADMIUM–NIOBIUM–OXYGEN–SULFUR

$CdS-CdO-Nb_2O_5$. The phase diagram is not constructed. The $Cd_2Nb_2O_6S$ quaternary compound is formed in this system (Bernard et al. 1971). It crystallizes in the cubic structure with the lattice parameter $a = 1067.4 \pm 0.2$ pm. This compound was obtained by the interaction of CdS and $CdNb_2O_6$ at 1050°C.

4.102   CADMIUM–OXYGEN–SELENIUM–SULFUR

$CdS-SeO_2$. This section is a nonquasibinary section of the Cd–O–Se–S quaternary system. The main products of CdS and SeO_2 (H_2SeO_3) interaction are CdSe, $CdSO_4$, and $CdSeO_3$ and the main selenium-containing phase is CdSe the output of which reaches 89% at 700°C (Markovski and Smirnova 1961, Smirnova and Markovski 1962, Goriayev et al. 1973). $CdSeO_3$ emerges in the solid reaction products only at the maximum contents of CdSe and the minimum contents of CdS. It was supposed that CdS and SeO_2 interact according to the next reactions:

$$CdS + SeO_2 = CdSe + SO_2;$$

$$CdS + 2SeO_2 = CdSO_4 + 2Se.$$

$CdSeO_3$ is formed as the result of CdSe and SeO_2 interaction.

4.103   CADMIUM–OXYGEN–MOLYBDENUM–SULFUR

$CdS-MoO_3$. This section is a nonquasibinary section of the Cd–O–Mo–S quaternary system. The main product of CdS and MoO_3 interaction at 450°C–700°C for 2.5 h in the nitrogen atmosphere is a hardly separable $MoO_2 + MoS_2$ (Mo_2S_3) mixture (Bhuiyan et al. 1979). The MoO_2 content in the mixture increases as the temperature increases. Pure MoS_2 also does not form at 400°C–600°C in the HCl atmosphere. The S/Mo atomic ratio in the reaction products changes within the interval of 0.38–0.81. The products with S/Mo = 0.21–0.42 are formed in the NH_3 atmosphere at 500°C–600°C for 2 h. Interaction of CdS and MoO_3 in the presence of $HCl + NH_3$ mixtures leads to formation of reaction products with S/Mo $\approx$ 0.4. Addition of solid NH_4Cl to the CdS + MoO_3 mixture gives the possibility to obtain at 525°C for 2 h pure MoS_2 with an outcome of 96.17%. The products of the solid-state interaction of CdS and MoO_3 were investigated by XRD and measuring of density.

4.104   CADMIUM–SELENIUM–TELLURIUM–SULFUR

CdS–CdSe–CdTe. The liquidus surface of this quasiternary system is shown in Figure 4.73 (Mizetskaya et al. 1986). It includes two fields of primary crystallization: solid solutions based on CdTe and CdS_xSe_{1-x} solid solutions.

Wide regions of solid solutions based on CdTe with sphalerite structure (α-solid solutions) and CdS_xSe_{1-x} solid solutions with wurtzite structure (β-solid solutions) exist in this system within 800°C and solidus temperatures (Figure 4.74) (Budionnaya et al. 1974a,b, 1979, Mizetskaya et al. 1986). One-phase regions are divided by the narrow α + β two-phase region. The region of α-solid solutions increases with the temperature decreasing. Calculated miscibility gap in this system at 600°C and 800°C (Ohtani et al. 1992) practically coincides with the obtained experimental data (Mizetskaya et al. 1986).

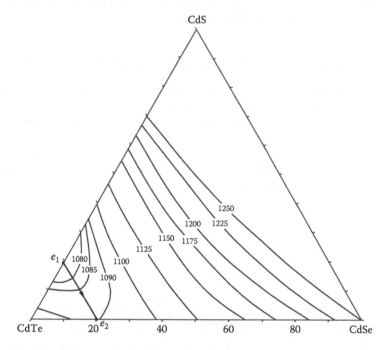

FIGURE 4.73 Liquidus surface of the CdS–CdSe–CdTe quasiternary system. (From Mizetskaya, I.B. et al., *Fiziko-khimicheskiye osnovy sinteza monokristallov poluprovodnikovyh tverdyh rastvorov soyedineniy $A^{II}B^{VI}$* [in Russian], Nauk. Dumka Publish., Kiev, Ukraine, 1986, 160pp.)

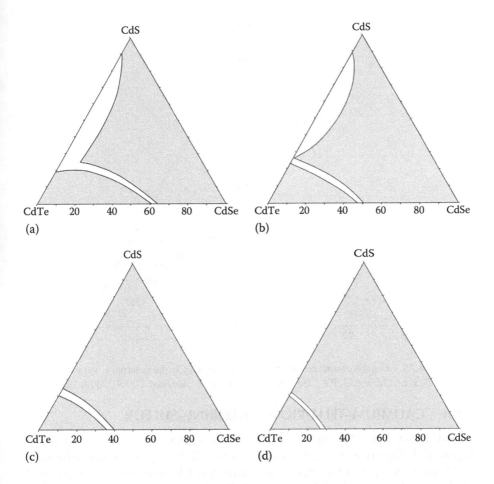

FIGURE 4.74 Isothermal sections of the CdS–CdSe–CdTe quasiternary system at (a) 700°C, (b) 750°C, (c) 900°C, and (d) 1000°C. (From Mizetskaya, I.B. et al., *Fiziko-khimicheskiye osnovy sinteza monokristallov poluprovodnikovyh tverdyh rastvorov soyedineniy A^{II}B^{VI}* [in Russian], Nauk. Dumka Publish., Kiev, Ukraine, 1986, 160pp.)

This system was investigated through DTA, XRD, metallography, and measuring of microhardness. The ingots were annealing at 800°C (Budionnaya et al. 1974a,b, 1979, Mizetskaya et al. 1986).

4.105   CADMIUM–SELENIUM–CHLORINE–SULFUR

CdS–CdSe–CdCl$_2$. The liquidus surface of this quasiternary system includes two fields of primary crystallization: CdCl$_2$ and CdSe$_x$S$_{1-x}$ solid solutions (Figure 4.75) (Oleinik and Mizetski 1984). This system was investigated through DTA and using mathematical simulation of experiment.

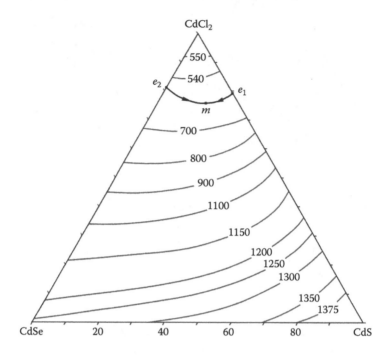

FIGURE 4.75 Liquidus surface of the CdS–CdSe–CdCl$_2$ quasiternary system. (From Oleinik, G.S. and Mizetski, P.A., *Izv. AN SSSR. Neorgan. materialy*, 20(9), 1490, 1984.)

4.106 CADMIUM–TELLURIUM–CHLORINE–SULFUR

CdS–CdTe–CdCl$_2$. The liquidus surface of this quasiternary system is shown in Figure 4.76 (Oleinik et al. 1986). It includes two fields of primary crystallization: CdCl$_2$ and CdS$_x$Te$_{1-x}$ solid solutions. The solubility of Cl$^-$ ion in the CdS$_x$Te$_{1-x}$ ($x \leq 0.1$) single crystals is equal to 0.3–0.4 mass.%. This system was investigated through DTA, metallography, and XRD.

4.107 CADMIUM–FLUORINE–MANGANESE–SULFUR

CdS–CdF$_2$–MnF$_2$. The phase diagram is not constructed. The Cd$_2$Mn$_2$F$_6$S quaternary compound is formed in this system (Pannetier et al. 1972). It crystallizes in the cubic structure with the lattice parameter $a = 1104$ pm. This compound was obtained by the annealing of the mixtures from binary compounds at 700°C for 4 days or by their quick heating up to 1000°C during 0.5 h.

4.108 CADMIUM–FLUORINE–IRON–SULFUR

CdS–CdF$_2$–FeF$_2$. The phase diagram is not constructed. The Cd$_2$Fe$_2$F$_6$S quaternary compound is formed in this system (Pannetier et al. 1972). It crystallizes in the cubic structure with the lattice parameter $a = 1089.5 \pm 0.5$ pm. This compound was obtained by the annealing of the mixtures from binary compounds at 700°C for 4 days or by their quick heating up to 1000°C during 0.5 h.

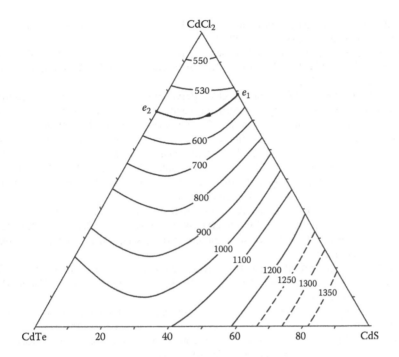

FIGURE 4.76 Liquidus surface of the CdS–CdTe–CdCl$_2$ quasiternary system. (From Oleinik, G.S. et al., *Izv. AN SSSR. Neorgan. materialy*, 22(1), 164, 1986.)

4.109 CADMIUM–FLUORINE–COBALT–SULFUR

CdS–CdF$_2$–CoF$_2$. The phase diagram is not constructed. The Cd$_2$Co$_2$F$_6$S quaternary compound is formed in this system (Pannetier et al. 1972). It crystallizes in the cubic structure with the lattice parameter $a = 1068.6 \pm 0.5$ pm. This compound was obtained by the annealing of the mixtures from binary compounds at 700°C for 4 days or by their quick heating up to 1000°C during 0.5 h.

4.110 CADMIUM–FLUORINE–NICKEL–SULFUR

CdS–CdF$_2$–NiF$_2$. The phase diagram is not constructed. The Cd$_2$Ni$_2$F$_6$S quaternary compound is formed in this system (Pannetier et al. 1972). It crystallizes in the cubic structure with the lattice parameter $a = 1052.3 \pm 0.2$ pm. This compound was obtained by the annealing of the mixtures from binary compounds at 700°C for 4 days or by their quick heating up to 1000°C during 0.5 h.

REFERENCES

Abdullayev A.G., Kyazumov M.G. An electron diffraction study of the Cd$_x$In$_y$Al$_z$S$_4$ (0.9 ≤ x ≤ 1, 1 ≤ y ≤ 1.3, 0.7 ≤ z ≤ 1) crystal structure, *Thin Solid Films*, **190**(2), 303–307 (1989).

Agaev A.B., Aliev V.O., Aliev O.M. Synthesis, X-ray investigations and physical properties of the compounds derived from the Y$_5$S$_7$ structure [in Russian], *Zhurn. neorgan. khimii*, **41**(2), 319–325 (1996).

Aliev I.I., Aliev O.M. Phase equilibria in the As_2Te_3–CdS system [in Russian], *Zhurn. neorgan. khimii*, **36**(9), 2383–2386 (1991).

Asadov M.M. Dependences of physico-chemical properties of solid solutions in the systems $GeSe_2$–A^2B^6 (A^2 = Hg, Cd; B^6 = S, Te) versus composition [in Russian], *Azerb. khim. zhurn.*, (2), 77–81 (2006).

Atuchin V.V., Pankevich V.Z., Parasyuk O.V., Pervukhina N.V., Pokrovsky L.D., Remesnik V.G., Uvarov V.N., Pekhnyo V.I. Structural and optical properties of noncentrosymmetric quaternary crystals $AgCd_2GaS_4$, *J. Cryst. Growth*, **292**(2), 494–499 (2006).

Atuchin V.V., Parasyuk O.V., Pervukhina N.V. Redetermination of the quaternary phase silver dicadmium gallium tetrasulfide, $AgCd_2GaS_4$, *Acta Crystallogr.*, **E61**(5), i91–i93 (2005).

Axtell III E.A., Kanatzidis M.G. First examples of gold thiocadmates: $A_2Au_2Cd_2S_4$ (A = Rb, Cs) and $K_2Au_4CdS_4$: Bright photoluminescence from new alkali metal/gold thiocadmates, *Chem. Eur. J.*, **4**(12), 2435–2441 (1998).

Azizov T.H., Guseinov G.G., Kuliev A.S., Nagiev R.A. Growth and properties of $CdInSbS_4$ single crystals [in Russian], *Izv. AN SSSR. Neorgan. materialy*, **22**(5), 726–728 (1986).

Balić-Žunić T., Makovicky E. The crystal structure of kudriavite, $(Cd, Pb)Bi_2S_4$, *Can. Mineral.*, **45**(3), 437–443 (2007).

Baranov I.Yu., Dolgih V.A., Popovkin B.A. Investigation of phase correlations in the La_2O_2X–MX (M = Zn, Cd, Hg; X = S, Se) systems [in Russian], *Zhurn. neorgan. khimii*, **41**(11), 1916–1919 (1996).

Barnier S., Guittard M., Julien C. Glass formation and structural studies of chalcogenide glasses in the CdS–Ga_2S_3–GeS_2 system, *Mater. Sci. Eng.*, **B7**(3), 209–214 (1990).

Bernard D., Le Montangner S., Pannetier J., Lucas J. Un nouveaux composé ferroelectrique $Cd_2Nb_2O_6S$, *Mater. Res. Bull.*, **6**(2), 75–80 (1971).

Bhuiyan N.H., Ahmed A., Begum S. Synthesis of molybdenum (IV) sulphide. Part II. Reaction of molybdenum (VI) oxide with (a) zinc sulphide and (b) cadmium sulphide in the presence of ammonium chloride, *J. Chem. Technol. Biotechnol.*, **29**(3), 169–174 (1979).

Blachnik R., Dreisbach H.A. Neue quaternäre Chalkogenohalogenide, *Monatsh. Chem.*, **117**(3), 305–311 (1986).

Black S.J., Einstein F.W.B., Hayes P.C., Kumar R., Tuck D.G. Reactions of $Cd(SR)_2$ (R = *n*-butyl, phenyl) and the molecular structure of the 2,2′-bipyridine adduct of bis(*n*-butyl thioxanthato)cadmium(II), *Inorg. Chem.*, **25**(23), 4181–4184 (1986).

Bochmann M., Bwembya G., Webb K.J., Malik M.A., Walsh J.R., O'Brien P. Arene chalcogenolato complexes of zinc and cadmium, *Inorg. Synthes.*, **31**, 19–24 (1997).

Bochmann M., Webb K. Novel precursors for the deposition of II-VI semiconductor films, *Mater. Res. Symp. Proc.*, **204**, 149–154 (1991).

Bochmann M., Webb K., Harman M., Hursthouse M.B. Synthesis, structure, and gas-phase decomposition of $[Cd(EC_6H_2tBu_3)_2]_2$ (E = S, Se): First examples of low-coordinate volatile cadmium chalcogenolato complexes, *Angew. Chem. Int. Ed. Engl.*, **29**(6), 638–639 (1990a).

Bochmann M., Webb K., Harman M., Hursthouse M.B. Synthese, Struktur und Gasphasenthermolyse von $[Cd(EC_6H_2tBu_3)_2]_2$ (E = S, Se); erste Beispiele für niedrigkoordinierte, flüchtige Chalcogenolatocadmiumkomplexe, *Angew. Chem.*, **102**(6), 703–704 (1990b).

Bochmann M., Webb K., Hursthouse M.B., Mazid M. Sterically hindered chalcogenolato complexes. Mono- and di-meric thiolates and selenolates of zinc and cadmium; structure of $[\{Cd(SeC_6H_2Bu^t_3-2,4,6)_2\}_2]$, the first three-co-ordinate cadmium–selenium complex, *J. Chem. Soc., Dalton Trans.*, (9), 2317–2323 (1991).

Brunetta C.D., Minstermann W.C. III, Lake C.H., Aitken J.A. Cation ordering and physicochemical characterization of the quaternary diamond-like semiconductor Ag_2CdGeS_4, *J. Solid State Chem.*, **187**, 177–185 (2012).

Bubenzer A., Nitsche R., Grieshaber E. The crystal structure of the cubic cadmium phosphorus sulphide iodide $Cd_{13}P_4S_{22}I_2$, *Acta Crystallogr.*, **B32**(10), 2825–2829 (1976).

Budionnaya L.D., Mitina L.A., Mizetskiy P.A. Formation of solid solutions based on CdTe in the CdTe–CdSe–CdS semiconductor system [in Russian]. In: *Fiz. processy v geterostrukturah i nekot. soyedineniah.* Kishinev, Republic of Moldova: Shtiintsa Publish., pp. 126–129 (1974a).

Budionnaya L.D., Mitina L.A., Mizetskiy P.A., Sharkina E.V. About possibility of obtaining of single crystals based on CdTe in the CdTe–CdSe–CdS semiconductor system [in Russian], *Poluprovodn. tehnika i mikroelektronika*, (17), 46–50 (1974b).

Budionnaya L.D., Mizetskaya I.B., Malinko V.N., Pidlisnyi E.V. Determination of energy gap of solid solution single crystals in the CdTe–CdSe–CdS system [in Russian], *Izv. AN SSSR. Neorgan. materialy*, **15**(5), 770–774 (1979).

Buzhor V.P., Lialikova R.Yu., Radautsan S.I., Ratseev S.V., Tezlevan V.E. Growth of In_2S_3 and $CdIn_2S_4$ single crystals in the solutions from the melts [in Russian], *Izv. AN SSSR. Neorgan. materialy*, **22**(3), 497–498 (1986).

Chaplygin I.V., Mozgova N.N., Magazina L.O., Kuznetsova O.Yu., Safonov Yu.G., Bryzgalov I.A., Makovicky E., Balić-Žunić T. Kudriavite, (Cd, Pb)Bi_2S_4, a new mineral species from Kudriavy volcano, Iturup Island, Kurile arc, Russia, *Can. Mineral.*, **43**(2), 695–701 (2005).

Chapuis G., Niggli A. Die idealisierte Kristallstruktur von Cu_2CdSiS_4, *Naturwissenschaften*, **55**(9), 441–442 (1968).

Chapuis G., Niggli A. The crystal structure of the "normal tetrahedral" compound Cu_2CdSiS_4, *Acta Crystallogr. B*, **28**(5), 1626–1628 (1972).

Chykhriy S.I., Parasyuk O.V., Halka V.O. Crystal structure of the new quaternary phase $AgCd_2GaS_4$ and phase diagram of the quasibinary system $AgGaS_2$–CdS, *J. Alloys Compd.*, **312**(1–2), 189–195 (2000a).

Chykhriy S.I., Sysa L.V., Parasyuk O.V., Piskach L.V. Crystal structure of the $Cu_2CdSn_3S_8$ compound, *J. Alloys Compd.*, **307**(1–2), 124–126 (2000b).

Coates G.E., Lauder A. Some alkoxy- and alkylthio-derivatives of methylcadmium, *J. Chem. Soc. A: Inorg. Phys. Theor.*, 264–267 (1966).

Craig D., Dance I.G., Garbutt R. The three-dimensionally non-molecular polyadamantanoid structure of $Cd(SPh)_2$, *Angew. Chem. Int. Ed. Engl.*, **25**(2), 165–166 (1986a).

Craig D., Dance I.G., Garbutt R. Die dreidimensionale, nichtmolekulare, polyadamantanartige Struktur von $Cd(SPh)_2$, *Angew. Chem.*, **98**(2), 178–179 (1986b).

Dance I.G., Garbutt R.G., Craig D.C. Applications of cadmium NMR to polycadmium compounds. 5. Diagonally coordinated halide within an octametal cage. Crystal and solution structures of $[XCd_8(SCH_2CH_2OH)_{12}X_3]$, X = Cl, Br, I, *Inorg. Chem.*, **26**(22), 3732–3740 (1987a).

Dance I.G., Garbutt R.G., Craig D.C. Scudder M.L. The different nonmolecular polyadamantanoid crystal structures of cadmium benzenethiolate and 4-methylbenzenethiolate. Analogies with microporous aluminosilicate frameworks, *Inorg. Chem.*, **26**(24), 4057–4064 (1987b).

Datsenko A.M., Razvazhnoy E.M., Chashchin V.A. Investigation of interaction in the Bi_2Te_3–(Cd,Pb)(S,Se,Te) systems [in Russian], *Tr. Mosk. khim. tehnol. in-t*, (120), 32–34 (1981).

Davydyuk G.Ye., Olekseyuk I.D., Parasyuk O.V., Piskach L.V., Semenyuk S.A., Kevshyn A.G., Pekhnyo V.I. Obtaining and investigation of the physical properties of the single crystals of the Cu_2CdGeS_4 and Cu_2CdSnS_4 compounds [in Ukrainian], *Nauk. visnyk Volyns'k. derzh. un-tu im. Lesi Ukrainky*, (1), 25–29 (2005).

Davydyuk G.Ye., Parasyuk O.V. Romanyuk Ya.E., Semenyuk S.A., Zaremba V.I., Piskach L.V., Kozioł J.J., Halka V.O. Single crystal growth and physical properties of the Cu_2CdGeS_4 compound, *J. Alloys Compd.*, **339**(1–2), 40–45 (2002).

Deng B., Chan G.H., Huang F.Q., Gray D.L., Ellis D.E., Van Duyne R.P., Ibers J.A. Synthesis, structure, optical properties and electronic structure of $NaLiCdS_2$, *J. Solid State Chem.*, **180**(2), 759–764 (2007).

Devi M.S., Vidyasagar K. First examples of sulfides in the quasiternary A/Cd/Sn/S (A = Li, Na) systems: Molten flux synthesis and single crystal X-ray structures of Li_2CdSnS_4, Na_2CdSnS_4 and $Na_6CdSn_4S_{12}$, *J. Chem. Soc. Dalton Trans.*, (9), 2092–2096 (2002).

Drobyazko V.P., Kuznetsova S.T. Interaction of indium arsenide with cadmium chalcogenides [in Russian], *Zhurn. neorgan. khimii*, **28**(11), 2929–2933 (1983).

Dubrovin I.V., Budionnaya L.D., Mizetskaya I.B., Sharkina E.V. Interaction according to the SnTe–CdS section of the SnTe + CdS ⇔ SnS + CdTe ternary mutual system [in Russian], *Izv. AN SSSR. Neorgan. materialy*, **21**(11), 1873–1878 (1985).

Farneth W.E., Herron N., Wang Y. Bulk semiconductors from molecular solids: A mechanistic investigation, *Chem. Mater.*, **4**(4), 916–922 (1992).

Filonenko V.V., Nechiporuk B.D., Novoseletski N.E., Yuhimchuk V.A., Lavorik Yu.F. Obtaining and some properties of Cu_2CdGeS_4 crystals [in Russian], *Izv. AN SSSR. Neorgan. materialy*, **27**(6), 1166–1168 (1991).

Fleischer M., Chao G.Y., Mandarino J.A. New mineral names, *Am. Mineral.*, **64**(5–6), 652–659 (1979).

Fouassier C., Levasseur A., Joubert J.C., Muller J., Hagenmuller P. Les systèmes B_2O_3–MO–MS. Boracites M-S (M = Mg, Mn, Fe, Cd) et sodalites M-S (M = Co, Zn), *Z. anorg. und allg. Chem.*, **375**(2), 202–208 (1970).

Goriayev V.M., Pechkovski V.V., Pinaev G.F. Interaction of gaseous SeO_2 with CdS [in Russian], *Izv. AN SSSR. Neorgan. materialy*, **9**(7), 1126–1131 (1973).

Grieshaber E., Nitsche R., Bubenzer A. New compounds with icosahedral structures of the type $Cd_{14-2x}P_4S_{24-x}I_{2x}$, *Mater. Res. Bull.*, **11**(9), 1169–1177 (1976).

Gu S., Hu H., Tao H., Zhao X. Properties and microstructure of GeS_2–Ga_2S_3–CdS glasses, *J. Wuhan Univ. Technol. Mater. Sci. Ed.*, **20**(4), 120–122 (2005).

Gulay L.D., Olekseyuk I.D., Parasyuk O.V. Crystal structures of the $Ag_4HgGe_2S_7$ and $Ag_4CdGe_2S_7$ compounds, *J. Alloys Compd.*, **340**(1–2), 157–166 (2002).

Haeuseler H., Cansiz A., Nimmrich M., Jung M. Materials with layered structures: Crystal structure of $Zn_{1.25}In_{2.5}S_3Se_2$, a new polytype of $Zn_2In_2S_5$ and the isotypic compounds $Cd_{0.5}Ga_2InS_5$ and $Hg_{0.8}Ga_{1.6}In_{1.2}S_5$, *J. Solid State Chem.*, **74**(1), 171–175 (1988).

Hahn H., Schulze H. Über quaternäre Chalkogenide des Germanium und Zinns, *Naturwissenschaften*, **52**(14) 426 (1965).

Imaoka M., Jamadzaki T. [in Japan]. *Monthly J. Inst. Industr. Sci. Univ. Tokyo*, **19**(9), 261–262 (1967).

Iyer R.G., Kanatzidis M.G. $[Mn_2(AsS_4)_4]^{8-}$ and $[Cd_2(AsS_4)_2(AsS_5)_2]^{8-}$: Discrete clusters with negative charge from alkali metal polythioarsenate fluxes, *Inorg. Chem.*, **43**(12), 3656–3662 (2004).

Kambas K. Preparation and optical investigation of the $CdInGaS_4$ compound, *Phys. Status Solidi (a)*, **76**(2), 735–741 (1983).

Kirovskaya I.A., Filatova T.N. New semiconductor InSb–CdS system [in Russian], *Sovrem. naukoemk. tekhnol.*, (8), 29–31 (2007).

Kirovskaya I.A., Timoshenko O.T. Peculiarities of the InP–CdS system and its obtaining [in Russian], *Sovrem. naukoemk. tekhnol.*, (6), 97–98 (2007).

Kirovskaya I.A., Zemtsov A.E. Chemical composition and acid-base properties of the surface of GaAs–CdS solid solutions [in Russian], **81**(1), 101–106 (2007).

Kirovskaya I.A., Zemtsov A.E., Shedenko A.V. Solid solutions of heterovalent substitution based on the GaAs–$A^{II}B^{VI}$ system [in Russian], *Sovremen. naukoemk. tekhnol.*, (8), 33–35 (2007).

Kislinskaya G.E. Investigation of coprecipitation of cadmium, indium and gallium with sulfides of some metals [in Russian], *Avtoref. dis. ... kand. khim. nauk*, Kiev, 27 pp. (1974).

Kissin S.A., Owens D.R., Roberts W.L. Černýite, a copper–cadmium–tin sulfide with the stannite structure, *Canad. Mineralog.*, **16**(2), 139–146 (1978).

Kozer V.R., Fedorchuk A., Olekseyuk I.D., Parasyuk O.V. Phase equilibria in the quasiternary system Ag_2S–In_2S_3–CdS at 870 K, *J. Alloys Compd.*, **480**(2), 360–364 (2009a).

Kozer V.R., Parasyuk O.V. Phase equilibria in the quasiternary system $Cu_2S–In_2S_3–CdS$ [in Ukrainian], *Chem. Met. Alloys*, **2**(1–2), 102–107 (2009b).

Kyazimov M.G., Amiraslanov I.R. The crystal structure of two packet $CdInGaS_4$ [in Russian], *Dokl. AN AzSSR*, **39**(2), 19–21 (1983).

Lee S., Colombet P., Ouvrard G., Brec R. General trends observed in the substituted thiophosphate family. Synthesis and structure of $AgScP_2S_6$ and $CdFeP_2S_6$, *Inorg. Chem.*, **27**(7), 1291–1294 (1988).

Lekse J.W., Moreau M.A., McNerny K.L., Yeon J., Halasyamani P.S., Aitken J.A. Second-harmonic generation and crystal structure of the diamond-like semiconductors Li_2CdGeS_4 and Li_2CdSnS_4, *Inorg. Chem.*, **48**(16), 7516–7518 (2009).

Leonov V.V., Chunarev E.N. Distribution of potassium and zinc chlorides at the crystallization of eutectics of cadmium chalcogenides with its chloride [in Russian]. In: *Physico-khim. processy v geterogennyh systemah*. Krasnoyarsk, Russia: Krasnoyar. Un-t Publish., pp. 59–64 (1977).

Leute V., Böttner H. Phase boundary processes in the system (Cd, Pb) (S, Se), *Ber. Bunsenges. Phys. Chem.*, **82**(3), 302–306 (1978).

Li H., Malliakas C.D., Peters J.A., Liu Z., Im J., Jin H., Morris C.D. et al. $CsCdInQ_3$ (Q = Se, Te): New photoconductive compounds as potential materials for hard radiation detection, *Chem. Mater.*, **25**(10), 2089–2099 (2013).

Lucas C.R., Peach M.E. Metal derivatives of pentachlorothiophenol, *Inorg. Nucl. Chem. Lett.*, **5**(2), 73–76 (1969).

Lucas C.R., Peach M.E. Reactions of pentachlorothiophenol. I. Preparation of some simple metallic and non-metallic derivatives, *Can. J. Chem.*, **48**(12), 1869–1875 (1970).

Lutz H.D., Bertram W.W., Oft B., Hauseler H. Phase relationships in the systems $MS–Cr_2S_3–In_2S_3$ (M = Co, Cd, Hg), *J. Solid State Chem.*, **46**(1), 56–63 (1983).

Markovski L.Ya., Smirnova R.I. About chemistry of cadmium sulfide and selenious acid interaction [in Russian], *Zhurn. neorgan. khimii*, **6**(4), 948–956 (1961).

Matsuhita H., Ichikawa T., Katsui A. Structural, thermodynamic and optical properties of Cu_2–II–IV–VI_4 quaternary compounds, *J. Mater. Sci.*, **40**(8), 2003–2005 (2005).

Mehtiev T.R., Nani R.H., Guseinov G.G. Crystal structure of the $CdInGaS_4$ single crystals [in Russian]. In: *Nekot. voprosy eksperiment. i teor. fiziki*. Baku, Azerbaijan: Elm Publish., pp. 201–202 (1977).

Mirzoyev A.J., Asadov M.M., Aliyev O.M. Phase equilibria in the $CdS–GeSe_2$ system and formed phases properties [in Azerbaijanian], *Zhurn. khim. problem*, (2), 322–325 (2006).

Mizetskaya I.B., Oleinik G.S., Budionnaya L.D., Tomashik V.N., Oleinik N.D. *Fiziko-khimicheskiye osnovy sinteza monokristallov poluprovodnikovyh tverdyh rastvorov soyedineniy $A^{II}B^{VI}$* [in Russian]. Kiev, Ukraine: Nauk. Dumka Publish., 160pp. (1986).

Moldovyan N.A., Markus M.M. Photoelectric properties of the $CdInAlS_4$ layered single crystals [in Russian], *Izv. AN MSSR. Ser. fiz.-tekhn. i mat. n.*, (2), 64–65 (1986).

Morris C.D., Li H., Jin H., Malliakas C.D., Peters J.A., Trikalitis P.N., Freeman A.J., Wessels B.W., Kanatzidis M.G. $Cs_2M^{II}M^{IV}_3Q_8$ (Q = S, Se, Te): An extensive family of layered semiconductors with diverse band gaps, *Chem. Mater.*, **25**(16), 3344–3356 (2013).

Nitsche R., Sargent D.F., Wild P. Crystal growth of quaternary 1_2246_4 chalcogenides by iodine vapor transport, *J. Cryst. Growth*, **1**(1), 52–53 (1967).

Odin I.N. Physico-chemical analysis of ternary and ternary mutual systems, containing cadmium, zinc, silicon, bismuth chalcogenides and properties of ingots in these systems [in Russian], *Zhurn. neorgan. khimii*, **41**(6), 941–953 (1996).

Odin I.N. *T-x-y* diagrams for reciprocal systems $PbX + CdI_2 = CdX + PbI_2$ (X = S, Se, Te) [in Russian], *Zhurn. neorgan. khimii*, **46**(10), 1733–1738 (2001).

Odin I.N., Chukichev M.V. Physico-chemical analysis of the Cd–Sb(Bi)–S systems and properties of photosensitivity solid solutions based on cadmium sulfide compounds [in Russian], *Zhurn. neorgan. khimii*, **45**(2), 255–260 (2000).

Odin I.N., Chukichev M.V. Metastable phases crystallizing from the melts in the PbX + CdI$_2$ = CdX + PbI$_2$ (X = S, Se, Te) mutual systems [in Russian], *Zhurn. neorgan. khimii*, **46**(12), 2083–2087 (2001).

Odin I.N., Kozlovskiy V.F., Grin'ko V.V. New quaternary compounds in the system Cu$_7$PS$_6$–Cd$_7$P$_2$S$_{12}$ [in Russian], *Zhurn. neorgan. khimii*, **50**(6), 1024–1026 (2005a).

Odin I.N., Presnyakov I.A., Pokholok K.V., Rubina M.E., Mikhaylina A.V. Investigation of the Cd$_3$In$_2$S$_2$Te$_4$ and Cd$_5$Ga$_2$S$_2$Te$_6$ quaternary compounds structure using probe Mossbauer spectroscopy on ^{119}Sn nuclei, *Zhurn. neorgan. khimii*, **50**(1), 87–89 (2005b).

Odin I.N., Rubina M.E. Cd$_3$S$_3$ + Ga$_2$Te$_3$ = Cd$_3$Te$_3$ + Ga$_2$S$_3$ system [in Russian], *Zhurn. neorgan. khimii*, **48**(7), 1206–1208 (2003).

Odin I.N., Rubina M.E., Demidova E.D. Phase relations in the Cd$_3$S$_3$ + In$_2$Te$_3$ = Cd$_3$Te$_3$ + In$_2$S$_3$ system [in Russian], *Zhurn. neorgan. khimii*, **49**(5), 848–851 (2004).

Odin I.N., Rubina M.E., Gapanovich M.V., Demidova E.D. *T-x-y* phase diagrams of the GaS + CdTe = CdS + GaTe, CdS–CdTe–InTe systems [in Russian], *Zhurn. neorgan. khimii*, **50**(4), 714–716 (2005c).

Odin I.N., Visitskiy E.V., Rubina M.E. *T-x-y* phase diagrams of the CdS + ZnTe ⇔ ZnS + CdTe, CdS–CdIn$_2$S$_4$, ZnS–In$_2$Te$_3$–ZnTe [in Russian], *Zhurn. neorgan. khimii*, **50**(6), 1018–1023 (2005d).

Ohtani H., Kojima K., Ishida K., Nishizawa T. Miscibility gap in II-VI semiconductor systems, *J. Alloys Compd.*, **182**(1), 103–114 (1992).

Oleinik G.S., Mizetski P.A. Liquidus surface of the CdCl$_2$–CdSe–CdS system [in Russian], *Izv. AN SSSR. Neorgan. materialy*, **20**(9), 1490–1493 (1984).

Oleinik G.S., Mizetski P.A., Nuzhnaya T.P. Liquidus surface of CdCl$_2$–CdS–CdTe ternary system [in Russian], *Izv. AN SSSR. Neorgan. materialy*, **22**(1), 164–165 (1986).

Olekseyuk I.D., Davidyuk H.Ye., Parasyuk O.V., Voronyuk S.V., Halka V.O., Oksyuta V.A. Phase diagram and electric transport properties of samples of the quasibinary system CuInS$_2$–CdS, *J. Alloys Compd.*, **309**(1–2), 39–44 (2000).

Olekseyuk I.D., Parasyuk O.V., Halka V.O., Piskach L.V., Pankevych V.Z., Romanyuk Ya.E. Phase equilibria in the quasiternary system Ag$_2$S–CdS–Ga$_2$S$_3$, *J. Alloys Compd.*, **325**(1–2), 167–179 (2001).

Olekseyuk I.D., Parasyuk O.V., Yurchenko O.M., Pankevych V.Z., Zaremba V.I., Valiente R., Romanyuk Y.E. Single-crystal growth and properties of AgCd$_2$GaS$_4$, *J. Cryst. Growth*, **279**(1–2), 140–145 (2005).

Olekseyuk I.D., Piskach L.V. Phase equilibria in the Cu$_2$SnX$_3$–CdX (X = S, Se, Te) systems [in Russian], *Zhurn. neorgan. khimii*, **42**(2), 331–333 (1997).

Olekseyuk I.D., Piskach L.V., Parasyuk O.V., Gorgut G.P., Volkov S.V., Pekhnyo V.I. Solid-liquid equilibria in the quasiternary system CdS–Ga$_2$S$_3$–GeS$_2$, *J. Alloys Compd.*, **421**(1–2), 91–97 (2006).

Osakada K., Yamamoto T. Formation of ZnS and CdS by thermolysis of homoleptic thiolato compounds [M(SMe)$_2$]$_n$ (M = Zn, Cd), *J. Chem. Soc. Chem. Commun.*, (14), 1117–1118 (1987).

Osakada K., Yamamoto T. Preparation of ZnS and CdS by thermal degradation of (methanethiolato)zinc and -cadmium complexes, [M(SMe)$_2$]$_n$ (M = Zn, Cd), *Inorg. Chem.*, **30**(10), 2328–2332 (1991).

Palchik O., Iyer R.G., Canlas C.G., Weliky D.P., Kanatzidis M.G. K$_{10}$M$_4$M′$_4$S$_{17}$ (M = Mn, Fe, Co, Zn; M′ = Sn, Ge) and Cs$_{10}$Cd$_4$Sn$_4$S$_{17}$: Compounds with a discrete supertetrahedral cluster, *Z. anorg. und allg. Chem.*, **630**(13–14), 2237–2247 (2004).

Pannetier J., Calaga Y., Lucas J. Nouvelles pyrochlores Cd$_2$M$_2$F$_6$S (M^{II} = Mn, Fe, Co, Ni, Cu, Zn), *Mater. Res. Bull.*, **7**(1), 57–62 (1972).

Paracchini C., Parasini A., Tarricone L. CdIn$_2$S$_2$Se$_2$: A new semiconducting compound, *J. Solid State Chem.*, **65**(1), 40–44 (1986).

Parasyuk O.V., Olekseyuk I.D., Piskach L.V., Volkov S.V., Pekhnyo V.I. Phase relations in the Ag$_2$S–CdS–SnS$_2$ system and the crystal structure of the compounds, *J. Alloys Compd.*, **399**(1–2), 173–177 (2005a).

Parasyuk O.V., Piskach L.V. Phase equilibria in the Ag$_2$SiS$_3$–CdS system [in Russian], *Zhurn. neorgan. khimii*, **44**(6), 1032–1033 (1999).

Parasyuk O.V., Piskach L.V. The Ag$_2$SnS$_3$–CdS system, *Pol. J. Chem.*, **72**(5), 966–968 (1998).

Parasyuk O.V., Piskach L.V., Olekseyuk I.D., Pekhnyo V.I. The quasiternary system Ag$_2$S–CdS–GeS$_2$ and the crystal structure of Ag$_2$CdGeS$_4$, *J. Alloys Compd.*, **397**(1–2), 95–98 (2005b).

Parasyuk O.V., Piskach L.V., Romanyuk Y.E., Olekseyuk I.D., Zaremba V.I., Pekhnyo V.I. Phase relations in the quasibinary Cu$_2$GeS$_3$–ZnS and quasiternary Cu$_2$S–Zn(Cd)S–GeS$_2$ systems and crystal structure of Cu$_2$ZnGeS$_4$, *J. Alloys Compd.*, **397**(1–2), 85–94 (2005c).

Parthé E., Yvon K., Deitch R.H. The crystal structure of Cu$_2$CdGeS$_4$ and other quaternary normal tetrahedral structure compounds, *Acta Crystallogr. B*, **25**(6), 1164–1174 (1969).

Peach M.E. Some reactions of pentafluorothiophenol. Preparation of some pentafluorophenylthio metal derivatives, *Can. J. Chem.*, **46**(16), 2699–2706 (1968).

Perez G., Darriet-Duale M., Hagenmuller P. Les systèmes ternaires MS$_2$–CdS–Ln$_2$S$_3$ à 1050°C (M = Si, Ge), (Ln = La… Gd), *J. Solid State Chem.*, **2** (1), 42–48 (1970).

Pervukhina N.V., Atuchin V.V., Parasyuk O.V. Redetermination of the quaternary phase silver dicadmium gallium tetrasulfide, AgCd$_2$GaS$_4$, *Acta Crystallogr.*, **E61**(5), i91–i93 (2005).

Piskach L.V., Olekseyuk I.D., Parasyuk O.V. Physico–chemical peculiarities of the Cu$_2$CdC^{IV}X$_4$ (C^{IV} – Si, Ge, Sn; X – S, Se, Te) quaternary phase formations [in Ukrainian], *Fizyka kondens. vysokomolek. system. Nauk. zap. Rivnens'kogo pedinstytutu*, (3), 153–157 (1997).

Piskach L.V., Parasyuk O.V. The Ag$_2$GeS$_3$–CdS system, *Pol. J. Chem.*, **72**(6), 1112–1115 (1998).

Piskach L.V., Parasyuk O.V., Olekseyuk I.D. The phase equilibria in the quasiternary Cu$_2$S–CdS–SnS$_2$ system, *J. Alloys Compd.*, **279**(2), 142–152 (1998).

Piskach L.V., Parasyuk O.V., Olekseyuk I.D. The Cu$_2$SiS$_3$–CdS system [in Russian], *Zhurn. neorgan. khimii*, **44**(5), 823–824 (1999).

Piskach L.V., Parasyuk O.V., Olekseyuk I.D., Romanyuk Y.E., Volkov S.V., Pekhnyo V.I. Interaction of argyrodite family compounds with the chalcogenides of II-b elements. *J. Alloys Compd.*, **421**(1–2), 98–104 (2006).

Piskach L.V., Parasyuk O.V., Romanyuk Ya.E. The phase equilibria in the quasibinary Cu$_2$GeS$_3$(Se$_3$)–CdS(Se) systems, *J. Alloys Compd.*, **299**(1–2), 227–231 (2000).

Pivneva S.P., Gavrilov V.V., Gurvich A.M., Postolova L.M., Borovitova M.P. Investigation of the processes which take place at the calcination of CdS with NaCl [in Russian], *Izv. AN SSSR. Neorgan. materialy*, **13**(2), 233–236 (1977).

Razzetti C., Lottici P.P., Zanotti L. Ternary and pseudoternary AB$_2$X$_4$ compounds (A = Zn, Cd; B = Ga, In; X = S, Se), *Mater. Chem. Phys.*, **11**(1), 65–83 (1984).

Robbins M., Lambrecht, Jr. V.G. Preparation and phase relationships in systems of the type CdS–M^{I}M^{III}S$_2$ where M^I = Ag, Cu and M^{III} = Al, Ga, In, *J. Solid State Chem.*, **6**(3), 402–405 (1973).

Ronis Ya.V., Krasnikov V.V., Bondars B.Ya., Vitola A.A., Miller T.N. Crystal structure of cadmium-phosphorus thionitride Cd$_4$P$_6$N$_{12}$S [in Russian], *Izv. AN LatvSSR. Ser. khim.*, (6), 643–646 (1988b).

Ronis Ya.V., Krasnikov V.V., Bondars B.Ya., Vitola A.A., Miller T.N. X-ray investigation of the metal-phosphorus thionitrides Me$_4$P$_6$N$_{12}$S (Me – Mg, Zn, Mn, Fe, Co) [in Russian], *Izv. AN LatvSSR. Ser. khim.*, (2), 139–144 (1989).

Ronis Ya.V., Vitola A.A., Lange I.Ya., Miller T.N. Cadmium–phosphorus thionitride Cd$_4$P$_6$N$_{12}$S [in Russian], *Izv. AN LatvSSR. Ser. khim.*, (4), 418–421 (1988a).

Ronis Ya.V., Vitola A.A., Miller T.N. Synthesis and properties of $Me_4P_6N_{12}S$ (Me – Zn, Cd, Mg) thionitrides [in Russian], *Tr. Mosk. khim.-tehnol. in-ta*, (150), 12–14 (1988c).

Rustamov P.G., Aliev I.I. The As_2Se_3–CdS system [in Russian], *Zhurn. neorgan. khimii*, **31**(3), 771–774 (1986).

Safarov M.G., Abilov Ch.I. Properties of solid solutions based on Sb_2Se_3 in the Sb_2Se_3–CdS system [in Russian], *Izv. AN SSSR. Neorgan. materialy*, **23**(1), 25–27 (1987).

Safarov M.G., Aliev F.G., Mamedov V.N., Aliev A.M., Kuliev A.G. Investigation of the Sb_2Se_3–CdS system [in Russian], *Zhurn. neorgan. khimii*, **34**(6), 1583–1585 (1989a).

Safarov M.G., Orudzhev N.M., Aliev I.I., Aliev I.P., Bagirova E.M. Synthesis and investigation of optical and photoelectrical properties of the complex sulfides of cadmium and V_2VI_3 compounds [in Russian]. In: *6th Resp. konf. "Fiz.-khim. analiz i neorgan. materialoved."* Baku, Azerbaijan, pp. 157–159 (2000).

Safarov M.G., Orudzhev N.M., Mehrabov A.O. The Sb_2Te_3–CdS system [in Russian], *Zhurn. neorgan. khimii*, **34**(7), 1906–1908 (1989b).

Safarov M.G., Orudzhev N.M., Mehrabov A.O. Physical properties of the solid solutions based on CdS in the Sb_2Se_3–CdS system [in Russian], *Izv. AN SSSR. Neorgan. materialy*, **26**(11), 2269–2271 (1990).

Said F.F., Tuck D.G. The direct electrochemical synthesis of some thiolates of zinc, cadmium and mercury, *Inorg. Chim. Acta*, **59**, 1–4 (1982).

Schäfer W., Nitsche R. Tetrahedral quaternary chalcogenides of the type Cu_2–II–IV–$S_4(Se_4)$, *Mater. Res. Bull.*, **9**(5), 645–654 (1974).

Schäfer W., Nitsche R. Zur Systematik tetraedrischer Verbindungen vom Typ $Cu_2Me^{II}Me^{IV}Me_4^{VI}$(Stannite und Wurtzstannite), *Z. Kristallogr.*, **145**(5–6), 356–370 (1977).

Schwer H., Keller E., Krämer V. Crystal structure and twinning of $KCd_4Ga_5S_{12}$ and some isotypic $AB_4C_5X_{12}$ compounds, *Z. Kristallogr.*, **204**(Pt. 2), 203–213 (1993).

Shand W.A. A new semiconducting alloy in the CdS–In_2S_3–Ga_2S_3 system, *Phys. Status Solidi (a)*, **3**(1), K77–K79 (1970).

Sirota M.I., Simonov M.A., Yegorov-Tismenko Yu.K., Simonov V.I., Belov N.V. Crystal structure of $CdSb_6S_8I_4$ [in Russian], *Kristallografiya*, **21**(1), 64–68 (1976).

Smirnova R.N., Markovski L.Ya. About reactions which take place at the interaction of cadmium sulfide dry powder and selenious anhydride [in Russian], *Zhurn. neorgan. khimii*, **7**(6), 1366–1369 (1962).

Szymański J.T. The crystal structure of černýite, Cu_2CdSnS_4, a cadmium analogue of stannite, *Canad. Mineralog.*, **16**(2), 147–151 (1978)

Teske Chr.L. Darstellung und Kristallstruktur von Barium-Cadmium-Thiostannat (IV) $BaCdSnS_4$, *Z. anorg. und allg. Chem.*, **460**(1), 163–168 (1980a).

Teske Chr.L. Zur Kenntnis von $BaCdGeS_4$ mit einem Beitrag zur Kristallchemie von Verbindungen des Typs $BaABS_4$, *Z. anorg. und allg. Chem.*, **468**(9), 27–34 (1980b).

Teske Chr.L., Darstellung und Kristallstruktur von $Ba_3CdSn_2S_8$ mit einer Anmerkung über $Ba_6CdAg_2Sn_4S_{16}$, *Z. anorg. und allg. Chem.*, **522**(3), 122–130 (1985).

Tomashik Z.F., Oleinik G.S., Tomashik V.N., Nizkova A.I. Phase equilibria in the PbSe–CdS system [in Russian], *Izv. AN SSSR. Neorgan. materialy*, **17**(12), 2155–2158 (1981).

Tomashik Z.F., Tomashik V.N. The CdS + PbSe ⇔ CdSe + PbS ternary mutual system [in Russian], *Izv. AN SSSR. Neorgan. materialy*, **20**(4), 568–570 (1984).

Tomashik Z.F., Tomashik V.N. Liquidus surface of the PbTe + CdS ⇔ PbS + CdTe ternary mutual system [in Russian], *Izv. AN SSSR. Neorgan. materialy*, **23**(12) 1981–1984 (1987).

Trishchuk L.I., Oleinik G.S., Mizetskaya I.B. Phase equilibria in the Ag_2Te–CdS system [in Russian], *Izv. AN SSSR. Neorgan. materialy*, **25**(7), 1081–1084 (1989).

Voitsehivski A.V., Drobiazko V.P. About solid solutions in the InAs–CdS system [in Ukrainian], *Ukr. fiz. zhurn.*, **12**(3), 460–461 (1967).

Voitsehovski A.V., Drobiazko V.P., Mitiurev V.K., Vasilenko V.P. Solid solutions in the InAs–CdS and InAs–CdSe systems [in Russian], *Izv. AN SSSR. Neorgan. materialy*, **4**(10), 1681–1684 (1968).

Voitsehovski A.V., Pashun A.D., Mitiurev V.K. About interaction of gallium arsenide with II-VI compounds [in Russian], *Izv. AN SSSR. Neorgan. materialy*, **6**(2), 379–380 (1970).

Wang L., Hung Y.-C., Hwu S.-J., Koo H.-J., Whangbo M.-H. Synthesis, structure, and properties of a new family of mixed-framework chalcohalide semiconductors: $CdSbS_2X$ (X = Cl, Br), $CdBiS_2X$ (X = Cl, Br), and $CdBiSe_2X$ (X = Br, I), *Chem. Mater.*, **18**(5), 1219–1225 (2006).

Wang X., Gu S., Yu J., Zhao X., Tao H. Formation and properties of chalcogenides glasses in the $CdS–Ga_2S_3–GeS_2$ system, *Mater. Chem. Phys.*, **83**(2–3), 284–288 (2004).

Wu Y., Bensch W. Synthesis, crystal structures and optical properties of $NaCdPnS_3$ (Pn – As, Sb), *J. Alloys Compd.*, **511**(1), 35–40 (2012).

Yang Y., Ibers J.A. Syntheses and structure of the new quaternary compounds $Ba_4Nd_2Cd_3S_{10}$ and $Ba_4Ln_2Cd_3S_{10}$ (Ln = Sm, Gd, Tb), *J. Solid State Chem.*, **149**(2), 384–390 (2000).

Yao G.-Q., Shen H.-S., Honig E.D., Kershaw R., Dwight K., Wold A. Preparation and characterization of the quaternary chalcogenides $Cu_2B(II)C(IV)X_4$[B(II) = Zn, Cd; C(IV) = Si, Ge; X = S, Se], *Solid State Ionics*, **24**(3), 249–252 (1987).

Zou J.-P., Peng Q., Luo S.-L., Tang X.-H., Zhang A.-Q., Zeng G.-S., Guo G-C. Synthesis, crystal and band structures, and optical properties of a novel quaternary mercury and cadmium chalcogenide halide: $(Hg_2Cd_2S_2Br)B$, *CrystEngComm*, **13**(11), 3862–3867 (2011).

5 Systems Based on CdSe

5.1 CADMIUM–HYDROGEN–CARBON–SELENIUM

Some compounds are formed in this quaternary system.

Cd(SeC$_6$H$_5$)$_2$ or C$_{12}$H$_{10}$Se$_2$Cd. To obtain this compound, C$_6$H$_5$SeH (22 mmol) in toluene (5 mL) was added to CdMe$_2$ (11 mmol) in toluene (5 mL) under inert atmosphere (Brennan et al. 1989). A white solid formed immediately. The reaction mixture was stirred for 4 h, and the white precipitate was collected, washed repeatedly with heptanes, and dried in vacuo to give a white solid Cd(SeC$_6$H$_5$)$_2$. Recrystallization from pyridine/heptane gives colorless needles. This compound melts at the temperature higher than 304°C. Zinc-blende nanometer-sized clusters of CdSe could be obtained as the result of Cd(SeC$_6$H$_5$)$_2$ thermolysis.

MeCd(SeC$_6$H$_2$Pr$_3^i$-2.4.6) or C$_{16}$H$_{26}$SeCd. For obtaining this compound, HSeC$_6$H$_2$Me$_3$-2,4,6 (8.65 mmol) was added to a solution of CdMe$_2$ (10.2 mmol) in light petroleum (10 mL) at −78°C (Bochmann et al. 1992). On warming the clear yellow solution to room temperature, a gas evolved and a white precipitate formed, which dissolved on stirring for another 2 h. Cooling to −20°C produced colorless crystals of C$_{16}$H$_{26}$SeCd. This compound melts at 180°C with decomposition. All operations were carried out under Ar using standard vacuum-line techniques.

Cd(SeC$_6$H$_2$Me$_3$-2,4,6)$_2$ or C$_{18}$H$_{22}$Se$_2$Cd. This compound was obtained by pyrolysis of Cd[N(SiMe$_3$)$_2$]$_2$ with 2,4,6-Me$_3$C$_6$H$_2$SeH as polymeric solids, which dissolved only in coordinating solvents (Bochmann et al. 1991). It melts at 400°C (Bochmann and Webb 1991).

The experiment was carried out under Ar atmosphere using standard vacuum-line techniques. Crystals suitable for x-ray diffraction (XRD) could not be grown (Bochmann et al. 1991).

Cd(SeC$_6$H$_2$Pr$_3^i$-2,4,6)$_2$ or C$_{30}$H$_{46}$Se$_2$Cd. The reaction of Cd[N(SiMe$_3$)$_2$]$_2$ with 2,4,6-Pr$_3^i$C$_6$H$_2$SeH gives C$_{30}$H$_{46}$Se$_2$Cd as a fine crystalline white solid, which is poorly soluble in light petroleum but dissolves in warm toluene (Bochmann et al. 1991). This compound melts at the temperature more than 250°C and sublimes at 230°C (0.13 Pa) (Bochmann et al. 1991, Bochmann and Webb 1991).

The experiment was carried out under Ar atmosphere using standard vacuum-line techniques (Bochmann et al. 1991).

Cd(SeC$_6$H$_2$Bu$_3^t$-2,4,6)$_2$ or C$_{36}$H$_{58}$Se$_2$Cd. To obtain this compound, tri-*t*-butylbenzeneselenol (4.62 mmol) and light petroleum (40 mL) are injected into a 100 mL, three-necked flask, equipped with a magnetic stirring bar and connected to the inert gas supply of a vacuum line via a stopcock adaptor (Bochmann et al. 1990a,b, 1991, 1997). Cd[N(SiMe$_3$)$_2$]$_2$ (2.3 mmol) is added to the stirred solution via syringe.

The solution turns bright yellow and the product begins to precipitate after a few minutes. Stirring is continued for 3 h. The solvent is removed in vacuo, and the yellow residue is washed with light petroleum (2 × 20 mL) and dissolved in hot (100°C–110°C) toluene (90 mL). Bright yellow crystals are obtained on cooling the solution overnight to 10°C; these are filtered off, washed with petroleum, and dried in vacuo. More products were obtained from the filtrate. $C_{36}H_{58}Se_2Cd$ sublimes at 240°C (0.13 Pa), melts at 300°C with decomposition, and deposits CdSe films from the vapor phase at 360°C (Bochmann et al. 1990a,b, 1997, Bochmann and Webb 1991). It crystallizes in the monoclinic structure with the lattice parameters $a = 1003.8 ± 0.2$, $b = 2010.2 ± 0.3$, $c = 1824.5 ± 0.2$ pm, and $\beta = 92.76° ± 0.01°$ and calculated density 1.37 g·cm^{-3} (Bochmann et al. 1991).

5.2   CADMIUM–HYDROGEN–OXYGEN–SELENIUM

$CdSeO_3 · H_2O$ compound is formed in this quaternary system. It crystallizes in the orthorhombic structure with the lattice parameters $a = 1318.0 ± 0.2$, $b = 589.0 ± 0.1$, and $c = 505.6 ± 0.1$ pm and calculated density 4.355 g·cm^{-3} (Engelen et al. 1996). It was obtained by the crystallization from aqueous solutions of $CdSeO_3$ or $Cd(HSeO_3)_2$.

5.3   CADMIUM–POTASSIUM–GALLIUM–SELENIUM

CdSe–K$_2$Se–Ga$_2$Se$_3$. The phase diagram is not constructed. The $KCd_4Ga_5Se_{12}$ quaternary compound is formed in this system, which crystallizes in the hexagonal structure with the lattice parameters $a = 1435.09 ± 0.30$ and $c = 974.36 ± 0.55$ pm ($a = 890.43 ± 0.14$ and $\alpha = 107.498° ± 0.008°$ in rhombohedral settings) and calculated density 5.111 g·cm^{-3} (Schwer et al. 1993). In order to produce $KZn_4Ga_5S_{12}$, $ZnGa_2S_4$ was solved in KBr in molar ratio 1:9 in quartz-glass ampoule, which was sealed under vacuum and heated to 900°C. After 1 day, the furnace was switched off and the sample cooled down slowly. The reguli were treated with hot water to remove the unreacted halide.

5.4   CADMIUM–POTASSIUM–TIN–SELENIUM

CdSe–K$_2$Se–SnSe$_2$. The phase diagram is not constructed. The $K_2CdSnSe_4$, $K_6Cd_4Sn_3Se_{13}$, and $K_{14}Cd_{15}Sn_{12}Se_{46}$ quaternary compounds are formed in this system. $K_6Cd_4Sn_3Se_{13}$ crystallizes in the trigonal structure with the lattice parameters $a = 1503.38 ± 0.11$ and $c = 1651.2 ± 0.2$ pm, calculated density 3.186 g·cm^{-3}, and energy gap of 2.33 eV (Ding et al. 2004).

$K_6Cd_4Sn_3Se_{13}$ decomposes at 280°C to $K_2CdSnSe_4$ and CdSe and is accessible only through hydrothermal synthesis (Ding et al. 2004). Several attempts to prepare it with a solid-state direct combination reaction of $K_2Se/Cd/Sn/Se$ yielded mixtures of $K_2CdSnSe_4$ and CdSe. Interestingly, however, when a mixture of $K_2CdSnSe_4$ and CdSe, having the proper nominal composition, was treated under hydrothermal conditions at 115°C for 96 h, $K_6Cd_4Sn_3Se_{13}$ formed in excellent yield.

$K_{14}Cd_{15}Sn_{12}Se_{46}$ crystallizes in the cubic structure with the lattice parameter $a = 2309.0 ± 0.3$ pm and calculated density 3.933 g·cm^{-3} (Ding and Kanatzidis 2006).

To prepare this compound, K_2Se, Cd, Sn, and Se were combined in an evacuated and flame-sealed fused silica tube and melted at 900°C. The obtained product was loaded into a Pyrex tube with deionized water, and the tube was evacuated, heated up to 180°C, and kept at this temperature for 6 days.

5.5 CADMIUM–POTASSIUM–PHOSPHORUS–SELENIUM

$K_2CdP_2Se_6$ quaternary compound exists in this system (Chondroudis and Kanatzidis 1998). This compound melts congruently at 640°C, is a wide-gap semiconductor ($E_g = 2.58$ eV), and crystallizes in the monoclinic structure. To obtain this compound, a mixture of Cd, P_2Se_5, K_2Se, and Se was sealed under vacuum in a Pyrex tube and heated to 500°C for 4 days followed by cooling to 150°C at 2°C/h.

5.6 CADMIUM–POTASSIUM–CHLORINE–SELENIUM

$CdSe–KCl–CdCl_2$. The phase diagram is not constructed. Using oriented zone recrystallization and radioactive isotopes, there was determined equilibrium coefficient of KCl distribution in the $CdSe–CdCl_2$ eutectic: $k_0 = 0.56$ (Leonov and Chunarev 1977).

5.7 CADMIUM–RUBIDIUM–GALLIUM–SELENIUM

$CdSe–Rb_2Se–Ga_2Se_3$. The phase diagram is not constructed. The $RbCd_4Ga_5Se_{12}$ quaternary compound is formed in this system, which crystallizes in the hexagonal structure with the lattice parameters $a = 1435.64 \pm 0.49$ and $c = 977.43 \pm 0.29$ pm ($a = 891.58 \pm 0.32$ and $\alpha = 107.453° \pm 0.016°$ in rhombohedral settings) and calculated density 5.218 g·cm^{-3} (Schwer et al. 1993). This compound was obtained if $CdGa_2Se_4$ was kept in RbBr at the temperature above 150°C of its melting point for 1 day and then cooled down slowly.

5.8 CADMIUM–RUBIDIUM–TIN–SELENIUM

$CdSe–Rb_2Se–SnSe_2$. The phase diagram is not constructed. The $Rb_{14}Cd_{15}Sn_{12}Se_{46}$ quaternary compound is formed in this system. This compound could be obtained if the Rb^+ ions replace the K^+ ions in the $K_{14}Cd_{15}Sn_{12}Se_{46}$ (Ding and Kanatzidis 2006).

5.9 CADMIUM–RUBIDIUM–PHOSPHORUS–SELENIUM

$Rb_2CdP_2Se_6$ and $Rb_8[Cd_4(Se_2)_2(PSe_4)_4]$ quaternary compounds exist in this system (Chondroudis and Kanatzidis 1997, 1998). $Rb_2CdP_2Se_6$ melts congruently at 720°C and crystallizes in the monoclinic structure with the lattice parameters $a = 664.0 \pm 0.1$, $b = 1272.9 \pm 0.2$, $c = 777.8 \pm 0.1$ pm, and $\beta = 98.24° \pm 0.01°$, calculated density 4.180 g·cm^{-3}, and energy gap 2.58 eV (2.66 eV from single-crystal transmission data). To obtain this compound, a mixture of Cd, P_2Se_5, Rb_2Se, and Se was sealed under vacuum in a Pyrex tube and heated to 500°C for 4 days followed by cooling to 150°C at 2°C/h.

$Rb_8[Cd_4(Se_2)_2(PSe_4)_4]$ melts incongruently at 456°C and crystallizes in the tetragonal structure with the lattice parameters $a = 1756.4 \pm 0.3$ and $c = 727.5 \pm 0.2$ pm and energy gap 2.57 eV (Chondroudis and Kanatzidis 1997). This compound was synthesized from a mixture of Cd, P_2Se_5, Rb_2Se, and Se heated to 550°C for 4 days followed by cooling to 150°C at 2°C/h.

5.10   CADMIUM–RUBIDIUM–BISMUTH–SELENIUM

$CdSe–Rb_2Se–Bi_2Se_3$. The phase diagram is not constructed. $Rb_2CdBi_6Se_{11}$ quaternary compound is formed in this system (Kim et al. 2006). It crystallizes in the orthorhombic structure with the lattice parameters $a = 1238.5 \pm 0.3$, $b = 2383.9 \pm 0.6$, and $c = 411.24 \pm 0.10$ pm at 173 K, energy gap 0.74 eV, and calculated density 6.581 g·cm^{-3}. $Rb_2CdBi_6Se_{11}$ was synthesized from direct combination reactions of the corresponding chemical elements targeting this compound. The thoroughly mixed elements were sealed in an evacuated, fused silica tube and heated at 750°C for 72 h followed by cooling to 550°C at a rate of 5°C/h and then to 50°C in 10 h.

5.11   CADMIUM–CESIUM–GALLIUM–SELENIUM

$CdSe–Cs_2Se–Ga_2Se_3$. The phase diagram is not constructed. The $CsCd_4Ga_5Se_{12}$ quaternary compound is formed in this system, which crystallizes in the hexagonal structure with the lattice parameters $a = 1439.98 \pm 0.29$ and $c = 979.14 \pm 0.24$ pm ($a = 893.19 \pm 0.17$ and $\alpha = 107.438° \pm 0.010°$ in rhombohedral settings) and calculated density 5.322 g·cm^{-3} (Schwer et al. 1993). This compound was obtained if $CdGa_2Se_4$ was kept in CsBr at the temperature above 150°C of its melting point for 1 day and then cooled down slowly.

5.12   CADMIUM–CESIUM–INDIUM–SELENIUM

$CdSe–Cs_2Se–In_2Se_3$. The phase diagram is not constructed. $CsCdInSe_3$ quaternary compound is formed in this system. This compound, which appears to melt congruently, crystallizes in the monoclinic structure with the lattice parameters $a = 1170.8 \pm 0.2$, $b = 1171.2 \pm 0.2$, $c = 2305.1 \pm 0.5$ pm, and $\beta = 97.28° \pm 0.03°$, calculated density 5.059 g·cm^{-3}, and energy gap 2.40 eV (Li et al. 2013). To obtain this compound, a mixture of Cs_2Se_3 (0.3 mmol), Cd (0.6 mmol), In (0.6 mmol), and Se (0.9 mmol) was loaded into a carbon-coated quartz tube in an N_2-filled glove box. The tube was flame-sealed under a vacuum of <10 kPa and placed in a temperature-controlled furnace. The mixture was heated slowly to 200°C and kept at this temperature for 6 h, then heated to 800°C in 8 h and kept there for 1 day, and finally cooled to room temperature for 1 day. The single crystals of $CsCdInSe_3$ were grown by the Bridgman method.

5.13   CADMIUM–CESIUM–YTTRIUM–SELENIUM

$CdSe–Cs_2Se–Y_2Se_3$. The phase diagram is not constructed. $CsCdYSe_3$ quaternary compound is formed in this system (Mitchell et al. 2003). It crystallizes in the orthorhombic structure with the lattice parameters $a = 423.86 \pm 0.05$, $b = 1581.9 \pm 0.5$, and

$c = 1106.39 \pm 0.14$ pm at 153 ± 2 K, energy gap 2.48 eV for (010) crystal face and 2.54 eV for (001) crystal face, and calculated density 5.114 g·cm^{-3}. This compound was prepared from the mixture of Cs_2Se_3, Y, Cd, Se, and CsI as a flux. The sample was heated to 850°C in 24 h, kept at this temperature for 96 h, and cooled to 200°C in 96 h, and then the furnace was turned off.

5.14   CADMIUM–CESIUM–CERIUM–SELENIUM

CdSe–Cs$_2$Se–Ce$_2$Se$_3$. The phase diagram is not constructed. CsCdCeSe$_3$ quaternary compound is formed in this system (Mitchell et al. 2003). It crystallizes in the orthorhombic structure with the lattice parameters $a = 438.83 \pm 0.05$, $b = 1575.76 \pm 0.18$, and $c = 1142.36 \pm 0.13$ pm at 153 ± 2 K, energy gap 2.40 eV, and calculated density 5.233 g·cm^{-3}. CsCdCeSe$_3$ was prepared from the mixture of Cs$_2$Se$_3$, Ce, Cd, Se, and CsI as a flux. The sample was heated to 850°C in 24 h, kept at this temperature for 96 h, and cooled to 200°C in 96 h, and then the furnace was turned off.

5.15   CADMIUM–CESIUM–PRASEODYMIUM–SELENIUM

CdSe–Cs$_2$Se–Pr$_2$Se$_3$. The phase diagram is not constructed. CsCdPrSe$_3$ quaternary compound is formed in this system (Mitchell et al. 2003). It crystallizes in the orthorhombic structure with the lattice parameters $a = 435.56 \pm 0.09$, $b = 1574.7 \pm 0.3$, and $c = 1134.8 \pm 0.3$ pm at 153 ± 2 K and calculated density 5.318 g·cm^{-3}. This compound was prepared from the mixture of Cs$_2$Se$_3$, Pr, Cd, Se, and CsI as a flux. The sample was heated to 850°C in 24 h, kept at this temperature for 96 h, and cooled to 200°C in 96 h, and then the furnace was turned off.

5.16   CADMIUM–CESIUM–NEODYMIUM–SELENIUM

CdSe–Cs$_2$Se–Nd$_2$Se$_3$. The phase diagram is not constructed. CsCdNdSe$_3$ quaternary compound is formed in this system (Mitchell et al. 2003). It was prepared from the mixture of Cs$_2$Se$_3$, Nd, Cd, Se, and CsI as a flux. The sample was heated to 850°C in 24 h, kept at this temperature for 96 h, and cooled to 200°C in 96 h, and then the furnace was turned off.

5.17   CADMIUM–CESIUM–SAMARIUM–SELENIUM

CdSe–Cs$_2$Se–Sm$_2$Se$_3$. The phase diagram is not constructed. CsCdSmSe$_3$ quaternary compound is formed in this system (Mitchell et al. 2003). It crystallizes in the orthorhombic structure with the lattice parameters $a = 429.95 \pm 0.04$, $b = 1577.93 \pm 0.15$, and $c = 1120.10 \pm 0.10$ pm at 153 ± 2 K, energy gap 2.47 eV for (010) crystal face and 2.45 eV for (001) crystal face, and calculated density 5.529 g·cm^{-3}. CsCdSmSe$_3$ was prepared from the mixture of Cs$_2$Se$_3$, Sm, Cd, Se, and CsI as a flux. The sample was heated to 850°C in 24 h, kept at this temperature for 96 h, and cooled to 200°C in 96 h, and then the furnace was turned off.

5.18　CADMIUM–CESIUM–GADOLINIUM–SELENIUM

CdSe–Cs₂Se–Gd₂Se₃. The phase diagram is not constructed. CsCdGdSe₃ quaternary compound is formed in this system (Mitchell et al. 2003). It crystallizes in the orthorhombic structure with the lattice parameters $a = 427.84 \pm 0.04$, $b = 1583.22 \pm 0.15$, and $c = 1115.48 \pm 0.11$ pm at 153 ± 2 K, energy gap 2.60 eV for (010) crystal face and 2.54 eV for (001) crystal face, and calculated density 5.621 g·cm⁻³. CsCdGdSe₃ was prepared from the mixture of Cs₂Se₃, Gd, Cd, Se, and CsI as a flux. The sample was heated to 850°C in 24 h, kept at this temperature for 96 h, and cooled to 200°C in 96 h, and then the furnace was turned off.

5.19　CADMIUM–CESIUM–TERBIUM–SELENIUM

CdSe–Cs₂Se–Tb₂Se₃. The phase diagram is not constructed. CsCdTbSe₃ quaternary compound is formed in this system (Mitchell et al. 2003). It crystallizes in the orthorhombic structure with the lattice parameters $a = 426.12 \pm 0.03$, $b = 1583.19 \pm 0.12$, and $c = 1110.96 \pm 0.09$ pm at 153 ± 2 K, energy gap 2.67 eV for (010) crystal face and 2.60 eV for (001) crystal face, and calculated density 5.682 g·cm⁻³. CsCdTbSe₃ was prepared from the mixture of Cs₂Se₃, Tb, Cd, Se, and CsI as a flux. The sample was heated to 850°C in 24 h, kept at this temperature for 96 h, and cooled to 200°C in 96 h, and then the furnace was turned off.

5.20　CADMIUM–CESIUM–DYSPROSIUM–SELENIUM

CdSe–Cs₂Se–Dy₂Se₃. The phase diagram is not constructed. CsCdDySe₃ quaternary compound is formed in this system (Mitchell et al. 2003). It crystallizes in the orthorhombic structure with the lattice parameters $a = 424.19 \pm 0.03$, $b = 1582.74 \pm 0.12$, and $c = 1108.28 \pm 0.08$ pm at 153 ± 2 K, energy gap 2.57 eV for (010) crystal face and 2.51 eV for (001) crystal face, and calculated density 5.755 g·cm⁻³. CsCdDySe₃ was prepared from the mixture of Cs₂Se₃, Dy, Cd, Se, and CsI as a flux. The sample was heated to 850°C in 24 h, kept at this temperature for 96 h, and cooled to 200°C in 96 h, and then the furnace was turned off.

5.21　CADMIUM–CESIUM–GERMANIUM–SELENIUM

CdSe–Cs₂Se–GeSe₂. The phase diagram is not constructed. The Cs₂CdGe₃Se₈ quaternary compound is formed in this system. It crystallizes in the orthorhombic structure with the lattice parameters $a = 770.48 \pm 0.03$, $b = 1268.22 \pm 0.06$, and $c = 1782.08 \pm 0.06$ pm, calculated density 4.683 g·cm⁻³, and energy gap 2.49 eV (Morris et al. 2013).

5.22　CADMIUM–CESIUM–TIN–SELENIUM

CdSe–Cs₂Se–SnSe₂. The phase diagram is not constructed. The Cs₂CdSn₃Se₈ quaternary compound is formed in this system. It crystallizes in the orthorhombic structure with the lattice parameters $a = 793.35 \pm 0.04$, $b = 1273.36 \pm 0.09$, and $c = 1853.87 \pm 0.10$ pm, calculated density 4.845 g·cm⁻³, and energy gap 2.16 eV

(Morris et al. 2013). To obtain $Cs_2CdSn_3Se_8$, a mixture of Cs_2Se_2, Zn, Sn, and Se in the appropriate ratios was added to a fused silica tube inside a nitrogen-filled glove box. The tube was evacuated, flame-sealed, and used in a flame-melting reaction. After cooling in air, a phase pure ingot containing orange platelike crystals was obtained.

5.23 CADMIUM–CESIUM–PHOSPHORUS–SELENIUM

$Cs_2CdP_2Se_6$ quaternary compound exists in this system (Chondroudis and Kanatzidis 1998). This compound melts congruently at 773°C, is a wide-gap semiconductor ($E_g = 2.63$ eV), and crystallizes in the monoclinic structure. To obtain this compound, a mixture of Cd, P_2Se_5, Cs_2Se, and Se was sealed under vacuum in a Pyrex tube and heated to 500°C for 4 days followed by cooling to 150°C at 2°C/h.

5.24 CADMIUM–CESIUM–BISMUTH–SELENIUM

$CdSe–Cs_2Se–Bi_2Se_3$. The phase diagram is not constructed. $CsCdBi_3Se_6$ quaternary compound is formed in this system (Kim et al. 2006). It crystallizes in the orthorhombic structure with the lattice parameters $a = 2651.2 \pm 0.8$, $b = 411.92 \pm 0.13$, and $c = 1239.6 \pm 0.4$ pm, energy gap 0.40 eV, and calculated density 6.604 g·cm⁻³. $CsCdBi_3Se_6$ was prepared by a direct combination of chemical elements. The mixture, sealed in evacuated fused silica tube, was heated at 750°C for 2 h with rocking, followed by cooling to 550°C at a rate of 20°C/h and then to room temperature in 10 h. Highly oriented polycrystalline ingots of this compound have been prepared using a vertical Bridgman growth technique.

5.25 CADMIUM–COPPER–ALUMINUM–SELENIUM

$CdSe–Cu_2Se–Al_2Se_3$. The phase diagram is not constructed. According to the data of Parthé et al. (1969), the $CuCd_2AlSe_4$ quaternary compound is formed in this system. It crystallizes in the hexagonal structure with the lattice parameters $a = 410.6 \pm 0.5$ and $c = 675.2 \pm 0.5$ pm, but this compound can be one of the solid solution compositions. This compound was obtained by the heating of mixtures from the chemical elements at 600°C–800°C.

5.26 CADMIUM–COPPER–GALLIUM–SELENIUM

$CdSe–Cu–Ga$. The phase diagram is not constructed. Using metallography and measuring of microhardness, it was determined that the solubility of Cu in CdSe at the presence of Ga reaches 1.3 at.% (Gusachenko et al. 1977a,b).

$CdSe–CuGaSe_2$. As far as the $CuGaSe_2$ compound forms by a peritectic process, this section is not quasibinary in the full concentration range (Figure 5.1) (Olekseyuk et al. 1997a). $CuCd_2GaSe_4$ quaternary compound is formed in this system, which melts incongruently at 1047°C [~1000°C (Hirai et al. 1967)]. Liquidus of this system consists from four branches of primary crystallization (solid solutions based on γ-Ga_2Se_3, $CuCdSe_2$, CdSe, and $CuCd_2GaSe_4$).

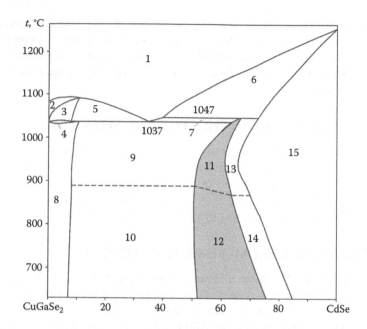

FIGURE 5.1 Phase relations in the CdSe–CuGaSe$_2$ system: 1, L; 2, L + (γ-Ga$_2$Se$_3$); 3, L + (CuGaSe$_2$) + (γ-Ga$_2$Se$_3$); 4, (CuGaSe$_2$) + (γ-Ga$_2$Se$_3$); 5, L + (CuGaSe$_2$); 6, L + (CdSe); 7, L + (β-CuCd$_2$GaSe$_4$); 8, (CuGaSe$_2$); 9, (CuGaSe$_2$) + (β-CuCd$_2$GaSe$_4$); 10, (CuGaSe$_2$) + (α-CuCd$_2$GaSe$_4$); 11, (β-CuCd$_2$GaSe$_4$); 12, (α-CuCd$_2$GaSe$_4$); 13, (CdSe) + (β-CuCd$_2$GaSe$_4$); 14, (CdSe) + (α-CuCd$_2$GaSe$_4$); and 15, (CdSe). (From Olekseyuk, I.D. et al., *Pol. J. Chem.*, 71(7), 893, 1997a.)

Peritectic horizontal at 1047°C in the field of 40–72 mol.% CdSe approaches the eutectic line very closely at 1037°C (Olekseyuk et al. 1997a). The eutectic point corresponds to 35 mol.% CdSe. The solubility of CdSe in CuGaSe$_2$ at 600°C reaches 7.5 mol.% and the solubility of CuGaSe$_2$ in CdSe at this temperature is equal to 12.5 mol.%. The maximum solubility of CuGaSe$_2$ in CdSe (35 mol.%) takes place at 950°C. At room temperature, the solubility of CuGaSe$_2$ in CdSe reaches 10 mol.% [1.3 mol.% (Gusachenko et al. 1977a,b)] and the solubility of CdSe in CuGaSe$_2$ is not higher than 1 mol.% (Dovletov et al. 1990). The homogeneity region of CuCd$_2$GaSe$_4$ at 600°C is within the interval from 52 to 78 mol.% CdSe and the parameter of cubic cell a goes from 575.0 ± 0.1 to 590.6 ± 0.2 pm.

CuCd$_2$GaSe$_4$ has no polymorphic transformation at 890°C from cubic (a = 580 ± 30 pm) to hexagonal modification (a = 410 and c = 673 pm) as indicated by Garbato and Manca (1974). Small thermal effects at 880°C–900°C are the result of partial ordering, which do not change neither the content nor the structure of this compound (Olekseyuk et al. 1997a).

This system was investigated through differential thermal analysis (DTA), metallography, XRD, and measuring of microhardness and density (Hirai et al. 1967, Dovletov et al. 1990, Olekseyuk et al. 1997a). The ingots were annealed at 600°C for 300 h (Olekseyuk et al. 1997a) [at 530°C–580°C for 120 h (Dovletov et al. 1990)]. Gusachenko et al. (1977a,b) introduced CuGaSe$_2$ into CdSe directly at the single-crystal growth.

5.27 CADMIUM–COPPER–INDIUM–SELENIUM

CdSe–CuInSe$_2$. The phase diagram of this system was constructed in Garbato et al. (1985), Manca and Garbato (1986), Zmiy et al. (1997), Vovk et al. (1999, 2000), and Olekseyuk et al. (2006), and the most reliable of them is shown in Figure 5.2 (Olekseyuk et al. 2006). A peritectic process L + β ⇔ γ takes place in the CdSe–CuInSe$_2$ system (nonvariant point coordinates are 2 mol.% CdSe and 987°C, respectively). The position of liquidus and solidus lines agrees well with the results of Garbato et al. (1985) and Manca and Garbato (1986). The eutectic suggested in Zmiy et al. (1997) and Vovk et al. (1999, 2000) [the coordinates of the eutectic point are 25 mol.% CdSe, 950°C (Zmiy et al. 1997), and 957°C (Vovk et al. 1999, 2000)] was not observed by Olekseyuk et al. (2006). In the region near CuInSe$_2$, the curve of the solid solution crystallization exhibits a minimum. The point of which liquidus and solidus lines converge is between 1 and 2 mol.% CdSe.

The solubility in β-solid solution based on CdSe reaches a maximum at the peritectic temperature (~33 mol.% CdSe) and substantially decreases with temperature decreasing: β-solid solution exists in 84–100 mol.% CdSe range at 600°C and in 89–100 mol.% CdSe range at 350°C (Olekseyuk et al. 2006). γ-Solid solution forms in three ways: immediately from the melt (0–2 mol.% CdSe), as a product of the peritectic process (2 to ~33 mol.% CdSe), and during the decomposition of β-solid solution. The solubility limits for γ-solid solution range from 26 to 76 mol.% CdSe at 600°C and from 47 to 83 mol.% CdSe at 350°C. α-Solid solution based on CuInSe$_2$ is localized in a concentration range of 0–6 mol.% CdSe at 600°C and from 0 to 8 mol.% CdSe at 350°C.

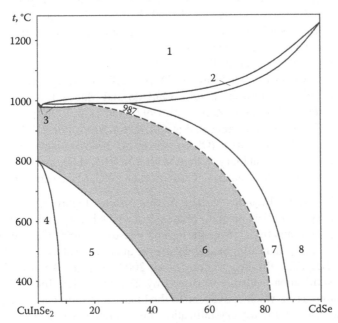

FIGURE 5.2 Phase relations in the CdSe–CuInSe$_2$ system: 1, L; 2, L + (CdSe); 3, L + γ; 4, (CuInSe$_2$); 5, (CuInSe$_2$) + γ; 6, γ; 7, (CdSe) + γ; and 8, (CdSe). (From Olekseyuk, I.D. et al., *J. Solid State Chem.*, 179(1), 315, 2006.)

The results of earlier investigations of this system differ significantly from those obtained by Olekseyuk et al. (2006). Intermediate phase with changing composition is formed in the $CdSe-CuInSe_2$ system, which melts incongruently at 1005°C [at 1010°C (Vovk et al. 1999, 2000) and at 1050°C (Hirai et al. 1967)]. Peritectic point corresponds to 40 mol.% CdSe (Zmiy et al. 1997, Vovk et al. 1999, 2000). The solubility of $CuInSe_2$ in CdSe reaches 25 mol.% at the peritectic temperature and decreases to ≤9 mol.% at 600°C [according to the data of Dovletov et al. (1992), the solubility of $CuInSe_2$ in CdSe reaches 10 mol.%]. The solubility of CdSe in $CuInSe_2$ at the eutectic and eutectoid temperatures is 8 and 15 mol.% [20 mol.% (Vovk et al. 1999, 2000)], respectively, and not higher than 7 mol.% at 600°C. Solid solutions based on $CuInSe_2$ decompose according to a eutectoid reaction at 710°C [at 635°C (Vovk et al. 1999, 2000)], and the eutectoid point contains 20 mol.% CdSe [22 mol.% CdSe (Vovk et al. 1999, 2000)] (Zmiy et al. 1997).

The $CuCdInSe_3$ quaternary compound, which melts at 1080°C and crystallizes in the cubic structure with the lattice parameter $a = 1140.7$ pm, calculation and experimental density 4.698 and 4.693 g·cm^{-3}, respectively, and energy gap $E_g = 1.66$ eV (Guseinov et al. 1972), was not found in the $CdSe-CuInSe_2$ system (Zmiy et al. 1997, Olekseyuk et al. 2006). The second quaternary compound $CuCd_2InSe_4$ that crystallizes in the cubic structure of sphalerite type with the lattice parameter $a = 593.4$ pm (Parthé et al. 1969) [$a = 593 \pm 6$ pm (Garbato and Manca 1974)] corresponds to the γ-solid solution (Zmiy et al. 1997, Olekseyuk et al. 2006). Vovk et al. (1999, 2000) noted the possibility of polymorphic transformation in this compound, which is absent on the phase diagram according to the data of Zmiy et al. (1997) and Olekseyuk et al. (2006). Energy gap of the $CuCd_2InSe_4$ depends on its composition and changes within the interval of 0.6–0.7 eV for the nonstoichiometric crystals and is equal to 0.4 eV for stoichiometric composition (Vovk et al. 1999, 2000).

This system was investigated through DTA, XRD, metallography, and measuring of microhardness (Hirai et al. 1967, Garbato and Manca 1974, Garbato et al. 1985, Manca and Garbato 1986, Zmiy et al. 1997, Vovk et al. 1999, 2000, Olekseyuk et al. 2006). The ingots were annealed at 600°C for 500 h [for 250 h (Zmiy et al. 1997) and at 350°C for 1440 h (Olekseyuk et al. 2006)]. Single crystals of the $(2CdSe)_x(CuInSe_2)_{1-x}$ quasibinary system with $x = 0.05$, 0.50, and 0.70 have been grown by using CdI_2 as flux material (Garbato et al. 1985, Manca and Garbato 1986).

$CdSe-Cu_2Se-In_2Se_3$. Liquidus surface of this system (Figure 5.3) consists of 11 fields of primary crystallization of the phases, 3 of which belong to the binary components Cu_2Se, CdSe, and In_2Se_3; 6 are the ternary phases Cu_3InSe_3, $CuInSe_2$, $CuIn_2Se_4$, $CdIn_6Se_{10}$, $CuIn_5Se_8$, and $CuIn_{11}Se_{17}$; and 2 are the quaternary phases $CuCd_2InSe_4$ and $Cu_{0.6}Cd_{0.7}In_6Se_{10}$ ($CuCdIn_{10}Se_{16.5}$) (Zmiy et al. 2004, Ivashchenko et al. 2005). The primary crystallization fields are separated by 23 monovariant lines. There are 23 invariant points, 11 of which correspond to the binary and 12 to the ternary invariant process. The fields of primary crystallization are separated by monovariant lines and invariant points (Table 5.1). It is necessary to note that constructed liquidus surface of this system must be reconstructed as it did not include the new results of Olekseyuk et al. (2006), concerning the $CdSe-CuInSe_2$ phase diagram.

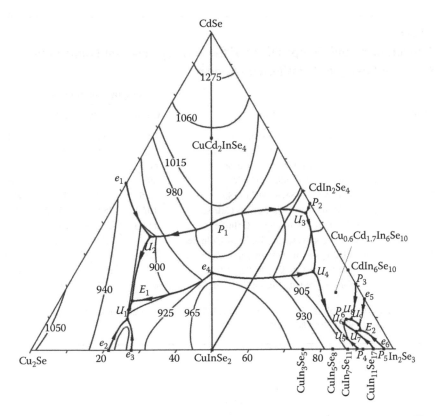

FIGURE 5.3 Liquidus surface of the $CdSe–Cu_2Se–In_2Se_3$ quasiternary system. (From Zmiy, O.F. et al., *J. Alloys Compd.*, 367(1–2), 49, 2004; Ivashchenko, I.A. et al., *J. Alloys Compd.*, 394(1–2), 186, 2005.)

$Cu_{0.6}Cd_{0.7}In_6Se_{10}$ quaternary compound crystallizes in the hexagonal structure with the lattice parameters $a = 404.98 \pm 0.02$ and $c = 328.86 \pm 0.02$ pm and calculated density 5.6550 ± 0.0009 g·cm⁻³ (Ivashchenko et al. 2005).

The isothermal section of this system at 550°C is shown in Figure 5.4 (Zmiy et al. 2004, Ivashchenko et al. 2005). It is divided into two separate subsystems: $CdSe–Cu_2Se–CuInSe_2$ and $CdSe–CuInSe_2–In_2Se_3$. This system was investigated through DTA and XRD and the homogenizing annealing was carried out at 550°C for 300 h.

5.28 CADMIUM–COPPER–SILICON–SELENIUM

$CdSe–Cu_2SiSe_3$. This section is nonquasibinary of the Cd–Cu–Si–Se quaternary system since Cu_2SiSe_3 melts incongruently (Figure 5.5) (Piskach et al. 1997, Olekseyuk et al. 1998). The $Cu_2CdSiSe_4$ quaternary compound is formed in this system, which melts incongruently at 912°C [at 921°C $\pm$ 5°C (Schäfer and Nitsche 1977)], and has a polymorphous transformation, and its low-temperature modification crystallizes in the orthorhombic structure with the lattice parameters $a = 797.9 \pm 0.4$, $b = 684.1 \pm 0.3$, and $c = 653.9 \pm 0.5$ pm [$a = 799.0$, $b = 682.4$, and $c = 656.4$ pm (Schäfer and Nitsche 1974, 1977)] and experimental density 4.96 g·cm⁻³ (Olekseyuk et al. 1998).

TABLE 5.1

Temperatures and Compositions of the Ternary Invariant Points in the CdSe–Cu$_2$Se–In$_2$Se$_3$ Quasiternary System

Invariant Points	Reaction	Composition, mol.%			t, °C
		Cu$_2$Se	CdSe	In$_2$Se$_3$	
E_1	L $\Leftrightarrow$ β-CuInSe$_2$ + CuCd$_2$InSe$_4$	64.5	15.5	20	865
E_2	L $\Leftrightarrow$ Cu$_{0.6}$Cd$_{0.7}$In$_6$Se$_{10}$ + In$_2$Se$_3$ + CuIn$_{11}$Se$_{17}$	6.25	6.75	87	795
U_1	L + Cu$_3$InSe$_3$ $\Leftrightarrow$ β-CuInSe$_2$ + Cu$_2$Se	68.5	21.5	10	907
U_2	L + CdSe $\Leftrightarrow$ CuCd$_2$InSe$_4$ + Cu$_2$Se	49	15	36	890
U_3	L + CdSe $\Leftrightarrow$ CdIn$_2$Se$_4$ + CuCd$_2$InSe$_4$	3.75	44.25	52	897
U_4	L + CuCd$_2$InSe$_4$ $\Leftrightarrow$ CdIn$_2$Se$_4$ + β-CuInSe$_2$	10.75	25.5	63.75	885
U_5	L + β-CuInSe$_2$ $\Leftrightarrow$ CuIn$_5$Se$_8$ + CdIn$_2$Se$_4$	11.25	4.25	84.5	850
U_6	L + CdIn$_2$Se$_4$ $\Leftrightarrow$ Cu$_{0.6}$Cd$_{0.7}$In$_6$Se$_{10}$ + CuIn$_5$Se$_8$	10	8.75	81.25	825
U_7	L + CuIn$_5$Se$_8$ $\Leftrightarrow$ Cu$_{0.6}$Cd$_{0.7}$In$_6$Se$_{10}$ + CuIn$_{11}$Se$_{17}$	7.75	6.25	86	810
U_8	L + CdIn$_2$Se$_4$ $\Leftrightarrow$ Cu$_{0.6}$Cd$_{0.7}$In$_6$Se$_{10}$ + CdIn$_6$Se$_{10}$	7.5	10	82.5	830
U_9	L + CdIn$_6$Se$_{10}$ $\Leftrightarrow$ Cu$_{0.6}$Cd$_{0.7}$In$_6$Se$_{10}$ + In$_2$Se$_3$	6.25	7.75	86	815

Source: Zmiy, O.F. et al., *J. Alloys Compd.*, 367(1–2), 49, 2004; Ivashchenko, I.A. et al., *J. Alloys Compd.*, 394(1–2), 186, 2005.

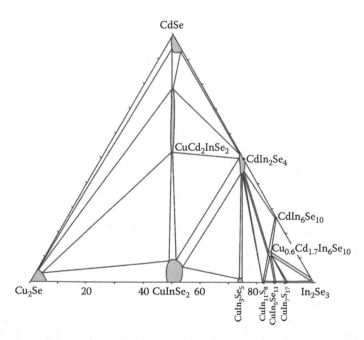

FIGURE 5.4 Isothermal section of the CdSe–Cu$_2$Se–In$_2$Se$_3$ quasiternary system at 550°C. (From Zmiy, O.F. et al., *J. Alloys Compd.*, 367(1–2), 49, 2004; Ivashchenko, I.A. et al., *J. Alloys Compd.*, 394(1–2), 186, 2005.)

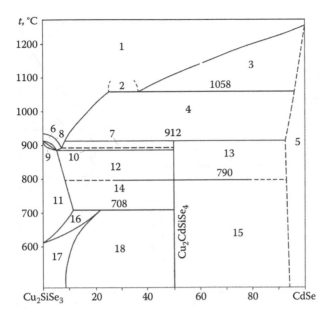

FIGURE 5.5 Phase relations in the CdSc–Cu$_2$SiSe$_3$ system: 1, L; 2, L$_1$ + L$_2$; 3, L$_1$ + (CdSe); 4, L$_2$ + (CdSe); 5, (CdSe); 6, L$_2$ + Cu$_8$SiSe$_6$; 7, L$_2$ + β-Cu$_2$CdSiSe$_4$; 8, L$_2$ + (β-Cu$_2$SiSe$_3$) + Cu$_8$SiSe$_6$, 9, (β-Cu$_2$SiSe$_3$) + Cu$_8$SiSe$_6$, 10, L$_2$ + Cu$_8$SiSe$_6$ + β-Cu$_2$CdSiSe$_4$; 11, (β-Cu$_2$SiSe$_3$); 12, (β-Cu$_2$SiSe$_3$) + β-Cu$_2$CdSiSe$_4$; 13, β-Cu$_2$CdSiSe$_4$ + (CdSe); 14, (β-Cu$_2$SiSe$_3$) + α-Cu$_2$CdSiSe$_4$; 15, α-Cu$_2$CdSiSe$_4$ + (CdSe); 16, (α-Cu$_2$SiSe$_3$) + (β-Cu$_2$SiSe$_3$); 17, (α-Cu$_2$SiSe$_3$); and 18, (α-Cu$_2$SiSe$_3$) + α-Cu$_2$CdSiSe$_4$. (From Piskach, L.V. et al., *Fizyka kondens. vysokomolek. system. Nauk. zap. Rinens'kogo pedinstytutu*, (3), 153, 1997; Olekseyuk, I.D. et al., *Zhurn. neorgan. khimii*, 43(3), 516, 1998.)

Cu$_8$SiSe$_6$, β-Cu$_2$CdSiSe$_4$ and solid solutions based on CdSe primarily crystallize in this system. The narrow immiscibility region exists at 1058°C. The solubility of Cu$_2$SiSe$_3$ in CdSe at peritectic temperature is not higher than 3 mol.% and the solubility of CdSe in Cu$_2$SiSe$_3$ is greater. The α-Cu$_2$SiSe$_3$ → β-Cu$_2$SiSe$_3$ peritectoid transformation takes place at 708°C. Thermal effects at 790°C correspond to the polymorphous transformation of Cu$_2$CdSiSe$_4$ and at 885°C and 892°C to the peritectic reaction L + Cu$_8$SiSe$_6$ ⇔ β-Cu$_2$CdSiSe$_4$ + α-Cu$_2$SiSe$_3$ and to the L + Cu$_8$SiSe$_6$ + β-Cu$_2$CdSiSe$_4$ secondary crystallization, respectively (Piskach et al. 1997, Olekseyuk et al. 1998).

This system was investigated through DTA, metallography, XRD, and measuring of microhardness and density. The ingots were annealed at 580°C for 1000 h (Piskach et al. 1997, Olekseyuk et al. 1998). The crystals of Cu$_2$CdSiSe$_4$ were grown by the horizontal gradient freezing or iodine transport method (Matsuhita et al. 2005).

5.29 CADMIUM–COPPER–GERMANIUM–SELENIUM

CdSe–Cu$_2$GeSe$_3$. The phase diagram is shown in Figure 5.6 (Piskach et al. 2000). Besides the quaternary compound Cu$_2$CdGeSe$_4$, possessing a narrow homogeneity region and forming incongruently at 830°C (Piskach et al. 2000, Gulay et al. 2002)

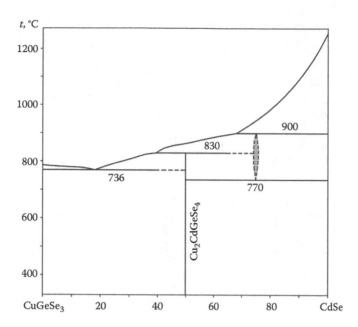

FIGURE 5.6 The CdSe–Cu$_2$GeSe$_3$ phase diagram. (From Piskach, L.V. et al., *J. Alloys Compd.*, 299(1–2), 227, 2000.)

[at 824°C ± 5°C (Schäfer and Nitsche 1977); at 852°C (Piskach et al. 1997); at 840°C ± 5°C (Zhukov et al. 1982a, 1984, Matsuhita et al. 2000, 2005)], another quaternary phase with the approximate composition Cu$_2$Cd$_3$GeSe$_6$ was revealed. Cu$_2$Cd$_3$GeSe$_6$ forms according to the peritectic reaction at 900°C and decomposes at 736°C according to the eutectoid reaction. The eutectic between Cu$_2$CdGeSe$_4$ and Cu$_2$GeSe$_3$ melts at 770°C and contains 18 mol.% CdSe [at 652°C and 25 mol.% CdSe (Piskach et al. 1997) and at 660°C ± 5°C and 25 mol.% CdSe (Zhukov et al. 1984)]. The solid solubility ranges of the components of this section do not exceed 2 mol.% (Piskach et al. 2000). According to the data of Zhukov et al. (1984), the solubility of CdSe in Cu$_2$GeSe$_3$ reaches 12 mol.% and the solubility of CdSe in Cu$_2$GeSe$_3$ is not higher than 7 mol.%. The increasing of Cu$_2$GeSe$_3$ contents in the solid solutions based on CdSe leads to the monotonous changing of electrophysical properties of forming solid solutions (Dovletov et al. 1987).

Cu$_2$CdGeSe$_4$ has two polymorphic modifications. α-Cu$_2$CdGeSe$_4$ crystallizes in the tetragonal structure of the stannite type with the lattice parameters $a = 574.82 \pm 0.02$ and $c = 1105.33 \pm 0.03$ pm (Gulay et al. 2002) [$a = 574.89 \pm 0.05$ and $c = 1195.5 \pm 0.1$ pm (Piskach et al. 2000); $a = 574.7$ and $c = 1105.9$ pm (Schäfer and Nitsche 1977); $a = 565$ and $c = 1096$ pm (Zhukov et al. 1984); $a = 565.7$ and $c = 1098.8$ pm (Schäfer and Nitsche 1974)], experimental density 5.45 g · cm^{-3} (Zhukov et al. 1984), and energy gap $E_g = 1.29$ eV (Konstantinova et al. 1989) [$E_g = 1.20$ eV (Mkrtchian et al. 1988, Matsuhita et al. 2000, 2005)].

β-Cu$_2$CdGeSe$_4$ crystallizes in the orthorhombic structure with the lattice parameters $a = 809.68 \pm 0.09$, $b = 689.29 \pm 0.06$, and $c = 662.64 \pm 0.06$ pm [$a = 808.8$, $b = 687.5$, and $c = 656.4$ pm (Matsuhita et al. 2000, 2005); $a = 806.2$, $b = 687.1$, and

$c = 659.7$ pm (Schäfer and Nitsche 1977); $a = 770.24 \pm 0.03$, $b = 654.86 \pm 0.02$, and $c = 629.28 \pm 0.03$ pm (Parasyuk et al. 2005)] and experimental density 5.638 ± 0.002 g·cm^{-3} (4.6067 ± 0.0006 g·cm^{-3} (Parasyuk et al. 2005)] (Gulay et al. 2002).

According to the data of Parthé et al. (1969), $Cu_2CdGeSe_4$ crystallizes in the cubic structure of sphalerite type with the lattice parameter $a = 565.7 \pm 0.5$ pm.

This system was investigated through DTA, XRD, metallography, and measuring of microhardness and density (Zhukov et al. 1982a, 1984, Piskach et al. 2000). The ingots were annealed at 800°C for 600 h (Zhukov et al. 1984). The $Cu_2CdGeSe_4$ compound was obtained at 1030°C–1080°C for 2 h and annealed at 530°C–580°C for 600 h (Mkrtchian et al. 1988). β-$Cu_2CdGeSe_4$ was produced by cooling of the respective melt without any annealing by directly quenching in water, while annealing of the alloy at 400°C leads to the obtaining of α-$Cu_2CdGeSe_4$ (Gulay et al. 2002). Single crystals of this compound were grown by the vertical Bridgman method with the next annealed for 100 h at 600°C and slowly cooling to room temperature (Parasyuk et al. 2005), or by the chemical transport reactions (Parthé et al. 1969, Schäfer and Nitsche 1974, Matsuhita et al. 2005), or by the oriented crystallization (Konstantinova et al. 1989), or by using the horizontal gradient freezing (Matsuhita et al. 2000, 2005).

CdSe–Cu$_8$GeSe$_6$. This section is not quasibinary section of the Cd–Cu–Ge–Se quaternary system (Figure 5.7) (Olekseyuk et al. 2000). It crosses the fields of

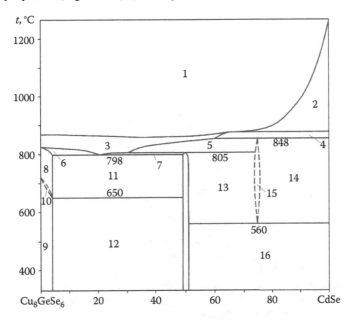

FIGURE 5.7 Phase relations in the CdSe–Cu$_8$GeSe$_6$ system: 1, L; 2, L + CdSe; 3, L + (Cu$_2$Se); 4, L + CdSe + (Cu$_2$Se); 5, L + (Cu$_2$Se) + Cu$_2$Cd$_3$GeSe$_6$; 6, L + (Cu$_2$Se) + (β-Cu$_8$GeSe$_6$); 7, L + (Cu$_2$Se) + Cu$_2$CdGeSe$_4$; 8, (Cu$_2$Se) + (β-Cu$_8$GeSe$_6$); 9, (Cu$_2$Se) + (α-Cu$_8$GeSe$_6$); 10, (Cu$_2$Se) + (α-Cu$_8$GeSe$_6$) + (β-Cu$_8$GeSe$_6$); 11, (Cu$_2$Se) + (β-Cu$_8$GeSe$_6$) + Cu$_2$CdGeSe$_4$; 12, (Cu$_2$Se) + (α-Cu$_8$GeSe$_6$) + Cu$_2$CdGeSe$_4$; 13, (Cu$_2$Se) + Cu$_2$Cd$_3$GeSe$_6$ + Cu$_2$CdGeSe$_4$; 14, CdSe + Cu$_2$Cd$_3$GeSe$_6$ + Cu$_2$CdGeSe$_4$; 15, (Cu$_2$Se) + Cu$_2$Cd$_3$GeSe$_6$; and 16, CdSe + (Cu$_2$Se) + Cu$_2$CdGeSe$_4$. (From Olekseyuk, I.D. et al., *J. Alloys Compd.*, 298(1–2), 203, 2000.)

primary crystallization of the solid solutions based on Cu_2Se and CdSe. The horizontal at 650°C corresponds to the polymorphic transformation of Cu_8GeSe_6 and that at 560°C corresponds to the decomposition of the $Cu_2Cd_3GeSe_6$ phase. This section was investigated through DTA, XRD, and metallography, and the alloys were annealed for 500 h at 400°C with next quenching in cold water.

CdSe–Cu_2Se–GeSe$_2$. Liquidus surface of this quasiternary system is shown in Figure 5.8 (Olekseyuk et al. 2000). It consists of eight fields of primary crystallization of phases. Three of them belong to the crystallization of Cu_2Se, CdSe, and GeSe$_2$. Another three belong to the crystallization of the ternary phases Cu_8GeSe_6, Cu_2GeSe_3, and Cd_4GeSe_6, and two belong to the crystallization of the quaternary phases $Cu_2CdGeSe_4$ and $Cu_2Cd_3GeSe_6$. The various fields of primary crystallization are divided by 19 monovariant lines and by 18 invariant points, 10 corresponding to binary invariant processes and 8 corresponding to the ternary ones. The types, the temperatures, and the coordinates of the ternary invariant points are presented in Table 5.2.

There is only one quaternary compound $Cu_2CdGeSe_4$ at 400°C presented in the CdSe–Cu_2Se–GeSe$_2$ quasiternary system (Figure 5.9) (Olekseyuk et al. 2000). It is in equilibrium with all existing binary and ternary compound. The $Cu_2Cd_3GeSe_6$ phase decomposes at higher temperatures and does not become apparent at this temperature.

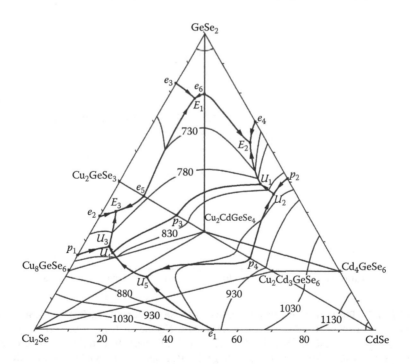

FIGURE 5.8 Liquidus surface of the CdSe–Cu_2Se–GeSe$_2$ quasiternary system. (From Olekseyuk, I.D. et al., *J. Alloys Compd.*, 298(1–2), 203, 2000.)

TABLE 5.2
Temperatures and Compositions of the Ternary Invariant Points in the CdSe–Cu₂Se–GeSe₂ Quasiternary System

Invariant Points	Reaction	Composition, mol.%			t, °C
		Cu₂Se	CdSe	GeSe₂	
E_1	L ⇔ Cu₂CdGeSe₄ + Cu₂GeSe₃ + GeSe₂	14	8	78	692
E_2	L ⇔ Cd₄GeSe₆ + GeSe₂ + Cu₂CdGeSe₄	5	32	63	686
E_3	L ⇔ Cu₂GeSe₃ + Cu₂CdGeSe₄ + β-Cu₈GeSe₆	56	5	39	754
U_1	L + Cu₂Cd₃GeSe₆ ⇔ Cu₂CdGeSe₄ + Cd₄GeSe₆	9	42	49	789
U_2	L + CdSe ⇔ Cu₂Cd₃GeSe₆ + Cd₄GeSe₆	6.5	47.5	46	850
U_3	L + Cu₂Se ⇔ β-Cu₈GeSe₆ + Cu₂CdGeSe₄	64	8	28	798
U_4	L + Cu₂Cd₃GeSe₆ ⇔ Cu₂CdGeSe₄ + Cu₂Se	64	10	26	805
U_5	L + CdSe ⇔ Cu₂Cd₃GeSe₆ + Cu₂Se	58	24	18	848

Source: Olekseyuk, I.D. et al. *J. Alloys Compd.*, 298(1–2), 203, 2000.

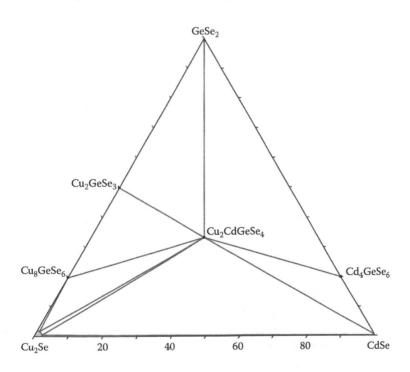

FIGURE 5.9 Isothermal section of the CdSe–Cu₂Se–GeSe₂ quasiternary system at 400°C. (From Olekseyuk, I.D. et al., *J. Alloys Compd.*, 298(1–2), 203, 2000.)

5.30 CADMIUM–COPPER–TIN–SELENIUM

CdSe–Cu₂SnSe₃. The phase diagram is shown in Figure 5.10 (Olekseyuk and Piskach 1997, Piskach et al. 1997). The eutectic contains 23 mol.% CdSe and crystallizes at 657°C [640°C ± 5°C (Zhukov et al. 1984)]. The $Cu_2CdSnSe_4$ quaternary compound is formed in this system. It melts incongruently at 782°C (Olekseyuk and Piskach 1997) [780°C (Matsuhita et al. 2000, 2005), 795°C ± 5°C (Zhukov et al. 1982a,b), and 768°C ± 5°C (Schäfer and Nitsche 1977)] and crystallizes in the tetragonal structure with the lattice parameters a = 583.37 ± 0.02 and c = 1140.39 ± 0.04 pm (Olekseyuk et al. 2002a) [a = 583.2 and c = 1138.9 pm (Matsuhita et al. 2000, 2005); a = 581.88 ± 0.06 and c = 1113.83 ± 0.02 pm (Olekseyuk and Piskach 1997); a = 581.4 and c = 1147 pm (Hahn and Schulze 1965); a = 582.6 and c = 1139.0 pm (Schäfer and Nitsche 1977); and a = 572 and c = 1112 pm (Zhukov et al. 1982b)], calculation and experimental density 5.77 [5.78 ± 0.02 (Olekseyuk et al. 2002a)] and 5.75 [5.76 (Olekseyuk and Piskach 1997); 5.77 (Zhukov et al. 1982b)] g·cm⁻³ (Hahn and Schulze 1965), and energy gap E_g = 1.30 eV (Mkrtchian et al. 1988) [E_g = 0.96 eV (Matsuhita et al. 2000, 2005); E_g = 0.89 eV (Konstantinova et al. 1989)].

At 530°C, the solubility of CdSe in Cu₂SnSe₃ reaches 20 mol.% and the solubility of Cu₂SnSe₃ in CdSe is not higher than 7 mol.% (Olekseyuk and Piskach 1997). According to the data of Zhukov et al. (1982b), at room temperature, the solubility

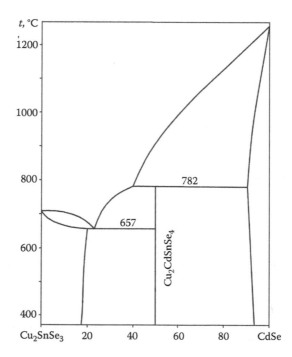

FIGURE 5.10 The CdSe–Cu₂SnSe₃ phase diagram. (From Olekseyuk, I.D., and Piskach, L.V., *Zhurn. neorgan. khimii*, 42(2), 331, 1997; Piskach, L.V. et al., *Fizyka kondens. vysokomolek. system. Nauk. zap. Rinens'kogo pedinstytutu*, (3), 153, 1997.)

of CdSe in Cu_2SnSe_3 is equal to 15 mol.% and Cu_2SnSe_3 in CdSe is not higher than 12 mol.%. The increasing of Cu_2SnSe_3 contents in the solid solutions based on CdSe leads to the monotonous changing of electrophysical properties of forming solid solutions (Dovletov et al. 1987).

This system was investigated through DTA, XRD, metallography, and measuring of microhardness and density (Zhukov et al. 1982a,b, Olekseyuk and Piskach 1997, Piskach et al. 1997). The ingots were annealed at 550°C–750°C for 500–600 h (Zhukov et al. 1982b). The $Cu_2CdSnSe_4$ compound was obtained at 1030°C–1080°C for 2 h and annealed at 530°C–580°C for 600 h (Mkrtchian et al. 1988). Its single crystals were grown by the oriented crystallization (Konstantinova et al. 1989), or using a solution-fusion method (Olekseyuk et al. 2002a), or using the horizontal gradient method from the melt (Matsuhita et al. 2000).

CdSe–Cu₂Se–SnSe₂. Liquidus surface of this quasiternary system is shown in Figure 5.11 (Parasyuk et al. 1999). It consists of five fields of primary crystallization of phases: CdSe, Cu_2Se, $SnSe_2$, Cu_2SnSe_3, and $Cu_2CdSnSe_4$. These fields of primary crystallization are divided by 10 monovariant lines and by 10 invariant points. The types, the temperatures, and the coordinates of the ternary invariant points are presented in Table 5.3.

Isothermal section of the CdSe–Cu₂Se–SnSc₂ quasiternary system is shown in Figure 5.12 (Parasyuk et al. 1999). The ingots were annealed at this temperature for 250 h.

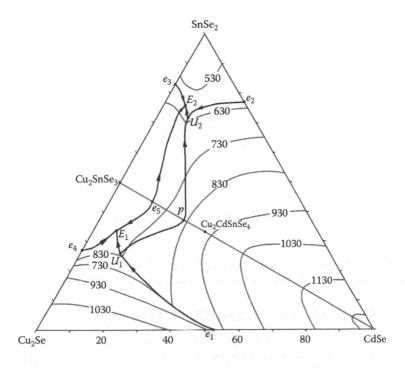

FIGURE 5.11 Liquidus surface of the CdSe–Cu₂Se–SnSe₂ quasiternary system. (From Parasyuk, O.V. et al., *Zhurn. neorgan. khimii*, 44(8), 1363, 1999.)

TABLE 5.3

Temperatures and Compositions of the Ternary Invariant Points in the CdSe–Cu₂Se–SnSe₂ Quasiternary System

Invariant Points	Reaction	Composition, mol.%			t, °C
		Cu_2Se	CdSe	$SnSe_2$	
E_1	L ⇔ Cu_2Se + $Cu_2CdSnSe_4$ + Cu_2SnSe_3	59	34	7	635
E_2	L ⇔ Cu_2SnSe_3 + $SnSe_2$ + $Cu_2CdSnSe_4$	17.5	76	6.5	575
U_1	L + CdSe ⇔ Cu_2Se + $Cu_2CdSnSe_4$	62	26	12	715
U_2	L + CdSe ⇔ $Cu_2CdSnSe_4$ + $SnSe_2$	19	71.5	9.5	605

Source: Parasyuk, O.V. et al., *Zhurn. neorgan. khimii*, 44(8), 1363, 1999.

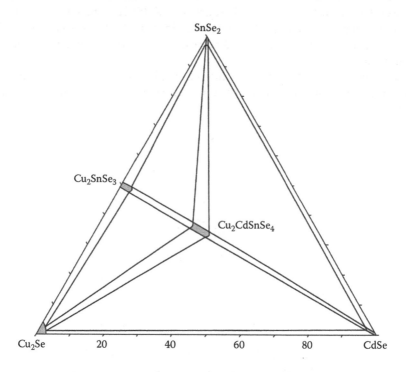

FIGURE 5.12 Isothermal section of the CdSe–Cu₂Se–SnSe₂ quasiternary system at 550°C. (From Parasyuk, O.V. et al., *Zhurn. neorgan. khimii*, 44(8), 1363, 1999.)

5.31 CADMIUM–COPPER–CHLORINE–SELENIUM

CdSe–CuCl₂. This section is nonquasibinary section of the Cd–Cu–Cl–Se quaternary system. The mixtures containing 0.1 mol.% $CuCl_2$ do not distinguish from the pure CdSe, while at the more $CuCl_2$ contents, CuSe and $CdCl_2$ were determined (Reisman and Berkenblit 1962). This indicates that CdSe and $CuCl_2$ interact according to the next reaction—$CdSe + CuCl_2 \Leftrightarrow CuSe + CdCl_2$—and forming CuSe partially dissolves into CdSe. The ingots containing 0.1, 1.0, 5.0, and 20 mol.% anhydrous $CuCl_2$ were mixed with CdSe, and such mixtures were annealed at 550°C and 900°C, quenched, and investigated through XRD.

5.32 CADMIUM–SILVER–GALLIUM–SELENIUM

CdSe–AgGaSe₂. Phase diagram of this system is of the eutectic type with eutectic point coordinates of ~33 mol.% CdSe and 815°C (Figure 5.13) (Olekseyuk et al. 2002b). The solubility of CdSe in $AgGaSe_2$ at the eutectic temperature is equal to ~23 and 9 mol.% at 600°C. The solubility of $AgGaSe_2$ in CdSe at the eutectic temperature reaches 59 mol.%. Decreasing of the temperature leads to three types of transformation depending on the $AgGaSe_2$ content. In the middle part of the solid solutions based on CdSe, a new phase occurs, the structure of which is a superstructure of wurtzite. The maximum ordering corresponds to the $AgCd_2GaSe_4$ composition, which crystallizes in the orthorhombic structure with the lattice parameters $a = 840.49 \pm 0.04$, $b = 719.34 \pm 0.04$, and $c = 684.34 \pm 0.04$ pm [in the hexagonal structure of wurtzite type with the lattice parameters $a = 425.1 \pm 0.5$ and $c = 695.6 \pm 0.5$ pm (Parthé et al. 1969)] and calculated density 5.763 ± 0.006 g·cm⁻³. The energy gap of this composition was estimated to be ≈1.7 eV at 293 K and ≈1.74 eV at 77 K (Bozhko et al. 2011). The solid solutions based on CdSe and enriched in $AgGaSe_2$ decompose with decreasing temperature according to the eutectoid reaction at 708°C (Olekseyuk et al. 2002b).

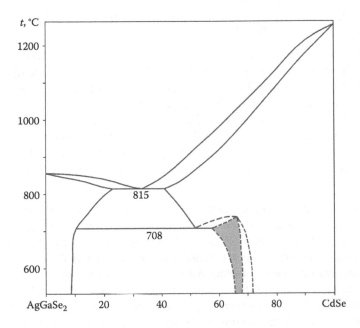

FIGURE 5.13 The CdSe–AgGaSe₂ phase diagram. (From Olekseyuk, I.D. et al., *J. Alloys Compd.*, 343(1–2), 125, 2002b.)

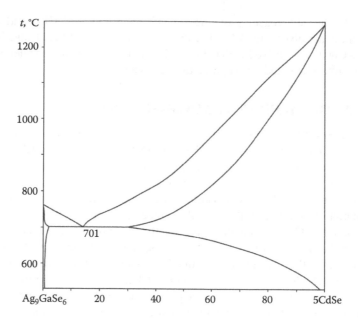

FIGURE 5.14 The 5CdSe–Ag$_9$GaSe$_6$ phase diagram. (From Kadykalo, E.M. et al., *J. Phase Equilib. Diff.*, 34(5), 403, 2013.)

This system was investigated through DTA, XRD, and metallography, and the ingots were annealing at 600°C for 500 h (Olekseyuk et al. 2002b). Single crystals of the AgCd$_2$GaSe$_4$ were grown using the method of directed crystallization, and the ingots were annealed at 600°C for 600 h (Bozhko et al. 2011).

CdSe–Ag$_9$GaSe$_6$. The phase diagram of this system is of the eutectic type with eutectic point coordinates 14 mol.% 5CdSe and 701°C (Figure 5.14) (Kadykalo et al. 2013) [45 mol.% CdSe and 704°C (Kadykalo et al. 2000)]. The subsolidus region features solid solutions based on CdSe in the range of 30–100 mol.% 5CdSe at the eutectic temperature and small (below 2 mol.% 5CdSe) solid solubility of Ag$_9$GaSe$_6$. As temperature decreases, the solid solution range of CdSe narrows and becomes negligible at 550°C. This system was investigated through DTA, XRD, and metallography, and the ingots were annealed at 550°C for 300–500 h (Kadykalo et al. 2000, 2013).

CdSe–Ag$_2$Se–Ga$_2$Se$_3$. Liquidus surface of this system consists of seven fields of the primary crystallization of the solid solutions based on CdSe, Ag$_2$Se, Ga$_2$Se$_3$, AgGaSe$_3$, Ag$_9$GaSe$_6$, α- and β-CdGa$_2$Se$_4$ (Figure 5.15) (Kadykalo et al. 2013). The fields of primary crystallization of the solid solutions are separated by 16 monovariant lines and 16 invariant points, of which 6 are ternary and 10 are binary or quasibinary.

The isothermal section of this system at 550°C was first constructed by Kadykalo et al. (2000) and then reconstructed by these authors and is shown in Figure 5.16 (Kadykalo et al. 2013). It contains seven single-phase regions, of which three are the solid solution ranges of the system components α-Ag$_2$Se, CdSe, and β-Ga$_2$Se$_3$; three are the solid solution ranges of the ternary compounds AgGaSe$_2$, α-CdGa$_2$Se$_4$,

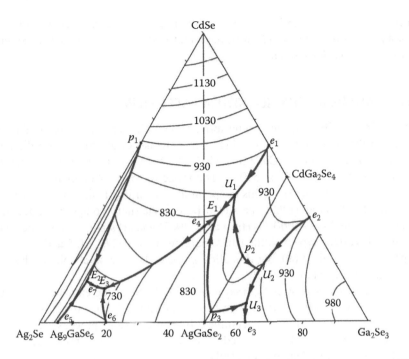

FIGURE 5.15 Liquidus surface of the CdSe–Ag₂Se–Ga₂Se₃ quasiternary system. (From Kadykalo, E.M. et al., *J. Phase Equilib. Diff.*, 34(5), 403, 2013.)

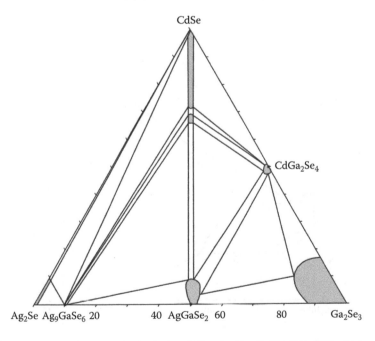

FIGURE 5.16 Isothermal section of the CdSe–Ag₂Se–Ga₂Se₃ quasiternary system at 550°C. (From Kadykalo, E.M. et al., *J. Phase Equilib. Diff.*, 34(5), 403, 2013.)

and β-Ag$_9$GaSe$_6$; and one is the solid solution range of the quaternary compound AgGa$_2$GaSe$_4$. Additionally, 12 two-phase and 6 three-phase regions exist in the CdSe–Ag$_2$Se–Ga$_2$Se$_3$ system.

5.33 CADMIUM–SILVER–INDIUM–SELENIUM

CdSe-AgInSe$_2$. This system is quasibinary with limited solubility of the components in the solid state (Figure 5.17) (Mishchenko et al. 2001). Coordinates of eutectic point are 10 mol.% CdSe and 750°C. The region of solid solution based on α-AgInSe$_2$ is 4 mol.% CdSe at 550°C. The range of solid solutions based on CdSe reaches 40 mol.% α-AgInSe$_2$ at this temperature. According to the data of Hirai et al. (1967) and Parthé et al. (1969), the AgCd$_2$InSe$_4$ quaternary compound is formed in this system. It melts at ~1000°C (Hirai et al. 1967) and crystallizes in the hexagonal structure of wurtzite type with the lattice parameters a = 427.7 ± 0.5 and c = 698.8 ± 0.5 pm (Parthé et al. 1969), but this alloy is located in the solid solution based on CdSe (Mishchenko et al. 2001).

The solubility based on β-AgInSe$_2$ is 5 mol.% CdSe at the eutectic temperature and increases with decreasing temperature to 17 mol.% CdSe. At 625°C, the solid solutions based on β-AgInSe$_2$ decompose by the eutectoid process. At the same temperature, the solid solutions based on α-AgInSe$_2$ have the largest extent: up to 12 mol.% CdSe (Mishchenko et al. 2001).

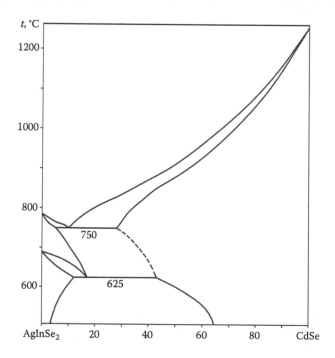

FIGURE 5.17   The CdSe–AgInSe$_2$ phase diagram. (From Mishchenko, I.A. et al., *Pol. J. Chem.*, 75(10), 1407, 2001.)

This system was investigated through DTA, XRD, metallography, and measuring of microhardness, and the ingots were annealed at 550°C for 300 h (Mishchenko et al. 2001).

CdSe–Ag$_2$Se–In$_2$Se$_3$. The liquidus surface of this quasiternary system is shown in Figure 5.18 (Ivashchenko et al. 2008b). It consists of 10 fields of primary crystallization of the phases, of which 3 belong to the crystallization of the solid solutions of the system components (Ag$_2$Se, CdSe, and In$_2$Se$_3$). Six fields belong to the crystallization of the ternary phases—α-AgInSe$_2$, β-AgInSe$_2$, CdIn$_2$Se$_4$, CdIn$_6$Se$_{10}$, AgIn$_5$Se$_8$, and AgIn$_{11}$Se$_{17}$—and one field belongs to the crystallization of the quaternary phase, Ag$_{0.4}$Cd$_{0.4}$In$_{6.3}$Se$_{10}$. These fields of primary crystallization are separated by 20 monovariant lines and 21 invariant points, of which 12 correspond to binary invariant processes and 9 to the ternary ones. The coordinates of the ternary invariant points, the character, and temperature of the corresponding invariant processes are listed in Table 5.4.

The isothermal section of the CdSe–Ag$_2$Se–In$_2$Se$_3$ quasiternary system at 550°C is given in Figure 5.19 (Ivashchenko et al. 2008a). Wide regions of two-phase equilibria of the solid solutions based on β-Ag$_2$Se, CdSe, α-AgInSe$_2$, α-AgIn$_5$Se$_8$,

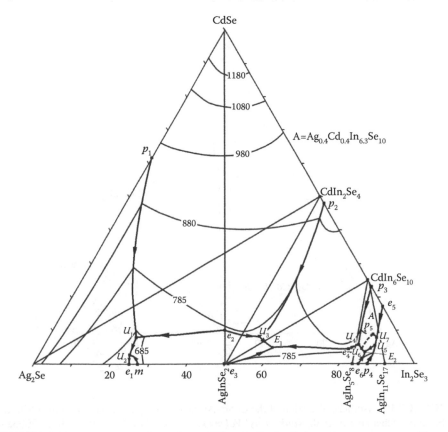

FIGURE 5.18 Liquidus surface of the CdSe–Ag$_2$Se–In$_2$Se$_3$ quasiternary system. (From Ivashchenko, I.A. et al., *Chem. Met. Alloys*, 1(3–4), 274, 2008b.)

TABLE 5.4
Temperatures and Compositions of the Ternary Invariant Points in the CdSe–Ag$_2$Se–In$_2$Se$_3$ Quasiternary System

Invariant Points	Reaction	Composition, mol.%			t, °C
		Ag$_2$Se	CdSe	In$_2$Se$_3$	
E_1	L $\Leftrightarrow$ β-AgInSe$_2$ + CdIn$_2$Se$_4$ + β-AgIn$_5$Se$_8$	3.5	5	60	727
E_2	L $\Leftrightarrow$ Ag$_{0.4}$Cd$_{0.4}$In$_{6.3}$Se$_{10}$ + β-AgIn$_5$Se$_8$ + AgIn$_{11}$Se$_{17}$	11.5	3.5	85	732
U_1	L + CdSe $\Leftrightarrow$ β-Ag$_2$Se + β-AgInSe$_2$	69	8	23	677
U_2	L + β-AgInSe$_2$ $\Leftrightarrow$ β-Ag$_2$Se + α-AgInSe$_2$	73	3	24	662
U_3	L + CdSe $\Leftrightarrow$ CdIn$_2$Se$_4$ + β-AgInSe$_2$	37	8	55	737
U_4	L + CdIn$_2$Se$_4$ $\Leftrightarrow$ CdIn$_6$Se$_{10}$ + β-AgIn$_5$Se$_8$	12	6	82	762
U_5	L + δ-In$_2$Se$_3$ $\Leftrightarrow$ AgIn$_{11}$Se$_{17}$ + Ag$_{0.4}$Cd$_{0.4}$In$_{6.3}$Se$_{10}$	8	5.5	86.5	757
U_6	L + CdIn$_6$Se$_{10}$ $\Leftrightarrow$ Ag$_{0.4}$Cd$_{0.4}$In$_{6.3}$Se$_{10}$ + β-AgIn$_5$Se$_8$	11	6	83	747
U_7	L + CdIn$_6$Se$_{10}$ $\Leftrightarrow$ Ag$_{0.4}$Cd$_{0.4}$In$_{6.3}$Se$_{10}$ + δ-In$_2$Se$_3$	9.5	4.5	86	772

Source: Ivashchenko, I.A. et al., *Chem. Met. Alloys*, 1(3–4), 274, 2008a.

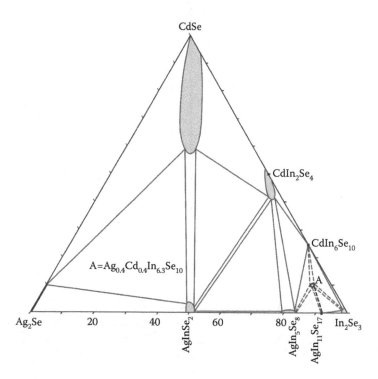

FIGURE 5.19 Isothermal section of the CdSe–Ag$_2$Se–In$_2$Se$_3$ quasiternary system at 550°C. (From Ivashchenko, I.A. et al., *Nauk. visnyk Volyns'k. un-tu im. Lesi Ukrainky*, (13), 27, 2008a.)

and $CdIn_2Se_4$, and of one of the polymorphous modifications of In_2Se_3, separate this system into quasiternary subsystem of various sizes. There is one quaternary compound $Ag_{0.4}Cd_{0.4}In_{6.3}Se_{10}$ in the $CdSe–Ag_2Se–In_2Se_3$ system. The two-phase equilibria of the compounds $AgIn_5Se_8$, $AgIn_{11}Se_{17}$, $Ag_{0.4}Cd_{0.4}In_{6.3}Se_{10}$, $CdIn_6Se_{10}$, and In_2Se_3, which split the investigated system in the In_2Se_3-rich region, are given by dotted lines because the crystal structure of the quaternary phase had not been determined.

5.34   CADMIUM–SILVER–SILICON–SELENIUM

$CdSe–Ag_8SiSe_6$. The phase diagram is shown in Figure 5.20 (Piskach et al. 2006). The peritectic composition and temperature are 86 ± 2 mol.% CdSe and 1002°C, respectively. The solubility of CdSe in Ag_8SiSe_6 at the eutectic temperature is 96 mol.%. This system was investigated by DTA, XRD, and metallography.

$CdSe–Ag_2Se–SiSe_2$. The isothermal section of this quasiternary system at room temperature is shown in Figure 5.21 and no quaternary phases were found (Parasyuk et al. 2003). The solid solubility range of Ag_8SiSe_6 is lower than 5 mol.% CdSe. The ingots were annealed at 400°C for 250 h and further cooling was done by turning off the furnace.

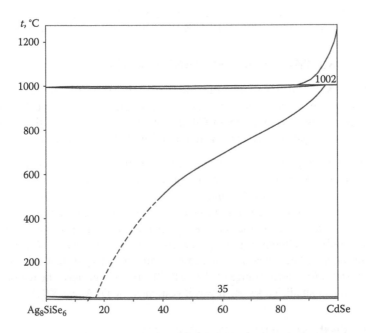

FIGURE 5.20 The $CdSe–Ag_8SiSe_6$ phase diagram. (From Piskach, L.V. et al., *J. Alloys Compd.*, 421(1–2), 98, 2006.)

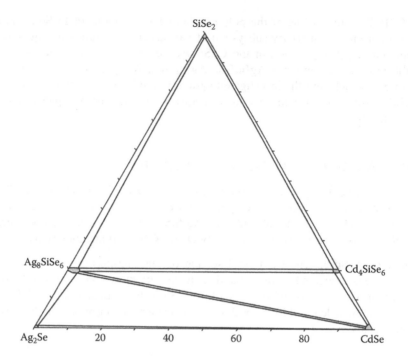

FIGURE 5.21 Isothermal section of the CdSe–Ag₂Se–SiSe₂ quasiternary system at room temperature. (From Parasyuk, O.V. et al., *J. Alloys Compd.*, 354(1–2), 138, 2003.)

5.35 CADMIUM–SILVER–GERMANIUM–SELENIUM

CdSe–Ag₈GeSe₆. The phase diagram is a eutectic type (Figure 5.22) (Piskach et al. 2006). The eutectic composition and temperature are 77 ± 1 mol.% CdSe and 872°C, respectively. The solubility of CdSe in Ag₈GeSe₆ at the eutectic temperature is 76 mol.%. This system was investigated by DTA, XRD, and metallography.

CdSe–Ag₂Se–GeSe₂. The glass-formation region in this system is shown in Figure 5.23 (Olekseyuk et al. 2009). Such region in the Ag₂Se–GeSe₂ boundary system is limited to the range 53–56 mol.% GeSe₂. The maximum amount of CdSe that can be introduced in the glass is 12 mol.%. The maximum GeSe₂ content is 62 mol.% at 8 mol.% CdSe. Characteristic temperatures of the glassy alloys, namely, the glass transition temperature (T_g), the crystallization temperature (T_c), and the melting temperature (T_m) of the crystallized alloys, have been measured. It was determined that T_g of the CdSe-containing glasses lies in a fairly narrow range (253°C ± 8°C), probably because the region of glass existence is rather small. This system was investigated by DTA and XRD.

5.36 CADMIUM–SILVER–TIN–SELENIUM

CdSe–"Ag₂SnSe₃". This section is not quasibinary one of the Cd–Ag–Sn–Se quaternary system (Figure 5.24) (Olekseyuk et al. 1997d). Ag₂CdSnSe₄ quaternary compound is formed in this section. It melts incongruently at 635°C and crystallizes

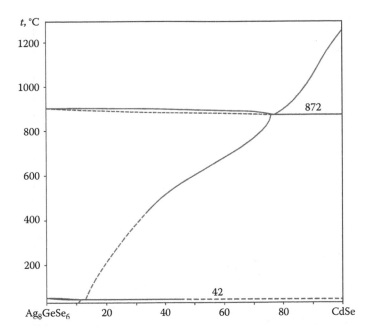

FIGURE 5.22   The CdSe–Ag$_8$GeSe$_6$ phase diagram. (From Piskach, L.V. et al., *J. Alloys Compd.*, 421(1–2), 98, 2006.)

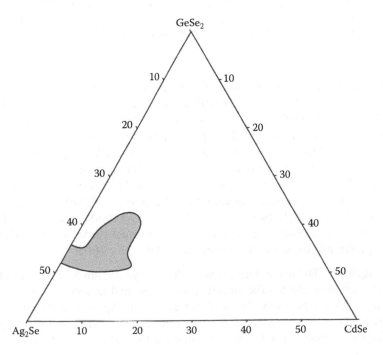

FIGURE 5.23   The glass-formation region in the CdSe–Ag$_2$Se–GeSe$_2$ quasiternary system. (From Olekseyuk, I.D. et al., *Chem. Met. Alloys*, 2(3–4), 146, 2009.)

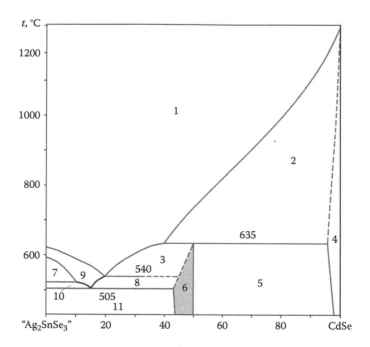

FIGURE 5.24 Phase relations in the CdSe–"Ag₂SnSe₃" system: 1, L; 2, L + (CdSe); 3, L + (Ag₂CdSnSe₄); 4, (CdSe); 5, (Ag₂CdSnSe₄) + (CdSe); 6, (Ag₂CdSnSe₄); 7, L + Ag₈SnSe₆ + Ag₁₋ₓSnₓSe; 8, L + Ag₈SnSe₆ + (Ag₂CdSnSe₄); 9, L + Ag₈SnSe₆; 10, L + Ag₈SnSe₆ + SnSe₂; and 11, Ag₈SnSe₆ + (Ag₂CdSnSe₄) + SnSe₂. (From Olekseyuk. I.D. et al., *Pol. J. Chem.*, 71(6), 721, 1997d.)

in the orthorhombic structure with the lattice parameters $a = 426.43 \pm 0.03$, $b = 731.75 \pm 0.04$, and $c = 698.48 \pm 0.05$ pm [$a = 426.3 \pm 0.3$, $b = 737.2 \pm 0.2$, and $c = 681.1 \pm 0.2$ pm (Olekseyuk et al. 1997d); $a = 426.2 \pm 0.5$, $b = 731.4 \pm 0.5$, and $c = 697.9 \pm 0.5$ pm (Parthé et al. 1969)] and calculated density 5.821 ± 0.03 g·cm⁻³ (Parasyuk et al. 2002). The homogeneity region of Ag₂CdSnSe₄ at 480°C is within the interval from 43.5 to 50 mol.% CdSe. The solid solutions based on CdSe do not exceed 2 mol.% "Ag₂SnSe₃" at 480°C. Alloys in the region 0–43.5 mol.% CdSe are three-phase and contain Ag₂CdSnSe₄, Ag₈SnSe₆, and SnSe₂. In the region 50–98 mol.% CdSe, only two phases (CdSe and Ag₂CdSnSe₄) are in equilibrium (Olekseyuk et al. 1997d). This system was investigated through DTA, XRD, and metallography, and the alloys were annealed at 480°C for 400 h.

CdSe–Ag₈SnSe₆. The phase diagram is a eutectic type (Figure 5.25) (Parasyuk et al. 2002, Piskach et al. 2006). The eutectic composition and temperature are 65 mol.% CdSe and 703°C, respectively. The solubility of CdSe in Ag₈SiSe₆ at the eutectic temperature is 42 mol.%. This system was investigated by DTA, XRD, and metallography. The alloys were annealed at 400°C over 250 h and then were quenched in cold water.

CdSe–Ag₂Se–SnSe₂. Isothermal section of this system at 400°C is shown in Figure 5.26 (Parasyuk et al. 2002). The Ag₂CdSnSe₄ compound is only quaternary

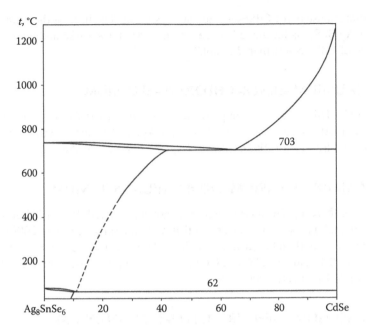

FIGURE 5.25 The CdSe–Ag$_8$SnSe$_6$ phase diagram. (From Piskach, L.V. et al., *J. Alloys Compd.*, 421(1–2), 98, 2006.)

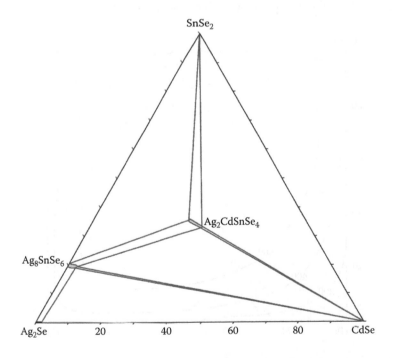

FIGURE 5.26 Isothermal section of the CdSe–Ag$_2$Se–SnSe$_2$ quasiternary system at room temperature. (From Parasyuk, O.V. et al., *J. Alloys Compd.*, 335(1–2), 176, 2002.)

294   Quaternary Alloys Based on II-VI Semiconductors

intermediate phase in the CdSe–Ag₂Se–SnSe₂ system. The limit of the solid solution region of Ag₈SnSe₆ is localized between 5 and 100 mol.% CdSe and the solid solution range of CdSe is less than 2.5 mol.%.

5.37 CADMIUM–SILVER–CHLORINE–SELENIUM

CdSe–AgCl. This section is nonquasibinary section of the Cd–Ag–Cl–Se quaternary system. At the interaction of CdSe and AgCl, the next reaction takes place (Reisman and Berkenblit 1962): CdSe + 2AgCl = CdCl₂ + Ag₂Se.

5.38 CADMIUM–BARIUM–NEODYMIUM–SELENIUM

CdSe–BaSe–Nd₂Se₃. The phase diagram is not constructed. Ba₄Cd₃Nd₂Se₁₀ quaternary compound has been synthesized in this system (Yang and Ibers 2000). It crystallizes in the orthorhombic structure with the lattice parameters $a = 433.40 \pm 0.09$, $b = 1866.7 \pm 0.4$, and $c = 2737.6 \pm 0.6$ pm. This compound was obtained at 900°C using BaBr₂/KBr eutectic flux.

5.39 CADMIUM–MERCURY–TELLURIUM–SELENIUM

CdSe + HgTe ⇔ CdTe + HgSe. The calculated miscibility gaps in the CdSe + HgTe ⇔ CdTe + HgSe ternary mutual system at 200°C–600°C are shown in Figure 5.27 (Ohtani et al. 1992). The position of the miscibility gap indicates that the phase

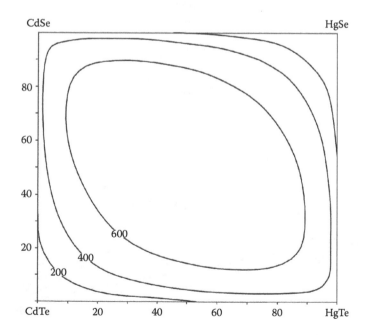

FIGURE 5.27 Calculated miscibility gaps in the CdSe + HgTe ⇔ CdTe + HgSe quasiternary system. (From Ohtani, H. et al., *J. Alloys Compd.*, 182(1), 103, 1992.)

separation occur toward the CdSe and HgTe ends that have the highest and lowest melting temperatures, respectively, among the four compounds (Leute et al. 1982, Ohtani et al. 1992).

A thermodynamic analysis of stability for the $Cd_xHg_{1-x}Te_ySe_{1-x}$ quaternary solid solutions has been developed by Kazakov et al. (2001). The locations of the unstable regions in (T-x) space are calculated using both the disordered solid solution approach and the short-range clustering approach. It was shown that such solid solutions decompose at low temperature.

5.40   CADMIUM–GALLIUM–INDIUM–SELENIUM

CdSe–Ga$_2$Se$_3$–In$_2$Se$_3$. The phase diagram is not constructed. $CdGa_3InSe_7$ quaternary compound has been synthesized in this system (Agaev et al. 1996). It melts at 847°C and crystallizes in the tetragonal structure with the lattice parameters $a = 936$ and $c = 625$ pm and energy band gap 0.88 eV. Single crystals of this compound were grown by the zone recrystallization.

5.41   CADMIUM–GALLIUM–PRASEODYMIUM–SELENIUM

CdSe–Ga$_2$Se$_3$–Pr$_2$Se$_3$. The phase diagram is not constructed. $CdGa_3PrSe_7$ quaternary compound has been synthesized in this system (Agaev et al. 1996). It melts at 897°C and crystallizes in the tetragonal structure with the lattice parameters $a = 950$ and $c = 628$ pm. Single crystals of this compound were grown by the chemical transport reactions.

5.42   CADMIUM–GALLIUM–NEODYMIUM–SELENIUM

CdSe–Ga$_2$Se$_3$–Nd$_2$Se$_3$. The phase diagram is not constructed. $CdGa_3NdSe_7$ quaternary compound has been synthesized in this system (Agaev et al. 1996). It melts at 972°C and crystallizes in the tetragonal structure with the lattice parameters $a = 964$ and $c = 625$ pm.

5.43   CADMIUM–GALLIUM–GERMANIUM–SELENIUM

CdSe–Ga$_2$Se$_3$–GeSe$_2$. Six fields of primary crystallization of the next phases— CdSe, Ga$_2$Se$_3$, GeSe$_2$, α- and β-CdGa$_2$Se$_4$, and Cd$_4$GeSe$_6$—exist on the liquidus surface of this system (Figure 5.28) (Olekseyuk and Parasyuk 1995). In the subsystem CdGa$_2$Se$_4$–Ga$_2$Se$_3$–GeSe$_2$, the crystallization ends in the ternary peritectic at 628°C. The crystallization in the CdSe–CdGa$_2$Se$_4$–GeSe$_2$ subsystem is more complex, and in the subsystem CdGa$_2$Se$_4$–Cd$_4$GeSe$_6$–GeSe$_2$, the crystallization ends at 642°C. In the subsystem CdSe–CdGa$_2$Se$_4$–Cd$_4$GeSe$_6$, the crystallization ends by the peritectic reaction at 730°C. This reaction with liquid excess is representative for other alloys from β-CdGa$_2$Se$_4$–U_1–p_1–CdSe peritectic tetragon. Temperatures and compositions of the ternary invariant points in the CdSe–Ga$_2$Se$_3$–GeSe$_2$ system are given in Table 5.5 (Olekseyuk and Parasyuk 1995).

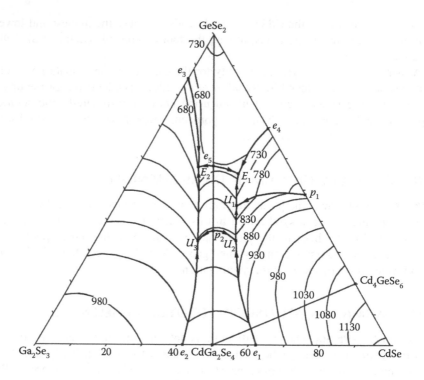

FIGURE 5.28 Liquidus surface of the $CdSe–Ga_2Se_3–GeSe_2$ quasiternary system. (From Olekseyuk, I.D., and Parasyuk, O.V., *Zhurn. neorgan. khimii*, 40(2), 315, 1995.)

TABLE 5.5

Temperatures and Compositions of the Ternary Invariant Points in the $CdSe–Ga_2Se_3–GeSe_2$ Quasiternary System

		Composition, mol.%			
Symbol	**Reaction**	**CdSe**	**Ga_2Se_3**	**$GeSe_2$**	**t, °C**
E_1	L ⇔ β-$CdGa_2Se_4$ + $GeSe_2$ + Cd_4GeSe_6	30	16.5	54.5	642
E_2	L ⇔ $GeSe_2$ + Ga_2Se_3 + β-$CdGa_2Se_4$	17	26	57	628
U_1	L + CdSe ⇔ Cd_4GeSe_6 + β-$CdGa_2Se_4$	35	20	45	730
U_2	L + CdSe ⇔ α-$CdGa_2Se_4$ + β-$CdGa_2Se_4$	41	26	33	830
U_3	L + Ga_2Se_3 ⇔ α-$CdGa_2Se_4$ + β-$CdGa_2Se_4$	28	38	34	830

Source: Olekseyuk, I.D. and Parasyuk, O.V., *Zhurn. neorgan. khimii*, 40(2), 315, 1995.

Isothermal section of the $CdSe–Ga_2Se_3–GeSe_2$ quasiternary system at 600°C is shown in Figure 5.29 (Olekseyuk et al. 1997c). The alloys were annealed at this temperature for 250 h.

There are two glass-forming regions in this system (Figure 5.30) (Olekseyuk et al. 1997b). The first of them begins from $GeSe_2$ and is elongated toward E_2 ternary eutectic and the second is situated around E_1 ternary eutectic.

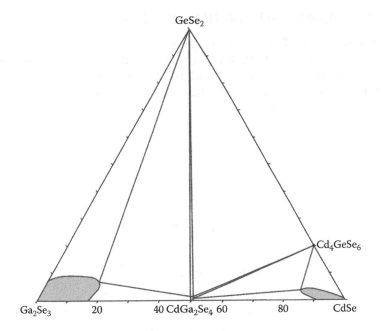

FIGURE 5.29 Isothermal section of the CdSe–Ga$_2$Se$_3$–GeSe$_2$ quasiternary system at 600°C. (From Olekseyuk, I.D. et al., *Pol. J. Chem.*, 71(6), 701, 1997c.)

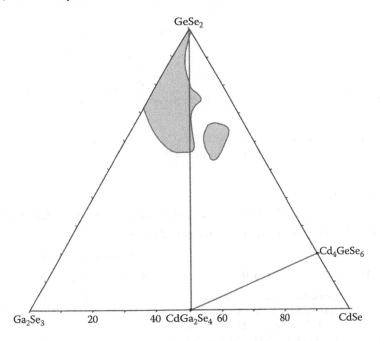

FIGURE 5.30 Glass-forming regions in the CdSe–Ga$_2$Se$_3$–GeSe$_2$ quasiternary system. (From Olekseyuk, I.D. et al., *Fizyka kondens. vysokomolek. system. Nauk. zap. Rinens'kogo pedinstytutu*, (3), 148, 1997b.)

5.44   CADMIUM–GALLIUM–TIN–SELENIUM

CdSe–Ga$_2$Se$_3$–SnSe$_2$. The liquidus surface of this quasiternary system is shown in Figure 5.31 (Piskach et al. 2002). It consists of five fields of primary crystallization of the phases CdSe, Ga$_2$Se$_3$, SnSe$_2$, α- and β-CdGa$_2$Se$_4$. Temperatures and compositions of the ternary invariant points in the CdSe–Ga$_2$Se$_3$–SnSe$_2$ system are given in Table 5.6 (Piskach et al. 2002).

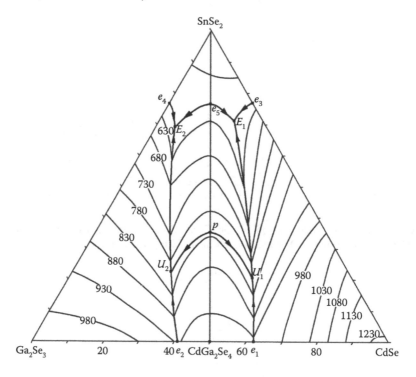

FIGURE 5.31   Liquidus surface of the CdSe–Ga$_2$Se$_3$–SnSe$_2$ quasiternary system. (From Piskach, L.V. et al., *Fiz. i khim. tv. tila*, 3(1), 25, 2002.)

TABLE 5.6
Temperatures and Compositions of the Ternary Invariant Points in the CdSe–Ga$_2$Se$_3$–SnSe$_2$ Quasiternary System

Symbol	Reaction	Composition, mol.%			*t*, °C
		CdSe	Ga$_2$Se$_3$	SnSe$_2$	
E_1	L ⇔ CdSe + SnSe$_2$ + α-CdGa$_2$Se$_4$	21	8	71	585
E_2	L ⇔ Ga$_2$Se$_3$ + SnSe$_2$ + α-CdGa$_2$Se$_4$	5	26	69	570
U_1	L + β-CdGa$_2$Se$_4$ ⇔ CdSe + α-CdGa$_2$Se$_4$	52	28	20	812
U_2	L + β-CdGa$_2$Se$_4$ ⇔ Ga$_2$Se$_3$ + α-CdGa$_2$Se$_4$	28	50	22	812

Source: Piskach, L.V. et al., *Fiz. i khim. tv. tila*, 3(1), 25, 2002.

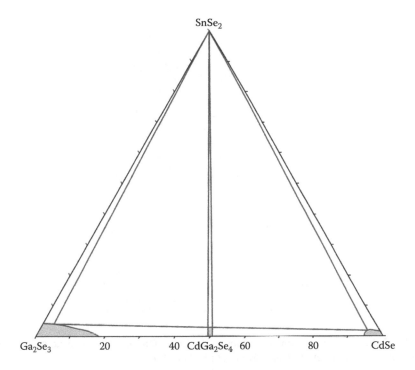

SnSe₂

Ga₂Se₃   20   40 CdGa₂Se₄ 60   80   CdSe

FIGURE 5.32   Isothermal section of the CdSe–Ga₂Se₃–SnSe₂ quasiternary system at 400°C system. (From Piskach, L.V. et al., *Fiz. i khim. tv. tila*, 3(1), 25, 2002.)

Isothermal section of the CdSe–Ga₂Se₃–SnSe₂ quasiternary system at 400°C is shown in Figure 5.32 (Piskach et al. 2002). The alloys were annealed at this temperature for 250 h.

The glass-formation region in this quasiternary system is shown in Figure 5.33 (Olekseyuk et al. 1999). The glass transition temperature (T_g), the crystallization temperature (T_c), and the melting temperature (T_m) were determined for obtained glasses. The glasses have considerable tendency for crystallization and they can be obtained only in the case of rigid hardening.

5.45   CADMIUM–GALLIUM–LEAD–SELENIUM

CdSe–Ga₂Se₃–PbSe. The liquidus surface of this system is shown in Figure 5.34 (Sosovska et al. 2008). It is divided into six fields of primary crystallization. Three of them belong to the system components (namely, to their solid solutions). The other three fields correspond to the primary crystallization of the ternary compounds α- and β-CdGa₂Se₄ and PbGa₂Se₄. The aforementioned fields are separated with 11 monovariant lines. There are seven binary and five ternary invariant points. Temperatures, compositions, and corresponding reaction of the ternary invariant points are listed in Table 5.7. The isothermal section of this system at 600°C is shown in Figure 5.35 (Sosovska et al. 2008). This system was investigated by DTA, XRD, and metallography. The alloys were annealed at 600°C over 250 h and then were rapidly quenched in cold water.

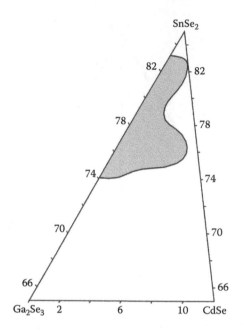

FIGURE 5.33 The glass-formation region in the CdSe–Ga₂Se₃–SnSe₂ quasiternary system. (From Olekseyuk, I.D. et al., *Funct. Mater.*, 6(3), 474, 1999.)

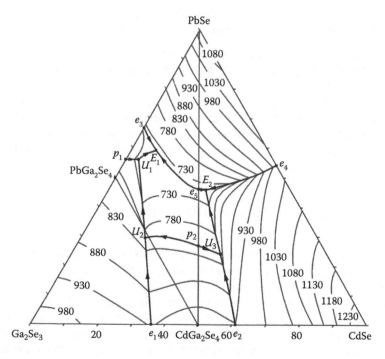

FIGURE 5.34 Liquidus surface of the CdSe–Ga₂Se₃–PbSe quasiternary system. (From Sosovska, S.M. et al., *J. Alloys Compd.*, 453(1–2), 115, 2008.)

TABLE 5.7

Temperatures and Compositions of the Ternary Invariant Points in the CdSe–Ga₂Se₃–PbSe Quasiternary System

		Composition, mol.%			
Symbol	Reaction	CdSe	Ga₂Se₃	PbSe	t, °C
E_1	L ⇔ β-CdGa₂Se₄ + PbGa₂Se₄ + PbSe	8	33	59	660
E_2	L ⇔ CdSe + β-CdGa₂Se₄ + PbSe	29	25	46	702
U_1	L + Ga₂Se₃ ⇔ β-CdGa₂Se₄ + PbGa₂Se₄	4	40	56	712
U_2	L + α-CdGa₂Se₄ ⇔ Ga₂Se₃ + β-CdGa₂Se₄	21	52	27	816
U_3	L + α-CdGa₂Se₄ ⇔ CdSe + β-CdGa₂Se₄	45	30	25	816

Source: Sosovska, S.M. et al., *J. Alloys Compd.*, 453(1–2), 115, 2008.

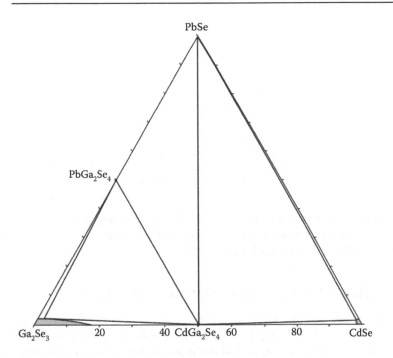

FIGURE 5.35 Isothermal section of the CdSe–Ga₂Se₃–PbSe quasiternary system at 600°C. (From Sosovska, S.M. et al., *J. Alloys Compd.*, 453(1–2), 115, 2008.)

5.46 CADMIUM–GALLIUM–ARSENIC–SELENIUM

CdSe–GaAs. The phase diagram is a eutectic type (Figure 5.36) (Glazov et al. 1979, 1983). The eutectic contains 67 mol.% CdSe and crystallizes at 1105°C (these data are taken from Figure 5.36). According to the data of Voitsehovski et al. (1970), the solubility of CdSe in GaAs reaches 10 mol.%. The solubility of GaAs in CdSe is also equal to 10 mol.% (Kirovskaya et al. 2007).

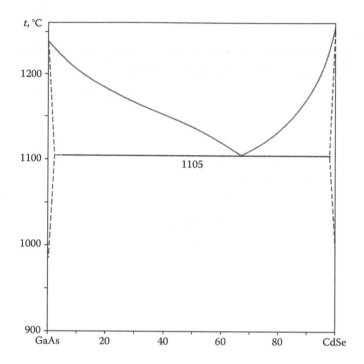

FIGURE 5.36 The CdSe–GaAs phase diagram. (From Glazov, V.M. et al., *Termodinamic-heskie svoistva metallicheskih rasplavov: Materialy 4-go Vsesoyuz. soveshch. po termodin-amike metal. splavov (rasplavov)*, Nauka, Alma-Ata, Kazakhstan, 1979, pp. 2, 26; Glazov, V.M. et al., *Izv. AN SSSR. Neorgan. materialy*, 19(2), 193, 1983.)

This system was investigated through DTA and metallography (Glazov et al. 1979, 1983). The ingots were obtained by the melting of mixtures from chemical elements at 1250°C (Voitsehovski et al. 1970).

5.47 CADMIUM–GALLIUM–ANTIMONY–SELENIUM

CdSe–Ga$_2$Se$_3$–Sb$_2$Se$_3$. The liquidus surface of this system is given in Figure 5.37 (Sosovska et al. 2007). It is represented by five fields of primary crystallization. Three of them belong to the solid solutions based on the system components: CdSe, Ga$_2$Se$_3$, and Sb$_2$Se$_3$. The other two fields correspond to the primary crystallization of the α- and β-CdGa$_2$Se$_4$. These fields are separated with nine monovariant lines. There are six binary and four ternary invariant points in this system, among them, two transition points and two ternary eutectics. Temperatures, compositions, and corresponding liquid–solid equilibria of the ternary invariant points are listed in Table 5.8. The polymorphous transformation of CdGa$_2$Se$_4$ appears on the liquidus surface as the curve U_1pU_2.

The isothermal section of this system at 400°C is shown in Figure 5.38 (Sosovska et al. 2007). This system was investigated by DTA, XRD, and metallography. The alloys were annealed at 400°C over 250 h and then were rapidly quenched in cold water.

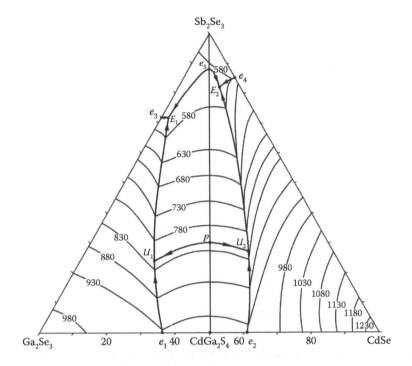

FIGURE 5.37 Liquidus surface of the CdSe–Ga$_2$Se$_3$–Sb$_2$Se$_3$ quasiternary system. (From Sosovska, S.M. et al., *Pol. J. Chem.*, 81(4), 505, 2007.)

TABLE 5.8

Temperatures and Compositions of the Ternary Invariant Points in the CdSe–Ga$_2$Se$_3$–Sb$_2$Se$_3$ Quasiternary System

Symbol	Reaction	Composition, mol.%			t, °C
		CdSe	Ga$_2$Se$_3$	Sb$_2$Se$_3$	
E_1	L ⇔ α-CdGa$_2$Se$_4$ + Ga$_2$Se$_3$ + Sb$_2$Se$_3$	2	26	72	527
E_2	L ⇔ CdSe + α-CdGa$_2$Se$_4$ + Sb$_2$Se$_3$	12	6	82	555
U_1	L + Ga$_2$Se$_3$ ⇔ α-CdGa$_2$Se$_4$+ Sb$_2$Se$_3$	22	54	24	812
U_2	L + β-CdGa$_2$Se$_4$ ⇔ CdSe + α-CdGa$_2$Se$_4$	48	25	27	811

Source: Sosovska, S.M. et al., *Pol. J. Chem.*, 81(4), 505, 2007.

5.48 CADMIUM–GALLIUM–BISMUTH–SELENIUM

CdSe–Ga$_2$Se$_3$–Bi$_2$Se$_3$. The liquidus surface of this system is given in Figure 5.39 (Sosovska et al. 2010). It is represented by six fields of primary crystallization. Three of them belong to the solid solutions based on the system components. The other three fields correspond to the primary crystallization of the ternary compounds α- and β-CdGa$_2$Se$_4$ and CdBi$_2$Se$_4$. These fields are separated by 11 monovariant lines. There are seven binary

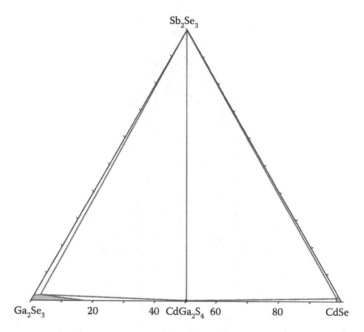

FIGURE 5.38 Isothermal section of the $CdSe–Ga_2Se_3–Sb_2Se_3$ quasiternary system at 400°C. (From Sosovska, S.M. et al., *Pol. J. Chem.*, 81(4), 505, 2007.)

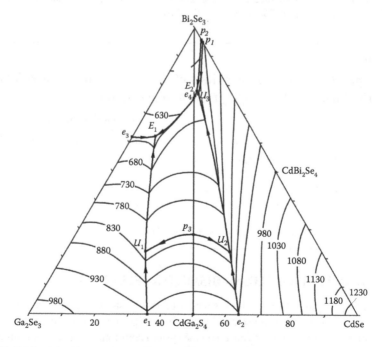

FIGURE 5.39 Liquidus surface of the $CdSe–Ga_2Se_3–Bi_2Se_3$ quasiternary system. (From Sosovska, S.M. et al., *Chem. Met. Alloys*, 3(1–2), 5, 2010.)

and five ternary invariant points in the system, among them, three transition points and two ternary eutectics. Temperatures, compositions, and corresponding reaction of the ternary invariant points are listed in Table 5.9. The polymorphous transformation of $CdGa_2Se_4$ appears on the liquidus surface as the invariant curve $U_1p_3U_2$.

The isothermal section of this system at 400°C is shown in Figure 5.40 (Sosovska et al. 2010). $CdBi_2Se_4$ is stable only in a limited high-temperature interval and does

TABLE 5.9
Temperatures and Compositions of the Ternary Invariant Points in the $CdSe-Ga_2Se_3-Bi_2Se_3$ Quasiternary System

Symbol	Reaction	Composition, mol.%			t, °C
		CdSe	Ga_2Se_3	Bi_2Se_3	
E_1	$L \Leftrightarrow \beta\text{-}CdGa_2Se_4 + Ga_2Se_3 + Bi_2Se_3$	7	31	62	619
E_2	$L \Leftrightarrow CdSe + \beta\text{-}CdGa_2Se_4 + Bi_2Se_3$	12	10	78	649
U_1	$L + \alpha\text{-}CdGa_2Se_4 \Leftrightarrow \beta\text{-}CdGa_2Se_4 + Ga_2Se_3$	24	54	22	816
U_2	$L + \alpha\text{-}CdGa_2Se_4 \Leftrightarrow \beta\text{-}CdGa_2Se_4 + CdSe$	28	50	22	815
U_3	$L + CdSe \Leftrightarrow \beta\text{-}CdGa_2Se_4 + CdBi_2Se_4$	13	10	77	669

Source: Sosovska, S.M. et al., *Chem. Met. Alloys*, 3(1–2), 5, 2010.

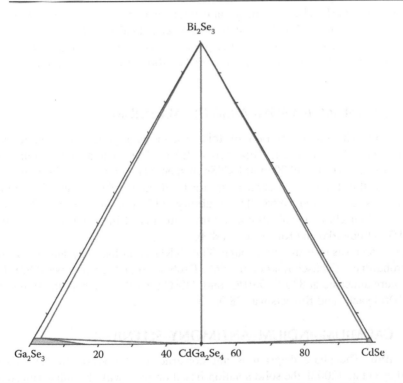

FIGURE 5.40 Isothermal section of the $CdSe-Ga_2Se_3-Bi_2Se_3$ quasiternary system at 400°C. (From Sosovska, S.M. et al., *Chem. Met. Alloys*, 3(1–2), 5, 2010.)

not exist at this temperature. This system was investigated by DTA, XRD, and metallography. The alloys were annealed at 400°C over 250 h and then were rapidly quenched in cold water.

5.49   CADMIUM–GALLIUM–CHLORINE–SELENIUM

$CdSe–Ga_2Se_3–GaCl_3$. The phase diagram is not constructed. The Cd_2GaSe_3Cl quaternary compound is formed in this system (Range and Handrick 1988). It crystallizes in the hexagonal structure with the lattice parameters $a = 418.38 \pm 0.03$ and $c = 683.64 \pm 0.09$ pm. This compound was obtained at 600°C–1100°C by the interaction of CdSe, $CdCl_2$, and Ga_2Se_3.

5.50   CADMIUM–GALLIUM–BROMINE–SELENIUM

$CdSe–Ga_2Se_3–GaBr_3$. The phase diagram is not constructed. The Cd_2GaSe_3Br quaternary compound is formed in this system (Range and Handrick 1988). It crystallizes in the hexagonal structure with the lattice parameters $a = 418.57 \pm 0.04$ and $c = 684.2 \pm 0.1$ pm. This compound was obtained at 600°C–1100°C by the interaction of CdSe, $CdBr_2$, and Ga_2Se_3.

5.51   CADMIUM–GALLIUM–IODINE–SELENIUM

$CdSe–Ga_2Se_3–GaI_3$. The phase diagram is not constructed. The Cd_2GaSe_3I quaternary compound is formed in this system (Range and Handrick 1988). It crystallizes in the hexagonal structure with the lattice parameters $a = 416.6 \pm 0.1$ and $c = 681.3 \pm 0.3$ pm. This compound was obtained at 600°C–1100°C by the interaction of CdSe, CdI_2, and Ga_2Se_3.

5.52   CADMIUM–INDIUM–ARSENIC–SELENIUM

CdSe–InAs. The phase diagram of this system belongs to the eutectic type (Figure 5.41) (Drobyazko and Kuznetsova 1983). The eutectic temperature and composition are 897°C and 30 mol.% CdSe, respectively. Liquidus and solidus temperatures within the concentration range up to 20 mol.% CdSe were determined also by Voitsehovski et al. (1968). The solubility of CdSe in InAs reaches 25 mol.% [30 mol.% (Voitsehovski et al. 1968)] and the solubility of InAs in CdSe is equal to 15 mol.% (Drobyazko and Kuznetsova 1983).

This system was investigated through DTA, XRD, metallography, and measuring of microhardness (Voitsehovski et al. 1968, Drobyazko and Kuznetsova 1983). The ingots were annealed at 830°C, 780°C, and 730°C for 120, 240, and 480 h, respectively (Drobyazko and Kuznetsova 1983).

5.53   CADMIUM–INDIUM–ANTIMONY–SELENIUM

CdSe–InSb. The phase diagram is not constructed. According to the data of Kirovskaya et al. (2002), the solid solution based on InSb with the sphalerite structure contains up to 10 mol.% CdSe.

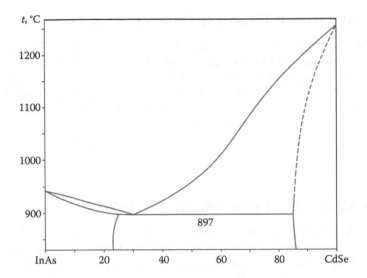

FIGURE 5.41 The CdSe–InAs phase diagram. (From Drobyazko, V.P. and Kuznetsova, S.T., *Zhurn. neorgan. khimii*, 28(11), 2929, 1983.)

5.54 CADMIUM–INDIUM–TELLURIUM–SELENIUM

$3CdSe + In_2Te_3 \Leftrightarrow 3CdTe + In_2Se_3$. The phase diagram is not constructed. There was determined the process of quaternary tetrahedral phase formations and solubility regions in this system (Figure 5.42) (Derid et al. 1964). It was shown that even small quantities of CdTe and CdSe significantly extend the homogeneity region. Thus, the addition of 1, 5, 10, and 25 mol.% (CdTe + CdSe) leads to increasing of solubility region up to 40, 50, 75, and 95 mol.%, respectively.

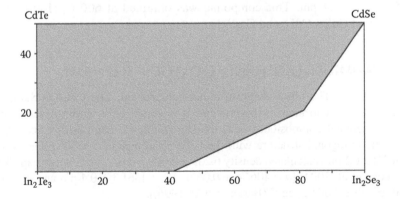

FIGURE 5.42 Formation of quaternary tetrahedral phases (shaded area) in the Cd–In–Te–Se system. (From Derid, O.P. et al., *Izv. AN SSSR. Ser. fiz.*, 28(6), 1053, 1964.)

5.55 CADMIUM–INDIUM–CHLORINE–SELENIUM

CdSe–In₂Se₃–InCl₃. The phase diagram is not constructed. The Cd_2InSe_3Cl quaternary compound is formed in this system (Range and Handrick 1988). It crystallizes in the hexagonal structure of the wurtzite type with the lattice parameters $a = 421.85 \pm 0.02$ and $c = 689.30 \pm 0.08$ pm. There was determined that this compound after annealing at 1400°C under 2.5 GPa (25 kbar) crystallizes in the defect cubic structure of NaCl type with the lattice parameter $a = 572.79 \pm 0.07$ pm. This compound was obtained at 600°C–1100°C by the interaction of CdSe, $CdCl_2$, and In_2Se_3.

5.56 CADMIUM–INDIUM–BROMINE–SELENIUM

CdSe–In₂Se₃–InBr₃. The phase diagram is not constructed. The Cd_2InSe_3Br quaternary compound is formed in this system (Range and Handrick 1988). It crystallizes in the hexagonal structure of the wurtzite type with the lattice parameters $a = 420.89 \pm 0.02$ and $c = 688.19 \pm 0.09$ pm. There was determined that this compound after annealing at 1400°C under 2.5 GPa (25 kbar) crystallizes in the defect cubic structure of NaCl type with the lattice parameter $a = 573.44 \pm 0.07$ pm. This compound was obtained at 600°C–1100°C by the interaction of CdSe, $CdBr_2$, and In_2Se_3.

5.57 CADMIUM–INDIUM–IODINE–SELENIUM

CdSe–In₂Se₃–InI₃. The phase diagram is not constructed. The Cd_2InSe_3I quaternary compound is formed in this system (Range and Handrick 1988). It crystallizes in the hexagonal structure of the wurtzite type with the lattice parameters $a = 421.05 \pm 0.03$ and $c = 688.83 \pm 0.06$ pm. There was determined that this compound after annealing at 1400°C under 2.5 GPa (25 kbar) crystallizes in the defect cubic structure of NaCl type with the lattice parameter $a = 575.65 \pm 0.04$ pm. This compound was obtained at 600°C–1100°C by the interaction of CdSe, CdI_2, and Ga_2Se_3.

5.58 CADMIUM–LANTHANUM–OXYGEN–SELENIUM

CdSe–La₂O₂Se. The phase diagram is not constructed. The $CdLa_2O_2Se_2$ quaternary compound is formed in this system. It is stable only within a very narrow range of chemical compositions near the stoichiometric composition and crystallizes in the tetragonal structure with the lattice parameters $a = 406.60 \pm 0.06$ and $c = 1863.4 \pm 0.1$ pm, calculated density 6.239 ± 0.009 g·cm⁻³, and energy gap ≈3.3 eV (Hiramatsu et al. 2004) [$a = 406.76 \pm 0.06$ and $c = 1861.3 \pm 0.4$ pm and calculated density 6.256 ± 0.003 g·cm⁻³ (Baranov et al. 1996)].

To obtain $CdLa_2O_2Se_2$, a mixture of CdSe, La_2O_3, and La_2Se_3 was sealed in an evacuated SiO_2 glass ampoule and heated to 950°C for 2 h (Hiramatsu et al. 2004). The ingots were annealed at 730°C–830°C for 250 h (Baranov et al. 1996).

5.59 CADMIUM–CERIUM–CHROMIUM–SELENIUM

A series of compounds with the general formula $Cd_xCe_yCr_2Se_4$ are obtained in this system. They crystallize in the cubic structure with the lattice parameter $a = 1075.42 \pm 0.01$ pm at $x = 0.96$ and $y = 0.03$ and $a = 1074.73 \pm 0.01$ pm at $x = 0.84$ and $y = 0.13$ (Rduch et al. 2012). Single crystals of $Cd_xCe_yCr_2Se_4$ were grown using the chemical vapor transport method with anhydrous $CrCl_3$ as the transport agent and with CdSe, Ce, and Se. Single-crystalline samples were not received with $y > 0.13$.

5.60 CADMIUM–GERMANIUM–ARSENIC–SELENIUM

$CdSe–GeSe_2–As_2Se_3$. The phase diagram is not constructed. A large glass-forming region exists in this system (Figure 5.43) (Zhao et al. 2005, 2008). The addition of Cd increases the crystallization trend of the glasses. Besides, the Cd content accommodated by the system to form glasses is found to be lower in the As_2Se_3-rich region than in the $GeSe_2$-rich region. With the introduction of CdSe density, microhardness and glass transition temperature (T_g) of glasses increase, whereas the thermal stability decreases. The T_g for the obtained glasses are relatively high, varying from 240°C to 340°C (Zhao et al. 2008). It is found that T_g increases with the content of Ge but decreases as As increases.

This system was investigated through differential scanning calorimetry (DSC), XRD, and measuring of density and microhardness and using IR transmission spectra (Zhao et al. 2005, 2008). The glasses were annealed at 300°C for 3 h.

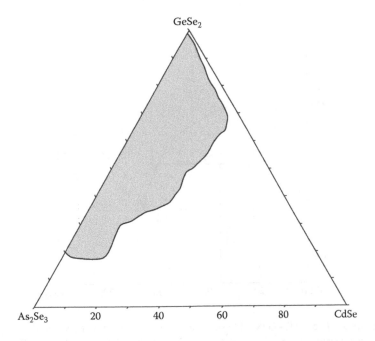

FIGURE 5.43 The glass-formation region in the $CdSe–GeSe_2–A_2Se_3$ quasiternary system. (From Zhao, D. et al., *J. Am. Ceram. Soc.*, 88(11), 3143, 2005.)

5.61 CADMIUM-GERMANIUM-TELLURIUM-SELENIUM

CdTe-GeSe$_2$. The phase diagram is a eutectic type (Figure 5.44) (Asadov et al. 2012). The eutectic composition and temperature are 66 mol.% GeSe$_2$ and 527°C, respectively. The Cd$_2$GeSe$_2$Te$_2$ quaternary compound, which melts incongruently at 647°C and crystallizes in the hexagonal structure with the lattice parameters $a = 569$ and $c = 1132$ pm, is formed in this system (Asadov 2006, Asadov et al. 2012). The solubility of CdTe in GeSe$_2$ and GeSe$_2$ in CdTe at 647°C is equal to 16 and 22 mol.%, respectively. The homogeneous glasses containing up to 5 mol.% CdTe have been found in this system (Vassilev et al. 2002).

This system was investigated through DTA, metallography, XRD, and measuring of microhardness and emf of concentrated chains (Asadov 2006, Asadov et al. 2012). Glass-forming region in the CdTe-GeSe$_2$ system has been determined by visual observation, XRD, and electron probe microanalysis (EPMA) (Vassilev et al. 2002).

5.62 CADMIUM-TIN-ARSENIC-SELENIUM

CdSe-CdSnAs$_2$. The phase diagram is not constructed. The solubility of CdSnAs$_2$ in 2CdSe reaches 6 mol.% (Dovletmuradov et al. 1971, 1972). This system was investigated through DTA, metallography, XRD, and measuring of microhardness.

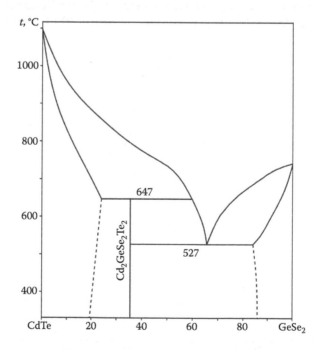

FIGURE 5.44 The CdTe-GeSe$_2$ phase diagram. (From Asadov, M.M. et al., *Konf. stran SNG po rostu kristallov. Tez. dokl. RK SNG-2012*, Kharkov, Ukraine, 2012, p. 99.)

5.63 CADMIUM–TIN–TELLURIUM–SELENIUM

CdSe–SnTe. This section is nonquasibinary section of the Cd–Sn–Te–Se quaternary system (Figure 5.45) (Dubrovin et al. 1986). The solubility of CdSe in SnTe at 780°C is equal to 14 mol.% and decreases to 3 mol.% at 400°C. The solubility of SnTe in CdSe is not higher than 1 mol.%. Solid solutions based on CdTe with the sphalerite structure (γ) and based on SnSe with the orthorhombic structure (δ) are formed at the interaction of CdSe and SnTe. Thermal effects at 540°C correspond to the polymorphous transformation of the δ-solid solutions. This system was investigated through DTA, metallography, and XRD. The ingots were annealed at 400°C, 700°C, and 750°C for 250, 120, and 100 h, respectively.

CdSe + SnTe ⇔ CdTe + SnSe. The isothermal sections of this ternary mutual system at 730°C and 530°C (Figure 5.46) have been constructed using EPMA and XRD (Leute and Menge 1992). The tie line fields for the structural miscibility gaps and the three-phase triangles were calculated from data, which mainly have been determined on the quasibinary edge systems. At an intermediate temperature between 530°C and 730°C, a region must exist, where four solid phases with different structures coexist in thermodynamic equilibrium. The ingots were equilibrated for 30 days at 730°C and for 70 days at 530°C.

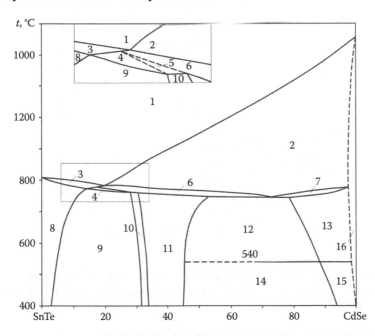

FIGURE 5.45 Phase relations in the CdSe–SnTe system: 1, L; 2, L + (CdSe); 3, L + (SnTe); 4, L + γ; 5, L + (CdSe) + γ; 6, L + (CdSe) + (SnTe); 7, L + (CdSe) + δ; 8, (SnTe); 9, (SnTe) + γ; 10, (CdSe) + (SnTe) + γ; 11, (CdSe) + (SnTe); 12, (CdSe) + (SnTe) + δ; 13, (CdSe) + δ; 14, (CdSe) + (SnTe) + δ′; 15, (CdSe) + δ′; and 16, (CdSe). (From Dubrovin, I.V. et al., *Izv. AN SSSR. Neorgan. materialy*, 22(4), 590, 1986.)

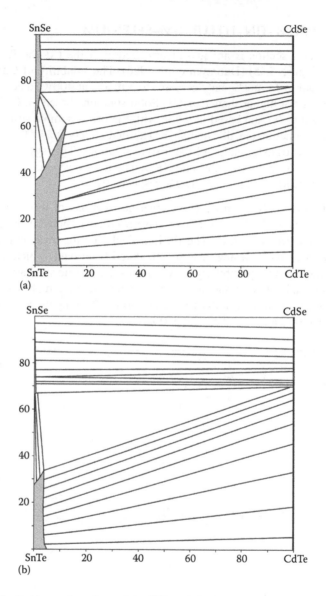

FIGURE 5.46 Isothermal sections of the CdSe + SnTe ⇔ CdTe + SnSe ternary mutual system at (a) 730°C and (b) 530°C. (From Leute, V., and Menge, D., *Z. Phys. Chem. (Munchen)*, 176(1), 65, 1992.)

5.64 CADMIUM–LEAD–TELLURIUM–SELENIUM

CdSe–PbTe. This section is nonquasibinary section of the Cd–Pb–Te–Se quaternary system (Figure 5.47) (Rayevski et al. 1983a,b). Three phases are in equilibrium in the solid state: α (PbSe$_x$Te$_{1-x}$ solid solutions with the cubic structure of NaCl type), β, and γ (CdSe$_x$Te$_{1-x}$ solid solutions with the sphalerite and wurtzite structure, respectively). The solubility of CdSe in PbTe at 870°C is equal to 16 mol.% [30 mol.% at

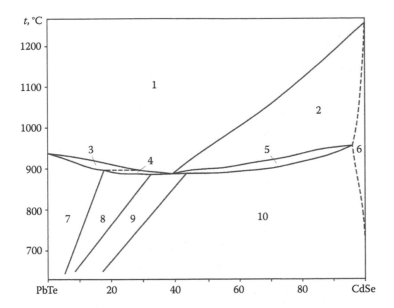

FIGURE 5.47 Phase relations in the CdSe–PbTe system: 1, L; 2, L + γ; 3, L + α; 4, L + α + β; 5, L + α + γ; 6, γ; 7, α; 8, α + β; 9, α + β + γ; and 10, α + γ. (From Rayevski, S.D. et al., *Izv. AN SSSR. Neorgan. materialy*, 19(6), 889, 1983a.)

600°C (Nikolič 1966)], and the solubility of PbTe in CdSe is not higher than 2 mol.%. Lattice parameters of solid solutions based on PbTe change linearly with composition (Rayevski et al. 1983a,b).

This system was investigated through DTA, metallography, XRD, and measuring of microhardness. The ingots were annealed at 650°C and 830°C for 340 and 120 h, respectively (Rayevski et al. 1983a,b).

CdTe–PbSe. This section is nonquasibinary section of the Cd–Pb–Te–Se quaternary system (Figure 5.48) (Rayevski et al. 1982). The $CdSe_xTe_{1-x}$ solid solutions undergo polymorphous transformation within the interval of 60–70 mol.% CdTe; therefore, three phases [solid solutions base on PbSe (α) and $CdSe_xTe_{1-x}$ solid solutions with the sphalerite and wurtzite structures (β and γ)] are in equilibrium in the narrow region of compositions (not higher than 3 mol.%). Three-phase equilibrium takes place at 885°C and 38 mol.% CdTe. The maximum solubility of CdTe in PbSe is equal to 27.5 mol.% at 880°C, and the solubility of PbSe in CdTe at 850°C is not higher than 2.5 mol.%. Lattice parameters of solid solutions based on PbSe change linearly with composition.

This system was investigated through DTA, metallography, XRD, and measuring of microhardness. The ingots were annealed at 590°C, 810°C, and 830°C (Rayevski et al. 1982).

CdSe + PbTc ⇔ CdTe + PbSe. The fields of primary crystallization of $CdSe_xTe_{1-x}$ and $PbSe_xTe_{1-x}$ solid solutions exist on the liquidus surface of this system (Figure 5.49) (Tomashik 1981). This system was investigated through DTA, metallography, and using mathematical planning of experiment.

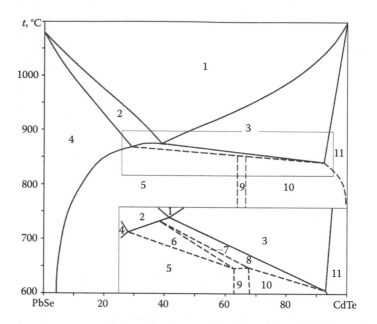

FIGURE 5.48 Phase relations in the CdTe–PbSe system (Rayevski et al. 1982): 1, L; 2, L + α; 3, L + γ; 4, α; 5, α + β; 6, L + α + β; 7, L + α + β + γ; 8, L + α + γ; 9, α + β + γ; 10, α + γ; and 11, γ. (From Rayevski, S.D. et al., *Izv. AN SSSR. Neorgan. materialy*, 18(8), 1267, 1982.)

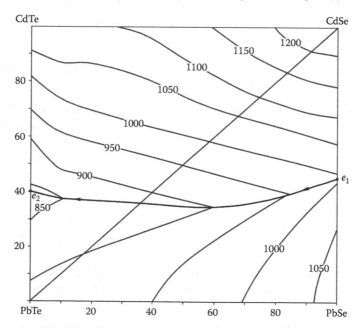

FIGURE 5.49 Liquidus surface of the CdSe + PbTe ⇔ CdTe + PbSe ternary mutual system. (From Tomashik, Z.F., *Izv. AN SSSR. Neorgan. materialy*, 17(9), 1575, 1981.)

5.65 CADMIUM–LEAD–IODINE–SELENIUM

CdSe–PbI$_2$. The phase diagram is a eutectic type (Figure 5.50) (Odin 2001). The eutectic composition and temperature are 19 ± 1 mol.% CdSe and 372°C, respectively. The mutual solubility of CdSe and PbI$_2$ is negligible: the solubility of PbI$_2$ in CdSe is equal to 0.15 mol.%. The crystallization of the melts from the PbI$_2$ side leads to the formation of two polytypic forms of this compound: 6R-PbI$_2$ and 2H-PbI$_2$ (Odin 2001, Odin and Chukichev 2001).

This system was investigated through DTA, XRD, and metallography, and the ingots were annealed for 1000 h (Odin 2001).

CdSe + PbI$_2$ ⇔ CdI$_2$ + PbSe. The fields of primary crystallization of solid solutions based on CdSe and PbSe and Cd$_x$Pb$_{1-x}$I$_2$ solid solutions exist on the liquidus surface of this ternary mutual system (Figure 5.51) (Odin 2001). The ternary eutectic E at 361°C is formed by the solid solutions based on CdSe and PbSe and Cd$_x$Pb$_{1-x}$I$_2$ solid solutions. A minimum m exists on the e_4e_2 line of the secondary crystallization. The isothermal section of this ternary mutual system at 350°C is shown in Figure 5.52 (Odin 2001).

Metastable phase can form at the crystallization of the melts in the CdSe–PbSe–PbI$_2$ system in the region enriched in PbI$_2$ (Odin 2001, Odin et al. 2005). This phase exists at room temperature for some months, but heating leads to its decomposition with formation of the phases according to the equilibrium phase diagram.

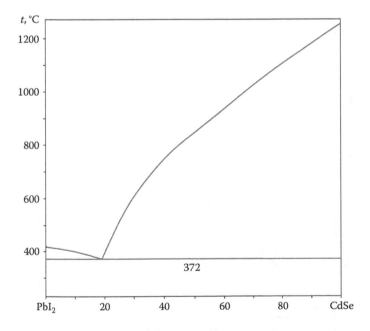

FIGURE 5.50 The CdSe–PbI$_2$ phase diagram. (From Odin, I.N., *Zhurn. neorgan. khimii*, 46(10), 1733, 2001.)

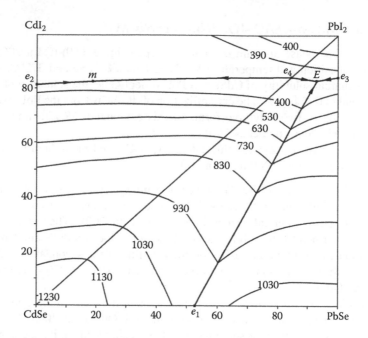

FIGURE 5.51 Liquidus surface of the $CdSe + PbI_2 \Leftrightarrow CdI_2 + PbSe$ ternary mutual system. (From Odin, I.N., *Zhurn. neorgan. khimii*, 46(10), 1733, 2001.)

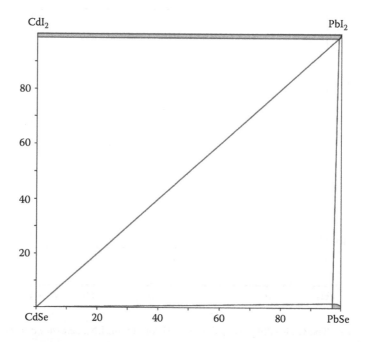

FIGURE 5.52 Isothermal section of the $CdSe + PbI_2 \Leftrightarrow CdI_2 + PbSe$ ternary mutual system at 350°C. (From Odin, I.N., *Zhurn. neorgan. khimii*, 46(10), 1733, 2001.)

5.66   CADMIUM–ARSENIC–TELLURIUM–SELENIUM

CdTe–As$_2$Se$_3$. The phase diagram is shown in Figure 5.53 (Aliev et al. 1988). The eutectic composition and temperature are 20 mol.% CdTe and 325°C, respectively. The immiscibility region exists in this system within the interval of 28–60 mol.% CdTe with monotectic temperature of 400°C. At the slow cooling, the glass-forming region based on As$_2$Se$_3$ contains up to 7 mol.% CdTe and up to 10 mol.% CdTe at the quenching (Aliev et al. 1988, Vassilev et al. 2002). The solubility of CdTe in As$_2$Se$_3$ is not higher than 1.5 mol.% and the solubility of As$_2$Se$_3$ in CdTe is equal to 2 mol.%.

This system was investigated through DTA, metallography, XRD, and measuring of microhardness and density (Aliev et al. 1988). Glass-forming region in the CdTe–GeSe$_2$ system has been determined by visual observation, XRD, and EPMA (Vassilev et al. 2002).

5.67   CADMIUM–ANTIMONY–TELLURIUM–SELENIUM

CdTe–Sb$_2$Se$_3$. The phase diagram is shown in Figure 5.54 (Safarov et al. 1989). The eutectic composition and temperature are 47 mol.% CdTe and 550°C, respectively. The Cd$_2$Sb$_2$Se$_3$Te$_2$ quaternary compound is formed in this system, which melts incongruently at 635°C and crystallizes in the cubic structure with the lattice parameter $a - 662$ pm. The solubility of CdTe in Sb$_2$Se$_3$ and Sb$_2$Se$_3$ in CdTe at room temperature reaches 8.0 and 7.5 mol.%, respectively. It is necessary to note that CdTe has no phase transformations in the investigated temperature region and the melting temperature of Sb$_2$Se$_3$ determined by the authors is too high. Therefore, this phase diagram raises some doubts.

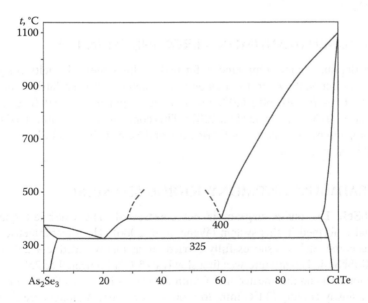

FIGURE 5.53 The CdTe–As$_2$Se$_3$ phase diagram. (From Aliev, I.I. et al., *Zhurn. neorgan. khimii*, 33(6), 1618, 1988.)

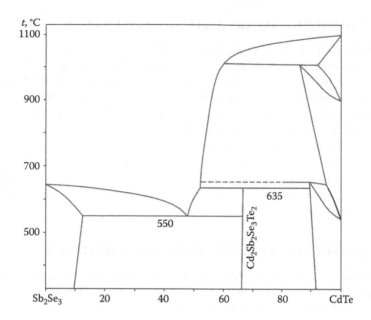

FIGURE 5.54   The CdTe–Sb$_2$Se$_3$ phase diagram. (From Safarov, M.G. et al., *Izv. AN SSSR. Neorgan. materialy*, 34(7), 1831, 1989.)

This system was investigated through DTA, metallography, XRD, and measuring of microhardness and density. The ingots containing up to 70 mol.% CdTe were annealed at 500°C for 500 h and containing more than 70 mol.% CdTe at 300°C for 400 h (Safarov et al. 1989).

5.68   CADMIUM–ANTIMONY–BROMINE–SELENIUM

CdSb$_2$Se$_3$Br$_2$ quaternary compound is formed in this system. It melts congruently at 509°C and crystallizes in the monoclinic structure with the lattice parameters $a = 2099.8 \pm 0.4$, $b = 402.60 \pm 0.08$, $c = 1124.9 \pm 0.2$ pm, and $\beta = 119.0° \pm 0.03°$ and energy gap 1.40 eV (Wang and Hwu 2007). This compound was synthesized by heating the stoichiometric mixture of CdBr$_2$, Sb, and Se or CdBr$_2$ and Sb$_2$Se$_3$ at 450°C followed by slow cooling.

5.69   CADMIUM–ANTIMONY–IODINE–SELENIUM

CdSe-SbSeI. The phase diagram is not constructed. The CdSbSe$_2$I quaternary compound is formed in this system (Wang et al. 2006). The stoichiometric synthesis of this compound was successfully carried out as follows: 3CdSe + 2Sb + SbI$_3$ + 3Se = 3CdSbSe$_2$I. The mixture was first slowly (0.5°C/min) heated to 250°C and isothermed for 10 h and then heated at 1°C/min to 400°C–430°C, isothermed for 4 days with next slowly cooling (0.1°C/min) to room temperature. A stoichiometric yield of polycrystalline product was acquired by employing different starting materials according to equation CdSe + 2Sb + CdI$_2$ + 3Se = 2CdSbSe$_2$I at temperature as low as 450°C.

5.70 CADMIUM–BISMUTH–TELLURIUM–SELENIUM

CdSe–Bi$_2$Te$_3$. The phase diagram is a eutectic type (Figure 5.55) (Datsenko et al. 1981). The eutectic is degenerated from the Bi$_2$Te$_3$-rich side and crystallizes at 583°C ± 3°C. The mutual solubility of CdSe and Bi$_2$Te$_3$ is insignificant. This system was investigated through DTA and XRD.

3CdSe + Bi$_2$Te$_3$ ⇔ 3CdTe + Bi$_2$Se$_3$. The phase diagram is not constructed. At 490°C, the Bi$_2$Se$_3$Te$_{3-3x}$ solid solutions with the tetradymite structure and CdSe$_x$Te$_{1-x}$ solid solutions with the sphalerite and wurtzite structure are in equilibrium (Figure 5.56) (Odin and Marugin 1991). The CdBi$_2$Se$_4$ is not stable at this temperature because it exists within the interval of 604°C–736°C. This system was investigated through XRD and measuring of electrophysical properties. The ingots were annealed at 490°C for 900 h.

5.71 CADMIUM–BISMUTH–BROMINE–SELENIUM

CdSe-BiSeBr. The phase diagram is not constructed. The CdBiSe$_2$Br quaternary compound is formed in this system. It crystallizes in the orthorhombic structure with the lattice parameters $a = 1002.5 ± 0.2$, $b = 411.90 ± 0.08$, and $c = 1314.3 ± 0.3$ pm (Wang et al. 2006). The stoichiometric synthesis of this compound was successfully carried out as follows: 3CdSe + 2Bi + BiBr$_3$ + 3Se = 3CdBiSe$_2$Br. The mixture was first slowly (0.5°C/min) heated to 250°C and isothermed for 10 h and then heated at 1°C/min to 400°C–430°C, isothermed for 4 days with next slowly cooling (0.1°C/min) to room temperature. A stoichiometric yield of polycrystalline product was acquired by employing different starting materials according to equation CdSe + 2Bi + CdBr$_2$ + 3Se = 2CdBiSe$_2$Br at temperature as low as 450°C. When the same reaction mixture

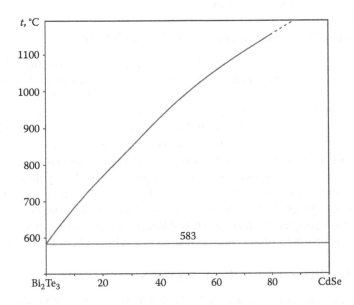

FIGURE 5.55 The CdSe–Bi$_2$Te$_3$ phase diagram. (From Datsenko, A.M. et al., *Tr. Mosk. khim. tehnol. in-t*, (120), 32, 1981.)

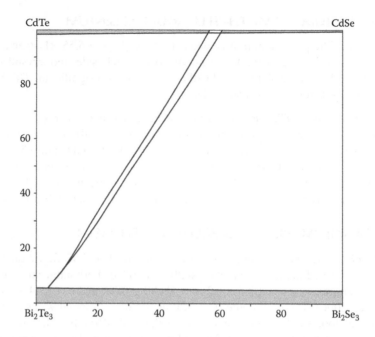

FIGURE 5.56 Isothermal section of the $3CdSe + Bi_2Te_3 \Leftrightarrow 3CdTe + Bi_2Se_3$ ternary mutual system at 490°C. (From Odin, I.N. and Marugin, V.V., *Zhurn. neorgan. khimii*, 36(7), 1865, 1991.)

was used, larger single crystals were grown by first heating the reactants to 900°C followed by slow cooling to room temperature. Single crystals of this compound can also be grown by a solid-state reaction.

5.72 CADMIUM–BISMUTH–IODINE–SELENIUM

CdSe-BiSeI. The phase diagram is not constructed. The $CdBiSe_2I$ quaternary compound is formed in this system. It crystallizes in the monoclinic structure with the lattice parameters $a = 1365.9 \pm 0.3$, $b = 419.20 \pm 0.08$, $c = 1019.3 \pm 0.2$ pm, and $\beta = 90.88° \pm 0.03°$ (Wang et al. 2006). The stoichiometric synthesis of this compound was successfully carried out as follows: $3CdSe + 2Bi + BiI_3 + 3Se = 3CdBiSe_2I$. The mixture was first slowly (0.5°C/min) heated to 250°C and isothermed for 10 h and then heated at 1°C/min to 400°C–430°C, isothermed for 4 days with next slowly cooling (0.1°C/min) to room temperature. A stoichiometric yield of polycrystalline product was acquired by employing different starting materials according to equation $CdSe + 2Bi + CdI_2 + 3Se = 2CdBiSe_2I$ at temperature as low as 450°C. When the same reaction mixture was used, larger single crystals were grown by first heating the reactants to 900°C followed by slow cooling to room temperature. Single crystals of this compound can also be grown by a solid-state reaction.

5.73 CADMIUM–VANADIUM–OXYGEN–SELENIUM

The phase diagram is not constructed. The $Cd_6V_2Se_5O_{21}$ [$Cd_6(V_2O_6)(SeO_3)_5$] quaternary compound is formed in this system. It is stable under an air atmosphere up to 350°C and crystallizes in the monoclinic structure with the lattice parameters

$a = 1530.1 \pm 0.6$, $b = 543.7 \pm 0.2$, $c = 2381.6 \pm 1.0$ pm, and $\beta = 93.950° \pm 0.004°$, calculated density 5.064 g·cm^{-3}, and energy gap 2.14 eV (Jiang et al. 2008). Red needle-shaped single crystals could be obtained by the solid-state reaction of a mixture composed of CdO, V_2O_5, and SeO_2 in a molar ratio of 2:1:1 or 1:1:1. The reaction mixture was thoroughly ground, pressed into a pellet, and sealed in an evacuated quartz tube. This tube was heated at 300°C for 1 day and at 680°C for 6 days, then slowly cooled to 280°C at 4°C/h, and finally cooled to room temperature for 10 h. Single-phase product was obtained quantitatively by reacting a mixture of CdO, V_2O_5, and SeO_2 in a molar ratio of 6:1:5 at 650°C for 6 days.

5.74 CADMIUM–OXYGEN–MANGANESE–SELENIUM

CdSe–MnSe–O. The phase diagram is not constructed. According to the data of thermodynamic calculations, an oxidation of the $Cd_xMn_{1-x}Se$ solid solutions begins from the oxidation of MnSe with formation of MnO and Se, and then CdSe is oxidized to form $CdSeO_3$ (Medvedev and Berchenko 1994).

5.75 CADMIUM–TELLURIUM–CHLORINE–SELENIUM

CdSe–CdTe–CdCl₂. Two fields of primary crystallizations of $CdCl_2$ and $CdSe_xTe_{1-x}$ solid solutions include the liquidus surface of this quasiternary system (Figure 5.57) (Luzhnaya et al. 1980). The lowest temperature from the CdSe–CdTe quasibinary

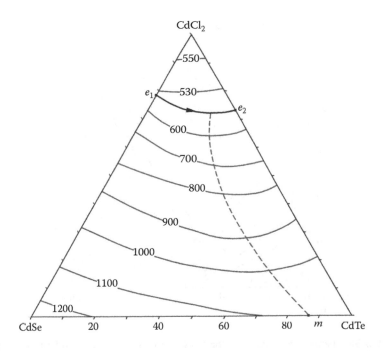

FIGURE 5.57 Liquidus surface of the CdSe–CdTe–CdCl₂ quasiternary system. (From Luzhnaya, N.P. et al., *Izv. AN SSSR. Neorgan. materialy*, 16(3), 541, 1980.)

system, becomes apparent on the liquidus surface also (its position is pointed by the dashed line in the field of $CdSe_xTe_{1-x}$ solid solutions' primary crystallization). This system was investigated through DTA.

5.76   CADMIUM–TELLURIUM–MANGANESE–SELENIUM

CdSe + MnTe ⇔ CdTe + MnSe. Isothermal sections of this ternary mutual system at 600°C and 960°C are shown in Figure 5.58 (Chebab Samir and Woolley 1985). This system was investigated through XRD.

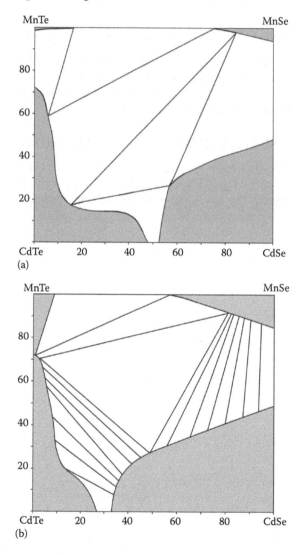

FIGURE 5.58   Isothermal sections of the CdSe + MnTe ⇔ CdTe + MnSe ternary mutual system at (a) 600°C and (b) 960°C. (From Chebab Samir, F. and Woolley, J.C., *J. Less Common Met.*, 106(1), 13, 1985.)

5.77 CADMIUM–CHROMIUM–TITANIUM–SELENIUM

A series of compounds with the general formula $Cd[Cr_xTi_y]Se_4$ are obtained in this system. They crystallize in the cubic structure with the lattice parameter $a = 1073.725 \pm 0.016$ pm for $Cd[Cr_{1.95}Ti_{0.07}]Se_4$, $a = 1074.468 \pm 0.017$ pm for $Cd[Cr_{1.62}Ti_{0.42}]Se_4$, and $a = 1075.00 \pm 0.03$ pm for $Cd[Cr_{1.49}Ti_{0.59}]Se_4$ (Rduch et al. 2013). Single crystals of $Cd[Cr_xTi_y]Se_4$ were grown using chemical vapor transport method with anhydrous $CrCl_3$ as a transporting agent and with CdSe and high-purity Ti and Se.

5.78 CADMIUM–CHROMIUM–CHLORINE–SELENIUM

3CdSe–2CrCl₃. This section is nonquasibinary section of the Cd–Cr–Cl–Se quaternary system (Luzhnaya et al. 1981). CdSe, $CrCl_3$, and Cr_2Se_3 primarily crystallize in this system. Simultaneous crystallization of $CrCl_3$ and Cr_2Se_3 takes place at 26 mol.% 3CdSe and 790°C, and simultaneous crystallization of CdSe and Cr_2Se_3 begins at 12 mol.% 3CdSe and 1140°C. Liquidus temperatures of this section are given in Table 5.10. This system was investigated through DTA. The ingots were annealed at 850°C–950°C for 5–6 days.

3CdSe + 2CrCl₃ ⇔ Cr₂Se₃ + 3CdCl₂. This section is not stable section of the Cd–Cr–Cl–Se quaternary system since the field of $CrCl_2$ primary crystallization exists on the liquidus surface (Figure 5.59) (Luzhnaya and Shabunina 1976, Luzhnaya et al. 1981). The main fields of the primary crystallization are the fields of CdSe, $CrCl_3$, and Cr_2Se_3. The field of $CdCr_2Se_4$ primary crystallization is situated as a narrow band near the $CdCl_2$-rich side. The E_1 and E_2 ternary eutectics crystallize at 510°C and 520°C, respectively. Three transition points exist in this system: U_1 where CdSe, $CdCr_2Se_4$, and Cr_2Se_3 simultaneously crystallize; U_2 where $CdCl_2$, Cr_2Se_3, and $CdCr_2Se_4$ simultaneously crystallize; and U_3 at 792°C where $CrCl_2$, $CrCl_3$, and Cr_2Se_3 simultaneously crystallize. This system was investigated through DTA and XRD.

TABLE 5.10

Liquidus Temperatures of the 3CdSe–CrCl₃ Section

3CdSe, mol.%	0	10	15	20	25	30	40	50
2CrCl₃, mol.%	100	90	85	80	75	70	60	50
t, °C		1150	1070	990	880	820	953	1045	1135
3CdSe, mol.%	60	65	75	80	85	90	100	
2CrCl₃, mol.%	40	35	25	20	15	10	0	
t, °C	1195	1186	1183	1178	1155	1158	1239	

Source: Luzhnaya, N.P et al., *Zhurn. neorgan. khimii*, 26(4), 1075, 1981.

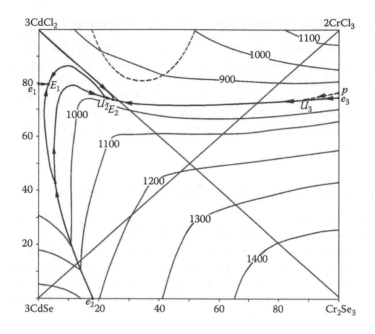

FIGURE 5.59　Liquidus surface of the $3CdSe + 2CrCl_3 \Leftrightarrow Cr_2Se_3 + 3CdCl_2$ ternary mutual system. (From Luzhnaya, N.P. et al., *Zhurn. neorgan. khimii*, 26(4), 1075, 1981.)

REFERENCES

Agaev A.B., Aliev V.O., Aliev O.M. Synthesis, x-ray investigations and physical properties of the compounds derived from the Y_5S_7 structure [in Russian], *Zhurn. neorgan. khimii*, **41**(2), 319–325 (1996).

Aliev I.I., Kuliev B.B., Gurshumova A.P., Nadzhafova Z.Z. Nature of chemical interaction in the As_2Se_3-CdTe system [in Russian], *Zhurn. neorgan. khimii*, **33**(6), 1618–1621 (1988).

Asadov M.M. Dependences of physico-chemical properties of solid solutions in the systems $GeSe_2–A^2B^6$ (A^2 = Hg, Cd); B^6 = S, Te) versus composition [in Russian], *Azerb. khim. zhurn.*, (2), 77–81 (2006).

Asadov M.M., Mamedov F.M., Mirzoev A.Ch., Aliev O.M. T-x phase diagrams of the FeSe–GeTe, $GeSe_2$–CdTe systems [in Russian], *Konf. stran SNG po rostu kristallov. Tez. dokl. RK SNG-2012*, Kharkov, Ukraine, p. 99 (2012).

Baranov I.Yu., Dolgih V.A., Popovkin B.A. Investigation of phase correlations in the La_2O_2X–MX (M = Zn, Cd, Hg; X = S, Se) systems [in Russian], *Zhurn. neorgan. khimii*, **41**(11), 1916–1919 (1996).

Bochmann M., Bwembya G., Webb K.J., Malik M.A., Walsh J.R., O'Brien P. Arene chalcogenolato complexes of zinc and cadmium, *Inorg. Synthes.*, **31**, 19–24 (1997).

Bochmann M, Coleman A.P., Powell A.K. Synthesis of some alkyl metal selenolato complexes of zinc, cadmium and mercury. X-ray crystal structure of Me, Hg, $Se(2,4,6-Pr^i_3C_6H_2)$, *Polyhedron*, **11**(5), 507–512 (1992).

Bochmann M., Webb K. Novel precursors for the deposition of II-VI semiconductor films, *Mater. Res. Symp. Proc.*, **204**, 149–154 (1991).

Bochmann M., Webb K., Harman M., Hursthouse M.B. Synthesis, structure, and gas-phase decomposition of $[Cd(EC_6H_2tBu_3)_2]_2$ (E = S, Se): First examples of low-coordinate volatile cadmium chalcogenolato complexes, *Angew. Chem. Int. Ed. Engl.*, **29**(6), 638–639 (1990a).

Bochmann M., Webb K., Harman M., Hursthouse M.B. Synthese, Struktur und Gasphasen-thermolyse von [Cd(EC$_6$H$_2t$Bu$_3$)$_2$]$_2$ (E = S, Se); erste Beispiele für niedrigkoordi-nierte, flüchtige Chalcogenolato cadmium komplexe, *Angew. Chem.*, **102**(6), 703–704 (1990b).

Bochmann M., Webb K., Hursthouse M.B., Mazid M. Sterically hindered chalcogenolato complexes. Mono- and di-meric thiolates and selenolates of zinc and cadmium; structure of [{Cd(SeC$_6$H$_2$Bu$_3^t$-2,4,6)$_2$}$_2$], the first three-co-ordinate cadmium–selenium complex, *J. Chem. Soc., Dalton Trans.*, (9), 2317–2323 (1991).

Bozhko V.V., Bulatetska L.V., Davydyuk G.Ye., Parasyuk O.V., Tretyak A.P., Vainorius N., Kažukauskas V. Growth and properties of the single AgCd$_2$GaSe$_4$ crystals, *J. Cryst. Growth*, **330**(1), 5–8 (2011).

Brennan J.G., Siegrist T., Carroll P.J., Stuczynski S.M., Brus L.E., Steigerwald M.L. The preparation of large semiconductor clusters via the pyrolysis of a molecular precursor, *J. Am. Chem. Soc.*, **111**(11), 4141–4143 (1989).

Chebab Samir F., Woolley J.C. Solid solution and lattice parameter values in the (Cd$_{1-x}$Mn$_x$)·(Te$_{1-y}$Se$_y$), *J. Less Common Met.*, **106**(1), 13–17 (1985).

Chondroudis K., Kanatzidis M.G. [M$_4$(Se$_2$)$_2$(PSe$_4$)$_4$]$^{8-}$: A novel, tetranuclear, cluster anion with a stellane-like core, *J. Chem. Soc. Chem. Commun.*, (4), 401–402 (1997).

Chondroudis K., Kanatzidis M.G. Group-10 and 12 one-dimensional selenodiphosphates: A$_2$MP$_2$Se$_6$ (A = K, Rb, Cs; M = Pd, Zn, Cd, Hg), *J. Solid State Chem.*, **138**(2), 321–328 (1998).

Datsenko A.M., Razvazhnoy E.M., Chashchin V.A. Investigation of interaction in the Bi$_2$Te$_3$–(Cd,Pb)(S,Se,Te) systems [in Russian], *Tr. Mosk. khim. tehnol. in-t*, (120), 32–34 (1981).

Derid O.P., Radautsan S.I., Mirgorodski V.M., Markus M.M. Physico-chemical properties of some alloys in the indium–selenium–tellurium–cadmium system [in Russian], *Izv. AN SSSR. Ser. fiz.*, **28**(6), 1053–1056 (1964).

Ding N., Chung D.-Y., Kanatzidis M.G. K$_6$Cd$_4$Sn$_3$S$_{13}$: A polar open-framework compound based on the partially destroyed super-tetrahedral [Cd$_4$Sn$_4$S$_{17}$]$^{10-}$ cluster, *Chem. Commun.*, (10), 1170–1171 (2004).

Ding N., Kanatzidis M.G. Acid-induced conversions in open-framework semiconductors: From [Cd$_4$Sn$_3$Se$_{13}$]$^{6-}$ to [Cd$_{15}$Sn$_{12}$Se$_{46}$]$^{14-}$, a remarkable disassembly/reassembly process, *Angew. Chem. Int. Ed.*, **45**(9), 1397–1401 (2006).

Dovletmuradov Ch., Dovletov K., Krzhivitskaya S.N., Mamaev S., Allanazarov A., Ashirov A. Obtaining and investigation of physico-chemical and electrical properties of CdSnAs$_2$–A^{II}B^{VI} (2CdSe, 2CdTe) solid solutions [in Russian]. In: *Troin. poluprovodniki A^{II}B^{IV}C$_2^V$ i A^{II}B$_2^{III}$C$_4^{VI}$*. Kishinev, Republic of Moldova: Shtiintsa Publish., pp. 96–98 (1972).

Dovletmuradov Ch., Krzhivitskaya S.N., Dovletov K., Mamaev S., Allanazarov A., Ashirov A. Some properties of CdSnAs$_2$–A^{II}B^{VI} (CdSe, CdTe) solid solutions [in Russian], *Izv. AN TurkmSSR. Ser. fiz.-tehn., khim. i geol. nauk*, (5), 111–114 (1971).

Dovletov K., Mkrtchian S.A., Zhukov E.G., Melikdzhanian A.G. Electrophysical properties of (CdSe)$_x$[Cu$_2$Ge(Sn)Se$_3$]$_{1-x}$ solid solutions [in Russian], *Izv. AN SSSR. Neorgan. materialy*, **23**(5), 857–860 (1987).

Dovletov K.O., Hanberdiev Ya.A., Nuryev C., Aleksanian S.N. Electrophysical properties of the (CdSe)$_{1-x}$(CuGaSe$_2$)$_x$ solid solutions [in Russian], *Izv. AN SSSR. Neorgan. materialy*, **26**(5), 939–941 (1990).

Dovletov K.O., Nuryev S., Garagulov O.Ch., Krzhivitskaya S.N., Dovletov S.K. Electrophysical and photoelectrical properties of the (CdSe)$_{1-x}$(CuInSe$_2$)$_x$ solid solutions [in Russian], *Izv. vuzov. Fiz.*, **35**(1), 41–43 (1992).

Drobyazko V.P., Kuznetsova S.T. Interaction of indium arsenide with cadmium chalcogenides [in Russian], *Zhurn. neorgan. khimii*, **28**(11), 2929–2933 (1983).

Dubrovin I.V., Budionnaya L.D., Mizetskaya I.B., Sharkina E.V. Phase diagram of the SnTe–CdSe section of the SnTe + CdSe ⇔ SnSe + CdTe ternary mutual system [in Russian], *Izv. AN SSSR. Neorgan. materialy*, **22**(4), 590–595 (1986).

Engelen B., Bäumer U., Hermann B., Müller H., Unterderweide K. Zur Polymorphie und Pseudosymmetrie der Hydrate $MSeO_3 \cdot H_2O$ (M = Mn, Co, Ni, Zn, Cd), *Z. anorg. und allg. Chem.*, **622**(11), 1886–1892 (1996).

Garbato L., Ledda F., Manca P., Rucci A., Spiga A. Phase diagram, growth and properties of quaternary diamond-like alloys in the $CuInSe_2$–CdSe system, *Progr. Cryst. Growth Charact.*, **10**(1–4), 199–205 (1985).

Garbato L., Manca P. Synthesis and characterization of some chalcogenides of $A^IB_2^{II}C^{III}D_4^{VI}$, *Mater. Res. Bull.*, **9**(4), 511–517 (1974).

Glazov V.M., Pavlova L.M., Perederiy L.I. Analysis of intermolecular interaction and thermodynamic properties of the GaAs–CdSe melts [in Russian]. In: *Termodinamicheskie svoistva metallicheskih rasplavov: Materialy 4-go Vsesoyuz. soveshch. po termodinamike metal. splavov (rasplavov)*. Alma-Ata, Kazakhstan: Nauka Publish., pp. 2, 26–29 (1979).

Glazov V.M., Pavlova L.M., Perederiy L.I. Phase equilibria and analysis of the intermolecular interaction in the GaAs–CdSe system [in Russian], *Izv. AN SSSR. Neorgan. materialy*, **19**(2), 193–196 (1983).

Gulay L.D., Romanyuk Ya.E., Parasyuk O.V. Crystal structure and low- and high-temperature modifications of $Cu_2CdGeSe_4$, *J. Alloys Compd.*, **347**(1–2), 193–197 (2002).

Gusachenko A.G., Atroshchenko L.V., Sysoev L.V., Lebed' N.B. Solubility of Cu and Ga in the CdSe single crystals [in Russian], *Izv. AN SSSR. Neorgan. materialy*, **13**(3), 530–531 (1977a).

Gusachenko A.G., Atroshchenko L.V., Sysoev L.V., Lebed' N.B. Solubility of copper and gallium in the cadmium selenide single crystals [in Russian]. In.: *Svoystva legirovan. poluprovodn.* Moscow, Russia: Nauka Publish., pp. 69–72 (1977b).

Guseinov G.D., Godzhaev E.M., Halilov H.Ya., Seidov F.M., Pashaev A.M. Complex semiconductor chalcogenides [in Russian], *Izv. AN SSSR. Neorgan. materialy*, **8**(9), 1569–1572 (1972).

Hahn H., Schulze H. Über quaternäre Chalkogenide des Germanium und Zinns, *Naturwissenschaften*, **52**(14) 426 (1965).

Hirai T., Kurata K., Takeda Y. Deviation of new semiconducting compounds by cross substitution for group IV semiconductors and their semiconducting and thermal properties, *Solid State Electron.*, **10**(10), 975–981 (1967).

Hiramatsu H., Ueda K., Kamiya T., Ohta H., Hirano M., Hosono H. Synthesis of single-phase layered oxychalcogenide $La_2CdO_2Se_2$: Crystal structure, optical and electrical properties, *J. Mater. Chem.*, **14**(19), 2946–2950 (2004).

Ivashchenko I.A., Gulay L.D., Zmiy O.F., Olekseyuk I.D. The quasiternary system Cu_2Se–CdSe–In_2Se_3 and the crystal structure of the $Cu_{0.6}Cd_{0.7}In_6Se_{10}$, *J. Alloys Compd.*, **394**(1–2), 186–193 (2005).

Ivashchenko I.A., Zmiy O.F., Olekseyuk I.D. Isothermal section of the Ag_2Se–CdSe–In_2Se_3 system at 820 K [in Ukrainian], *Nauk. visnyk Volyns'k. un-tu im. Lesi Ukrainky*, (13), 27–33 (2008a).

Ivashchenko I.A., Zmiy O.F., Olekseyuk I.D. Phase equilibria in the quasiternary system Ag_2Se–CdSe–In_2Se_3, *Chem. Met. Alloys*, **1**(3–4), 274–282 (2008b).

Jiang H.-L., Kong F., Fan Y., Mao J.-G. $ZnVSe_2O_7$ and $Cd_6V_2Se_5O_{21}$: New d^{10} transition-metal selenites with V(IV) or V(V) cations, *Inorg. Chem.*, **47**(16), 7430–7437 (2008).

Kadykalo E.M., Marushko L.P., Zmiy O.F., Olekseyuk I.D. Quasi-ternary system Ag_2Se–CdSe–Ga_2Se_3, *J. Phase Equilib. Diff.*, **34**(5), 403–415 (2013).

Kadykalo E., Zmiy O., Olekseyuk I. Isothermal section of the Ag_2Se–CdSe–Ga_2Se_3 quasiternary system at 820 K and phase diagram of the Ag_9GaSe_6–CdSe section [in Ukrainian], *Visnyk L'viv. un-tu. Ser. khim.*, (39), 67–71 (2000).

Kazakov A.I., Stepanov M.A., Shlikher V.A. Calculation of the immiscibility regions in the four-component $Cd_xHg_{1-x}Te_ySe_{1-y}$ and $Zn_xHg_{1-x}Te_ySe_{1-x}$ solid solutions [in Russian], *Trudy Odes. politekhn. un-ta*, (2), 145–150 (2001).

Kim J.H., Chung D.-Y., Kanatzidis M.G. A new chalcogenide homologous series $A_2[M_{5+n}Se_{9+n}]$ (A = Rb, Cs; M = Bi, Ag, Cd), *Chem. Commun.*, (15), 1628–1630 (2006).

Kirovskaya I.A., Azarova O.P., Shubenkova E.G., Dubina O.N. Synthesis and optical absorption of the solid solutions of the InSb–(II-VI) systems [in Russian], *Neorgan. materialy*, **38**(2), 135–138 (2002).

Kirovskaya I.A., Zemtsov A.E., Shedenko A.V. Solid solutions of heterovalent substitution based on the GaAs–$A^{II}B^{VI}$ system [in Russian], *Sovremen. naukoemk. tekhnol.*, (8), 33–35 (2007).

Konstantinova N.N., Medvedkin G.A., Polushina I.K., Smirnova A.D., Sokolova V.I., Tairov M.A. Optical and electrical properties of the $Cu_2CdSnSe_4$ and $Cu_2CdGeSe_4$ crystals [in Russian], *Izv. AN SSSR. Neorgan. materialy*, **25**(9), 1445–1448 (1989).

Leonov V.V., Chunarev E.N. Distribution of potassium and zinc chlorides at the crystallization of eutectics of cadmium chalcogenides with its chloride [in Russian]. In: *Physico-khim. processy v geterogennyh systemah*. Krasnoyarsk, Russia: Krasnoyar. Un-t Publish., pp. 59–64 (1977).

Leute V., Menge D. The quasiternary system $(Cd_kSn_{1-k})(Se_lTe_{1-l})$, *Z. Phys. Chem. (Munchen)*, **176**(1), 65–76 (1992).

Leute V., Schmidtke H.M., Stratmann W. The influence of a miscibility gap on the reaction behavior of the quasiternary system $Hg_kCd_{(1-k)}Te_lSe_{(1-l)}$, *Ber. Bunsenges. phys. Chem.*, **86**(8), 732–738 (1982).

Li H., Malliakas C.D., Peters J.A., Liu Z., Im J., Jin H., Morris C.D. et al. $CsCdInQ_3$ (Q = Se, Te): New photoconductive compounds as potential materials for hard radiation detection, *Chem. Mater.*, **25**(10), 2089–2099 (2013).

Luzhnaya N.P., Mizetski P.A., Oleinik G.S. Liquidus surface of the $CdCl_2$–CdSe–CdTe ternary system [in Russian], *Izv. AN SSSR. Neorgan. materialy*, **16**(3), 541–542 (1980).

Luzhnaya N.P., Shabunina G.G. Investigation of $CdCr_2Se_4$ crystallization in the $3CdSe + 2CrCl_3 \Leftrightarrow Cr_2Se_3 + 3CdCl_2$ ternary mutual system [in Russian]. In: *Troin. polupro-vodn. i ih primenenie*: *Tez. dokl.* Kishinev, Republic of Moldova: Shtiintsa Publish., pp. 86–87 (1976).

Luzhnaya N.P., Shabunina G.G., Kalinnikov V.T., Aminov T.G. The $3CdSe + 2CrCl_3 \Leftrightarrow Cr_2Se_3 + CdCl_2$ ternary mutual system [in Russian], *Zhurn. neorgan. khimii*, **26**(4), 1075–1080 (1981).

Manca P., Garbato L. Phase relationships, crystal growth and stoichiometry defects in $A^IC^{III}D_2^{YI}/B^{II}D^{VI}$ heterojunction-forming system, *Sol. Cells*, **16**, 101–121 (1986).

Matsuhita H., Ichikawa T., Katsui A. Structural, thermodynamic and optical properties of Cu_2–II–IV–VI_4 quaternary compounds, *J. Mater. Sci.*, **40**(8), 2003–2005 (2005).

Matsuhita H., Maeda T., Katsui A., Takizawa T. Thermal analysis and synthesis from the melts of Cu-based quaternary compounds Cu–III–IV–VI_4 and Cu_2–II–IV–VI_4 (II – Zn, Cd; III – Ga, In; IV – Ge, Sn; VI – Se), *J. Cryst. Growth*, **208**(1–4), 416–422 (2000).

Medvedev Yu.V., Berchenko N.N. Analysis of the phase compositions at the interface of $Mn_{1-x}A_x^{II}B^{VI}$(A^{II} = Zn, Cd, Hg; B^{VI} = Te, Se) solid solution with own oxides [in Russian], *Zhurn. neorgan. khimii*, **39**(5), 846–848 (1994).

Mishchenko I.A., Zmiy O.F., Olekseyuk I.D. Phase equilibrium in the $AgInSe_2$–CdSe system, *Pol. J. Chem.*, **75**(10), 1407–1411 (2001).

Mitchell K., Huang F.Q., McFarland A.D., Haynes Ch.L., Somers R.C., Van Duyne R.P., Ibers J.A. The $CsLnMSe_3$ semiconductors (Ln = rare-earth element, Y; M = Zn, Cd, Hg), *Inorg. Chem.*, **42**(13), 4109–4116 (2003).

Mkrtchian S.A., Dovletov K., Zhukov E.G., Melikdzhanian A.G., Nuryiev S. Electrophysical properties of the $Cu_2A^{II}B^{IV}Se_4$ (A^{II} – Cd, Hg; B^{IV} – Ge, Sn) compounds [in Russian], *Izv. AN SSSR. Neorgan. materialy*, **24**(7), 1094–1096 (1988).

Morris C.D., Li H., Jin H., Malliakas C.D., Peters J.A., Trikalitis P.N., Freeman A.J., Wessels B.W., Kanatzidis M.G. $Cs_2M^{II}M_3^{IV}Q_8$ (Q = S, Se, Te): An extensive family of layered semiconductors with diverse band gaps, *Chem. Mater.*, **25**(16), 3344–3356 (2013).

Nikolič P.M. Solid solutions of CdSe and CdTe in PbTe and their optical properties, *Br. J. Appl. Phys.*, **17**(3), 341–344 (1966).

Odin I.N. *T-x-y* diagrams for mutual systems $PbX + CdI_2 = CdX + PbI_2$ (X = S, Se, Te) [in Russian], *Zhurn. neorgan. khimii*, **46**(10), 1733–1738 (2001).

Odin I.N., Chukichev M.V. Metastable phases crystallizing from the melts in the $PbX + CdI_2 = CdX + PbI_2$ (X = S, Se, Te) mutual systems [in Russian], *Zhurn. neorgan. khimii*, **46**(12), 2083–2087 (2001).

Odin I.N., Grin'ko V.V., Kozlovskiy V.F. Stable and metastable phases in the $PbSe + AgI_2 \Leftrightarrow Ag_2Se + CdI_2$ and $PbSe + CdI_2 \Leftrightarrow CdSe + PbI_2$ mutual system [in Russian], *Zhurn. neorgan. khimii*, **50**(5), 843–847 (2005).

Odin I.N., Marugin V.V. Electrophysical properties of the alloys of the $Bi_2Se_3 + Cd_3Te_3 \Leftrightarrow Bi_2Te_3 + Cd_3Se_3$ system [in Russian], *Zhurn. neorgan. khimii*, **36**(7), 1865–1869 (1991).

Ohtani H., Kojima K., Ishida K., Nishizawa T. Miscibility gap in II-VI semiconductor systems, *J. Alloys Compd.*, **182**(1), 103–114 (1992).

Olekseyuk I.D., Gulay L.D., Dudchak I.V., Piskach L.V., Parasyuk O.V., Marchuk O.V. Single crystal preparation and crystal structure of the $Cu_2Zn(Cd,Hg)SnSe_4$ compounds, *J. Alloys Compd.*, **340**(1–2), 141–145 (2002a).

Olekseyuk I.D., Gulay L.D., Parasyuk O.V., Husak O.A., Kadykalo E.M. Phase diagram of the $AgGaSe_2–CdSe$ system and crystal structure of the $AgCd_2GaSe_4$ compound, *J. Alloys Compd.*, **343**(1–2), 125–131 (2002b).

Olekseyuk I.D., Kadykalo E.M., Zmiy O.F. The $CuGaSe_2–CdSe$ section of the quasiternary $Cu_2Se–CdSe–Ga_2Se_3$ system, *Pol. J. Chem.*, **71**(7), 893–897 (1997a).

Olekseyuk I.D., Kogut Yu.M., Parasyuk O.V., Piskach L.V., Gorgut G.P., Kus'ko O.P., Pekhnyo V.I., Volkov S.V. Glass-formation in the $Ag_2Se–Zn(Cd, Hg)Se–GeSe_2$, *Chem. Met. Alloys*, **2**(3–4), 146–150 (2009).

Olekseyuk I.D., Parasyuk O.V. The $CdSe–Ga_2Se_3–GeSe_2$ system [in Russian], *Zhurn. neorgan. khimii*, **40**(2), 315–319 (1995).

Olekseyuk I.D., Parasyuk O.V., Bozhko V.V., Galyan V.V., Petrus' I.I. Glass-forming in the $Zn(Cd,Hg)Se–Ga_2Se_3–GeSe_2$ systems [in Ukrainian], *Fizyka kondens. vysokomolek. system. Nauk. zap. Rinens'kogo pedinstytutu*, (3), 148–152 (1997b).

Olekseyuk I.D., Parasyuk O.V., Bozhko V.V., Petrus' I.I., Galyan V.V. Formation and properties of the quasiternary $Zn(Cd,Hg)Se–Ga_2Se_3–SnSe_2$ system glasses, *Funct. Mater.*, **6**(3), 474–477 (1999).

Olekseyuk I.D., Parasyuk O.V., Dzham O.A., Piskach L.V. The reciprocal $CuInS_2 + 2CdSe \Leftrightarrow CuInSe_2 + 2CdS$ system. Part. I. The quasibinary $CuInSe_2–CdSe$ system: Phase diagram and crystal structure of solid solutions, *J. Solid State Chem.*, **179**(1), 315–322 (2006).

Olekseyuk I.D., Parasyuk O.V., Sysa L.V., Yurchenko Yu.V. The $CdSe–Ga_2Se_3–GeSe_2$ system at 870 K, *Pol. J. Chem.*, **71**(6), 701–704 (1997c).

Olekseyuk I.D., Piskach L.V. Phase equilibria in the $Cu_2SnX_3–CdX$ (X = S, Se, Te) systems [in Russian], *Zhurn. neorgan. khimii*, **42**(2), 331–333 (1997).

Olekseyuk I.D., Piskach L.V., Parasyuk O.V. Phase equilibria in the $Cu_2SiSe_3(Te_3)–CdSe(Te)$ systems [in Russian], *Zhurn. neorgan. khimii*, **43**(3), 516–519 (1998).

Olekseyuk I.D., Piskach L.V., Parasyuk O.V. Phase equilibria of $Ag_{0.33}Sn_{16.7}Se_{50}–CdSe$ section of the quasiternary $Ag_2Se–CdSe–SnSe_2$ system, *Pol. J. Chem.*, **71**(6), 721–724 (1997d).

Olekseyuk I.D., Piskach L.V., Parasyuk O.V., Mel'nyk O.M., Lyskovets T.A. The $Cu_2Se–CdSe–GeSe_2$ system, *J. Alloys Compd.*, **298**(1–2), 203–212 (2000).

Parasyuk O.V. Romanyuk Ya.E., Olekseyuk I.D. Single-crystal growth of $Cu_2CdGeSe_4$, *J. Cryst. Growth*, **275**(1–2), e159–e162 (2005).

Parasyuk O.V., Gulay L.D., Piskach L.V., Olekseyuk I.D. The $Ag_2Se–CdSe–SnSe_2$ system at 670K and the crystal structure of the $Ag_2CdSnSe_4$ compound, *J. Alloys Compd.*, **335**(1–2), 176–180 (2002).

Parasyuk O.V., Olekseyuk I.D., Gulay L.D., Piskach L.V. Phase diagram of the $Ag_2Se-Zn(Cd)$ Se–SiSe$_2$ systems and crystal structure of the Cd_4SiSe_6 compound, *J. Alloys Compd.*, **354**(1–2), 138–142 (2003).

Parasyuk O.V., Piskach L.V., Olekseyuk I.D. The $Cu_2Se-CdSe-SnSe_2$ system [in Russian], *Zhurn. neorgan. khimii*, **44**(8), 1363–1367 (1999).

Parthé E., Yvon K., Deitch R.H. The crystal structure of Cu_2CdGeS_4 and other quaternary normal tetrahedral structure compounds, *Acta Crystallogr. B*, **25**(6), 1164–1174 (1969).

Piskach L.V., Olekseyuk I.D., Parasyuk O.V. Physico-chemical peculiarities of the $Cu_2CdC^{IV}X_4$ (C^{IV} – Si, Ge, Sn; X – S, Se, Te) quaternary phase formations [in Ukrainian], *Fizyka kondens. vysokomolek. system. Nauk. zap. Rinens'kogo pedinstytutu*, (3), 153–157 (1997).

Piskach L.V., Parasyuk O.V., Olekseyuk I.D., Halahan V.Ya. $CdSe-Ga_2Se_3-SnSe_2$ system [in Ukrainian], *Fiz. i khim. tv. tila*, **3**(1), 25–32 (2002).

Piskach L.V., Parasyuk O.V., Olekseyuk I.D., Romanyuk Y.E., Volkov S.V., Pekhnyo V.I. Interaction of argyrodite family compounds with the chalcogenides of II-b elements. *J. Alloys Compd.*, **421**(1–2), 98–104 (2006).

Piskach L.V., Parasyuk O.V., Romanyuk Ya.E. The phase equilibria in the quasibinary $Cu_2GeS_3(Se_3)-CdS(Se)$ systems, *J. Alloys Compd.*, **299**(1–2), 227–231 (2000).

Range K.-J., Handrick K. Synthese und Hochdruckverhalten quaternärer Chalkogenidhalogenide $M_2M'X_3Y$ (M = Zn, Cd; M' = Al, Ga, In; X = Se, Te; Y = Cl, Br, I), *Z. Naturforsch.*, **B43**(2), 153–158 (1988).

Rayevski S.D., Zbigli K.R., Kazak G.F., Prunich M.D. Phase diagram of the PbSe–CdTe system [in Russian], *Izv. AN SSSR. Neorgan. materialy*, **18**(8), 1267–1270 (1982).

Rayevski S.D., Zbigli K.R., Kazak G.F., Prunich M.D. Phase diagram of the PbTe–CdSe system [in Russian], *Izv. AN SSSR. Neorgan. materialy*, **19**(6), 889–892 (1983a).

Rayevski S.D., Zbigli K.R., Kazak G.F., Prunich M.D. Solid solutions of cadmium selenide in lead telluride [in Russian], *Izv. AN MSSR. Ser. fiz.-tehn. i mat. n.*, (2), 42–44 (1983b).

Rduch P., Duda H., Groń T., Skrzypek D., Malicka E., Gągor A. Influence of Ce substitution on the critical properties of 3D-Heisenberg $Cd_xCe_yCr_2Se_4$ ferromagnets, *Philos. Mag.*, **92**(18), 2382–2396 (2012).

Rduch P., Duda H., Guzik A., Malicka E., Groń T., Mazur S., Gągor A., Sitko R. Critical behaviour of the 3D-Ising ferromagnets $Cd[Cr_xTi_y]Se_4$, *J. Phys. Chem. Solids*, **74**(10), 1419–1425 (2013).

Reisman A., Berkenblit M. Impurity incorporation into CdSe and equilibria in the system $CdSe-CdCl_2$, *J. Electrochem. Soc.*, **109**(11), 1111–1113 (1962).

Safarov M.G., Aliev F.G., Mamedov V.N., Bairamov G.M. Interaction in the Sb_2Se_3-CdTe system [in Russian], *Izv. AN SSSR. Neorgan. materialy*, **34**(7), 1831–1833 (1989).

Schäfer W., Nitsche R. Tetrahedral quaternary chalcogenides of the type $Cu_2-II-IV-S_4(Se_4)$, *Mater. Res. Bull.*, **9**(5), 645–654 (1974).

Schäfer W., Nitsche R. Zur Systematik tetraedrischer Verbindungen vom Typ $Cu_2Me^{II}Me^{IV}Me_4^{VI}$(Stannite und Wurtzstannite), *Z. Kristallogr.*, **145**(5–6), 356–370 (1977).

Schwer H., Keller E., Krämer V. Crystal structure and twinning of $KCd_4Ga_5S_{12}$ and some isotypic $AB_4C_5X_{12}$ compounds, *Z. Kristallogr.*, **204**(Pt. 2), 203–213 (1993).

Sosovska S.M., Olekseyuk I.D., Parasyuk O.V. The quasiternary $CdSe-Ga_2Se_3-Sb_2Se_3$ system, *Pol. J. Chem.*, **81**(4), 505–513 (2007).

Sosovska S.M., Olekseyuk I.D., Parasyuk O.V. The $CdSe-Ga_2Se_3-PbSe$ system, *J. Alloys Compd.*, **453**(1–2), 115–120 (2008).

Sosovska S.M., Olekseyuk I.D., Parasyuk O.V. The quasiternary $CdSe-Ga_2Se_3-Bi_2Se_3$ system, *Chem. Met. Alloys*, **3**(1–2), 5–11 (2010).

Tomashik Z.F. Liquidus surface of the CdSe + PbTe ⇔ CdTe + PbSe ternary mutual system [in Russian], *Izv. AN SSSR. Neorgan. materialy*, **17**(9), 1575–1577 (1981).

Vassilev V.S., Boycheva S.V., Petkov P. Glass formation in the $GeSe_2(As_2Se_3)–Sb_2Se_3–CdTe$, *Mater. Lett.*, **52**(1–2), 126–129 (2002).

Voitsehovski A.V., Drobiazko V.P., Mitiurev V.K., Vasilenko V.P. Solid solutions in the InAs– CdS and InAs–CdSe systems [in Russian], *Izv. AN SSSR. Neorgan. materialy*, **4**(10), 1681–1684 (1968).

Voitsehovski A.V., Pashun A.D., Mitiurev V.K. About interaction of gallium arsenide with II– VI compounds [in Russian], *Izv. AN SSSR. Neorgan. materialy*, **6**(2), 379–380 (1970).

Vovk P., Davydyuk G., Mishchenko I., Zmiy O. Electrical and photoelectrical properties of the solid solutions based on the $CuCd_2InSe_4$ compound [in Ukrainian], *Nauk. visnyk Volyns'k. derzh. un-tu*, (14), 33–38 (1999).

Vovk P., Davydyuk G., Mishchenko I., Zmiy O. Electrical and photoelectrical properties of the solid solutions based on the $CuCd_2InSe_4$ compound [in Ukrainian], *Visnyk L'viv. un-tu. Ser. khim.*, (39), 167–172 (2000).

Wang L., Hung Y.-C., Hwu S.-J., Koo H.-J., Whangbo M.-H. Synthesis, structure, and prop-erties of a new family of mixed-framework chalcohalide semiconductors: $CdSbS_2X$ (X = Cl, Br), $CdBiS_2X$ (X = Cl, Br), and $CdBiSe_2X$ (X = Br, I), *Chem. Mater.*, **18**(5), 1219–1225 (2006).

Wang L., Hwu S.-J. A new series of chalcohalide semiconductors with composite $CdBr_2/ Sb_2Se_3$ lattices: Synthesis and characterization of $CdSb_2Se_3Br_2$ and indium derivatives $InSb_2S_4X$ (X = Cl and Br) and InM_2Se_4Br (M = Sb and Bi), *Chem. Mater.*, **19**(25), 6212–6221 (2007).

Yang Y., Ibers J.A. Syntheses and structure of the new quaternary compounds $Ba_4Nd_2Cd_3S_{10}$ and $Ba_4Ln_2Cd_3S_{10}$ (Ln = Sm, Gd, Tb), *J. Solid State Chem.*, **149**(2), 384–390 (2000).

Zhao D., Xia F., Chen G., Zhang X., Ma H., Adam J.L. Formation and properties of chalcogenide glasses in the $GeSe_2–As_2Se_3–CdSe$ system, *J. Am. Ceram. Soc.*, **88**(11), 3143–3146 (2005).

Zhao D., Zhang X., Wang H., Zeng H., Ma H., Adam J.L., Chen G. Thermal properties of chal-cogenide glasses in the $GeSe_2–As_2Se_3–CdSe$ system, *J. Non-Cryst. Solids*, **354**(12–13), 1281–1284 (2008).

Zhukov E.G., Mkrtchian S.A., Dovletov K.O., Kalinnikov V.T., Ashirov A.A. The $Cu_2SnSe_3–CdSe$ system [in Russian], *Zhurn. neorgan. khimii*, **27**(3), 761–762 (1982b).

Zhukov E.G., Mkrtchian S.A., Dovletov K.O., Kalinnikov V.T., Melikdzhanian A.G., Ashirov A.A. Thermal and X-ray analysis of the $Cu_2Sn(Ge)Se_3–CdSe$ systems [in Russian]. In: *Termicheski analiz. Tez. dokl. VIII Vses. konf*, Moskva-Kuibyshev, Nauka Publish., p. 120 (1982a).

Zhukov E.G., Mkrtchian S.A., Dovletov K.O., Melikdzhanian A.G., Kalinnikov V.T., Ashirov A.A. The $Cu_2GeSe_3–CdSe$ system [in Russian], *Zhurn. neorgan. khimii*, **29**(7), 1897–1898 (1984).

Zmiy O.F., Mishchenko I.A., Olekseyuk I.D. Phase equilibria in the quasiternary system $Cu_2Se–CdSe–In_2Se_3$, *J. Alloys Compd.*, **367**(1–2), 49–57 (2004).

Zmiy O.F., Parasyuk O.V., Mischenko I.A. Investigation of the $CuInSe_2–CdSe$ and $CuInSe_2– CdInSe_4$ sections in the $Cu_2Se–CdSe–In_2Se_3$ quasiternary system [in Ukrainian], *Fizyka kondens. vysokomolek. system. Nauk. zap. Rivnens'kogo pedinstytutu*, (3), 194–198 (1997).

6 Systems Based on CdTe

6.1 CADMIUM–HYDROGEN–CARBON–TELLURIUM

Some compounds are formed in this quaternary system.

Cd(TePhMe)$_2$ or C$_{14}$H$_{14}$Te$_2$Cd, bis(methylbenzenetellurato)cadmium. To obtain this compound, a slight excess of (4-methylphenyl)(trimethylsilyl)tellurium was added to CdCl$_2$ suspended in tetrahydrofuran, and the mixture was stirred at room temperature during 14 h (Steigerwald and Sprinkle 1987). The yellow solid was isolated, washed with pentane, and extracted with toluene/PMe$_3$. Evaporation gave C$_{14}$H$_{14}$Te$_2$Cd as a bright yellow solid. Pyrolysis of this compound in the solid state (200°C, 16 h) gives CdTe.

Cd(TeC$_6$H$_2$Me$_3$-2,4,6)$_2$ or C$_{18}$H$_{22}$Te$_2$Cd, bis(2,4,6-trimetylbenzenetellurato) cadmium. This compound was prepared from 2,4,6-Me$_3$C$_6$H$_2$TeH (2.2 mmol) and Cd[N(SiMe$_3$)$_2$]$_2$ (1.08 mmol) in petroleum ether at −20°C. After stirring for 20 min, the residue was allowed to warm to room temperature, filtered off, washed with 20 mL of petroleum ether, and recrystallized from dimethylformamide to give yellow crystals that melt with decomposition at 230°C–240°C (Bochmann et al. 1991a,b, 1997). C$_{18}$H$_{22}$Te$_2$Cd crystallizes in the tetragonal structure with the lattice parameters a = 1594.5 and c = 759.4 pm and calculated density 2.085 g·cm^{-3}. An identical product was obtained from CdCl$_2$ and PhMe$_3$TeSnBu$_3^n$. Heating a solution of C$_{18}$H$_{22}$Te$_2$Cd in mesitylene leads to the formation of CdTe.

6.2 CADMIUM–POTASSIUM–CHLORINE–TELLURIUM

CdTe–KCl–CdCl$_2$. The phase diagram is not constructed. Using oriented zone recrystallization and radioactive isotopes, the equilibrium coefficient of KCl distribution in the CdTe–CdCl$_2$ eutectic was determined: k_0 = 0.78 (Leonov and Chunarev 1977).

6.3 CADMIUM–CESIUM–GALLIUM–TELLURIUM

CdTe–Cs$_2$Te–Ga$_2$Te$_3$. The phase diagram is not constructed. The CsCd$_4$Ga$_5$Te$_{12}$ quaternary compound is formed in this system, which crystallizes in the hexagonal structure with the lattice parameters a = 1546.97 ± 0.15 and c = 1053.29 ± 0.41 pm (a = 959.73 ± 0.08 and α = 107.413° ± 0.009° in rhombohedral settings) and calculated density 5.619 g·cm^{-3} (Schwer et al. 1993). This compound was obtained if CdGa$_2$Te$_4$ was kept in CsBr at temperature above 150°C of its melting point for 1 day and then cooled down slowly.

6.4   CADMIUM–CESIUM–INDIUM–TELLURIUM

$CdTe–Cs_2Te–In_2Te_3$. The phase diagram is not constructed. $CsCdInTe_3$ quaternary compound is formed in this system. This compound, which appears to melt congruently, crystallizes in the monoclinic structure with the lattice parameters $a = 1252.3 \pm 0.3$, $b = 1517.2 \pm 0.3$, $c = 2444.1 \pm 0.5$ pm, and $\beta = 97.38° \pm 0.03°$; calculated density 4.810 $g \cdot cm^{-3}$; and energy gap 1.78 eV (Li et al. 2013). To obtain this compound, a mixture of Cs_2Te_3 (0.4 mmol), Cd (0.8 mmol), In (0.8 mmol), and Te (1.2 mmol) was loaded into a carbon-coated quartz tube in a N_2-filled glove box. The tube was flame-sealed under a vacuum of <10 kPa and placed in a temperature-controlled furnace. The mixture was heated to 650°C in 18 h and kept there for 2.5 days and then cooled to room temperature for 3 days. The single crystals of $CsCdInSe_3$ were grown by the Bridgman method.

6.5   CADMIUM–CESIUM–LANTHANUM–TELLURIUM

$CdTe–Cs_2Te–La_2Te_3$. The phase diagram is not constructed. $CsCdLaTe_3$ quaternary compound exists in this system (Liu et al. 2008). It crystallizes in the orthorhombic structure with the lattice parameters $a = 463.92 \pm 0.17$, $b = 1670.2 \pm 0.7$, and $c = 1216.8 \pm 0.5$ pm and calculated density 5.404 $g \cdot cm^{-3}$.

$CsCdLaTe_3$ compound was produced by the mixture of La, Cd, Te, and CsCl via the high-temperature solid-state reaction (Liu et al. 2008). CsCl works as a flux to assist the crystallization of this compound and also as a Cs source at the interaction. A mixture of the reagents is loaded into a short fused silica tube, which is subsequently situated inside a larger evacuated quartz tube. The tube is flame-sealed under a 10^{-3} Pa atmosphere and then placed into a temperature-controlled furnace. The heating and cooling profiles are as follows. The furnace was heated up to 850°C at a rate of 30°C/h, the temperature is held constant at 850°C for 96 h and then slowly cooled at 3°C/h to 200°C before the furnace is turned off, and then the sample is cooled radiatively to ambient temperature in the furnace. The raw product is washed first with distilled water to remove excess flux and the chloride by-products and dried with ethanol.

6.6   CADMIUM–CESIUM–PRASEODYMIUM–TELLURIUM

$CdTe–Cs_2Te–Pr_2Te_3$. The phase diagram is not constructed. $CsCdPrTe_3$ quaternary compound exists in this system (Liu et al. 2008). It crystallizes in the orthorhombic structure with the lattice parameters $a = 459.4 \pm 0.3$, $b = 1677.0 \pm 1.2$, and $c = 1202.3 \pm 0.5$ pm and calculated density 5.515 $g \cdot cm^{-3}$. $CsCdPrTe_3$ compound was produced by the same way as $CsCdLaTe_3$.

6.7   CADMIUM–CESIUM–NEODYMIUM–TELLURIUM

$CdTe–Cs_2Te–Nd_2Te_3$. The phase diagram is not constructed. $CsCdNdTe_3$ quaternary compound exists in this system (Liu et al. 2008). It crystallizes in the orthorhombic structure with the lattice parameters $a = 457.30 \pm 0.04$, $b = 1675.01 \pm 0.15$, and $c = 1196.75 \pm 0.10$ pm and calculated density 5.596 g·cm^{-3}. CsCdNdTe$_3$ compound was produced by the same way as CsCdLaTe$_3$.

6.8 CADMIUM–CESIUM–SAMARIUM–TELLURIUM

CdTe–Cs$_2$Te–Sm$_2$Te$_3$. The phase diagram is not constructed. CsCdSmTe$_3$ quaternary compound exists in this system (Liu et al. 2008). It crystallizes in the orthorhombic structure with the lattice parameters $a = 454.72 \pm 0.06$, $b = 1680.4 \pm 0.2$, and $c = 1189.47 \pm 0.18$ pm and calculated density 5.689 g·cm^{-3}. CsCdSmTe$_3$ compound was produced by the same way as CsCdLaTe$_3$.

6.9 CADMIUM–CESIUM–EUROPIUM–TELLURIUM

CdTe–Cs$_2$Te–Eu$_2$Te$_3$. The phase diagram is not constructed. CsCdEuTe$_3$ quaternary compound could not be synthesized by the same way as CsCdLaTe$_3$ (Liu et al. 2008).

6.10 CADMIUM–CESIUM–GADOLINIUM–TELLURIUM

CdTe–Cs$_2$Te–Gd$_2$Te$_3$. The phase diagram is not constructed. CsCdGdTe$_3$ quaternary compound exists in this system (Liu et al. 2008). It crystallizes in the orthorhombic structure with the lattice parameters $a = 451.22 \pm 0.17$, $b = 1678.3 \pm 0.7$, and $c = 1181.8 \pm 0.5$ pm and calculated density 5.829 g·cm^{-3}. CsCdGdTe$_3$ compound was produced by the same way as CsCdLaTe$_3$.

6.11 CADMIUM–CESIUM–TERBIUM–TELLURIUM

CdTe–Cs$_2$Te–Tb$_2$Te$_3$. The phase diagram is not constructed. CsCdTbTe$_3$ quaternary compound exists in this system (Liu et al. 2008). It crystallizes in the orthorhombic structure with the lattice parameters $a = 449.56 \pm 0.04$, $b = 1677.90 \pm 0.16$, and $c = 1178.04 \pm 0.11$ pm and calculated density 5.883 g·cm^{-3}. CsCdTbTe$_3$ compound was produced by the same way as CsCdLaTe$_3$.

6.12 CADMIUM–CESIUM–DYSPROSIUM–TELLURIUM

CdTe–Cs$_2$Te–Dy$_2$Te$_3$. The phase diagram is not constructed. CsCdDyTe$_3$ quaternary compound exists in this system (Liu et al. 2008). It crystallizes in the orthorhombic structure with the lattice parameters $a = 448.3 \pm 0.7$, $b = 1675 \pm 3$, and $c = 1174.5 \pm 0.8$ pm and calculated density 5.956 g·cm^{-3}. CsCdDyTe$_3$ compound was produced by the same way as CsCdLaTe$_3$.

6.13 CADMIUM–CESIUM–HOLMIUM–TELLURIUM

CdTe–Cs$_2$Te–Ho$_2$Te$_3$. The phase diagram is not constructed. CsCdHoTe$_3$ quaternary compound exists in this system (Liu et al. 2008). It crystallizes in the orthorhombic structure with the lattice parameters $a = 447.27 \pm 0.15$, $b = 1680.2 \pm 0.6$, and

$c = 1172.7 \pm 0.4$ pm and calculated density 5.977 g·cm^{-3}. CsCdHoTe$_3$ compound was produced by the same way as CsCdLaTe$_3$.

6.14 CADMIUM–CESIUM–ERBIUM–TELLURIUM

CdTe–Cs$_2$Te–Er$_2$Te$_3$. The phase diagram is not constructed. CsCdErTe$_3$ quaternary compound exists in this system (Liu et al. 2008). It crystallizes in the orthorhombic structure with the lattice parameters $a = 446.27 \pm 0.14$, $b = 1681.2 \pm 0.6$, and $c = 1170.6 \pm 0.4$ pm and calculated density 6.016 g·cm^{-3}. CsCdErTe$_3$ compound was produced by the same way as CsCdLaTe$_3$.

6.15 CADMIUM–CESIUM–THULIUM–TELLURIUM

CdTe–Cs$_2$Te–Tm$_2$Te$_3$. The phase diagram is not constructed. CsCdTmTe$_3$ quaternary compound exists in this system (Liu et al. 2008). It crystallizes in the orthorhombic structure with the lattice parameters $a = 445.7 \pm 0.2$, $b = 1682.4 \pm 0.8$, and $c = 1169.6 \pm 0.6$ pm; calculated density 6.036 g·cm^{-3}; and energy gap $E_g = 2.01$ eV. CsCdTmTe$_3$ compound was produced by the same way as CsCdLaTe$_3$.

6.16 CADMIUM–CESIUM–YTTERBIUM–TELLURIUM

CdTe–Cs$_2$Te–Yb$_2$Te$_3$. The phase diagram is not constructed. CsCdYbTe$_3$ quaternary compound could not be synthesized by the same way as CsCdLaTe$_3$ (Liu et al. 2008).

6.17 CADMIUM–CESIUM–LUTETIUM–TELLURIUM

CdTe–Cs$_2$Te–Lu$_2$Te$_3$. The phase diagram is not constructed. CsCdLuTe$_3$ quaternary compound exists in this system (Liu et al. 2008). It crystallizes in the orthorhombic structure with the lattice parameters $a = 444.4 \pm 0.4$, $b = 1684.1 \pm 1.7$, and $c = 1167.2 \pm 1.1$ pm and calculated density 6.106 g·cm^{-3}. CsCdLuTe$_3$ compound was produced by the same way as CsCdLaTe$_3$.

6.18 CADMIUM–COPPER–GALLIUM–TELLURIUM

CdTe–CuGaTe$_2$. The phase diagram is a eutectic type (Figure 6.1) (Dovletov et al. 1989). The eutectic temperature and composition are 877°C $\pm$ 5°C [850°C (Aresti et al. 1977)] and 28 mol.% CuGaTe$_2$, respectively. CuCdGaTe$_3$ quaternary compound is formed in this system. It melts incongruently at 897°C $\pm$ 5°C and crystallizes in the cubic structure of sphalerite type with the lattice parameter $a = 623$ pm [$a = 620 \pm 0$ pm (Garbato and Manca 1974)]. The solubility of CuGaTe$_2$ in CdTe and CdTe in CuGaTe$_2$ at room temperature is equal to 5 and 1 mol.%, respectively (Dovletov et al. 1989). According to the data of Aresti et al. (1977), the solubility of CuGaTe$_2$ in 2CdTe (the system considered as 2CdTe–CuGaTe$_2$) reaches 67 mol.%. It was indicated (Gallardo 1992) that it is possible to predict

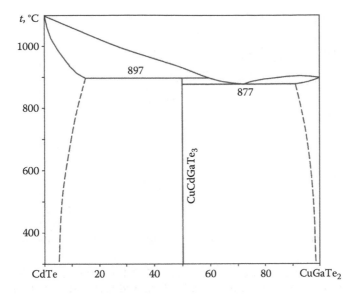

FIGURE 6.1 The CdTe–CuGaTe$_2$ phase diagram. (From Dovletov, K.O. et al., *Izv. AN SSSR. Neorgan. materialy*, 25(7), 1218, 1989.)

the value at which the chalcopyrite phase in this system disappears (x_c) and $T_c(x)$ behavior for the Cd$_{2x}$(CuGa)$_{1-x}$Te$_2$ alloys for which a chalcopyrite–sphalerite-like disordering transition with extensive terminal solid solution formation exists.

This system was investigated by differential thermal analysis (DTA), x-ray diffraction (XRD), metallography, and measuring of microhardness and density (Aresti et al. 1977, Dovletov et al. 1989).

6.19 CADMIUM–COPPER–INDIUM–TELLURIUM

CdTe–CuInTe$_2$. The most reliable phase diagram is shown in Figure 6.2 (Kadykalo et al. 2013). As the stoichiometric composition of CuInTe$_2$ lies beyond its homogeneity region at 400°C, a two-phase region of the coexistence of α-Cu$_2$Te and solid solutions based on CuInTe$_2$ is present in the phase diagram. Therefore, this section is not quasibinary below 668°C.

The existence of the two-phase region (chalcopyrite + sphalerite) in the solid state of this system was also indicated earlier by Cherniavski (1962), Voitsehivski (1964), and Cherniavski et al. (1969). The lattice parameters of forming solid solutions change linearly with composition. It was indicated (Gallardo 1992) that it is possible to predict the value at which the chalcopyrite phase in this system disappears (x_c) and $T_c(x)$ behavior for the Cd$_{2x}$(CuIn)$_{1-x}$Te$_2$ alloys for which a chalcopyrite–sphalerite-like disordering transition with extensive terminal solid solution formation exists.

According to the data of Guseinov et al. (1972), the CuCdInTe$_3$ and Cu$_7$Cd$_3$In$_7$Te$_{17}$ quaternary compounds are formed in this system. CuCdInTe$_3$ compound melts congruently at 915°C and crystallizes in the cubic structure with the lattice parameter $a = 1141.0$ pm; calculation and experimental densities of 6.053 and 6.057 g·cm^{-3},

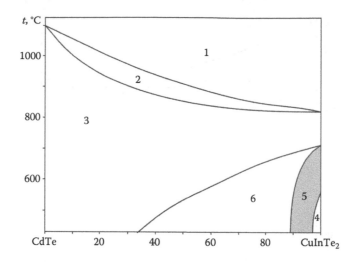

FIGURE 6.2 The 2CdTe–CuInTe$_2$ phase diagram: 1, L; 2, L + (CdTe); 3, (CdTe); 4, (α-Cu$_2$Te) + (CuInTe$_2$); 5, (CuInTe$_2$); and 6, (CuInTe$_2$) + (CdTe). (From Kadykalo, E.M. et al., *J. Phase Equilib. Diff.*, 34(3), 221, 2013.)

respectively; and energy gap $E_g = 0.70$ eV. According to the data of Hirai et al. (1967), Parthé et al. (1969), and Garbato and Manca (1974), the CuCd$_2$InTe$_4$ quaternary compound is formed in this system. It melts incongruently at 850°C (Hirai et al. 1967) and crystallizes in the cubic structure of sphalerite type with the lattice parameter $a = 633.5 \pm 0.5$ pm (Parthé et al. 1969) [$a = 637 \pm 2$ pm (Garbato and Manca 1974)]. All these compounds can be of the solid solution compositions.

This system was investigated by DTA, XRD, metallography, and measuring of microhardness (Cherniavski 1962, Hirai et al. 1967, Cherniavski et al. 1969, Kadykalo et al. 2013). The ingots were anneal at 400°C for 250–500 h (Kadykalo et al. 2013).

CdTe–CuIn$_5$Te$_8$. The liquidus of this vertical section is represented by the curve of the primary crystallization of the solid solutions based on CdTe (Figure 6.3) (Kadykalo et al. 2013). The alloys are single phase and contain only these solid solutions below solidus to 682°C.

CdTe–Cu$_2$Te–In$_2$Te$_3$. The liquidus surface of this system consists of four fields of primary crystallization (Figure 6.4): solid solutions based on CdTe, which occupy the largest share of the concentration triangle, and solid solutions based on β-Cu$_2$Te, CdIn$_2$Te$_4$, and CdIn$_8$Te$_{13}$ (Kadykalo et al. 2013). The fields of the primary crystallization are separated by four monovariant lines and six nonvariant points. The coordinates of the ternary transition point U are 4 mol.% CdTe, 3 mol.% Cu$_2$Te, and 93 mol.% In$_2$Te$_3$ and 692°C. The eutectoid process (β-In$_2$Te$_3$) ⇔ (α-In$_2$Te$_3$) + (CdIn$_8$Te$_{13}$) + (CdIn$_2$Te$_4$) takes place at 417°C in the CdTe–CuInTe$_2$–In$_2$Te$_3$ subsystem. The nonvariant process (β-Cu$_2$Te) + (β-In$_2$Te$_3$) ⇔ (α-Cu$_2$Te) + (α-CuInTe$_2$) takes place at 574°C in the CdTe–Cu$_2$Te–CuInTe$_2$ subsystem.

Isothermal section of the CdTe–Cu$_2$Te–In$_2$Te$_3$ system is shown in Figure 6.5 (Kadykalo et al. 2013).

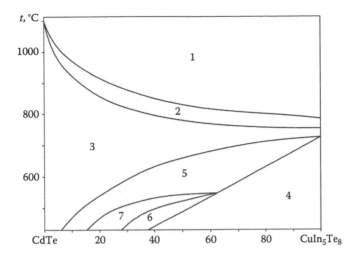

FIGURE 6.3 The CdTe–CuIn$_5$Te$_8$ vertical section: 1, L; 2, L + (CdTe); 3, (CdTe); 4, (CuIn$_5$Te$_8$); 5, (CdTe) + (CuIn$_5$Te$_8$); 6, (CuIn$_5$Te$_8$) + (CuInTe$_2$); and 7, (CdTe) +(CuIn$_5$Te$_8$) + (CuInTe$_2$). (From Kadykalo, E.M. et al., *J. Phase Equilib. Diff.*, 34(3), 221, 2013.)

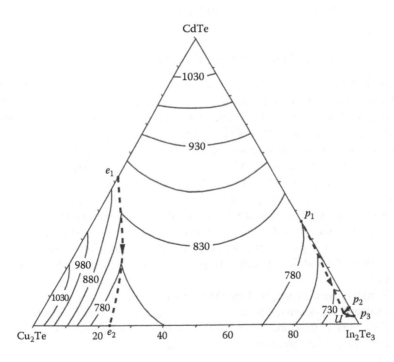

FIGURE 6.4 Liquidus surface of the CdTe–Cu$_2$Te–In$_2$Te$_3$ quasiternary system. (From Kadykalo, E.M., et al., *J. Phase Equilib. Diff.*, 34(3), 221, 2013.)

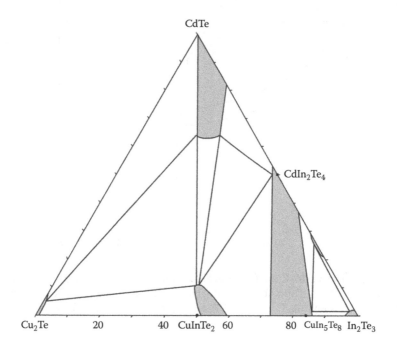

FIGURE 6.5   Isothermal section of the CdTe–Cu$_2$Te–In$_2$Te$_3$ quasiternary system at 400°C.
(From Kadykalo, E.M. et al., *J. Phase Equilib. Diff.*, 34(3), 221, 2013.)

6.20   CADMIUM–COPPER–SILICON–TELLURIUM

CdTe–Cu$_2$SiTe$_3$. This section is a nonquasibinary section of the Cd–Cu–Si–Te
quaternary system since Si$_2$Te$_3$ primarily crystallizes from the Cu$_2$SiTe$_3$-rich side
(Figure 6.6) (Piskach et al. 1997, Olekseyuk et al. 1998). The eutectic composition and temperature are 23 mol.% CdTe and 539°C, respectively. The immiscibility region within the interval of 46–56 mol.% CdTe exists in this system
with monotectic temperature 762°C. The Cu$_2$CdSiTe$_4$ quaternary compound is
formed in the CdTe–Cu$_2$SiTe$_3$ system. It melts incongruently at 657°C [at 650°C
(Matsuhita et al. 2005)] (the peritectic composition is 40 mol.% CdTe) and crystallizes in the tetragonal structure with the lattice parameters $a = 607.3 \pm 0.2$ and
$c = 1177.1 \pm 0.6$ pm [$a = 612$ and $c = 1179$ pm (Matsuhita et al. 2005; $a = 611.0 \pm 0.1$
and $c = 1181.1 \pm 0.3$ pm (Haeuseler et al. 1991)] and calculation and experimental densities 5.64 and 5.57 g·cm^{-3}, respectively (Piskach et al. 1997, Olekseyuk
et al. 1998).

This system was investigated by DTA, XRD, metallography, and measuring of
microhardness and density. The ingots were annealed at 500°C for 1000 h (Piskach
et al. 1997, Olekseyuk et al. 1998). Single crystals of Cu$_2$CdSiTe$_4$ were obtained by
the horizontal gradient freezing method (Matsuhita et al. 2005).

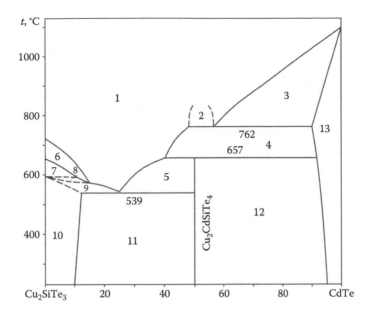

FIGURE 6.6 Phase relations in the CdTe–Cu$_2$SiTe$_3$ system: 1, L; 2, L$_1$ + L$_2$; 3, L$_2$ + (CdTe); 4, L + (CdTe); 5, L + Cu$_2$CdSiTe$_4$; 6, L + Si$_2$Te$_3$; 7, L + Si$_2$Te$_3$ + Cu$_2$Te; 8, L + Cu$_2$Te + (Cu$_2$SiTe$_3$); 9, L + (Cu$_2$SiTe$_3$); 10, (Cu$_2$SiTe$_3$); 11, (Cu$_2$SiTe$_3$) + Cu$_2$CdSiTe$_4$; 12, Cu$_2$CdSiTe$_4$ + (CdTe); and 13, (CdTe). (From Olekseyuk, I.D. et al., *Zhurn. neorgan. khimii*, 43(3), 516, 1998.)

6.21 CADMIUM–COPPER–GERMANIUM–TELLURIUM

CdTe–Cu$_2$GeTe$_3$. This section is a nonquasibinary one of the Cd–Cu–Ge–Te quaternary system since Cu$_2$Te and GeTe primarily crystallize from the Cu$_2$GeTe$_3$-rich side (Figure 6.7) (Olekseyuk et al. 1996, Piskach et al. 1997). The eutectic composition and temperature are 14 mol.% CdTe and 482°C [500°C ± 5°C (Piskach et al. 1988)], respectively. The Cu$_2$CdGeTe$_4$ quaternary compound is formed in this system. It melts incongruently at 532°C [545°C ± 5°C (Piskach et al. 1988)] and crystallizes in the tetragonal structure with the lattice parameters a = 611.4 ± 0.3 and c = 1190.6 ± 0.3 pm [a = 612.7 ± 0.1 and c = 1191.9 ± 0.3 pm (Haeuseler et al. 1991); a = 614 and c = 1196 pm (Piskach et al. 1988)] and calculation and experimental densities 6.02 and 6.05 g·cm^{-3}, respectively (Olekseyuk et al. 1996). Matsuhita et al. (2005) noted that Cu$_2$CdGeTe$_4$ was not synthesized and had the mixed phases of CdTe and Cu$_2$GeTe$_3$.

Solubility of CdTe in Cu$_2$GeTe$_3$ and Cu$_2$GeTe$_3$ in CdTe at 430°C is equal to 13 and 8 mol.%, respectively (Olekseyuk et al. 1996).

This system was investigated by DTA, XRD, metallography, and measuring of microhardness and density (Olekseyuk et al. 1996, Piskach et al. 1997, 1988). The ingots were annealed at 430°C for 300–500 h (Olekseyuk et al. 1996) [at 200°C for 100 h (Piskach et al. 1988)].

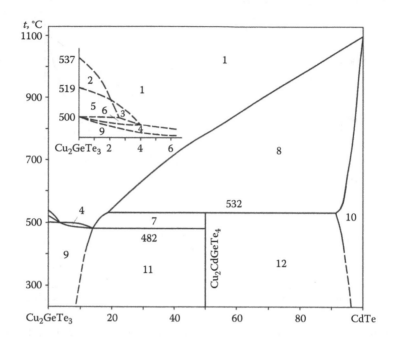

FIGURE 6.7 Phase relations in the CdTe–Cu$_2$GeTe$_3$ system: 1, L; 2, L + Cu$_2$Te; 3, L + GeTe; 4, L + (Cu$_2$GeTe$_3$); 5, L + Cu$_2$Te + GeTe; 6, L + (Cu$_2$GeTe$_3$) + GeTe; 7, L + Cu$_2$CdGeTe$_4$; 8, L + (CdTe); 9, (Cu$_2$GeTe$_3$); 10, (CdTe); 11, (Cu$_2$GeTe$_3$) + Cu$_2$CdGeTe$_4$; and 12, Cu$_2$CdGeTe$_4$ + (CdTe). (From Olekseyuk, I.D. et al., *Zhurn. neorgan. khimii*, 41(9), 1420, 1996; Piskach, L.V. et al., *Fizyka kondens. vysokomolek. system. Nauk. zap. Rinens'kogo pedinstytutu*, (3), 153, 1997.)

6.22 CADMIUM–COPPER–TIN–TELLURIUM

CdTe–Cu$_2$SnTe$_3$. This section is a nonquasibinary section of the Cd–Cu–Sn–Te quaternary system since Cu$_2$SnTe$_3$ melts incongruently (Figure 6.8) (Olekseyuk and Piskach 1997, Piskach et al. 1997). The eutectic composition and temperature are 24 mol.% CdTe and 405°C, respectively. SnTe primarily crystallizes within the interval of 0–7 mol.% CdTe. The Cu$_2$CdSnTe$_4$ quaternary compound is formed in this system. It melts incongruently at 470°C (the peritectic composition is 17 mol.% CdTe) and crystallizes in the monoclinic structure with the lattice parameters $a = 425.6 \pm 0.3$, $b = 427.7 \pm 0.3$, $c = 1042.1 \pm 0.4$, and $\beta = 119.35° \pm 0.05°$ [in the tetragonal structure with the lattice parameters $a = 619.8 \pm 0.1$ and $c = 1225.6 \pm 0.3$ pm (Haeuseler et al. 1991)] and experimental density 6.14 g · cm^{-3} (Olekseyuk and Piskach 1997).

The solubility of CdTe in Cu$_2$SnTe$_3$ and Cu$_2$SnTe$_3$ in CdTe at 350°C is equal to 12 and 4 mol.%, respectively (Olekseyuk and Piskach 1997). Matsuhita et al. (2005) noted that Cu$_2$CdSnTe$_4$ was not synthesized and had the mixed phases of CdTe and Cu$_2$SnTe$_3$.

This system was investigated by DTA, XRD, metallography, and measuring of microhardness and density (Olekseyuk and Piskach 1997, Piskach et al. 1997).

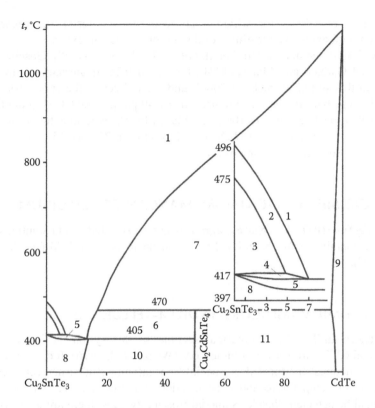

FIGURE 6.8 Phase relations in the CdTe–Cu$_2$SnTe$_3$ system: 1, L; 2, L + SnTe; 3, L + Cu$_3$Te$_2$; 4, L + (Cu$_2$SnTe$_3$) + Cu$_3$Te$_2$; 5, L + (Cu$_2$SnTe$_3$); 6, L + Cu$_2$CdSnTe$_4$; 7, L + (CdTe); 8, (Cu$_2$SnTe$_3$); 9, (CdTe); 10, (Cu$_2$SnTe$_3$) + Cu$_2$CdSnTe$_4$; and 11, Cu$_2$CdSnTe$_4$ + (CdTe). (From Olekseyuk, I.D. and Piskach, L.V., *Zhurn. neorgan. khimii*, 42(2), 331, 1997.)

6.23 CADMIUM–SILVER–GALLIUM–TELLURIUM

CdTe–AgGaTe$_2$. The phase diagram is not constructed. According to the data of Parthé et al. (1969), the AgCd$_2$GaTe$_4$ quaternary compound is formed in this system. It crystallizes in the cubic structure of sphalerite type with the lattice parameter a = 637.5 ± 0.5 pm, but this compound can be one of the solid solution compositions.

6.24 CADMIUM–SILVER–INDIUM–TELLURIUM

CdTe–AgInTe$_2$. The phase diagram is not constructed. One phase with the sphalerite structure exists in this system within the interval of 0–50 mol.% AgInTe$_2$ (annealing temperature is 600°C) (Woolley and Williams 1966). The structure becomes chalcopyrite at the AgInTe$_2$ contents more than 60 mol.%. It was indicated (Gallardo 1992) that it is possible to predict the value at which the chalcopyrite phase in this system disappears (x_c) and $T_c(x)$ behavior for the

$Cd_{2x}(AgIn)_{1-x}Te_2$ alloys for which a chalcopyrite–sphalerite-like disordering transition with extensive terminal solid solution formation exists.

According to the data of Hirai et al. (1967), Parthé et al. (1969), Guseinov et al. (1972), and Garbato and Manca (1974), the $AgCd_2InTe_4$ quaternary compound is formed in this system. It melts at 700°C and crystallizes in the cubic structure of sphalerite type with the lattice parameter $a = 1149$ pm [$a = 643.8 \pm 0.5$ pm (Parthé et al. 1969); $a = 645 \pm 7$ pm (Garbato and Manca 1974)], experimental density 6.269 $g \cdot cm^{-3}$, and energy gap $E_g = 0.47$ eV (Guseinov et al. 1972), but this compound can be one of the solid solution compositions.

6.25 CADMIUM–MAGNESIUM–MANGANESE–TELLURIUM

CdTe–MgTe–MnTe. The phase diagram is not constructed. Solid solutions based on CdTe contain up to 25 mol.% MgTe and up to 50 mol.% MnTe (Averkieva et al. 1992).

6.26 CADMIUM–MERCURY–INDIUM–TELLURIUM

CdTe–HgTe–In$_2$Te$_3$. The isothermal sections of this system at 330°C, 430°C, 530°C, and 630°C are shown in Figure 6.9 (Weitze et al. 1996). It can be seen that ordered structures are formed at special concentrations of structural vacancies. Depending on temperature and concentration, continuous order–disorder transitions could be detected. Besides spinodal miscibility gaps, there are also structural miscibility gaps and several three-phase regions. For $t < 450$°C, one can observe an ordered phase with a narrow existence region. This phase with stoichiometry $(Cd_{1-y}Hg_y)_5In_2Te_8$ is separated from the surrounding phase by a miscibility gap that increases with decreasing temperature.

Interdiffusion experiments along the quasibinary section $(Cd_{3(1-m-n)}Hg_{3m}In_{2n}V_n)Te_3$ have shown that the structural vacancies (V) determine the diffusion properties of such alloys in different ways (von Wensierski et al. 1997). At low vacancy densities, for example, at 0.01 mol.% In_2Te_3, the structural vacancies behave like usual defects introduced by doping. At higher vacancy densities, the diffusion coefficient increases much less than the concentration of the structural vacancies because a great part of them are bound in associates. At special stoichiometric conditions ($n = 3/8$ and 3/4), the structural vacancies take strictly ordered positions and behave no more as defects, but as interstitial sites of an ordered superstructure of the zinc blende lattice. Near the points of transition from the ordered phases to the more or less disordered alloys, the diffusion coefficient increases continuously.

The $CdTe–HgTe–In_2Te_3$ quasiternary system was investigated by DTA, XRD, and electron probe microanalysis, and the annealing temperatures of the mixtures ranged between 330°C and 630°C and annealing times between 30 and 280 days (Weitze et al. 1996).

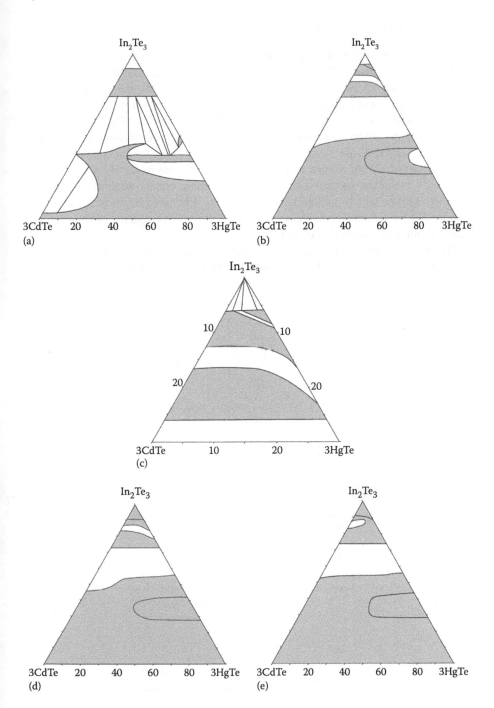

FIGURE 6.9 Isothermal sections of the CdTe–HgTe–In$_2$Te$_3$ quasiternary system at (a) 330°C, (b, c) 430°C, (d) 530°C, and (e) 630°C. (From Weitze, D. et al., *J. Alloys Compd.*, 239(2), 117, 1996.)

6.27 CADMIUM–MERCURY–OXYGEN–TELLURIUM

The predominance areas of the condensed phases in equilibrium with the gas phase in the system Cd–Hg–O–Te at 330°C, 430°C, and 530°C are shown in Figure 6.10 (Diehl and Nolaeng 1984). The left parts of these diagrams are occupied by the predominance area of $Cd_xHg_{1-x}Te$ solid solutions. Depending on the cadmium activity in the system, $Cd_xHg_{1-x}Te$ has phase boundaries with CdO, $CdTeO_3$, $CdTe_2O_5$, or TeO_2. For higher and medium values of a_{Cd}, the phase relationships are similar to those in the Cd–O–Te system, and for low values of a_{Cd}, the sequence of oxide phases with increasing oxygen activity coincides with that in the Hg–O–Te system for a $a_{Hg} \approx 1$. For the special value range of the independent variables, Cd_3TeO_6 is the only tellurate with a predominance area in the diagrams. Regardless of the a_{Cd}, reactions of the various oxide phases with a gas phase of increasing oxygen activity finally result in the formation of HgO.

Of particular interest for the dry oxidation of $Cd_xHg_{1-x}Te$ solid solutions is the situation around the point $Cd_xHg_{1-x}Te$–CdO–$CdTeO_3$ where three condensed phases

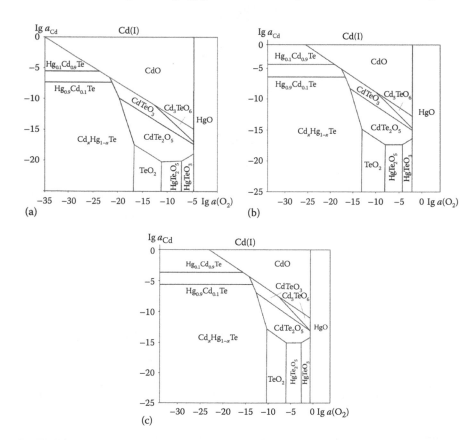

FIGURE 6.10 Predominance areas of the condensed phases in equilibrium with the gas phase in the system Cd–Hg–O–Te in saturated mercury vapor at (a) 330°C, (b) 430°C, and (c) 530°C. (From Diehl, R. and Nolaeng, B.I., *J. Cryst. Growth*, 66(1), 91, 1984.)

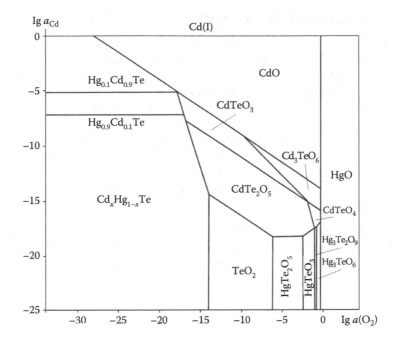

FIGURE 6.11 Predominance areas of the condensed phases in equilibrium with the gas phase in the system Cd–Hg–O–Te at 430°C and $a_{Hg} = 0.1$. (From Diehl, R. and Nolaeng, B.I., *J. Cryst. Growth*, 66(1), 91, 1984.)

coexist with the gas phase. Solid solutions with x up to around 0.6 react with O_2 to form CdO, whereas solid solutions with x in excess of 0.6 are in equilibrium with $CdTeO_3$.

Slight differences in the oxidation of $Cd_xHg_{1-x}Te$ are noticeable in a gas phase with reduced a_{Hg} (Figure 6.11) (Diehl and Nolaeng 1984). All isocompositional lines in the $Cd_xHg_{1-x}Te$ predominance area ranging from $x = 0.9$ to 0.1 meet the left-hand boundary of the $CdTeO_3$ predominance area, that is, oxidation of $Cd_xHg_{1-x}Te$ in the specified compositional range yields $CdTeO_3$ as the only oxide phase in equilibrium with $Cd_xHg_{1-x}Te$. A further difference in comparison to Figure 6.10b is the appearance of small predominance areas of the tellurates $CdTeO_4$, $Hg_3Te_2O_9$, and Hg_3TeO_6.

A quaternary diagram has been constructed by Rhiger and Kvaas (1983) to describe thermodynamic equilibrium relationships among the low-temperature condensed phases of the Cd–Hg–O–Te system. Important inferences can be drawn from this diagram based on the stability relationships involving elemental Hg and $CdTeO_3$. Because Hg coexists with all of the oxides and tellurides, it will be present in most situations where oxidation is incomplete. In actual oxide films, Hg will probably be the last element to oxidize and the first to be reduced if oxygen is lost. Similarly, $CdTeO_3$ can coexist in equilibrium with all of the other oxides and tellurides. Thus, $CdTeO_3$ is probably the first oxide to form, and it remains stable.

According to the data of Brandt and Moritz (1985), the $Cd_3Te_2O_9$ structure is stabilized by 30 at.% Hg. The Cd–Hg–O–Te system was investigated using thermodynamic equilibrium simulations (Rhiger and Kvaas 1983, Diehl and Nolaeng 1984).

6.28   CADMIUM–MERCURY–IODINE–TELLURIUM

CdTe–HgI$_2$. This section is a nonquasibinary section of the Cd–Hg–I–Te quaternary system (Figure 6.12) (Liahovitskaya et al. 1986). Depending on binary compounds ratio, the stable phases are CdI$_2$, Cd$_{1-x}$Hg$_x$Te, Hg$_3$Te$_2$I$_2$, and solid solutions based on α-HgI$_2$. The temperature of α-β-transformation of the solid solution based on HgI$_2$ sharply increases at the CdTe contents increasing. The solubility of CdTe in β-HgI$_2$ is equal to 2.5 ± 0.5 mol.%, while the solubility of CdTe in α-HgI$_2$ is much more.

Three four-phase nonvariant equilibria exist in this section (Liahovitskaya et al. 1986): α-HgI$_2$ + β-HgI$_2$ + Hg$_3$Te$_2$I$_2$ + L at 243°C, α-HgI$_2$ + CdI$_2$ + Hg$_3$Te$_2$I$_2$ + L at 303°C, and HgTe + CdI$_2$ + Hg$_3$Te$_2$I$_2$ + L at 345°C.

This system was investigated by DTA, XRD, and x-ray fluorescence analysis (Liahovitskaya et al. 1986).

CdTe + HgI$_2$ ⇔ HgTe + CdI$_2$. The isothermal section of this system at room temperature is shown in Figure 6.13 (Liahovitskaya et al. 1986). The HgTe–CdI$_2$ and CdI$_2$–Hg$_3$Te$_2$I$_2$ are the stable sections in this ternary mutual system.

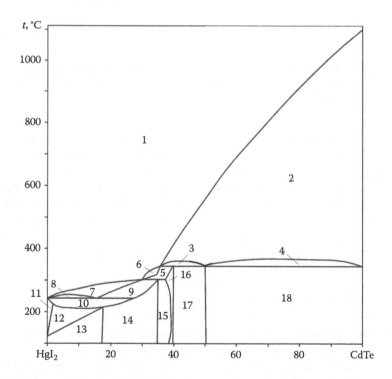

FIGURE 6.12   Phase relations in the CdTe–HgI$_2$ system: 1, L; 2, L + Cd$_{1-x}$Hg$_x$Te; 3, L + Hg$_3$Te$_2$I$_2$; 4, L + CdI$_2$ + Cd$_{1-x}$Hg$_x$Te; 5, L + CdI$_2$ + Hg$_3$Te$_2$I$_2$; 6, L + Hg$_3$Te$_2$I$_2$; 7, L + (α-HgI$_2$); 8, L + (α-HgI$_2$) + (β-HgI$_2$); 9, L + (α-HgI$_2$) + Hg$_3$Te$_2$I$_2$; 10, (α-HgI$_2$) + (β-HgI$_2$) + Hg$_3$Te$_2$I$_2$; 11, (β-HgI$_2$); 12, (α-HgI$_2$) + (β-HgI$_2$); 13, (α-HgI$_2$); 14, (α-HgI$_2$) + Hg$_3$Te$_2$I$_2$; 15, (α-HgI$_2$) + Hg$_3$Te$_2$I$_2$ + CdI$_2$; 16, Hg$_3$Te$_2$I$_2$ + CdI$_2$; 17, HgTe + Hg$_3$Te$_2$I$_2$ + CdI$_2$; and 18, CdI$_2$ + Cd$_{1-x}$Hg$_x$Te. (From Liahovitskaya, V.A. et al., *Zhurn. neorgan. khimii*, 31(4), 1020, 1986.)

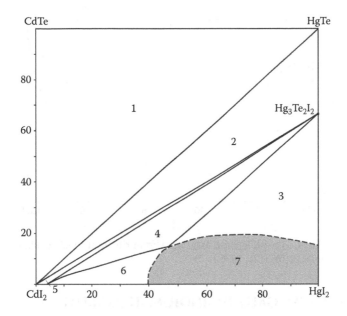

FIGURE 6.13 Isothermal section of the CdTe + HgI$_2$ ⇔ HgTe + CdI$_2$ ternary mutual system at room temperature: 1, CdI$_2$ + Cd$_{1-x}$Hg$_x$Te; 2, HgTe + CdI$_2$ + Hg$_3$Te$_2$I$_2$; 3, (α-HgI$_2$) + Hg$_3$Te$_2$I$_2$; 4, CdI$_2$ + (α-HgI$_2$) + Hg$_3$Te$_2$I$_2$; 5, CdI$_2$ + Hg$_3$Te$_2$I$_2$; 6, CdI$_2$ + (α-HgI$_2$); and 7, (α-HgI$_2$). (From Liahovitskaya, V.A. et al., *Zhurn. neorgan. khimii*, 31(4), 1020, 1986.)

6.29 CADMIUM–MERCURY–MANGANESE–TELLURIUM

CdTe–HgTe–MnTe. The phase diagram is not constructed. Solid solutions with the sphalerite structure are formed in most parts of this quasiternary system (Manhas et al. 1987). A two-phase region exists only from the MnTe-rich side and contains up to 20–25 mol.% (CdTe + HgTe). The Cd$_x$Hg$_y$Mn$_z$Te forming solid solutions have the lattice parameter $a = (648.1 - 1.9y - 14.4z)$ pm and energy gap $E_g = (1.46 - 1.62y + 1.33z)$ eV. This system was investigated by XRD and the ingots were annealed at 550°C for 1 month.

6.30 CADMIUM–ALUMINUM–ANTIMONY–TELLURIUM

CdTe–AlSb. The phase diagram is shown in Figure 6.14 (Kuz'mina 1976). Solid solutions over the entire range of concentration are formed in this system. However, there were determined thermal effects on the heating and cooling curves at nearly 450°C, which can correspond to the melting of indium antimonide. This assumption is confirmed by the data of XRD.

Nonequilibrium state of forming solid solutions exists at low temperatures (Kuz'mina and Khabarov 1969). Solid solutions based on CdTe crystallize in the hexagonal structure at high rates of crystallization. Homogeneous solid solutions were obtained by the high-temperature annealing with their next quenching or by the quenching from the temperatures closed to crystallization temperatures. This system was investigated by DTA and XRD (Kuz'mina 1976).

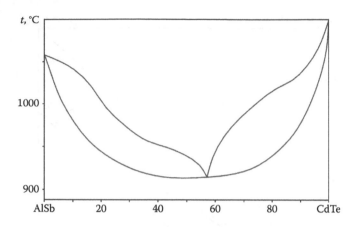

FIGURE 6.14   The CdTe–AlSb phase diagram. (From Kuz'mina, G.A. et al., *Izv. AN SSSR. Neorgan. materialy*, 12(6), 1121, 1976.)

6.31   CADMIUM–GALLIUM–INDIUM–TELLURIUM

CdTe–Ga$_2$Te$_3$–In$_2$Te$_3$. Figure 6.15 shows isothermal sections of this quasiternary system at 430°C, 530°C, 580°C, and 630°C (Leute and Bolwin 2001). It shows the calculated equilibrium boundaries of homogeneous phases and the calculated tie line fields of miscibility gaps. Characteristic for this system is that an ordering tendency near 25 mol.% 3CdTe and a spinodal miscibility gap around 80 mol.% 3CdTe occur simultaneously in the temperature region between 330°C and 620°C. Below 620°C, a spinodal miscibility gap extends from the 3CdTe–Ga$_2$Te$_3$ edge into the ternary region. For 586°C, this gap overlaps with the structural miscibility gap between the tetragonal ordered phase of the chalcopyrite type and the cubic disordered phase of the zinc blende type causing the formation of a three-phase region of two cubic regions and one tetragonal phase. With decreasing temperature, this three-phase region moves from the 3CdTe–Ga$_2$Te$_3$ system to the 3CdTe–In$_2$Te$_3$ system. At 430°C (Figure 6.15a), there remains only a very thin region of a cubic solid solution along the CdTe-rich part of the 3CdTe–In$_2$Te$_3$ system. The CdTe–Ga$_2$Te$_3$–In$_2$Te$_3$ system was investigated by DTA, XRD, and EPMA.

6.32   CADMIUM–GALLIUM–ARSENIC–TELLURIUM

CdTe–GaAs. The phase diagram is not constructed. The solubility of CdTe in GaAs is equal to 5 mol.% (Voitsehovski et al. 1970) and the solubility of GaAs in CdTe reaches 15 mol.% (Glazov et al. 1975b). Forming solid solutions crystallize in the cubic structure of sphalerite type.

This system was investigated by metallography (Voitsehovski et al. 1970, Glazov et al. 1975b). The ingots were obtained by melting mixtures of chemical elements in stoichiometric ratio at 1250°C (Voitsehovski et al. 1970).

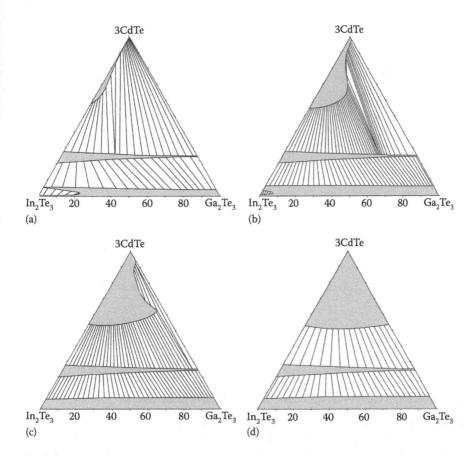

FIGURE 6.15 Isothermal sections of the CdTe–Ga$_2$Te$_3$–In$_2$Te$_3$ quasiternary system at (a) 430°C, (b) 530°C, (c) 580°C, and (d) 630°C. (From Leute, V. and Bolwin, H. et al., *Solid State Ionics*, 141–142, 279, 2001.)

6.33 CADMIUM–GALLIUM–ANTIMONY–TELLURIUM

CdTe–GaSb. The phase diagram is a eutectic type (Figure 6.16) (Glazov et al. 1975d). The eutectic composition and temperature are 95 mol.% GaSb and 670°C, respectively. The solubility of GaSb in CdTe is not higher than 10 mol.% (Glazov et al. 1975c,d) [solubility reaches 15 mol.% (Burdiyan and Makeichik 1966, Kirovskaya et al. 2007a); 20 mol.% (Burdiyan 1970, 1971)] and the solubility of CdTe in GaSb is equal to 10 mol.% (Kirovskaya et al. 2007a). Energy gap of forming solid solutions changes near linearly with composition (Burdiyan 1970, 1971).

This system was investigated by DTA, metallography, and measuring of micro-hardness (Glazov et al. 1975c,d).

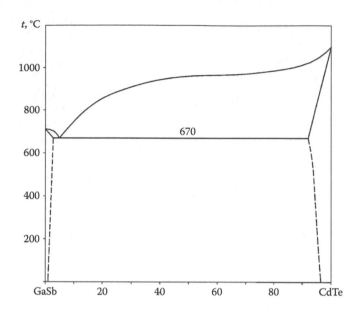

FIGURE 6.16   The CdTe–GaSb phase diagram. (From Glazov, V.M. et al., *Izv. AN SSSR. Neorgan. materialy*, 11(3), 418, 1975d.)

6.34   CADMIUM–GALLIUM–BROMINE–TELLURIUM

CdTe–Ga$_2$Te$_3$–GaBr$_3$. The phase diagram is not constructed. The Cd$_2$GaTe$_3$Br quaternary compound is formed in this system (Range and Handrick 1988). It crystallizes in the cubic structure with the lattice parameter a = 628.36 ± 0.03 pm. This compound was obtained at 600°C–1100°C by the interaction of CdTe, CdBr$_2$, and Ga$_2$Te$_3$.

6.35   CADMIUM–GALLIUM–IODINE–TELLURIUM

CdTe–Ga$_2$Te$_3$–GaI$_3$. The phase diagram is not constructed. The Cd$_2$GaTe$_3$I quaternary compound is formed in this system (Range and Handrick 1988). It crystallizes in the cubic structure with the lattice parameter a = 627.94 ± 0.05 pm. This compound was obtained at 600°C–1100°C by the interaction of CdTe, CdI$_2$, and Ga$_2$Te$_3$.

6.36   CADMIUM–INDIUM–PHOSPHORUS–TELLURIUM

CdTe–InP. The phase diagram is a eutectic type (Figure 6.17) (Glazov et al. 1972, 1973, 1975a). The eutectic composition and temperature are 50 mol.% InP and 410°C ± 10°C, respectively. The solubility of CdTe in InP is equal to 3 mol.% and the solubility of InP in CdTe reaches 20–25 mol.%. This system was investigated by DTA and metallography.

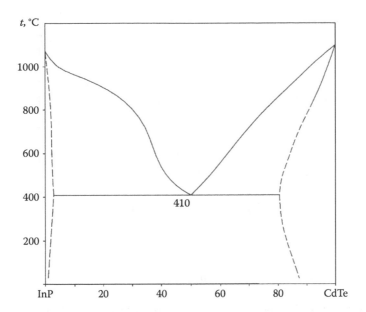

FIGURE 6.17 The CdTe–InP phase diagram. (From Glazov, V.M. et al., *Izv. AN SSSR. Neorgan. materialy*, 9(11), 1883, 1973.)

6.37 CADMIUM–INDIUM–ARSENIC–TELLURIUM

CdTe–InAs. The phase diagram is a eutectic type (Figure 6.18) (Morozov and Chernov 1979). The eutectic composition and temperature are 33 ± 2 mol.% CdTe and $874°C \pm 3°C$ (Buzevich 1972, Bazhenova et al. 1974, Balagurova et al. 1975, Morozov and Chernov 1979) [882°C (Drobyazko and Kuznetsova 1983; 870°C (Stuckes and Chasmar 1964)], respectively. The solubility of CdTe in InAs at the eutectic temperature reaches 30 mol.% (Voitsehovski and Goriunova 1962, Kleshchinski et al. 1964, Bazhenova et al. 1974, Balagurova et al. 1975, 1976) [26 mol.% (Drobyazko and Kuznetsova 1983)] and the solubility of InAs in CdTe at the same temperature is equal to 5 mol.% (Balagurova et al. 1975, 1976) [35 mol.% (Drobyazko and Kuznetsova 1983)].

Temperature dependence of CdTe solubility in InAs within the interval of 780°C–874°C is satisfactorily described by the following equation (Balagurova et al. 1975): $\ln x_{CdTe} = -13{,}080/T + 9.02$. The solubility of CdTe in InAs saturated by In is given in Table 6.1 (Buzevich 1972). Within the interval of 0–1 mol.% CdTe, InTe and CdAs complexes include in crystal matrix, which are formed as the result of the following reaction: $CdTe + InAs \Leftrightarrow CdAs + InTe$. Thus, this system is nonquasibinary in the region of small CdTe contents and turns into InTe–InAs–CdAs system.

The coefficient of CdTe distribution in the CdTe–InAs solid solutions obtained by the Bridgman method is equal to 0.25 (Kalashnikova et al. 1975).

This system was investigated by DTA, XRD, emission spectrum analysis, and measuring of microhardness (Buzevich 1972, Bazhenova et al. 1974, Balagurova et al. 1975,

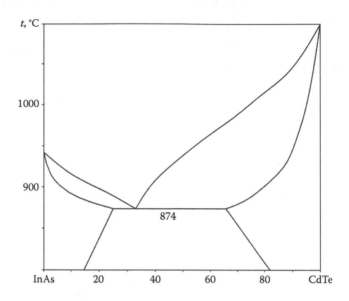

FIGURE 6.18 The CdTe–InAs phase diagram. (From Morozov, V.N. and Chernov, V.G., *Izv. AN SSSR. Neorgan. materialy*, 15(8), 1324, 1979.)

TABLE 6.1
Solubility of CdTe in InAs Saturated by In

t, °C	x_{CdTe}, mol.%	t, °C	x_{CdTe}, mol.%
400	1.47	550	7.90
450	2.75	600	13.50
500	4.30	650	16.00

Source: Buzevich, G.I., *Avtoref. dis. ... kand. fiz.-mat. nauk*, Irkutsk, Russia, 1972, 17 pp.

1976, Morozov and Chernov 1979, Drobyazko and Kuznetsova 1983). The ingots were annealing at 830°C, 780°C, and 730°C during 120, 240, and 480 h, respectively (Drobyazko and Kuznetsova 1983).

CdTe–In–InAs. The liquidus surface of this quasiternary system (Figure 6.19) was constructed using DTA and mathematical simulation of experiment (Riazantsev and Telegina 1978).

6.38 CADMIUM–INDIUM–ANTIMONY–TELLURIUM

CdTe–InSb. The phase diagram is a eutectic type (Figure 6.20) (Morozov et al. 1974). The eutectic composition and temperature are 6 ± 1 mol.% CdTe and 510°C ± 1°C, respectively. The solubility of CdTe in InSb reaches 5 mol.% (Goriunova et al. 1961, Khabarov and Sharavski 1963, 1964, Khabarova et al. 1963, Iniutkin et al. 1964, Morozov et al. 1974, Kirovskaya and Mironova 2006, 2007,

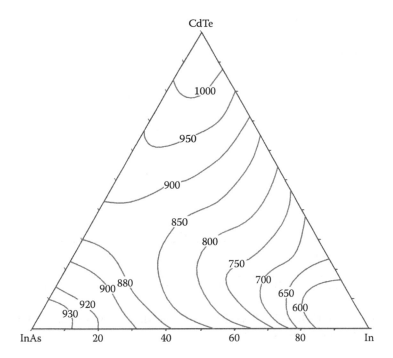

FIGURE 6.19 Liquidus surface of the CdTe–In–InAs quasiternary system. (From Riazantsev, A.A. and Telegina, M.P., *Zhurn. neorgan. khimii*, 23(8), 2211, 1978.)

Kirovskaya et al. 2007b) and the solubility of InSb in CdTe at eutectic temperature is equal to 6 mol.% and decreases to 3 mol.% at 730°C (Riazantsev et al. 1980, Kirovskaya and Mironova 2006, 2007). The coefficient of CdTe distribution in the CdTe–InAs solid solutions obtained by the Bridgman method is equal to 0.3 (Kalashnikova et al. 1975).

Some maximums exist on the dependence of the lattice parameter from the composition within the interval of 0–5 mol.% CdTe (Brodovoy et al. 1997). An opinion was expressed about the formation of donor–acceptor complexes in the solid solutions based on InSb. The band gap of solid solution based on InSb is characterized by a minimum (0.17 eV) at 3 mol.% CdTe (Kirovskaya and Mironova 2006, Kirovskaya et al. 2007b).

This system was investigated by DTA, XRD, metallography, and measuring of microhardness (Goriunova et al. 1961, Khabarov and Sharavski 1963, 1964, Khabarova et al. 1963, Iniutkin et al. 1964, Morozov et al. 1974, Kirovskaya and Mironova 2006, Riazantsev et al. 1980, Kirovskaya et al. 2007b). The ingots were annealed at 510°C for 350 h (Morozov et al. 1974). The melts were homogenized at 1100°C for 100 h and then the ingots were annealed at 500°C for 100–150 h (Brodovoy et al. 1997).

CdTe–Cd–InSb. The liquidus surface of this quasiternary system (Figure 6.21) was constructed using DTA and mathematical simulation of experiment (Riazantsev et al. 1980).

CdTe–In–InSb. The liquidus surface of this quasiternary system (Figure 6.22) was constructed using DTA and mathematical simulation of experiment (Riazantsev et al. 1980).

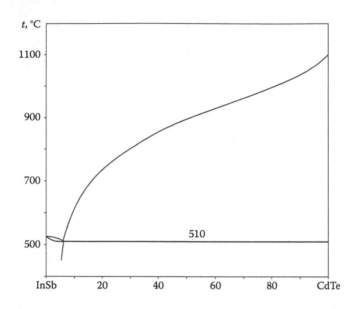

FIGURE 6.20 The CdTe–InSb phase diagram. (From Morozov, V.N. et al., *Izv. Sib. otd. AN SSSR. Ser. khim. nauk*, 4(9), 52, 1974.)

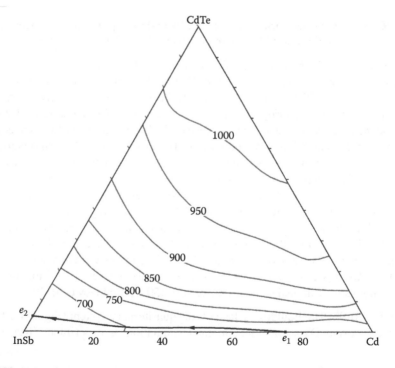

FIGURE 6.21 Liquidus surface of the CdTe–Cd–InSb quasiternary system. (From Riazantsev, A.A. and Telegina, M.P., *Zhurn. neorgan. khimii*, 23(8), 2211, 1978.)

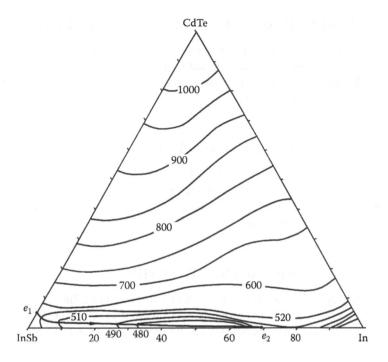

FIGURE 6.22 Liquidus surface of the CdTe–In–InSb quasiternary system. (From Riazantsev, A.A. and Telegina, M.P., *Zhurn. neorgan. khimii*, 23(8), 2211, 1978.)

6.39 CADMIUM–INDIUM–BROMINE–TELLURIUM

CdTe–In$_2$Te$_3$–InBr$_3$. The phase diagram is not constructed. The Cd$_2$InTe$_3$Br quaternary compound is formed in this system (Range and Handrick 1988). It crystallizes in the defect zinc blende structure with the lattice parameter $a = 632.17 \pm 0.07$ pm. It was determined that this compound after annealing at 1400°C under 4.0 GPa (40 kbar) crystallizes in the defect cubic structure of NaCl type. This compound was obtained at 600°C–1100°C by the interaction of CdTe, CdBr$_2$, and In$_2$Te$_3$.

6.40 CADMIUM–INDIUM–IODINE–TELLURIUM

CdTe–In$_2$Te$_3$–InI$_3$. The phase diagram is not constructed. The Cd$_2$InTe$_3$I quaternary compound is formed in this system (Range and Handrick 1988). It crystallizes in the defect zinc blende structure with the lattice parameter $a = 634.59 \pm 0.04$ pm. It was determined that this compound after annealing at 1400°C under 4.0 GPa (40 kbar) crystallizes in the defect cubic structure of NaCl type with the lattice parameter $a = 615.1 \pm 0.1$ pm. This compound was obtained at 600°C–1100°C by the interaction of CdTe, CdI$_2$, and In$_2$Te$_3$.

6.41   CADMIUM–THALLIUM–GERMANIUM–TELLURIUM

$CdTl_2GeTe_4$ quaternary compound is formed in this system. It crystallizes in the tetragonal structure with the lattice parameters $a = 838.25 \pm 0.19$ and $c = 707.75 \pm 0.18$ pm (McGuire et al. 2005). This compound was synthesized by the reaction of the elements in 1:2:1:4 molar ratio.

6.42   CADMIUM–THALLIUM–TIN–TELLURIUM

$CdTl_2SnTe_4$ quaternary compound is formed in this system. It crystallizes in the tetragonal structure with the lattice parameters $a = 842.50 \pm 0.04$ and $c = 721.71 \pm 0.05$ pm (McGuire et al. 2005). This compound was synthesized by the reaction of the elements in 1:2:1:4 molar ratio.

6.43   CADMIUM–TIN–ARSENIC–TELLURIUM

CdTe–CdSnAs₂. The phase diagram is not constructed. The solubility of $2CdTe$ in $CdSnAs_2$ is equal to 6 mol.% (Dovletmuradov et al. 1971, 1972). This system was investigated by DTA, XRD, metallography, and measuring of microhardness.

6.44   CADMIUM–LEAD–IODINE–TELLURIUM

CdTe–PbI₂. The phase diagram is a eutectic type (Figure 6.23) (Odin 2001). The eutectic composition and temperature are 15 ± 1 mol.% CdTe and 384°C, respectively. Mutual solubility of CdSe and PbI_2 is negligible: the solubility of PbI_2 in CdTe is equal to 0.2 mol.%.

The crystallization of the melts from the PbI_2 side leads to the formation of $2H-PbI_2$ metastable phase (Odin 2001, Odin and Chukichev 2001). This system was investigated by DTA, XRD, and metallography, and the ingots were annealed during 1000 h (Odin 2001).

CdTe + PbI₂ ⇔ CdI₂ + PbTe. The fields of primary crystallization of the solid solutions based on CdSe and PbTe and $Cd_xPb_{1-x}I_2$ solid solutions exist on the liquidus surface of this ternary mutual system (Figure 6.24) (Odin 2001). The ternary eutectic E at 368°C is formed by the solid solutions based on CdTe and PbTe and $Cd_xPb_{1-x}I_2$ solid solutions. A minimum m exists on the e_4e_2 line of the secondary crystallization.

The isothermal section of this ternary mutual system at 350°C is analogous to one of the CdSe + PbI₂ ⇔ CdI₂+PbSe ternary mutual system (Figure 5.52) (Odin 2001).

Metastable phases of $2H-PbI_2$ and $6R-PbI_2$ can form at the crystallization of the melts in the CdTe–PbTe–PbI₂ quasiternary system in the region enriched in PbI_2 (Odin and Chukichev 2001).

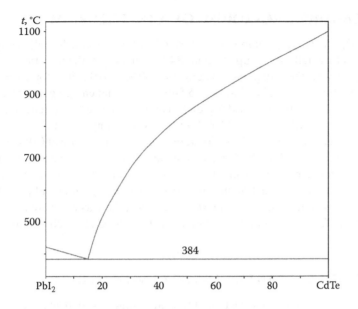

FIGURE 6.23 The CdTe–PbI$_2$ phase diagram. (From Odin, I.N., *Zhurn. neorgan. khimii*, 46(10), 1733, 2001.)

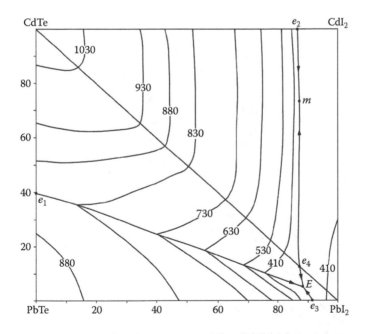

FIGURE 6.24 Liquidus surface of the CdTe + PbI$_2$ ⇔ CdI$_2$ + PbTe ternary mutual system. (From Odin, I.N., *Zhurn. neorgan. khimii*, 46(10), 1733, 2001.)

6.45   CADMIUM–VANADIUM–OXYGEN–TELLURIUM

$Cd_4V_2Te_3O_{15}$ quaternary compound is formed in this system. It melts congruently at 751°C, is thermally stable up to about 840°C, and crystallizes in the orthorhombic structure with the lattice parameters $a = 539.93 \pm 0.04$, $b = 1604.8 \pm 0.1$, and $c = 1623.5 \pm 0.1$ pm; calculated density 5.545 g·cm^{-3}; and energy gap 2.66 eV (Jiang et al. 2008). The synthesis of $Cd_4V_2Te_3O_{15}$ can be expressed by a reaction $4CdO + V_2O_5 + 3TeO_2 = Cd_4V_2Te_3O_{15}$ at 610°C. Pure powder samples of $Cd_4V_2Te_3O_{15}$ could be prepared quantitatively by the solid-state reaction of a mixture of $CdO/V_2O_5/TeO_2$ in a molar ratio of 4:1:3 at 610°C for 6 days. Its single crystals (needle in shape and light yellow in color) were obtained using CdO (1.1 mmol), V_2O_5 (0.55 mmol), and TeO_2 (0.55 mmol). The reaction mixture was thoroughly ground and pressed into a pellet, which was then put into a quartz tube. This tube was evacuated, sealed, and heated at 720°C for 6 days and was then cooled to 320°C at 4°C/h before switching off the furnace.

6.46   CADMIUM–TANTALUM–OXYGEN–TELLURIUM

CdTe–Ta–O. The isothermal section of this quasiternary system at 25°C is shown in Figure 6.25 (Cordes and Schmid-Fetzer 1995). All system compositions on this section are composed only of the phases present on this section. Solid-state equilibria in this system were calculated thermodynamically.

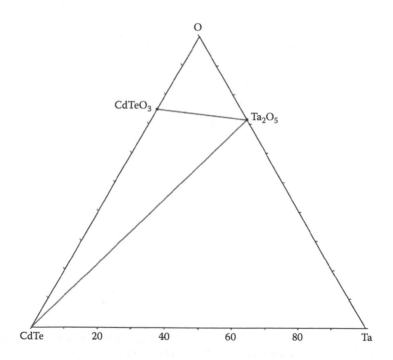

FIGURE 6.25   Isothermal section of the CdTe–Ta–O quasiternary system at 25°C. (From Cordes, H. and Schmid-Fetzer, R., *Z. Metallk.*, 86(5), 304, 1995.)

6.47 CADMIUM–OXYGEN–MOLYBDENUM–TELLURIUM

$CdMoTeO_6$ quaternary compound is formed in this system. It melts congruently at approximately 766°C [at 760°C (Forzatti and Trifirò 1977, Forzatti and Tieghi 1978)] and crystallizes in the tetragonal structure with the lattice parameters $a = 528.60 \pm 0.07$ and $c = 906.60 \pm 0.18$ pm (Zhao et al. 2013) [$a = 528.40 \pm 0.01$ and $c = 905.95 \pm 0.02$ pm (Laligant 2001); $a = 527.9 \pm 0.2$ and $c = 905.6 \pm 0.2$ pm, calculated and experimental densities 5.682 and 5.59 $g \cdot cm^{-3}$, respectively (Forzatti and Tieghi 1978)]. Calculated and experimental energy gaps of $CdMoTeO_6$ are equal to 2.91 and 3.59 eV, respectively (Zhao et al. 2013). According to the data of Botto and Baran (1980), this compound crystallizes in the orthorhombic structure with the lattice parameters $a = 553 \pm 1$, $b = 510 \pm 1$, and $c = 908 \pm 1$ pm and calculated and experimental densities 5.60 and 5.5 $g \cdot cm^{-3}$, respectively.

Polycrystalline samples of $CdMoTeO_6$ were synthesized by the solid-state reaction techniques. The starting chemicals were CdO, TeO_2, and MoO_3 or $CdMoO_4$ and TeO_2 or $CdMoO_4$ and H_6TeO_6. The equimolar mixture of CdO, TeO_2, and MoO_3 was thoroughly ground, slowly heated to 550°C at a rate of 20°C/h, and then sintered at this temperature for 72 h with several intermediate grindings (Zhao et al. 2013). $CdMoTeO_6$ could be also prepared from a mixture of $CdMoO_4$ and TeO_2 in the ratio 1:1, which was mixed in an agate mortar (Forzatti and Tieghi 1978, Cannizzaro et al. 1981, Laligant 2001). The mixture was placed in an alumina crucible and heated progressively to 600°C [to 500°C (Forzatti and Trifirò 1977)] for 20 h and then slowly cooled to room temperature. This compound could be obtained at 425°C when orthorhombic TeO_2 was used, while using tetragonal TeO_2, the synthesis temperature increased up to 470°C and up to 490°C when $CdMoO_4$ interacts with H_6TeO_6 (Forzatti and Tieghi 1978).

Single crystals of this compound have been grown from its stoichiometric melts by a spontaneous nucleation approach (Forzatti and Tieghi 1978, Zhao et al. 2013).

CdTe–O–Mo. The isothermal section of this quasiternary system at 25°C is shown in Figure 6.26 (Cordes and Schmid-Fetzer 1995). All system compositions on this section are composed only of the phases present on this section. Solid-state equilibria in this system were calculated thermodynamically.

6.48 CADMIUM–OXYGEN–TUNGSTEN–TELLURIUM

CdTe–O–W. The isothermal section of this quasiternary system at 25°C is shown in Figure 6.26 (Cordes and Schmid-Fetzer 1995). All system compositions on this section are composed only of the phases present on this section. Solid-state equilibria in this system were calculated thermodynamically.

6.49 CADMIUM–OXYGEN–MANGANESE–TELLURIUM

CdTe–MnTe–O. The phase diagram is not constructed. According to the data of thermodynamic calculations, an oxidation of the $Cd_xMn_{1-x}Te$ solid solutions begins from the oxidation of MnTe with the formation of MnO and Te, and then CdTe oxidized to form $CdTeO_3$ (Medvedev and Berchenko 1994).

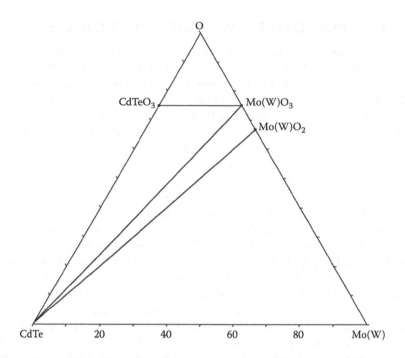

FIGURE 6.26 Isothermal section of the CdTe–O–Mo(W) quasiternary system at 25°C. (From Cordes, H. and Schmid-Fetzer, R., *Z. Metallk.*, 86(5), 304, 1995.)

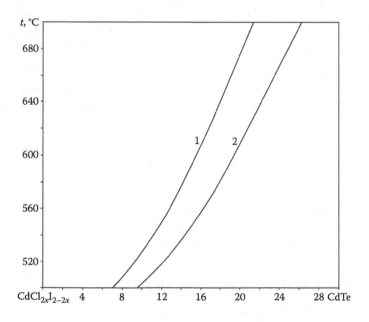

FIGURE 6.27 Temperature dependence of CdTe solubility in the $CdCl_{2x}I_{2-2x}$ melts: 1, 31 and 2, 60 mol.% $CdCl_2$. (From Polistanski, G.D. et al., *Deposited in "Tsvetmetinformatsia"*, № 442Dep. M. 1978.)

6.50 CADMIUM–CHLORINE–IODINE–TELLURIUM

$CdTe–CdCl_2–CdI_2$. The phase diagram is not constructed. The solubility of CdTe in the (31 mol.% $CdCl_2$ + 69 mol.% CdI_2) and (60 mol.% $CdCl_2$ + 40 mol.% CdI_2) melts was determined within the interval of 500°C–700°C (Figure 6.27) (Polistanski et al. 1978). The time of the melt saturation by CdTe is not higher than 3 h. The solubility was determined by the filtration method.

REFERENCES

Aresti A., Garbato L., Manca P., Rucci A. Phase diagram of the $(CuGaTe_2)_{1-x}$–$(2CdTe)_x$ system and semiconducting properties of $CuCd_2GaTe_4$, *J. Electrochem. Soc.*, **124**(5), 766–768 (1977).

Averkieva G.K., Boiko M.E., Konstantinova N.N., Popova T.B., Prochuhan V.D., Rud' Yu.V. Investigation of solid solutions in the CdTe–MnTe–MgTe system [in Russian], *Fiz. tv. tela*, **34**(7), 2284–2286 (1992).

Balagurova E.A., Bazhenova G.I., Riazantsev A.A., Khabarov E.N. *T-x*-projection of InAs–CdTe phase diagram [in Russian]. In: *Protsessy rosta i sinteza poluprovodnikovyh kristallov i plenok.* Novosibirsk, Russia: Nauka Publish., Pt. 2, pp. 236–239 (1975).

Balagurova E.A., Morozov V.N., Khabarov E.N. Microhardness dependence of the solid solutions crystals from components interaction nature [in Russian], *Izv. AN SSSR. Neorgan. materialy*, **12**(1), 105–107 (1976).

Bazhenova G.I., Balagurova E.A., Riazantsev A.A., Khabarov E.N. Solid solutions in the InAs–CdTe system [in Russian], *Izv. AN SSSR. Neorgan. materialy*, **10**(10), 1770–1773 (1974).

Bochmann M., Bwembya G., Webb K.J., Malik M.A., Walsh J.R., O'Brien P. Arene chalcogenolato complexes of zinc and cadmium, *Inorg. Synthes.*, **31**, 19–24 (1997).

Bochmann M., Coleman A.P., Webb K.J., Hursthouse M.B., Mazid M. Synthese und Eigenschaften sterisch anspruchsvoller Tellurophenole; Struktur von $[Cd(\mu\text{-}TeC_6H_2Me_3)_2]_\infty$, *Angew. Chem.*, **103**(8), 975–976 (1991a).

Bochmann M., Coleman A.P., Webb K.J., Hursthouse M.B., Mazid M. Synthesis of sterically hindered tellurophenols and the structure of $[Cd(\mu\text{-}TeC_6H_2Me_3)_2]_\infty$, *Angew. Chem. Int. Ed. Engl.*, **30**(8), 973–975 (1991b).

Botto I.L., Baran E.J. Röntgenographische und spektroskopische Untersuchung einiger Telluromolybdate, *Z. anorg. und allg. Chem.*, **468**(1), 221–227 (1980).

Brandt G., Moritz R. Compound formation in the mercury–cadmium–tellurium–oxygen system, *Mater. Res. Bull.*, 1985, **20**(1), 49–56.

Brodovoy V.A., Vialyi N.G., Knorozok L.M., Markiv I.Ya. Peculiarities of lattice parameter changing of the $(ZnSb)_{1-x}(CdTe)_x$ solid solutions [in Russian], *Neorgan. materialy*, **33**(3), 303–304 (1997).

Burdiyan I.I. The solid solutions of gallium antimonide with cadmium, zinc and mercury tellurides [in Russian]. In: *Nekot. voprosy khimii i fiz. poluprovodn. slozhn. sostava.* Uzhgorod, Ukraine: Uzhgorod. Un-t Publish., pp. 190–195 (1970).

Burdiyan I.I. Investigation of solid solutions of GaSb with cadmium, zinc and mercury tellurides [in Russian], *Izv. AN SSSR. Neorgan. materialy*, **7**(3), 414–416 (1971).

Burdiyan I.I., Makeichik A.I. Solid solutions in the GaSb–CdTe system [in Russian], *Uchen. zap. Tirasp. ped. in-ta*, **16**, 125–126 (1966).

Buzevich G.I. Investigation of growth and physical properties of epitaxial layers of InAs–CdTe solid solutions [in Russian], *Avtoref. dis. ... kand. fiz.-mat. nauk*, Irkutsk, Russia, 17pp., (1972).

Cannizzaro M., Forzatti P., Tittarelli P. Synthesis of $CdTeWO_6$ and its catalytic behaviour in the allylic oxidation of propylene, *Gazz. Chim. Ital.*, **111**, 469–472 (1981).

Cherniavski V.P. Solid solutions of the mCuInTe$_2$–(1–m)2CdTe system [in Russian]. In: *Fizika*. L.: Publish. of Leningr. inzh.-stroit. in-t, pp. 10–12 (1962).

Cherniavski V.P., Goriunova N.A., Borshchevski A.S. About some properties of the CuInTe$_2$–CdTe alloys of the Cu–Cd–In–Te system [in Russian]. In: *Khim. sviaz' v kristallah*. Minsk, Bielorussia: Nauka i tehnika Publish., pp. 413–416 (1969).

Cordes H., Schmid-Fetzer R. Thermochemically stable metal contacts on CdTe: Tungsten, molybdenum and tantalum, *Z. Metallk.*, **86**(5), 304–311 (1995).

Diehl R., Nolaeng B.I. Dry oxidation of Hg$_{1-x}$Cd$_x$Te: Calculation of predominance area diagrams of the oxide phases, *J. Cryst. Growth*, **66**(1), 91–105 (1984).

Dovletmuradov Ch., Dovletov K., Krzhivitskaya S.N., Mamaev S., Allanazarov A., Ashirov A. Obtaining and investigation of physico-chemical and electrical properties of CdSnAs$_2$–A^{II}B^{VI} (2CdSe, 2CdTe) solid solutions [in Russian]. In: *Troin. poluprovodniki A^{II}B^{IV}C$_2^V$ i A^{II}B$_2^{III}$C$_4^{VI}$*. Kishinev, Republic of Moldova: Shtiintsa Publish., pp. 96–98 (1972).

Dovletmuradov Ch., Krzhivitskaya S.N., Dovletov K., Mamaev S., Allanazarov A., Ashirov A. Some properties of CdSnAs$_2$–A^{II}B^{VI} (CdSe, CdTe) solid solutions [in Russian], *Izv. AN TurkmSSR. Ser. fiz.-tehn., khim. i geol. nauk*, (5), 111–114 (1971).

Dovletov K.O., Nuryev S., Hanberdiev Ya., Garagulov O., Ragimov F.D., Aleksanian S.N. Obtaining and electrophysical properties of the (CdTe)$_{1-x}$(CuGaTe$_2$)$_x$ solid solutions [in Russian], *Izv. AN SSSR. Neorgan. materialy*, **25**(7), 1218–1220 (1989).

Drobyazko V.P., Kuznetsova S.T. Interaction of indium arsenide with cadmium chalcogenides [in Russian], *Zhurn. neorgan. khimii*, **28**(11), 2929–2933 (1983).

Forzatti P., Tieghi G. Solid state reactions to CdTeMoO$_6$ and its structural characterization, *J. Solid State Chem.*, **25**(4), 387–390 (1978).

Forzatti P., Trifirò F. Synthesis and characterization of cadmium tellurium molybdate, *Gazz. Chim. Ital.*, **107**(1–2), 35–37 (1977).

Gallardo P.G. Order-disorder phase transitions in D$_{2x}^{II}$(A^{I}B^{III})$_{1-x}$C$_2^{VI}$ alloys systems, *Phys. Status Solidi (a)*, **134**(1), 119–125 (1992).

Garbato L., Manca P. Synthesis and characterization of some chalcogenides of A^{I}B$_2^{II}$C^{III}D$_4^{VI}$, *Mater. Res. Bull.*, **9**(4), 511–517 (1974).

Glazov V.M., Krestovnikov A.N., Nagiev V.A., Rzaev F.R. Investigation of phase equilibria in the InP–ZnTe and InP–CdTe quasibinary system [in Russian], *Elektron. tehnika. Ser. 6: Materialy*, (4), 127–129 (1972).

Glazov V.M., Krestovnikov A.N., Nagiev V.A., Rzaev F.R. Phase equilibria in the InP–ZnTe and InP–CdTe quasibinary system [in Russian], *Izv. AN SSSR. Neorgan. materialy*, **9**(11), 1883–1889 (1973).

Glazov V.M., Krestovnikov A.N., Nagiev V.A., Rzaev F.R. Investigation of phase equilibrium and analysis of intermolecular interaction in the quasibinary systems [in Russian]. In: *Termodin. svoistva metallicheskih splavov*. Baku, Azerbaijan: Elm Publish., pp. 380–386 (1975a).

Glazov V.M., Pavlova L.M., Griazeva N.L. Investigation of phase equilibria and analysis of intermolecular interaction in the GaSb–Zn(Cd)Te quasibinary systems [in Russian]. In: *Termodinamicheskiye svoistva metallicheskih splavov*. Baku, Azerbaijan: Elm Publish., pp. 368–371 (1975c).

Glazov V.M., Pavlova L.M., Griazeva N.L. Phase equilibria and analysis of intermolecular interaction in the GaSb–Zn(Cd)Te quasibinary systems [in Russian], *Izv. AN SSSR. Neorgan. materialy*, **11**(3), 418–423 (1975d).

Glazov V.M., Pavlova L.M., Lebedeva L.V. Thermodynamic analysis of interaction of gallium arsenide with zinc and cadmium tellurides [in Russian]. In: *Termodinamicheskiye svoistva metallicheskih splavov*. Baku, Azerbaijan: Elm Publish., pp. 372–375 (1975b).

Goriunova N.A., Averkieva G.K., Sharavski P.V., Tovpentsev Yu.K. Investigation of quaternary alloys based on indium antimonide and cadmium telluride [in Russian]. In: *Fizika i khimia*. L.: Publish. of Leningr. inzh.-stroit. in-t, p. 10 (1961).

Guseinov G.D., Godzhaev E.M., Halilov H.Ya., Seidov F.M., Pashaev A.M. Complex semiconductor chalcogenides [in Russian], *Izv. AN SSSR. Neorgan. materialy*, **8**(9), 1569–1572 (1972).

Haeuseler H., Ohrendori F., Himmrich M. Zur Kenntnis quaternärer Telluride $Cu_2MM'Te_4$ mit Tetrahederstrukturen, *Z. Naturforsch.*, **B46**(8), 1049–1052 (1991).

Hirai T., Kurata K., Takeda Y. Deviation of new semiconducting compounds by cross substitution for group IV semiconductors and their semiconducting and thermal properties, *Solid State Electron.*, **10**(10), 975–981 (1967).

Iniutkin A., Kolosov E., Osnach L., Habarova V., Habarov E., Sharavski P. Some investigations of solid solutions based on $A^{III}B^V$ and $A^{II}B^{VI}$ compounds [in Russian], *Izv. AN SSSR. Ser. fiz.*, **28**(6), 1010–1016 (1964).

Jiang H.-L., Huang S.-P., Fan Y., Mao J.-G., Cheng W.-D. Explorations of new types of second-order nonlinear optical materials in $Cd(Zn)$-V^V-Te^{IV}-O systems, *Chem. Eur. J.*, **14**(6), 1972–1981 (2008).

Kadykalo E.M., Marushko L.P., Ivashchenko I.A., Zmiy O.F., Olekseyuk I.D. Quasi-ternary system Cu_2Te–$CdTe$–In_2Te_3, *J. Phase Equilib. Diff.*, **34**(3), 221–228 (2013).

Kalashnikova E.V., Korzhov V.I., Morozov V.N., Petrov A.A., Semikolenova N.A., Skorobogatova L.A., Finogenova V.K., Khabarova V.A., Khabarov E.N. Obtaining of the $A^{III}B^V$–$A^{II}B^{VI}$ solid solutions closed to intrinsic [in Russian]. In: *Processy rosta i sinteza poluprovodnikovyh kristallov i plenok*. Novosibirsk, Russia: Nauka Publish., Pt. 2, pp. 232–236 (1975).

Khabarova V.A., Khabarov E.N., Sharavski P.V. Determination of CdTe solubility limit in InSb [in Russian], *Izv. vuzov. Ser. Fizika*, (6), 62–64 (1963).

Khabarov E.N., Sharavski P.V. About intcratomic constraint forces in the InSb–CdTe solid solutions [in Russian]. In: *Fizika*. L.: Publish. of Leningr. inzh.-stroit. in-t, p. 31 (1963).

Khabarov E.N., Sharavski P.V. Investigation of properties of InSb·CdTe limited solid solutions [in Russian], *Dokl. AN SSSR*, **155**(3), 542–544 (1964).

Kirovskaya I.A., Mironova E.V. Obtaining and identification of substitutional solid solutions of the system InSb–CdTe [in Russian], *Zhurn. neorgan. khimii*, **51**(4), 701–705 (2006).

Kirovskaya I.A., Mironova E.V. Synthesis and properties of new materials—Solid solutions $(InSb)_x(CdTe)_{1-x}$ [in Russian], *Dokl. AN Vysshey shkoly Rosii*, (1), 34–43 (2007).

Kirovskaya I.A., Novgorodtseva L.V., Vasina M.V., Baranovskaya M.V., Chkalova A.L., Kuznetsova I.Yu. New materials based on the GaSb–$A^{II}B^{VI}$ systems [in Russian], *Sovrem. naukoemk. tekhnol.*, (6), 96–97 (2007a).

Kirovskaya I.A., Shubenkova E.G., Mironova E.V., Rud'ko T.A., Bykova E.I. Obtaining and properties of the solid solutions of the InSb–$A^{II}B^{VI}$ system [in Russian], *Sovremen. naukoemk. tekhnol.*, (8), 31–33 (2007b).

Kleshchinski L.I., Habarov E.N., Sharavski P.V. Determination of solid solution limits in the InAs–CdTe system [in Russian]. In: *Fizika*. L.: Publish. of Leningr. inzh.-stroit. in-t, pp. 12–15 (1964).

Kuz'mina G.A. Investigation of the AlSb–CdTe and InAs–HgTe phase diagram [in Russian], *Izv. AN SSSR. Neorgan. materialy*, **12**(6), 1121–1122 (1976).

Kuz'mina G.A., Khabarov E.N. Obtaining of the solid solutions in the AlSb–CdTe system [in Russian], *Izv. AN SSSR. Neorgan. materialy*, **5**(1), 30–32 (1969).

Laligant Y. X-ray and TEM studies of $CdTeMoO_6$ and $CoTeMoO_6$: A new superstructure of fluorite type with cation and anion deficiencies ($\blacksquare CoTeMo)(\square_2O_6)$, *J. Solid State Chem.*, **160**(2), 401–408 (2001).

Leonov V.V., Chunarev E.N. Distribution of potassium and zinc chlorides at the crystallization of eutectics of cadmium chalcogenides with its chloride [in Russian]. In: *Physico-khim. processy v geterogennyh systemah*. Krasnoyarsk, Russia: Krasnoyar. Un-t Publish., pp. 59–64 (1977).

Leute V., Bolwin H. Ordering and demixing in the system $In_2Te_3/Ga_2Te_3/Cd_3Te_3$, *Solid State Ionics*, **141–142**, 279–287 (2001).

Li H., Malliakas C.D., Peters J.A., Liu Z., Im J., Jin H., Morris C.D. et al. CsCdInQ$_3$ (Q = Se, Te): New photoconductive compounds as potential materials for hard radiation detection, *Chem. Mater.*, **25**(10), 2089–2099 (2013).

Liahovitskaya V.A., Zhmurova Z.I., Fedorov P.P. Interaction of mercury iodide with cadmium telluride in the condensed phases [in Russian], *Zhurn. neorgan. khimii*, **31**(4), 1020–1023 (1986).

Liu Y., Chen L., Wu L.-M., Chan G.H., Van Duyne R.P. Syntheses, crystal and band structures and magnetic and optical properties of new CsLnCdTe$_3$ (Ln = La, Pr, Nd, Sm, Gd-Tm and Lu), *Inorg. Chem.*, **47**(3), 855–862 (2008).

Manhas S., Khulbe K.C., Beckett D.J.S., Lamarche G., Woolley J.C. Lattice parameters, energy gaps, and magnetic properties of the Cd$_x$Hg$_y$Mn$_z$Te alloy system, *Phys. Status Solidi (b)*, **143**(1), 267–274 (1987).

Matsuhita H., Ichikawa T., Katsui A. Structural, thermodynamic and optical properties of Cu$_2$–II–IV–VI$_4$ quaternary compounds. *J. Mater. Sci.*, **40**(8), 2003–2005 (2005).

McGuire M.A., Scheidemantel T.J., Badding J.V., DiSalvo F.J. Tl$_2$AXTe$_4$ (A = Cd, Hg, Mn; X = Ge, Sn): Crystal structure, electronic structure, and thermoelectric properties, *Chem. Mater.*, **17**(24), 6186–6191 (2005).

Medvedev Yu.V., Berchenko N.N. Analysis of the phase compositions at the interface of Mn$_{1-x}$A$_x^{II}$B^{VI}(A^{II} = Zn, Cd, Hg; B^{VI} = Te, Se) solid solution with own oxides [in Russian], *Zhurn. neorgan. khimii*, **39**(5), 846–848 (1994).

Morozov V.N., Chernov V.G. Phase equilibria in the InAs–CdTe system [in Russian], *Izv. AN SSSR. Neorgan. materialy*, **15**(8), 1324–1329 (1979).

Morozov V.N., Karnauhova E.N., Skorobogatova L.A., Riazantsev A.A. Apparatus for differential thermal analysis and investigation of InSb–CdTe phase diagram [in Russian], *Izv. Sib. otd. AN SSSR. Ser. khim. nauk*, (4[9]), 52–56 (1974).

Odin I.N. *T-x-y* diagrams for mutual systems PbX + CdI$_2$ = CdX + PbI$_2$ (X = S, Se, Te) [in Russian], *Zhurn. neorgan. khimii*, **46**(10), 1733–1738 (2001).

Odin I.N., Chukichev M.V. Metastable phases crystallizing from the melts in the PbX + CdI$_2$ = CdX + PbI$_2$ (X = S, Se, Te) mutual systems [in Russian], *Zhurn. neorgan. khimii*, **46**(12), 2083–2087 (2001).

Olekseyuk I.D., Piskach L.V. Phase equilibria in the Cu$_2$SnX$_3$–CdX (X = S, Se, Te) systems [in Russian], *Zhurn. neorgan. khimii*, **42**(2), 331–333 (1997).

Olekseyuk I.D., Piskach L.V., Parasiuk O.V. Phase equilibria in the Cu$_2$SiSe$_3$(Te$_3$)–CdSe(Te) systems [in Russian], *Zhurn. neorgan. khimii*, **43**(3), 516–519 (1998).

Olekseyuk I.D., Piskach L.V., Sysa L.V. The Cu$_2$GeTe$_3$–CdTe system and structure of the Cu$_2$CdGeTe$_4$ compound [in Russian], *Zhurn. neorgan. khimii*, **41**(9), 1420–1422 (1996).

Parthé E., Yvon K., Deitch R.H. The crystal structure of Cu$_2$CdGeS$_4$ and other quaternary normal tetrahedral structure compounds, *Acta Crystallogr. B*, **25**(6), 1164–1174 (1969).

Piskach L.V., Olekseyuk I.D., Parasyuk O.V. Physico–chemical peculiarities of the Cu$_2$CdC^{IV}X$_4$ (C^{IV} – Si, Ge, Sn; X – S, Se, Te) quaternary phase formations [in Ukrainian], *Fizyka kondens. vysokomolek. system. Nauk. zap. Rinens'kogo pedinstytutu*, (3), 153–157 (1997).

Piskach L.V., Zmiy O.F., Olekseyuk I.D., Mokraya I.R. Phase equilibria in the Cu$_2$GeTe$_3$–CdTe system [in Russian], *Vestn. L'vov. un-ta. Ser. khim.*, (29), 40–42 (1988).

Polistanski G.D., Ivanov Yu.M., Vaniukov A.V., Surkov E.P. Investigation of solubility and cadmium telluride growth in the salt melts [in Russian], *Deposited in "Tsvetmetinformatsia"*, № 442Dep. M. (1978).

Range K.-J., Handrick K. Synthese und Hochdruckverhalten quaternärer Chalkogenidhalogenide M$_2$M'X$_3$Y (M = Zn, Cd; M' = Al, Ga, In; X = Se, Te; Y = Cl, Br, I), *Z. Naturforsch.*, **B43**(2), 153–158 (1988).

Rhiger D.R., Kvaas R.E. Solid state quaternary phase equilibrium diagram for the Hg–Cd–Te–O system, *J. Vac. Sci. Technol.*, **A1**(3), 1712–1718 (1983).

Riazantsev A.A., Karnauhova E.N., Kuz'mina G.A. Phase diagrams and solubility of components in the A^{3}B^5–CdTe systems [in Russian], *Zhurn. neorgan. khimii*, **25**(3), 802–805 (1980).

Riazantsev A.A., Telegina M.P. Liquidus surfaces and thermodynamic analysis of InSb–CdTe–In, InSb–CdTe–Cd and InAs–CdTe–In phase diagrams [in Russian], *Zhurn. neorgan. khimii*, **23**(8), 2211–2216 (1978).

Schwer H., Keller E., Krämer V. Crystal structure and twinning of KCd$_4$Ga$_5$S$_{12}$ and some isotypic AB$_4$C$_5$X$_{12}$ compounds, *Z. Kristallogr.*, **204**(Pt. 2), 203–213 (1993).

Steigerwald M.L., Sprinkle C.R. Organometallic synthesis of II-VI semiconductors. 1. Formation and decomposition of bis(organotelluro)mercury and bis(organotelluro) cadmium compounds, *J. Am. Chem. Soc.*, **109**(23), 7200–7201 (1987).

Stuckes A.D., Chasmar R.P. Electrical and thermal properties of alloys of InAs and CdTe, *J. Phys. Chem. Sol.*, **25**(5), 469–476 (1964).

Voitsehivski A.V. Some four-component semiconductor phases [in Ukrainian], *Ukr. fiz. zhurn.*, **9**(7), 796–797 (1964).

Voitsehovski A.V., Goriunova N.A. Solid solutions in some quaternary semiconductor systems [in Russian]. In: *Fizika*. L.: Publish. of Leningr. inzh.-stroit. in-t, pp. 12–14 (1962).

Voitsehovski A.V., Pashun A.D., Mitiurev V.K. About interaction of gallium arsenide with II–VI compounds [in Russian], *Izv. AN SSSR. Neorgan. materialy*, **6**(2), 379–380 (1970).

von Wensierski H., Weitze D., Leute V. Ordering and diffusion in II-VI/III-VI alloys with structural vacancies, *Solid State Ionics*, **101–103**(Pt. 1), 479–487 (1997).

Weitze D., Schmidtke H.M., Leute V. Ordering phenomena and demixing in the quasiternary system HgTe/CdTe/In$_2$Te$_3$, *J. Alloys Compd.*, **239**(2), 117–123 (1996).

Woolley J.C., Williams E. W. Some cross-substitutional alloys of CdTe, *J. Electrochem. Soc.*, **113**(9), 899–901 (1966).

Zhao S., Jiang X., He R., Zhang S.-Q., Sun Z., Luo J., Lin Z., Hong M. A combination of multiple chromophores enhances second-harmonic generation in a nonpolar noncentrosymmetric oxide: CdTeMoO$_6$, *J. Mater. Chem. C*, **1**(16), 2906–2912 (2013).

7 Systems Based on HgS

7.1 MERCURY–HYDROGEN–CARBON–SULFUR

Some compounds are formed in this quaternary system.

MeHgSMe or C$_2$H$_6$SHg, methyl(methylthio)mercury. This compound melts at 24°C–25°C (Nyquist and Mann 1972, Bach and Weibel 1976). To obtain this compound, chloromethylmercury (2.5 g) was dissolved in absolute EtOH, and an equivalent amount of NaOH (2 mL of a 2 N solution) was added with stirring (Nyquist and Mann 1972). The solution was heated to the boiling point (~78°C), allowed to cool to room temperature, and finally cooled in an ice bath. The cold solution was rapidly filtered to remove the precipitated NaCl. The filtered solution was maintained at 0°C until a slight excess of methanethiol was added with stirring. The reaction mixture was allowed to come to room temperature at which it remained for several hours. As water was slowly added to the methanol, turbidity was obtained, and upon cooling, C$_2$H$_6$SHg precipitated from solution. This was filtered using an ice-cooled, sintered glass funnel. The precipitate was recrystallized from an McOH–H$_2$O solvent giving large white platelets.

This compound could be also obtained if methylmercaptan (39 mmol) was condensed into a small tube immersed in dry ice and subsequently distilled through a gas bubbler into a solution of 36 mmol of MeHgCl in 19 mL of 2 N aqueous NaOH (38 mmol) and 60 mL of absolute MeOH (Bach and Weibel 1976). The mixture was allowed to stand overnight at 0°C, affording a white precipitate, which was a clear viscous oil at room temperature. Water (~10 mL) was added to the mixture, which then was extracted with pentane (4 × 50 mL). Evaporation of pentane from combined extracts afforded C$_2$H$_6$HgS. Further purification was achieved by high-vacuum distillation.

(HgMe)$_2$S or C$_2$H$_6$SHg$_2$, bis(methylmercury) sulfide. This compound melts at 141°C–143°C with decomposition (Bach and Weibel 1976). The crude product was recrystallized from benzene to give a white solid.

Hg(MeS)$_2$ or C$_2$H$_6$S$_2$Hg, bis(methanethiolato)mercury. This compound melts at 127°C–138°C [at 185°C (Bradley and Kunchur 1964), at 175°C–178°C (Wertheim 1929, Biscarini et al. 1974)] with decomposition (Canty and Tyson 1978) and crystallizes in the orthorhombic structure with the lattice parameters a = 1980.0 ± 0.5, b = 758 ± 2, and c = 780 ± 2 pm and experimental density 3.34 g·cm^{-3} (Bradley and Kunchur 1964).

To obtain C$_2$H$_6$S$_2$Hg, imidazole (C$_3$H$_4$N$_2$) (0.39 mmol) in H$_2$O (5 mL) was added to a solution of MeSHgO$_2$CMe (0.39 mmol) in H$_2$O (10 mL). A white powder that formed immediately was collected, washed with water, and dried in vacuo over P$_2$O$_5$ (Canty and Tyson 1978). This compound can be also obtained by adding six drops of CH_3SH to about 2 mL of EtOH, subsequently adding an excess of mercuric cyanide solution (10%) (Wertheim 1929). After shaking and cooling the mixture for a few minutes, the resulting precipitates were separated by filtration and dried in the air for a short time on a porous plate. They were purified by recrystallization from the ethanol. Bis(methanthiolato)mercury was prepared also by the reaction of slight excess of CH_3SH with a solution of $HgCl_2$ in absolute EtOH and was recrystallized from EtOH (Bowmaker et al. 1984).

$Hg(EtS)_2$ or $C_4H_{10}S_2Hg$, bis(ethanthiolato)mercury. This compound melts at 72°C–73°C (Biscarini et al. 1974) [at 76°C (Wertheim 1929, Bradley and Kunchur 1965)] and crystallizes in the monoclinic structure with the lattice parameters $a = 734.2 \pm 0.2$, $b = 470.2 \pm 0.2$, $c = 2313.1 \pm 0.6$ pm, and $\beta = 101.83° \pm 0.01°$ and experimental density 2.74 g·cm^{-3} (Fraser et al. 1995)[$a = 754 \pm 2$, $b = 487 \pm 1$, $c = 2380 \pm 4$ pm, and $\beta = 85°$ and calculation and experimental densities 2.45 and 2.47 g·cm^{-3}, respectively (Bradley and Kunchur 1965)]. $C_4H_{10}S_2Hg$ was recrystallized from EtOH to produce very thin flaky plates. Good-quality single crystals were obtained by very slow evaporation of a saturated ethanolic solution at room temperature (Fraser et al. 1995). This compound can be also obtained by adding six drops of C_2H_5SH to about 2 mL of EtOH, subsequently adding an excess of mercuric cyanide solution (10%) (Wertheim 1929). After shaking and cooling the mixture for a few minutes, the resulting precipitates were separated by filtration and dried in the air for a short time on a porous plate. They were purified by recrystallization from the ethanol.

$HgMeBu^tS$ or $C_5H_{12}SHg$, methylmercury t-butylmercaptide. This compound melts at 41°C–42°C. To obtain $C_5H_{12}SHg$, methylmercury chloride (10 mmol) and t-butylmercaptan (10 mmol) were dissolved in a methanolic solution containing 5 mL of 3 M aqueous NaOH (Bach and Weibel 1976). Distilled water (~15 mL) was added until a lasting cloudiness developed. After the mercurial solution was extracted with pentane (2 × 25 mL), 20 mL of water and 15 mL of 3 N NaOH were added to the aqueous MeOH and it was again extracted with pentane (4 × 25 mL). The pentane was removed, affording a crude solid, which on sublimation (25°C, 5.3 kPa) afforded a white granular solid.

$HgBu^tMeS$ or $C_5H_{12}SHg$, t-butylmercury methylmercaptide. This compound melts at 56°C–57°C. It was prepared from 2 mmol of Me_3CHgCl, 1 mL of 3 N NaOH, and 0.13 mmol of MeSH (2.4 mmol) in 40 mL of MeOH at −78°C (Bach and Weibel 1976). The reaction mixture was allowed to stir for 30 min at 25°C and the product was recovered by extraction into pentane. The pentane was dried ($MgSO_4$) and concentrated, and on standing at 0°C, large clear prisms were obtained.

$Hg(Pr^nS)_2$ and $Hg(Pr^iS)_2$ or $C_6H_{14}S_2Hg$, bis(n-propylthiolato)mercury and bis(i-propylthiolato)mercury. $Hg(Pr^nS)_2$ melts at 67°C–69°C (Biscarini et al. 1974) [at 71°C–72°C (Wertheim 1929)] and $Hg(Pr^iS)_2$ melts at 62°C–63°C (Wertheim 1929, Biscarini et al. 1974). These compounds can be obtained by adding six drops of n-C_3H_7SH or i-C_3H_7SH to about 2 mL of EtOH, subsequently adding an excess of mercuric cyanide solution (10%) (Wertheim 1929). After shaking and cooling the mixture for a few minutes, the resulting precipitates were separated by filtration and dried in the air for a short time on a porous plate. They were purified by recrystallization from the ethanol.

HgMeSPh or C_7H_8SHg, (benzenethiolato)methylmercury (methylmercury phenylmercaptide). This compound that was obtained for the first time by Märcker (1865) melts at 91°C–92°C (Sytsma and Kline 1973, Bach and Weibel 1976) [at 83°C (Canty and Kishimoto 1977)]. To obtain C_7H_8SHg, a solution of 8.8 mmol of MeHgCl in 50 mL of MeOH was combined with equivalent amount of aqueous 2 N NaOH (4.4. mL) (Bach and Weibel 1976). After stirring for 1 h, the solution was filtered and 8.8 mL of PhSH was added to the filtrate. On addition of mercaptan, a white precipitate was formed immediately. The solution was stored at 5°C for 2 h and then filtered. Recrystallization from methylene chloride afforded white needles.

C_7H_8SHg could be also prepared if a solution containing methylmercury nitrate (0.5 mmol), thiophenol (0.5 mmol), and triethylamine (0.5 mmol) in MeOH (10 mL) was stirred for 2 h at room temperature (Block et al. 1990). The solution was filtered and the filtrate concentrated to about 90% of its original volume. After several days at ~20°C, colorless needles were collected and washed with distilled H_2O to yield a colorless crystalline product.

For preparing C_7H_8SHg, it is possible to use an aqueous solution of methylmercury hydroxide prepared from methylmercury iodide and excess amount of freshly precipitated silver oxide (Sytsma and Kline 1973). The methylmercury hydroxide solution, from which AgI and excess Ag_2O have been removed by filtration, was added to a methanolic solution of thiophenol. The crude product was purified by the recrystallization in water/acetone.

HgMeC$_6$H$_{11}$S or $C_7H_{14}SHg$, methylcyclohexylmercaptomercury. This compound melts at 65°C. To obtain it, methylmercury acetate was dissolved in water and added to cyclohexylmercaptan dissolved in MeOH or methanolic KOH solution. The crude product was purified by the recrystallization in water/acetone (Sytsma and Kline 1973).

HgMe(p-MePhS) and HgMe(o-MePhS) or $C_8H_{10}SHg$, methyl(p-methylthiophenolato)mercury and methyl(o-methylthiophenolato)mercury. HgMe(p-MePhS) and HgMe(o-MePhS) melt at 75°C. To obtain these compounds, methylmercury acetate was dissolved in water and added to p-methylthiophenol or p-methylthiophenol dissolved in MeOH or methanolic KOH solution. The crude products were purified by the recrystallization in hexane (Sytsma and Kline 1973).

HgEtPhS or $C_8H_{10}SHg$, ethanthiolato(phenyl)mercury. This compound melts at 50°C (Canty and Kishimoto 1977). Crystals of $C_8H_{10}SHg$ were obtained by recrystallization from EtOH/chloroform.

MeHgPhCH$_2$S or $C_8H_{10}SHg$, methylmercury benzylmercaptide. To obtain this compound, 2.0 mmol of PhCH$_2$SH was added to a solution of MeHgCl (2.0 mmol) and 0.08 g of NaOH in 30 mL of MeOH (Bach and Weibel 1976). After stirring for

15 min, 10 mL of H_2O was added and the product was extracted with three 30 mL portions of pentane. The removal of solvent afforded crude product as clear dense oil.

Bu^{t}HgBu^{t}S or C$_8$H$_{18}$SHg, t-butylmercury t-butylmercaptide. This compound melts at 56°C–57°C. It was prepared if t-BuHgCl (2.0 mmol) was dissolved in 30 mL of MeOH and 4 mL of aqueous 1 N NaOH (4.0 mmol) (Bach and Weibel 1976). t-Butylmercaptan (2.2 mmol) was added and the mixture was stoppered and stirred for 1.5 h. After the crystals had begun to form, the solution was stored overnight in the refrigerator. The crude product was collected by filtration and recrystallized from ethyl ether to give a white solid.

Hg(Bu^{n}S)$_2$, Hg(Bu^{i}S)$_2$, and Hg(Bu^{t}S)$_2$ or C$_8$H$_{18}$S$_2$Hg, bis(n-butylthiolato)mercury, bis(i-butylthiolato)mercury, and bis(2-methylpropane-2-thiolato)mercury. Hg(Bu^{n}S)$_2$ melts at 81°C–83°C (Biscarini et al. 1974) [at 85°C–86°C (Wertheim 1929)], Hg(Bu^{i}S)$_2$ melts at 90°C–91°C (Biscarini et al. 1974) [at 94°C–95°C (Wertheim 1929)], and Hg(Bu^{t}S)$_2$ melts at 157°C–159°C (Biscarini et al. 1974). Hg(Bu^{n}S)$_2$ could be prepared electrochemically using a simple cell: Pt(–)/CH$_3$CN + n-C$_4$H$_9$SH/Hg(+) (Said and Tuck 1982). The anode was a pool of Hg in contact with a Pt wire sealed through the glass wall. When the electrolysis began, a white solid was formed at the anode almost immediately. The resulted solid was collected, washed several times with acetonitrile and petroleum ether, and dried. This compound is air stable. Hg(Bu^{n}S)$_2$ and Hg(Bu^{i}S)$_2$ can be also obtained by adding six drops of n-C$_4$H$_9$SH or i-C$_4$H$_9$SH to about 2 mL of EtOH, subsequently adding an excess of mercuric cyanide solution (10%) (Wertheim 1929). After shaking and cooling the mixture for a few minutes, the resulting precipitates were separated by filtration and dried in the air for a short time on a porous plate. They were purified by recrystallization from the ethanol. Hg(Bu^{t}S)$_2$ was prepared by the reaction of slight excess of t-C$_4$H$_9$SH with a solution of HgCl$_2$ in absolute EtOH and was recrystallized from EtOH (Bowmaker et al. 1984).

HgPr^{i}PhS or C$_9$H$_{12}$SHg, i-propanethiolato(phenyl)mercury. This compound melts at 54°C (Canty and Kishimoto 1977). Crystals of C$_9$H$_{12}$SHg were obtained by recrystallization from EtOH/chloroform.

MeHg(4-MePhCH$_2$)S or C$_9$H$_{12}$SHg, methylmercury 4-methylbenzylmercaptide. This compound melts at 71°C–73°C. The reaction of Bu^{t}HgCl (1 mmol) and HSCH(Me)Ph in basic MeOH afforded a crude product (Bach and Weibel 1976). Recrystallization from cyclopentane afforded white needles of Bu^{t}Hg(4-MePhCH$_2$)S.

HgMe$_3$C$_6$H$_3$CH$_2$S or C$_{10}$H$_{14}$SHg, methylmercury mesitylmercaptide. This compound melts at 122°C–124°C. It was prepared from the solutions of CH$_3$HgCl and NaOH with addition of mesitylmercaptan (Bach and Weibel 1976). Upon addition of mesitylmercaptan, a white precipitate was formed. Extraction of aqueous methanol solution with methylene chloride, removal of the solvent, and recrystallization from CH$_2$Cl$_2$/MeOH solution afforded a white fibrous solid.

HgBu^{t}PhS or C$_{10}$H$_{14}$SHg, t-butanethiolato(phenyl)mercury. This compound melts at 107°C–108°C (Canty and Kishimoto 1977). Crystals of C$_{10}$H$_{14}$SHg were obtained by recrystallization from EtOH/chloroform.

Hg(n-C₅H₁₁S)₂ and Hg(i-C₅H₁₁S)₂ or C₁₀H₂₂S₂Hg, bis(n-pentylthiolato)mercury and bis(i-pentylthiolato)mercury. $Hg(n\text{-}C_5H_{11}S)_2$ melts at 74°C–75°C and $Hg(i\text{-}C_5H_{11}S)_2$ melts at 100°C (Wertheim 1929). These compounds can be obtained by adding six drops of $n\text{-}C_5H_{11}SH$ or $i\text{-}C_5H_{11}SH$ to about 2 mL of EtOH, subsequently adding an excess of mercuric cyanide solution (10%) (Wertheim 1929). After shaking and cooling the mixture for a few minutes, the resulting precipitates were separated by filtration and dried in the air for a short time on a porous plate. They were purified by recrystallization from the ethanol.

Hg(2,2-Me₂PrS)₂ or C₁₀H₂₂S₂Hg, bis(2,2-dimethylpropanethiolato)mercury. To obtain this compound, 2,2-dimethylpropanethiol (4 mmol) in MeOH (10 mL) was added to a solution containing $HgCl_2$ (2.0 mmol) and triethylamine (2.0 mmol) in MeOH (5 mL) (Block et al. 1990). White solid was produced immediately, and the solution was stirred at room temperature for 1 h. The solid was collected by filtration and washed with MeOH and distilled water to yield, after solvent removal at 13 Pa, a white powder of $Hg(2,2\text{-}Me_2PrS)_2$.

HgMe(C₁₀H₇S) or C₁₁H₁₀SHg, methyl(2-mercaptonaphtalato)mercury and methyl(1-mercaptonaphtalato)mercury. These compounds melt at 93°C and 97°C, respectively. To obtain them, methylmercury acetate was dissolved in water and added to 2-mercaptonaphtalene or 1-mercaptonaphtalene dissolved in MeOH or methanolic KOH solution. The crude products were purified by the recrystallization in hexane (Sytsma and Kline 1973).

HgMe(p-BuʲCMe₃S) or C₁₁H₁₄SHg, methyl(p-t-butylthiophenolato)mercury. This compound melts at 89°C. To obtain it, methylmercury acetate was dissolved in water and added to p-t-butylthiophenol dissolved in MeOH or methanolic KOH solution. The crude product was purified by the recrystallization in hexane (Sytsma and Kline 1973).

BuʲHgPhCH₂S or C₁₁H₁₆SHg, t-butylmercury benzylmercaptide. This compound melts at 75°C–76°C. To obtain it, a solution of t-butylmercury chloride (2.0 mmol) in 50 mL of MeOH was combined with a solution consisting of benzylmercaptan (2.0 mmol), 12 mL of MeOH, and 1 mL of 3 N NaOH (Bach and Weibel 1976). After the resulting solution was stirred for 20 min, it was warmed and H_2O (5.5 mL) added until a permanent cloudiness persisted. The solution was heated until clear and then stored in the freezer. The crude product was collected by filtration and the product was recrystallized from pentane to give white needles. $Bu^tHgPhCH_2S$ is unstable in the presence of light.

PhSHgPh or C₁₂H₁₀SHg, benzenethiolato(phenyl)mercury. This compound that was obtained for the first time by Lecher (1915, 1920) melts at 102°C (Canty and Kishimoto 1977). Crystals of $C_{12}H_{10}SHg$ were obtained by recrystallization from EtOH/chloroform.

Hg(PhS)₂ or C₁₂H₁₀S₂Hg, bis(benzenethiolato)mercury. This compound that was obtained for the first time by Vogt (1861) melts at 150°C–153°C (Koton et al. 1950, Gregg et al. 1951, Canty and Kishimoto 1977). On addition of benzenethiol in EtOH to mercuric acetate or mercuric chloride in EtOH, a white precipitate of $Hg(PhS)_2$ was formed immediately. It was collected, washed with EtOH, and purified by recrystallization from EtOH or ethyl acetate. Preparation using MeOH as the reaction solvent yielded the same product as with EtOH (Gregg et al. 1951, Canty and Kishimoto 1977).

This compound could be also prepared by the same procedure as for bis(n-butylthiolato)mercury using C_6H_5SH instead of n-C_4H_9SH (Said and Tuck 1982) or by the interaction of $HgPh_2$ with PhSH at 130°C during 3 h (Koton et al. 1950).

Hg($C_6H_{13}S)_2$ or $C_{12}H_{26}S_2Hg$, bis(hexylthiolato)mercury. Hg($C_6H_{13}S)_2$ melts at 76°C–77°C (Wertheim 1929). This compound can be obtained by adding six drops of $C_6H_{13}SH$ to about 2 mL of EtOH, subsequently adding an excess of mercuric cyanide solution (10%) (Wertheim 1929). After shaking and cooling the mixture for a few minutes, the resulting precipitates were separated by filtration and dried in the air for a short time on a porous plate. They were purified by recrystallization from the ethanol.

PhHgS(PhMe$_2$–2,6) or $C_{14}H_{14}SHg$, (2,6-dimethylphenylthioato)phenylmercury. This compound crystallizes in the monoclinic structure with the lattice parameters $a = 982.9 \pm 0.5$, $b = 2509.0 \pm 0.2$, $c = 540.0 \pm 0.6$ pm, and $\beta = 92.95° \pm 0.07°$ and calculated density 2.08 g·cm^{-3} (Kuz'mina et al. 1974).

Hg(o-MePhS)$_2$, Hg(m-MePhS)$_2$, and Hg(p-MePhS)$_2$ or $C_{14}H_{14}S_2Hg$, bis(o-tolylthiolato)mercury, bis(m-tolylthiolato)mercury, and bis(p-tolylthiolato)mercury. Hg(o-MePhS)$_2$, Hg(m-MePhS)$_2$, and Hg(p-MePhS)$_2$ melt at 169°C–173°C, 126°C–130°C, and 162°C–166°C [158°C–160°C (Koton et al. 1950)], respectively (Gregg et al. 1951). To obtain these compounds, a solution containing $HgCl_2$ in EtOH was slowly added at room temperature to o-MePhSH, m-MePhSH, or p-MePhSH in EtOH. They could be also obtained by the interaction of the solution of $HgCl_2$ in EtOH with the solutions of triphenylaryl sulfides in EtOH. The mercaptides precipitate almost immediately and completely on cooling. They were purified by recrystallization from EtOH or ethyl acetate. Preparations using MeOH as the reaction solvent yielded the same products as with EtOH (Gregg et al. 1951). Hg(p-MePhS)$_2$ was also prepared by the interaction of $HgPh_2$ and p-MePhSH at 130°C during 2–4 h (Koton et al. 1950).

(2,4,6-Pr$_3^i$PhS)MeHg or $C_{16}H_{26}SHg$, (2,4,6-tri-i-propylbenzenethiolato)methylmercury. This compound crystallizes in the triclinic structure with the lattice parameters $a = 596.2 \pm 0.1$, $b = 964.9 \pm 0.1$, $c = 1595.1 \pm 0.2$ pm, $\alpha = 79.60° \pm 0.01°$, $\beta = 79.58° \pm 0.01°$, and $\gamma = 84.68° \pm 0.01°$ and calculated density 1.69 g·cm^{-3} (Block et al. 1990). To obtain this compound, a solution containing 0.35 mmol of methylmercury nitrate and 0.35 mmol of triethylamine in MeOH was slowly added to a solution of 0.35 mmol of 2,4,6-tri-i-propylthiophenol in 10 mL of MeOH at room temperature. After 2 h of stirring, the resultant solution was filtered, the filtrate was concentrated to ~5, and 5 mL of acetonitrile was added slowly. The solution was kept at −20°C for several days. The colorless block-shaped crystals were collected and recrystallized from CH_3OH/CH_3CN at −20°C.

Hg($SC_6H_2Me_3$–2,4,6)$_2$ or $C_{18}H_{22}S_2Hg$. This compound does not sublime and melts at 230°C with decomposition (Bochmann and Webb 1991a,b). The reaction of 2,4,6-trimethylbenzenethiol with mercury bis[bis(trimethylsilyl)amide] in light petroleum leads to the precipitation of $C_{18}H_{22}S_2Hg$ as a colorless fibrous crystalline solids (Bochmann and Webb 1991b). This compound could be also obtained if a solution of $Hg[N(SiMe_3)_2]_2$ (0.9 mmol) in light petroleum (5 mL) was added to a solution of 1.8 mmol of $HSC_6H_2Me_3-2,4,6$ in light petroleum (20 mL) via syringe at room temperature. The mixture was stirred for 2 h, filtered, and the residue recrystallized from toluene at −20°C to obtain $C_{18}H_{22}S_2Hg$ as white microcrystals.

$PhCH_2HgSCPh_3$ or $C_{18}H_{22}S_2Hg$, benzyltriphenylmethylthiomercury. This compound melts at 120°C with decomposition and crystallizes in the monoclinic structure with the lattice parameters $a = 1186.2 \pm 0.7$, $b = 1813.7 \pm 1.5$, $c = 1293.2 \pm 0.8$ pm, and $\beta = 128.93° \pm 0.03°$ (Bach et al. 1974). It was prepared by the treatment of benzylmercuric chloride with triphenylmethanethiol in methanolic NaOH. Recrystallization from benzenepentane gave white opaque crystals.

$Hg(Bu_2^tMeS)_2$ or $C_{18}H_{38}S_2Hg$, bis(di-t-buthylmethanthiolato)mercury. To obtain this compound, di-t-butylmethanethiol (2 mmol) was added to a solution containing $HgCl_2$ (2.0 mmol) and triethylamine (2.0 mmol) in MeOH (5 mL) (Block et al. 1990). White solid was produced immediately, and the solution was stirred at room temperature for 1 h. The solid was collected by filtration and washed with MeOH and distilled water to yield, after solvent removal at 13 Pa, a white powder of $Hg(Bu_2^tMeS)_2$.

$MeHgSPh_3Me$ or $C_{20}H_{18}SHg$, methylmercury triphenylmethylmercaptide. This compound melts at 167°C–169°C with decomposition. To obtain $C_{20}H_{18}SHg$, a solution of 5.0 mmol of MeHgCl was dissolved in 30 mL of MeOH (Bach and Weibel 1976). To this solution was added a slurry composed of 5.0 mmol of triphenylmethylmercaptide, 10 mL of aqueous 1 N NaOH (10 mmol), and 30 mL of MeOH. The resulting solution was stirred with warming until the formation of a gray suspension was noted. After 30 min at 25°C, the solvent was evaporated and the crude gray mass was dissolved in benzene, filtered through $MgSO_4$, concentrated, and cooled. The large clear crystals were obtained.

After stirring for 1 h, the solution was filtered and 8.8 mL of PhSH was added to the filtrate. On addition of mercaptan, a white precipitate was formed immediately. The solution was stored at 5°C for 2 h and then filtered. Recrystallization from methylene chloride afforded white needles.

$Hg(Bu_3^tS)_2$ or $C_{24}H_{54}S_2Hg$, bis(t-butylmercapto)mercury. This compound sublimes at 135°C under low pressure and crystallizes in the orthorhombic structure with the lattice parameters $a = 1836 \pm 3$, $b = 918 \pm 2$, and $c = 752 \pm 1$ pm and calculation and experimental densities 1.972 and 1.97 $g \cdot cm^{-3}$, respectively (Kunchur 1964).

$Hg(SC_6H_2Pr_3^i-2,4,6)_2$ or $C_{30}H_{46}S_2Hg$. This compound melts at 140°C and sublimes at 130°C (1.3 Pa) (Bochmann and Webb 1991a). The reaction of tris(isopropyl)benzenethiol with mercury bis[bis(trimethylsilyl)amide] in light petroleum leads to the precipitation of $C_{30}H_{46}S_2Hg$ as a colorless fibrous crystalline solids (Bochmann and Webb 1991b). This compound could be also obtained if a solution of $Hg[N(SiMe_3)_2]_2$ (0.9 mmol) in light petroleum (5 mL) was added to a solution of 1.8 mmol of $HSC_6H_2Pr_3^i-2,4,6$ in light petroleum (20 mL) via syringe at room temperature.

The mixture was stirred for 2 h and filtered and the residue recrystallized from toluene at $-20°C$ to obtain $C_{30}H_{46}S_2Hg$ as white microcrystals.

bis(2,4,6-Pr$_3^i$PhS)$_2$MeHg or $C_{31}H_{48}S_2Hg$, (2,4,6-tri-i-propylbenzenethiolato)methylmercury. This compound was obtained if to a solution containing $HgCl_2$ and triethylamine in MeOH was added 2,4,6-tri-i-propylbenzenethiol in MeOH (Block et al. 1990). White solid was produced immediately, and the solution was stirred for 1 h. The solid was collected by filtration and washed with MeOH and distilled water to yield, after solvent removal at 13 Pa, a white powder of $C_{31}H_{48}S_2Hg$. Recrystallization from CH_2Cl/CH_3CN afforded colorless crystals, which were collected and dried in air.

Hg(SC$_6$H$_2$Bu$_3^t$-2,4,6)$_2$ or $C_{36}H_{58}S_2Hg$, bis(2,4,6-tri-t-butylbenzenethiolato)mercury. This compound sublimes at 170°C (1.3 Pa) and melts at 270°C with decomposition (Bochmann and Webb 1991a,b, Bochmann et al. 1997). To obtain this compound, a solution of $Hg[N(SiMe_3)_2]_2$ (0.9 mmol) in light petroleum (5 mL) was added to a solution of 1.8 mmol of $HSC_6H_2Bu_3^t$-2,4,6 in light petroleum (20 mL) via syringe at room temperature (Bochmann and Webb 1991b, Bochmann et al. 1997). The mixture was stirred for 2 h and filtered and the residue recrystallized from toluene at $-20°C$ to obtain $C_{36}H_{58}S_2Hg$ as white microcrystals.

7.2   MERCURY–HYDROGEN–NITROGEN–SULFUR

HgS–N$_2$–NH$_3$. The phase diagram is not constructed. $HgN_2S \cdot NH_3$ quaternary compound is formed in this system. It crystallizes in the orthorhombic structure with the lattice parameters $a = 554.8$, $b = 1015.8$, and $c = 1491.9$ pm and calculated density 4.4 g·cm^{-3} (Martan and Weiss 1984).

7.3   MERCURY–HYDROGEN–OXYGEN–SULFUR

$Hg(OH)_2 \cdot 2HgSO_4 \cdot H_2O$ compound is formed in this system. It crystallizes in the monoclinic structure with the lattice parameters $a = 715.2 \pm 0.1$, $b = 891.9 \pm 0.2$, $c = 1448.8 \pm 0.2$ pm, and $\beta = 98.94° \pm 0.02°$ and calculation and experimental densities 6.17 and 6.18 g·cm^{-3}, respectively (Björnlund 1974). Crystals of this compound were formed and diluted H_2SO_4 was allowed to be in contact with solid, freshly precipitated, yellow HgO.

7.4   MERCURY–LITHIUM–RUBIDIUM–SULFUR

HgS–Li$_2$S–Rb$_2$S. The phase diagram is not constructed. $Li_{1.8}Rb_{0.2}Hg_6S_7$ quaternary compound is formed in this system (Axtell et al. 1998). It can be obtained by the ionic exchange of $Rb_2Hg_6S_7$ and LiI.

7.5   MERCURY–POTASSIUM–GERMANIUM–SULFUR

HgS–K$_2$S–GeS$_2$. The phase diagram is not constructed. $K_2Hg_3Ge_2S_8$ quaternary compound is formed in this system. It melts incongruently at ~576°C [580°C (Kanatzidis et al. 1997a,b)], recrystallizes at ~501°C, and has two polymorphic modifications (Liao et al. 2003). One of them crystallizes in the orthorhombic structure with the lattice parameters $a = 1908.2 \pm 0.2$, $b = 955.1 \pm 0.1$, and $c = 828.71 \pm 0.08$ pm [$a = 1918.8 \pm 0.3$, $b = 961.8 \pm 0.2$, and $c = 832.8 \pm 0.5$ pm (Kanatzidis et al. 1997a,b)]; calculated density 9.514 g $\cdot$ cm^{-3} [4.674 g $\cdot$ cm^{-3} (Kanatzidis et al. 1997a,b)]; and bandgap $E_g = 2.64$ eV (Kanatzidis et al. 1997a,b, Liao et al. 2003).

Other modification crystallizes in the monoclinic structure with the lattice parameters $a = 959.48 \pm 0.07$, $b = 836.08 \pm 0.06$, $c = 966.38 \pm 0.07$ pm, and $\beta = 94.637°$; calculated density 9.322 g $\cdot$ cm^{-3}; and bandgap $E_g = 2.70$ eV (Liao et al. 2003).

α-K$_2$Hg$_3$Ge$_2$S$_8$ was obtained by the reaction of the mixture of Ge, HgS, K$_2$S, and S, taking in stoichiometric ratio, in evacuated glass ampoule at 400°C for 96 h upon a cooling rate of 4°C/h (Kanatzidis et al. 1997a,b, Liao et al. 2003). β-K$_2$Hg$_3$Ge$_2$S$_8$ was synthesized by the reaction of Ge, HgS, K$_2$S, and excess S in evacuated glass ampoule at 520°C over 6 h with the next holding at that temperature for 1 h and cooling to 480°C at a rate of 1°C/h (Liao et al. 2003). This modification could be obtained also by heating the mixtures up to 400°C–500°C over a period of 8–12 h with the next holding at that temperatures for 48–96 h and cooling to about 200°C at a rate of 4°C/h. Relatively large (larger than a millimeter on edge) optical quality pieces of K$_2$Hg$_3$Ge$_2$S$_8$ crystals can be synthesized by recrystallization using slow cooling and/or temperature cycling of K$_2$Hg$_3$Ge$_2$S$_8$/K$_2$S$_8$ flux reaction (Kanatzidis et al. 1997a,b, Liao et al. 2003).

7.6 MERCURY–POTASSIUM–TIN–SULFUR

HgS–K$_2$S–SnS$_2$. The phase diagram is not constructed. K$_2$Hg$_3$Sn$_2$S$_8$ quaternary compound is formed in this system. It melts incongruently at ~545°C, recrystallizes at ~430°C, and has two polymorphic modifications (Kanatzidis et al. 1997a,b, Liao et al. 2003). α-K$_2$Hg$_3$Sn$_2$S$_8$ crystallizes in the orthorhombic structure with the lattice parameters $a = 1956.3 \pm 0.2$, $b = 985.3 \pm 0.1$, and $c = 846.7 \pm 0.1$ pm [$a = 1952.2 \pm 0.5$, $b = 983.5 \pm 0.3$, and $c = 843.1 \pm 0.2$ pm (Kanatzidis et al. 1997a,b)]; calculated density 9.555 g $\cdot$ cm^{-3} [4.816 g $\cdot$ cm^{-3} (Kanatzidis et al. 1997a,b)]; and bandgap $E_g = 2.40$ eV (Liao et al. 2003) [$E_g = 2.39$ eV (Kanatzidis et al. 1997a,b)].

Other modification crystallizes in the monoclinic structure with calculated density 9.322 g $\cdot$ cm^{-3} and bandgap $E_g = 2.50$ eV (Liao et al. 2003).

α-K$_2$Hg$_3$Sn$_2$S$_8$ was obtained by the reaction of the mixture of Sn, HgS, K$_2$S, and S, taking in stoichiometric ratio, at 400°C for 96 h upon a cooling rate of 4°C/h (Kanatzidis et al. 1997a,b, Liao et al. 2003). β-K$_2$Hg$_3$Sn$_2$S$_8$ was synthesized by the reaction of Sn, HgS, K$_2$S, and S in evacuated glass ampoule at 400°C for 96 h with the next cooling to 160°C at a rate of 80°C/h (Liao et al. 2003). This modification could be obtained also by heating the mixtures up to 400°C–500°C over a period of 8–12 h with the next holding at that temperatures for 48–96 h and cooling to about 200°C at a rate of 4°C/h. Relatively large (larger than a millimeter on edge) optical quality pieces of K$_2$Hg$_3$Sn$_2$S$_8$ crystals can be synthesized by recrystallization using slow cooling and/or temperature cycling of K$_2$Hg$_3$Sn$_2$S$_8$/K$_2$S$_8$ flux reaction (Kanatzidis et al. 1997a,b, Liao et al. 2003).

7.7   MERCURY–POTASSIUM–ANTIMONY–SULFUR

HgS–K$_2$S–Sb$_2$S$_3$. The phase diagram is not constructed. KHgSbS$_3$ quaternary compound is formed in this system. It crystallizes in the monoclinic structure with the lattice parameters a = 1708.4 ± 0.8, b = 864.7 ± 0.4, c = 895.1 ± 0.3 pm, and β = 105.78° ± 0.05° and calculated density 4.4 g·cm^{-3} (Imafuku et al. 1985, 1986). This compound was synthesized by hydrothermal heating of the powder mixture of HgS and Sb$_2$S$_3$ in the solution of KOH/K$_2$S = 1/5 at 180°C for 50 h in a Pyrex glass tube.

7.8   MERCURY–RUBIDIUM–GERMANIUM–SULFUR

HgS–Rb$_2$S–GeS$_2$. The phase diagram is not constructed. Rb$_2$Hg$_3$Ge$_2$S$_8$ quaternary compound is formed in this system (Kanatzidis et al. 1997a,b, Marking et al. 1998, Liao et al. 2003). It melts incongruently (Liao et al. 2003) and crystallizes in the monoclinic structure with the lattice parameters a = 993.8, b = 635.2, c = 1311.7 pm, and β = 97.33° (Marking et al. 1998) and energy gap 2.65 eV (Kanatzidis et al. 1997a,b).

Rb$_2$Hg$_3$Ge$_2$S$_8$ was prepared with a slight excess of HgS powder in the mixture Ge + HgS + Rb$_2$S + S by heating at 500°C for 96 h and cooling to 200°C at 4°C/h (Liao et al. 2003) or by the interaction of HgS with Ge in the Rb$_2$S$_x$ melting at 350°C (Marking et al. 1998).

7.9   MERCURY–RUBIDIUM–TIN–SULFUR

HgS–Rb$_2$S–SnS$_2$. The phase diagram is not constructed. Rb$_2$Hg$_3$Sn$_2$S$_8$ quaternary compound is formed in this system (Kanatzidis et al. 1997a,b, Marking et al. 1998, Liao et al. 2003). It melts incongruently (Liao et al. 2003) and crystallizes in the monoclinic structure with the lattice parameters a = 1013.2, b = 654.0, c = 1343.4 pm, and β = 97.93°; calculated density 4.770 g·cm^{-3} (Marking et al. 1998); and energy gap 2.40 eV (Kanatzidis et al. 1997a,b).

Rb$_2$Hg$_3$Sn$_2$S$_8$ was prepared with a slight excess of HgS powder in the mixture Sn + HgS + Rb$_2$S + S by the heating at 500°C for 96 h and cooling to 200°C at 4°C/h (Liao et al. 2003) or by the interaction of HgS with Sn in the Rb$_2$S$_x$ melt at 350°C (Marking et al. 1998).

7.10   MERCURY–RUBIDIUM–SELENIUM–SULFUR

HgS + Rb$_2$Se ⇔ HgSe + Rb$_2$S. The phase diagram is not constructed. Solid solution Rb$_2$Hg$_6$S$_{3.5}$Se$_{3.5}$ was obtained in this system by the reaction of Rb$_2$S, Rb$_2$Se, HgS, and HgSe under vacuum at 650°C for 48 h with the next cooling to 50°C and quenching to room temperature (Axtell et al. 1998). This product is air and water stable and is insoluble in common organic solvents.

7.11   MERCURY–CESIUM–INDIUM–SULFUR

HgS–Cs$_2$S–In$_2$S$_3$. The phase diagram is not constructed. CsHgInS$_2$ quaternary compound is formed in this system. It melts incongruently and crystallizes in the monoclinic structure with the lattice parameters a = 1125.03 ± 0.06, b = 1125.31 ± 0.05,

$c = 2228.84 \pm 0.14$ pm, and $\beta = 97.260° \pm 0.005°$ at 140 ± 2 K; calculated density
5.168 g·cm^{-3}; and energy gap 2.30 eV (Li et al. 2012). It was synthesized from a
mixture of Cs_2S, HgS, In, and S, which was loaded into a fused silica tube and flame
sealed under a vacuum. The tube was then heated to $600°C$ in 12 h, annealed at this
temperature for 24 h, followed by heating to $800°C$ in 3 h, held there for additional
8 h, and then cooled quickly to room temperature in 10 h. The Bridgman method
and a modified horizontal traveling heater method were used to grow the crystals.

7.12 MERCURY–CESIUM–GERMANIUM–SULFUR

HgS–Cs$_2$S–GeS$_2$. The phase diagram is not constructed. $Cs_2Hg_3Ge_2S_8$ quaternary
compound is formed in this system. It crystallizes in the triclinic structure with the
lattice parameters $a = 780.8$, $b = 916.4$, $c = 661.2$ pm, $\alpha = 92.02°$, $\beta = 108.65°$, and
$\gamma = 108.10°$ (Marking et al. 1998). This compound was obtained by the interaction of
HgS with Ge in the Cs_2S_x melt at $520°C$.

7.13 MERCURY–CESIUM–TIN–SULFUR

HgS–Cs$_2$S–SnS$_2$. The phase diagram is not constructed. $Cs_2Hg_3Sn_2S_8$ and
$Cs_2HgSn_3S_8$ quaternary compounds are formed in this system. $Cs_2Hg_3Sn_2S_8$ crystallizes in the triclinic structure with the lattice parameters $a = 787.8$, $b = 915.7$,
$c = 680.3$ pm, $\alpha = 92.96°$, $\beta = 109.45°$, and $\gamma = 107.81°$ and calculated density
5.207 g·cm^{-3} (Marking et al. 1998). This compound was obtained by the interaction
of HgS with Sn in the Cs_2S_x melt at $520°C$.

$Cs_2HgSn_3S_8$ crystallizes in the orthorhombic structure with the lattice parameters
$a = 766.72 \pm 0.04$, $b = 1234.05 \pm 0.10$, and $c = 1790.11 \pm 0.11$ pm; calculated density
4.231 g·cm^{-3}; and energy gap 2.72 eV (Morris et al. 2013).

7.14 MERCURY–CESIUM–SELENIUM–SULFUR

HgS + Cs$_2$Se $\Leftrightarrow$ HgSe + Cs$_2$S. The phase diagram is not constructed. The solid
solution $Cs_2Hg_6S_{5.25}Se_{1.75}$ was obtained in this system by the reaction of Cs_2S, Cs_2Se,
HgS, and HgSe under vacuum at $650°C$ for 48 h with the next cooling to $50°C$ and
quenching to room temperature (Axtell et al. 1998). This product is air and water
stable and is insoluble in common organic solvents.

7.15 MERCURY–COPPER–SILICON–SULFUR

HgS–Cu$_2$S–SiS$_2$. The phase diagram is not constructed. Cu_2HgSiS_4 and
$Cu_6Hg_{0.973}SiS_{5.973}$ quaternary compounds are formed in this system. Cu_2HgSiS_4 melts
incongruently at $859°C \pm 5°C$ and crystallizes in the orthorhombic structure with the
lattice parameters $a = 759.2$, $b = 648.4$, and $c = 626.9$ pm (Schäfer and Nitsche 1974,
1977). $Cu_6Hg_{0.973}SiS_{5.973}$ crystallizes in the cubic structure with the lattice parameter
$a = 989.38 \pm 0.01$ pm and calculated density 5.416 ± 0.002 g·cm^{-3} (Gulay et al. 2002b).

Cu_2HgSiS_4 was obtained by the chemical transport reactions (Schäfer and Nitsche
1974).

7.16   MERCURY–COPPER–GERMANIUM–SULFUR

HgS–CuGeS$_3$. The phase diagram of this system is shown in Figure 7.1 (Parasyuk et al. 2002b). Two quaternary intermediate phases exist in the HgS–CuGeS$_3$ system. Cu$_2$HgGeS$_4$ melts congruently at 936°C [at 907°C ± 5°C (Schäfer and Nitsche 1977)] and gives rise to a polymorphous transformation between 650°C and 677°C. The low-temperature modification crystallizes in the tetragonal structure with the lattice parameters a = 548.73 ± 0.04 and c = 1054.23 ± 0.02 pm [a = 549.0 and c = 1055.0 pm (Schäfer and Nitsche 1974, 1977)] and calculated density 5.5339 ± 0.0006 g·cm^{-3} (Parasyuk et al. 2002b).

The high-temperature modification of Cu$_2$HgGeS$_4$ crystallizes in the orthorhombic structure with the lattice parameters a = 768.11 ± 0.04, b = 655.46 ± 0.02, and c = 631.44 ± 0.04 pm (Parasyuk et al. 2002b) [a = 776.9, b = 652. 2, and c = 632.5 pm (Schäfer and Nitsche 1977)]. Cu$_2$Hg$_3$GeSe$_6$ melts incongruently at 826°C and decomposes according to a eutectoid reaction at 706°C.

The eutectic with a melting temperature of 924°C and a composition of 33 mol.% HgS forms between Cu$_2$GeS$_3$ and Cu$_2$HgGeS$_4$ (Parasyuk et al. 2002b). The second eutectic is located between Cu$_2$Hg$_3$GeS$_6$ and HgS. It melts at 737°C and corresponds to the composition with 93 mol.% HgS. The solid solubility ranges of the initial compounds are small.

This system was investigated by differential thermal analysis (DTA), x-ray diffraction (XRD), and metallography and the annealing of the ingots was carried out at 400°C during 250 h (Parasyuk et al. 2002b). After annealing, the ampoules with the samples were quenched in air. The growing of the Cu$_2$HgGeS$_4$ single crystals was carried out by chemical vapor transport reactions and I$_2$ as a transport agent (Schäfer and Nitsche 1974, Parasyuk et al. 2002b).

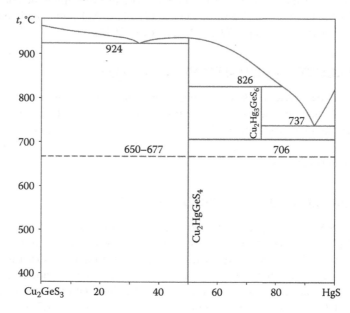

FIGURE 7.1   Phase diagram of HgS–Cu$_2$GeS$_3$ system. (From Parasyuk, O.V. et al., *J. Alloys Compd.*, 334(1–2), 143, 2002.)

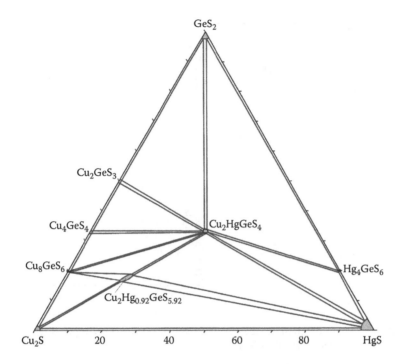

FIGURE 7.2 Isothermal section of the HgS–Cu$_2$S–GeS$_2$ quasiternary system at 400°C. (From Marchuk, O.V. et al., *J. Alloys Compd.*, 333(1–2), 143, 2002.)

HgS–Cu$_2$S–GeS$_2$. The isothermal section of this quasiternary system at 400°C is shown in Figure 7.2 (Marchuk et al. 2002). Equilibria exist between the quaternary compound Cu$_2$HgGeS$_4$ and the binary components HgS and GeS$_2$ and the ternary Cu$_2$GeS$_3$, Cu$_8$GeS$_6$, Hg$_4$GeS$_6$, and Cu$_4$GeS$_4$ phases and quaternary phase Cu$_6$Hg$_{0.92}$GeS$_{5.92}$. Cu$_6$Hg$_{0.92}$GeS$_{5.92}$ crystallizes in the cubic structure with the lattice parameter a = 999.88 ± 0.01 pm and calculated density 5.494 ± 0.001 g · cm^{-3}. This compound has a homogeneity region within the interval of 63–66 mol.% Cu$_2$S with changing of the lattice parameter from 1054.1 to 1048.5 pm. This system was investigated by DTA and metallography and the alloys were annealed at 400°C during 250 h (Marchuk et al. 2002).

7.17 MERCURY–COPPER–TIN–SULFUR

HgS–Cu$_2$SnS$_3$. The phase diagram is shown in Figure 7.3 (Olekseyuk et al. 2000). Cu$_2$HgSnS$_4$ quaternary compound is formed in this system. It melts congruently at 849°C [at 845°C ± 5°C (Schäfer and Nitsche 1977)] and crystallizes in the tetragonal structure with the lattice parameters a = 558.0 ± 0.2 and c = 1089.5 ± 0.3 pm (Olekseyuk et al. 2000) [a = 557.49 ± 0.06 and c = 1088.2 ± 0.1 pm (Kabalov et al. 1998), a = 557.5 and c = 1084.4 pm (Schäfer and Nitsche 1977), a = 556.6 and c = 1088 pm (Hahn and Schulze 1965)] and calculation and experimental densities 5.66 and 5.56 g · cm^{-3}, respectively (Hahn and Schulze 1965). The melting maximum of this quaternary phase is somewhat shifted from stoichiometric composition and it

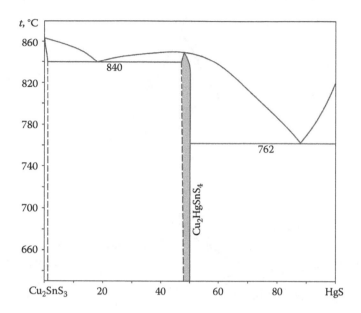

FIGURE 7.3 Phase diagram of HgS–Cu₂SnS₃ system. (From Olekseyuk, I. et al., *Visnyk L'viv. un-tu. Ser. khim.*, (39), 48, 2000.)

takes place at 47–49 mol.% HgS (Olekseyuk et al. 2000). Cu_2HgSnS_4 forms eutectics with the initial compounds of the system, which melt at 840°C (18 mol.% HgS) and at 762°C (88 mol.% HgS).

Cu_2HgSnS_4 quaternary compound corresponds to the mineral velikite (Kaplunnik et al. 1977, Mandarino 1998) [according to Kaplunnik et al. (1977), the composition of this mineral is $Cu_{3.75}Hg_{1.75}Sn_2S_8$], which crystallizes in the tetragonal structure with the lattice parameters $a = 556.0$ and $c = 1090.5$ pm [$a = 554.2 \pm 0.3$ and $c = 1090.8 \pm 0.7$ pm (Kaplunnik et al. 1977)] and calculation and experimental densities 5.42 and 5.45 g·cm⁻³ [5.48 and 5.59 g·cm⁻³ (Kaplunnik et al. 1977)], respectively (Mandarino 1998).

The alloys were annealed at 400°C during 250 h and this system was investigated by DTA, XRD, and metallography (Olekseyuk et al. 2000).

7.18 MERCURY–COPPER–PHOSPHORUS–SULFUR

HgS–Cu₂S–P₂S₅. The phase diagram is not constructed. $CuHgPS_4$ compound is formed in this system. It crystallizes in the orthorhombic structure with the lattice parameters $a = 1266.0$, $b = 734.98$, and $c = 609.43$ pm (Carrillo-Cabrera et al. 1992). This compound was obtained from the chemical elements at 900°C.

7.19 MERCURY–COPPER–ARSENIC–SULFUR

HgS–Cu₂S–As₂S₃. The phase diagram is not constructed. $Cu_6Hg_3As_4S_{12}$ quaternary compound (mineral aktashite) is formed in this system. It crystallizes in the tetragonal structure with the lattice parameters $a = 1373.0 \pm 0.3$ and $c = 932.9 \pm 0.1$ pm [in the cubic structure with the lattice parameter $a = 539.8$ pm (Vasil'ev 1968)] and calculation and experimental densities 5.72 and 5.5 g·cm^{-3}, respectively (Novatski 1982). According to the data of Vasil'ev (1968), aktashite contains up to 2.5 mass.% Sb.

7.20　MERCURY–COPPER–FLUORINE–SULFUR

HgS–CuF$_2$–HgF$_2$. The phase diagram is not constructed. Cu$_2$Hg$_2$SF$_6$ quaternary compound is formed in this system. It crystallizes in the cubic structure with the lattice parameter $a = 1068.82 \pm 0.02$ pm at 2.4 K, $a = 1069.43 \pm 0.02$ pm at 107 K, and $a = 1071.54 \pm 0.02$ pm at 298 K (Kawabata et al. 2007) [$a = 1073.47 \pm 0.13$ pm (Bernard et al. 1975)]. Cu$_2$Hg$_2$SF$_6$ has no structural phase transition down 3 K. Polycrystalline samples of this compound were prepared by a solid-state reaction.

7.21　MERCURY–COPPER–CHLORINE–SULFUR

HgS–CuCl. The phase diagram is not constructed. CuHgSCl and CuHg$_2$S$_2$Cl quaternary compounds are formed in this system. CuHgSCl crystallizes in the orthorhombic structure with the lattice parameters $a = 984.01 \pm 0.08$, $b = 1775.1 \pm 0.2$, and $c = 409.59 \pm 0.03$ pm and calculated density 6.16 g·cm^{-3} at 22°C and $a = 1001.2 \pm 0.8$, $b = 1824 \pm 2$, and $c = 412.1 \pm 0.4$ pm and calculated density 6.64 g·cm^{-3} at –120°C (Beck and Rompel 2003) [$a = 984.4$, $b = 1775.2$, and $c = 409.5$ pm; calculation and experimental densities 6.20 and 6.16 g·cm^{-3}, respectively; and energy gap $E_g = 1.52$ eV at room temperature (Guillo et al. 1979)]. This compound is characterized by an ionic conductivity at the temperatures >100°C (Moro'oka et al. 2003).

CuHg$_2$S$_2$Cl crystallizes in the tetragonal structure with the lattice parameters $a = 2644.4 \pm 0.3$ and $c = 409.2 \pm 0.1$ pm (Blachnik and Lytze 1990).

The crystal chemistry of CuHgSCl was discussed in Borisov et al. (2005), and it was shown that the degree of anion ordering in this compound substantially exceeds (by 15%–20%) the degree of cation ordering in the same interval of interplanar spacings d_{hkl}, which determines the structure arrangement.

CuHgSCl was synthesized by the hydrothermal reaction of CuCl with HgS in concentrated HCl as solvent at 400°C in sealed glass ampoules (Beck and Rompel 2003).

7.22　MERCURY–COPPER–BROMINE–SULFUR

HgS–CuBr. The phase diagram is not constructed. CuHgSBr quaternary compound is formed in this system. It crystallizes in the orthorhombic structure with the lattice parameters $a = 1001.3 \pm 0.8$, $b = 1827 \pm 3$, and $c = 412.3 \pm 0.5$ pm and calculated density 6.62 g·cm^{-3} at –70°C and $a = 1003.7 \pm 0.4$, $b = 1833.6 \pm 0.5$, and $c = 412.4 \pm 0.2$ pm and calculated density 6.58 g·cm^{-3} at 22°C (Beck and Rompel 2003) [$a = 1004.5$, $b = 1832.0$, and $c = 412.8$ pm and calculation and experimental densities 6.60 and 6.57 g·cm^{-3}, respectively, at room temperature (Guillo et al. 1979)]. At 50°C, CuHgSBr undergoes a second-order phase transition into a higher symmetric structure with halved b-axis ($a = 1009.2 \pm 0.3$, $b = 918.40 \pm 0.04$, and $c = 413.81 \pm 0.02$ pm and calculated density 6.57 g·cm^{-3} at 85°C) (Beck and Rompel 2003). This compound is characterized by an ionic conductivity at the temperatures >73°C (Moro'oka et al. 2003).

The crystal chemistry of CuHgSBr was discussed in Borisov et al. (2005), and it was shown that the degree of anion ordering in this compound substantially exceeds (by 15%–20%) the degree of cation ordering in the same interval of interplanar spacings d_{hkl}, which determines the structure arrangement.

CuHgSBr was synthesized by the hydrothermal reaction of CuBr with HgS in concentrated HBr as solvent at 400°C in sealed glass ampoules (Beck and Rompel 2003).

7.23 MERCURY–COPPER–IODINE–SULFUR

HgS–CuI. The phase diagram is not constructed. CuHgSI and $CuHg_2S_2I$ quaternary compounds are formed in this system (Blachnik and Dreisbach 1986, Keller and Wimbert 2004, Moro'oka et al. 2004b). CuHgSI crystallizes at low temperatures in the orthorhombic structure with the lattice parameters $a = 718.3 \pm 0.1$, $b = 834.3 \pm 0.2$, and $c = 698.9 \pm 0.1$ pm and calculation and experimental densities 6.71 $g \cdot cm^{-3}$ (Keller and Wimbert 2004). $CuHg_2S_2I$ also crystallizes in the orthorhombic structure with the lattice parameters $a = 1261.8 \pm 0.3$, $b = 722.4 \pm 0.1$, and $c = 693.7 \pm 0.1$ pm and calculated density 6.89 $g \cdot cm^{-3}$.

The crystal chemistry of CuHgSI and $CuHg_2S_2I$ was discussed in Borisov et al. (2005), and it was shown that the degree of anion ordering in these compounds substantially exceeds (by 15%–20%) the degree of cation ordering in the same interval of interplanar spacings d_{hkl}, which determines the structure arrangement.

CuHgSI has been synthesized by annealing mixtures of HgS and CuI (Blachnik and Dreisbach 1986, Moro'oka et al. 2004b) or by the hydrothermal reaction of CuI and α-HgS in diluted aqueous HI solution as a solvent at 180°C (Keller and Wimbert 2004). $CuHg_2S_2I$ was obtained by the hydrothermal reaction of CuI and α-HgS in diluted aqueous HI solution at 300°C (Keller and Wimbert 2004).

7.24 MERCURY–COPPER–IRON–SULFUR

$(Cu, Fe)_6Hg_2S_5$ quaternary compound (mineral gortdrumite) exists in this system. It crystallizes in the orthorhombic structure with the lattice parameters $a = 1496$, $b = 790$, and $c = 2410$ pm (Steed 1983, Dunn et al. 1984).

7.25 MERCURY–SILVER–SILICON–SULFUR

$HgS–Ag_2S–SiS_2$. The phase diagram is not constructed. $Ag_6Hg_{0.897}SiS_{5.897}$ quaternary compound is formed in this system. It crystallizes in the cubic structure with the lattice parameter $a = 1050.55 \pm 0.02$ pm (Gulay et al. 2002b).

7.26 MERCURY–SILVER–GERMANIUM–SULFUR

$HgS–Ag_2GeS_3$. The phase diagram is not constructed. Ag_2HgGeS_4, $Ag_2Hg_3GeS_6$, and $Ag_4HgGe_2S_7$ quaternary compounds are formed in this system. Ag_2HgGeS_4 possesses a narrow homogeneity range and crystallizes in the orthorhombic structure with the lattice parameters $a = 802.47 \pm 0.04$, $b = 686.84 \pm 0.03$, and

$c = 659.55 \pm 0.04$ pm [$a = 800.8$, $b = 687.1$, and $c = 659.3$ pm (Haeuseler and Himmrich 1989)] and calculated density 5.629 ± 0.002 g·cm^{-3} (Parasyuk et al. 2002a). $Ag_2Hg_3GeS_6$ crystallizes in the tetragonal structure with the lattice parameters $a = 1461.9 \pm 0.3$ and $c = 2079.6 \pm 0.5$ pm (Parasyuk et al. 2002a), and $Ag_4HgGe_2S_7$ crystallizes in the monoclinic structure with the lattice parameters $a = 1745.46 \pm 0.08$, $b = 680.93 \pm 0.02$, $c = 1053.42 \pm 0.03$ pm, and $\beta = 93.398° \pm 0.03°$ and calculated density 5.321 ± 0.003 g·cm^{-3} (Gulay et al. 2002a).

Ag_2HgGeS_4 was obtained by the interaction of chemical elements at 700°C (Haeuseler and Himmrich 1989). This system was investigated by DTA, XRD, and metallography (Parasyuk et al. 2002a).

$HgS–Ag_2S–GeS_2$. Isothermal section of this quasiternary system at 400°C is shown in Figure 7.4 (Parasyuk et al. 2002a). Besides the quaternary compounds, forming in the $HgS–Ag_2GeS_3$ section, one more quaternary compound $Ag_6Hg_{0.82}GeS_{5.82}$ is obtained in the $HgS–Ag_2S–GeS_2$ quasiternary system. It crystallizes in the cubic structure with the lattice parameter $a = 1055.47 \pm 0.02$ pm and calculated density 5.9930 ± 0.0007 g·cm^{-3} (Gulay and Parasyuk 2001). This quaternary compound possesses a homogeneity region along the $Ag_8GeS_6–Hg_4GeS_6$ section in the concentration interval 22–31 mol.% Hg_4GeS_6 and the lattice constant decreases from $a = 1055.84 \pm 0.07$ to 1054.04 ± 0.07 pm in the homogeneity range (Parasyuk et al. 2002a).

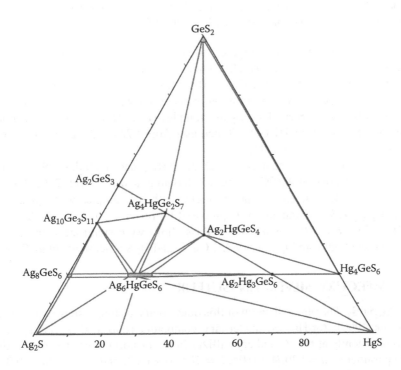

FIGURE 7.4 Isothermal section of the $HgS–Ag_2S–GeS_2$ quasiternary system at 400°C. (From Parasyuk, O.V. et al., *J. Alloys Compd.*, 336(1–2), 213, 2002.)

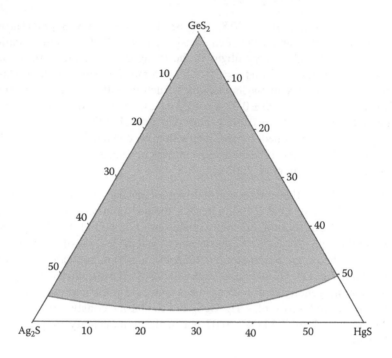

FIGURE 7.5   Glass-formation region of the HgS–Ag$_2$S–GeS$_2$ quasiternary system. (From Olekseyuk, I.D. et al., *Chem. Met. Alloys*, 2(1–2), 49, 2009.)

The glass-formation region (Figure 7.5) covers the entire concentration triangle for GeS$_2$ concentrations over 50 mol.% (Olekseyuk et al. 2009). The glass transition temperature, changing from 232°C to 340°C, gradually falls with decreasing GeS$_2$ content. For a steady GeS$_2$ concentration, it decreases with increasing HgS content. The optical bandgap energy for all glass samples varies from 1.7 to 2.3 eV. For the glassy sample with 50 mol.% GeS$_2$, it increases from 1.72 to 2.07 eV as HgS content increases from 16.7 to 33.3 mol.%.

This system was investigated by DTA, XRD, and metallography, and the alloys were annealed at 400°C during 250 h (Gulay and Parasyuk 2001, Parasyuk et al. 2002a). After annealing, the ampoules with the samples were quenched in air. Ag$_6$Hg$_{0.82}$GeS$_{5.82}$ compound was prepared by fusion of high-purity Ag, Ge, S, and HgS (Gulay and Parasyuk 2001). Glass alloys were synthesized by the melt-quenching method from high-purity Ag, Ge, S, and HgS (Olekseyuk et al. 2009).

7.27   MERCURY–SILVER–TIN–SULFUR

HgS–Ag$_2$SnS$_3$. The phase diagram of this quasibinary system is shown in Figure 7.6 (Parasyuk 1999). Ag$_2$HgSnS$_4$ quaternary compound is formed in this system. It melts congruently at 660°C and crystallizes in the orthorhombic structure with the lattice parameters $a = 820.74 \pm 0.04$, $b = 703.30 \pm 0.03$, and $c = 671.80 \pm 0.03$ pm [$a = 814.9 \pm 0.2$, $b = 700.8 \pm 0.2$, and $c = 670.4 \pm 0.3$ pm (Parasyuk 1999); $a = 820.3$, $b = 702.6$, and $c = 671.0$ pm (Haeuseler and Himmrich 1989)]; calculated density

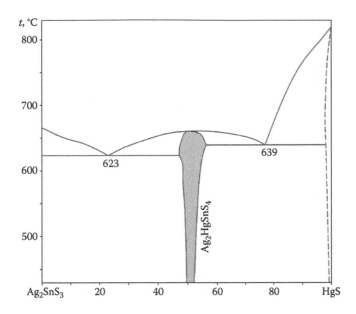

FIGURE 7.6 Phase diagram of the HgS–Ag$_2$SnS$_3$ quasibinary system. (From Parasyuk, O.V., *J. Alloys Compd.*, 291(1–2), 215, 1999.)

5.7740 ± 0.0008 g·cm^{-3}; and bandgap energy values 1.38 eV at room temperature and 1.45 eV at liquid N$_2$ temperature (Parasyuk et al. 2005). The highest point of melting is found for Ag$_2$HgSnS$_4$ situated in the interval 50–52.5 mol.% HgS. The homogeneity region does not exceed 3 mol.% from Ag$_2$SnS$_3$ side and 6 mol.% from HgS side and it decreases with decreasing temperature (Parasyuk 1999).

The coordinates of the eutectic points are 23 mol.% HgS and 623°C and 77 mol.% HgS and 639°C (Parasyuk 1999).

This system was investigated by DTA, XRD, and metallography and the ingots were annealed at 400°C during 250 h with further quenching in cold water (Parasyuk 1999). The Ag$_2$HgSnS$_4$ compound was obtained by the interaction of chemical elements at 700°C (Haeuseler and Himmrich 1989). Single crystals of this compound were grown using the Bridgman–Stockbarger technique (Parasyuk et al. 2005).

HgS–Ag$_8$SnS$_6$. The phase diagram of this quasibinary system (Figure 7.7) is of eutectic type, the eutectic coordinates being 89 mol.% HgS and 608°C (Parasyuk et al. 2005). Solid solution range based on β-Ag$_8$SnS$_6$ extends to ~57 mol.% HgS and significantly decreases with decreasing temperature. A eutectoid decomposition of this solid solution with the formation of the solid solution based on α-Ag$_8$SnS$_6$ and HgS occurs at 157°C. The horizontal line at 344°C corresponds to the phase transition of HgS. This system was investigated by DTA and XRD, and the ingots were annealed at 400°C during 250 h (Parasyuk et al. 2005).

HgS–Ag$_2$S–SnS$_2$. Isothermal section of this quasiternary system is shown in Figure 7.8 (Parasyuk et al. 2005). Only one quaternary compound, Ag$_2$HgSnS$_4$, exists in the HgS–Ag$_2$S–SnS$_2$ system at this temperature.

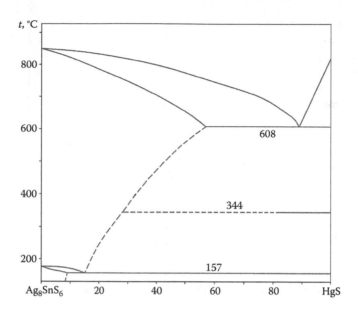

FIGURE 7.7 Phase diagram of the HgS–Ag$_8$SnS$_6$ quasibinary system. (From Parasyuk, O.V. et al., *J. Alloys Compd.*, 399(1–2), 32, 2005.)

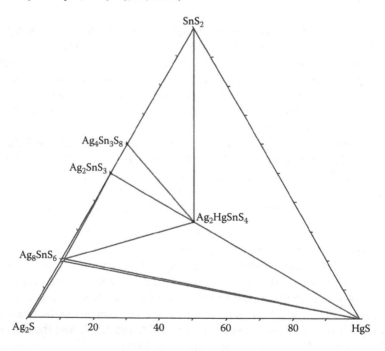

FIGURE 7.8 Isothermal section of the HgS–Ag$_2$S–SnS$_2$ quasiternary system at room temperature. (From Parasyuk, O.V. et al., *J. Alloys Compd.*, 399(1–2), 32, 2005.)

7.28 MERCURY–SILVER–ARSENIC–SULFUR

$HgS–AgAsS_2$. The phase diagram is not constructed. $AgHgAsS_3$ quaternary compound is formed in this system (mineral laffittite). It melts incongruently at 428°C ± 2°C (Il'yasheva et al. 1979) and crystallizes in the monoclinic structure with the lattice parameters $a = 773.2$, $b = 1128.5$, $c = 664.3$ pm, and $\beta = 115.16°$ (Nakai and Appleman 1983) [$a = 656.7 ± 0.3$, $b = 1402.0 ± 0.5$, $c = 638.8 ± 0.2$ pm, and $\beta = 119°05'$ (Johan et al. 1974, Fleischer and Mandarino 1975)] and calculated density 6.15 $g \cdot cm^{-3}$ (Johan et al. 1974) [in the pseudoorthorhombic structure with the lattice parameters $a = 1148.84 ± 0.05$, $b = 1402.0 ± 0.5$, and $c = 638.8 ± 0.2$ pm (Johan et al. 1974)]. The $AgHgAsS_2 + AgAsS_2$ eutectic crystallizes at 384°C ± 2°C and liquidus temperature corresponding to the $AgHgAsS_4$ compound is equal to 520°C ± 10°C (Il'yasheva et al. 1979).

This system was investigated by DTA and metallography. One-phase $AgHgAsS_3$ ingots were obtained by their annealing at 384°C–428°C for at least 2 months (Il'yasheva et al. 1979). Its single crystals were grown by the interaction of HgS, Ag_2S, and As_2S_3 at 900°C with next recrystallization from the mixture of Na_2S and $Na_2B_4O_7$ solutions (Nakai and Appleman 1983).

7.29 MERCURY–SILVER–CHLORINE–SULFUR

$HgS–AgCl$. The phase diagram is not constructed. $AgHgSCl$ and $AgHg_2S_2Cl$ quaternary compounds are formed in this system (Blachnik and Lytze 1990). The $AgHgSCl$ crystallizes in the orthorhombic structure with the lattice parameters $a = 1024 ± 1$, $b = 1273 ± 3$, and $c = 902.3 ± 0.5$ pm.

7.30 MERCURY–SILVER–BROMINE–SULFUR

$HgS–AgBr$. The phase diagram is not constructed. $AgHgSBr$ quaternary compound, which is Ag^+-ionic conductor, is formed in this system (Blachnik and Dreisbach 1986, Moro'oka et al. 2004a). It crystallizes in the orthorhombic structure with the lattice parameters $a = 964.8 ± 0.8$, $b = 466.1 ± 0.4$, and $c = 942.6 ± 0.6$ pm and calculated density 6.59 $g \cdot cm^{-3}$ (Beck et al. 2001). The Ag_2HgSBr_2 compound, which was obtained by the heating of the Ag_2S and $HgBr_2$ mixture in evacuated ampoules at 200°C for 15 h (Karataeva and Sviridov 1970), corresponds to the mixture of AgBr and $AgHgSBr$ (Blachnik and Dreisbach 1986).

The crystal chemistry of $AgHgSBr$ was discussed in Borisov et al. (2005), and it was shown that the degree of anion ordering in these compounds substantially exceeds (by 15%–20%) the degree of cation ordering in the same interval of interplanar spacings d_{hkl}, which determines the structure arrangement.

$AgHgSBr$ has been synthesized by a solid-state interaction of HgS and AgBr (Blachnik and Dreisbach 1986, Moro'oka et al. 2004a) or by the hydrothermal reaction of HgS and AgBr in half-concentrated HBr as a solvent at 300°C (Beck et al. 2001).

7.31 MERCURY–SILVER–IODINE–SULFUR

HgS–AgI. The phase diagram is shown in Figure 7.9 (Blachnik and Dreisbach 1986). AgHgSI and Ag_2HgSI_2 quaternary compounds are formed in this system. AgHgSI melts incongruently at 250°C ± 3°C and has a polymorphous transition at 305°C (Moro'oka et al. 2004a). Low-temperature modification crystallizes in the orthorhombic structure with the lattice parameters a = 1015.9 ± 0.2, b = 464.77 ± 0.05, and c = 984.9 ± 0.2 pm and calculated density 6.68 g·cm^{-3} (Beck et al. 2001). High-temperature modification also crystallizes in the orthorhombic structure with the lattice parameters a = 772.75, b = 846.95, and c = 707.10 pm (Moro'oka et al. 2004a) [a = 773.18 ± 0.02, b = 847.53 ± 0.02, and c = 707.47 ± 0.01 pm and calculation and experimental densities 6.70 and 6.67 ± 0.02 g·cm^{-3}, respectively (Beck et al. 2004); a = 772.3 ± 0.3, b = 847.7 ± 0.3, and c = 707.3 ± 0.3 pm (Blachnik and Dreisbach 1986)].

Ag_2HgSI_2 also crystallizes in the orthorhombic structure with the lattice parameters a = 1384.35 ± 0.06, b = 746.97 ± 0.03, and c = 710.29 ± 0.03 pm and calculated density 6.35 g·cm^{-3} (Keller and Wimbert 2003). It decomposes at 242°C with formation of AgHgSI and AgI. According to the data of Suchow and Pond (1954), this compound crystallizes in the cubic structure with the lattice parameter a = 495 pm.

The crystal chemistry of α- and β-AgHgSI and Ag_2HgSI_2 was discussed in Borisov et al. (2005), and it was shown that the degree of anion ordering in these compounds substantially exceeds (by 15%–20%) the degree of cation ordering in the same interval of interplanar spacings d_{hkl}, which determines the structure arrangement.

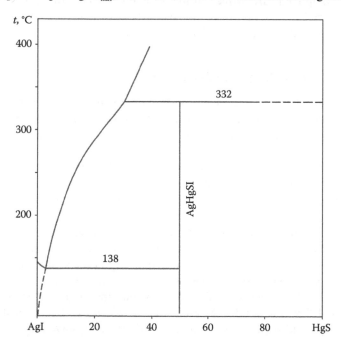

FIGURE 7.9 The HgS–AgI phase diagram within the interval of 70°C–370°C. (From Blachnik, R. and Dreisbach, H.A., *Monatsh. Chem.*, 117(3), 305, 1986.)

AgHgSI has been synthesized by a solid-state interaction of HgS and AgI (Blachnik and Dreisbach 1986, Moro'oka et al. 2004a) or by the hydrothermal reaction of HgS and AgI in half-concentrated HI as a solvent at 400°C (Beck et al. 2001, 2004). High-temperature modification of AgHgSI can be obtained at 273°C from low-temperature modification of this compound (Beck et al. 2004). Ag_2HgSI_2 was obtained by heating a mixture of AgI and α-HgS at 120°C (Keller and Wimbert 2003) or by heating a mixture of Ag_2S and HgI_2 at 200°C for 16 h in sealed evacuated tubes (Suchow and Pond 1954).

7.32 MERCURY–MAGNESIUM–FLUORINE–SULFUR

$HgS–MgF_2–HgF_2$. The phase diagram is not constructed. $Mg_2Hg_2SF_6$ quaternary compound is formed in this system. It crystallizes in the cubic structure with the lattice parameter $a = 1069.73 \pm 0.14$ pm (Bernard et al. 1975).

7.33 MERCURY–BARIUM–TIN–SULFUR

$HgS–BaS–SnS_2$. The phase diagram is not constructed. $BaHgSnS_4$ quaternary compound is formed in this system. It crystallizes in the orthorhombic structure with the lattice parameters $a = 1080.4$, $b = 1084.0$, and $c = 661.3$ pm and calculation and experimental densities 4.91 and 4.96 $g \cdot cm^{-3}$, respectively (Teske 1980). This compound was obtained using stoichiometric ratio of binary sulfides by their sintering at 550°C for 2 days and then at 650°C for 1 day.

7.34 MERCURY–GALLIUM–INDIUM–SULFUR

$HgS–Ga_2S_3–In_2S_3$. The phase diagram is not constructed. $HgGaInS_4$ and $Hg_{0.8}Ga_{1.6}In_{1.2}S_5$ quaternary compounds are formed in this system. $HgGaInS_4$ crystallizes in the hexagonal structure with the lattice parameters $a = 390$ and $c = 3140$ pm (Moldovyan and Metlinski 1987, Moldovyan et al. 1993) and energy gap $E_g = 2.41$ eV (Moldovyan and Chebotaru 1990, Moldovyan et al. 1993) [$E_g = 2.3$ eV (Moldovyan and Metlinski 1987)]. $Hg_{0.8}Ga_{1.6}In_{1.2}S_5$ crystallizes also in the hexagonal structure with the lattice parameters $a = 382$ and $c = 3092.9$ pm (Haeuseler et al. 1988).

HgGaInS$_4$ was obtained by the chemical transport reactions (Moldovyan and Metlinski 1987, Moldovyan and Chebotaru 1990), and $Hg_{0.8}Ga_{1.6}In_{1.2}S_5$ was synthesized from the binary sulfides at 600°C–800°C (Haeuseler et al. 1988).

7.35 MERCURY–GALLIUM–GERMANIUM–SULFUR

$HgS–Ga_2S_3–GeS_2$. No quaternary phases were found in this system. Seven fields of primary crystallization form the liquidus surface of the $HgS–Ga_2S_3–GeS_2$ quasiternary system (Figure 7.10) (Olekseyuk et al. 2006). Four of them correspond to the components of the system: HgS, γ-Ga_2S_3, δ-Ga_2S_3, and GeS_2. The other three fields belong to $HgGa_2S_4$, $HgGa_6S_{10}$, and Hg_4GeS_6. Eleven monovariant lines and five ternary invariant points separate the fields of primary crystallization. The types, the temperatures, and the coordinates of the ternary invariant points are presented in Table 7.1.

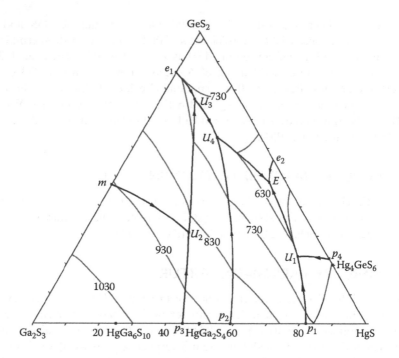

FIGURE 7.10 Liquidus surface of the $HgS–Ga_2S_3–GeS_2$ quasiternary system. (From Olekseyuk, I.D. et al., *J. Alloys Compd.*, 417(1–2), 131, 2006.)

TABLE 7.1
Temperatures and Compositions of the Ternary Invariant Points in the $HgS–Ga_2S_3–GeS_2$ Quasiternary System

Invariant Points	Reaction	Composition, mol.%			t, °C
		HgS	Ga₂S₃	GeS₂	
E	$L \Leftrightarrow HgGa_2S_4 + Hg_4GeS_6 + GeS_2$	47	5	48	556
U_1	$L + \alpha\text{-}HgS \Leftrightarrow HgGa_2S_4 + Hg_4GeS_6$	68	9	23	643
U_2	$L + \gamma\text{-}Ga_2S_3 \Leftrightarrow HgGa_6S_{10} + \delta\text{-}Ga_2S_3$	31	38	31	915
U_3	$L + \delta\text{-}Ga_2S_3 \Leftrightarrow HgGa_6S_{10} + GeS_2$	10	13	77	680
U_4	$L + HgGa_6S_{10} \Leftrightarrow HgGa_2S_4 + GeS_2$	23	13	64	667

Source: Olekseyuk, I.D. et al., *J. Alloys Compd.*, 417(1–2), 131, 2006.

Isothermal section of the $HgS–Ga_2S_3–GeS_2$ quasiternary system at 400°C is shown in Figure 7.11 (Olekseyuk et al. 2001). The existence of the $HgGe_2S_5$ ternary compound was not confirmed. The region of the solid solution based on HgS is not higher than 5 mol.% and the solubility of $HgGa_2S_4$ in Hg_4GeS_6 is equal to 2 mol.% at this temperature.

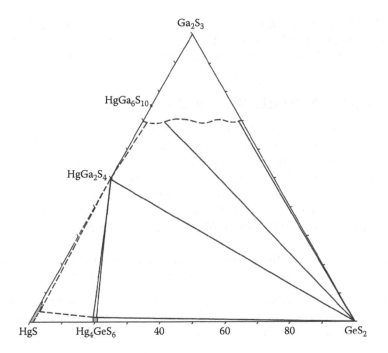

FIGURE 7.11 Isothermal section of the HgS–Ga$_2$S$_3$–GeS$_2$ quasiternary system at 400°C. (From Olekseyuk, I.D. et al., *Nauk. visnyk Volyn. derzh. un-tu. Khim. nauky*, (6), 38, 2001.)

This system was investigated by the XRD and metallography, and all alloys were annealed at 400°C for 250 h with the next quenching in air or in cold water (Olekseyuk et al. 2001, 2006).

7.36 MERCURY–INDIUM–CHROMIUM–SULFUR

HgS–In$_2$S$_3$–Cr$_2$S$_3$. The phase diagram is not constructed. Solid solutions with the spinel structure over the entire range of concentrations are formed in the subsystem HgIn$_2$S$_4$–HgCr$_2$S$_4$–In$_2$S$_3$ (Lutz et al. 1983). Two-phase regions exist in the HgS–HgIn$_2$S$_4$–HgCr$_2$S$_4$ and HgCr$_2$S$_4$–In$_2$S$_3$–Cr$_2$S$_3$ subsystems. This system was investigated by DTA, XRD, and thermogravimetry.

7.37 MERCURY–INDIUM–CHLORINE–SULFUR

HgS–HgCl$_2$–InCl$_3$. The phase diagram is not constructed. Hg$_4$In$_2$S$_3$Cl$_8$ and Hg$_7$InS$_6$Cl$_5$ quaternary compounds are formed in this system. Hg$_4$In$_2$S$_3$Cl$_8$ is stable up to 300°C and crystallizes in the orthorhombic structure with the lattice parameters $a = 1351.0 \pm 0.3$, $b = 734.00 \pm 0.15$, and $c = 1819.0 \pm 0.4$ pm; calculated density 5.199 g·cm^{-3}; and calculation and experimental energy gap 2.30 and 2.43 eV, respectively (Liu et al. 2012a). Hg$_7$InS$_6$Cl$_5$ is stable in the air at room temperature and crystallizes in the triclinic structure with the lattice parameters $a = 737.43 \pm 0.05$, $b = 1045.20 \pm 0.08$, $c = 1375.21 \pm 0.11$ pm, $\alpha = 109.415° \pm 0.03°$, $\beta = 93.746° \pm 0.003°$, and $\gamma = 105.246° \pm 0.003°$; calculated density 6.597 g·cm^{-3}; and energy gap 2.54 eV (Liu et al. 2012b).

$Hg_4In_2S_3Cl_8$ and $Hg_7InS_6Cl_5$ were prepared via the solid-state reaction from a mixture containing Hg_2Cl_2, In_2S_3, and S. Crystalline $Hg_4In_2S_3Cl_8$ can only be stable in air for about 4 days and then becomes amorphous (Liu et al. 2012a,b).

7.38 MERCURY–THALLIUM–SILICON–SULFUR

HgS–Tl_2SiS_3. The phase diagram is a eutectic type (Figure 7.12) (Olekseyuk et al. 2012). $HgTl_2SiS_4$ quaternary compound is formed in this system. It melts incongruently at 381°C, has polymorphous transformation at 312°C, and crystallizes in the tetragonal structure with the lattice parameters $a = 804.07 \pm 0.01$ and $c = 688.52 \pm 0.02$ pm. The eutectic composition and temperature are 36 mol.% HgS and 367°C, respectively. Thermal effects at 345°C correspond to the polymorphous transformation of HgS. Solubility of HgS in Tl_2SiS_3 is not higher than 5 mol.% at the eutectic temperature.

This system was investigated by DTA, XRD, and measuring of microhardness. The ingots were annealed at 250°C for 250 h (Olekseyuk et al. 2012).

HgS–Tl_2S–SiS_2. Isothermal section of this system at 250°C is shown in Figure 7.13 (Olekseyuk et al. 2012). $HgTl_2SiS_4$ and $HgTl_2Si_3S_8$ quaternary compounds exist in the HgS–Tl_2S–SiS_2 quasiternary system at this temperature.

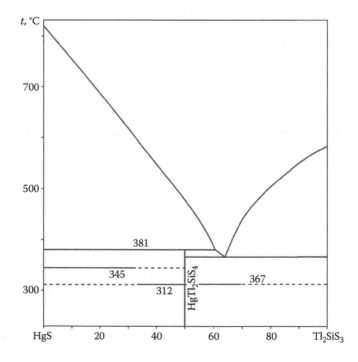

FIGURE 7.12   The HgS–Tl_2SiS_3 phase diagram. (From Olekseyuk, I.D. et al., *Nauk. visnyk Volyn. nats. un-tu im. Lesi Ukrainky. Khim. nauky*, [(17)242], 62, 2012.)

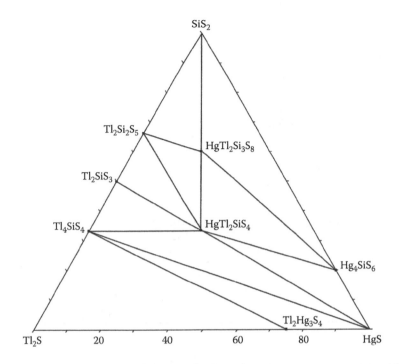

FIGURE 7.13 Isothermal section of the HgS–Tl_2S–SiS_2 quasiternary system at 250°C. (From Olekseyuk, I.D. et al., *Nauk. visnyk Volyn. nats. un-tu im. Lesi Ukrainky. Khim. nauky*, [(17)242], 62, 2012.)

7.39 MERCURY–THALLIUM–GERMANIUM–SULFUR

HgS–Tl_2GeS_3. The phase diagram is not constructed. $HgTl_2GeS_4$ quaternary compound is formed in this system (Mozolyuk et al. 2012).

7.40 MERCURY–THALLIUM–TIN–SULFUR

HgS–Tl_2S–SnS_2. Isothermal section of this system at 250°C is shown in Figure 7.14 (Olekseyuk et al. 2010). $HgTl_2SnS_4$ quaternary compound exists in the HgS–Tl_2S–SnS_2 quasiternary system at this temperature. It crystallizes in the tetragonal structure with the lattice parameters $a = 785.86 \pm 0.03$ and $c = 670.05 \pm 0.03$ pm. The ingots were annealed at 250°C for 250 h.

7.41 MERCURY–THALLIUM–ARSENIC–SULFUR

HgS–Tl_2S–As_2S_3. The phase diagram is not constructed. $HgTlAsS_3$ (mineral christite) and $HgTlAs_3S_6$ (mineral simonite) quaternary compounds are formed in this system. Both compounds crystallize in the monoclinic structure with the lattice parameters $a = 611.3 \pm 0.1$, $b = 1618.8 \pm 0.4$, $c = 611.1 \pm 0.1$ pm, and $\beta = 96.71° \pm 0.02°$ and calculation and experimental densities 6.37 and 6.2 ± 0.2 g·cm^{-3}, respectively, for $HgTlAsS_3$ (Brown and Dickson 1976, Radtke et al. 1977) and $a = 594.8$, $b = 1140.4$, $c = 1597.9$ pm, and $\beta = 90.15°$ and calculated density 5.036 g·cm^{-3} for $HgTlAs_3S_6$ (Engel et al. 1982).

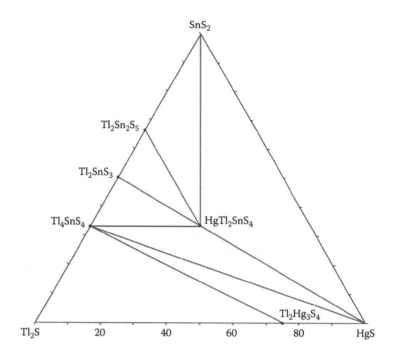

FIGURE 7.14 Isothermal section of the $HgS–Tl_2S–SnS_2$ quasiternary system at 250°C. (From Olekseyuk, I.D. et al., *Nauk. visnyk Volyn. nats. un-tu im. Lesi Ukrainky. Khim. nauky*, (30), 19, 2010.)

$HgTlAsS_3$ can be synthesized by reacting mixtures of HgS and $TlAsS_2$ in stoichiometric proportions in evacuated glass tubes at 290°C and below (Radtke et al. 1977). Heating above 300°C causes its thermal decomposition.

7.42 MERCURY–THALLIUM–ANTIMONY–SULFUR

$HgS–Tl_2S–Sb_2S_3$. The phase diagram is not constructed. $HgTlSb_4S_7$ (mineral vaughanite) quaternary compound is formed in this system. It crystallizes in the triclinic structure with the lattice parameters $a = 901.2 \pm 0.3$, $b = 1322.3 \pm 0.3$, $c = 590.6 \pm 0.2$ pm, $\alpha = 93.27° \pm 0.03°$, $\beta = 95.05° \pm 0.04°$, and $\gamma = 109.16° \pm 0.03°$ and calculated density 5.62 g·cm^{-3} (Harris and Roberts 1989).

7.43 MERCURY–THALLIUM–SELENIUM–SULFUR

$HgS–TlSe$. The phase diagram is a eutectic type (Figure 7.15) (Asadov et al. 1982). The eutectic composition and temperature are 20 mol.% HgS and 320°C, respectively. Solubility of HgS in TlSe reaches 18 mol.% at the eutectic temperature and decreases to 15 mol.% at 250°C. This system was investigated by DTA, XRD, and measuring of microhardness. The ingots were annealed at 280°C for 100 h.

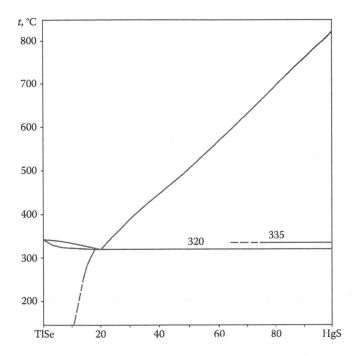

FIGURE 7.15 The HgS–TlSe phase diagram. (From Asadov, M.M. et al., *Azerb. khim. zhurn.*, (1), 102, 1982.)

HgS–Tl$_2$Se. The phase diagram is shown in Figure 7.16 (Kuliev et al. 1978). The eutectic composition and temperature are 43 mol.% HgS and 255°C, respectively. The Hg$_3$Tl$_2$SeS$_3$ quaternary compound is formed in this system. It melts incongruently at 400°C. Thermal effects at 335°C correspond to the polymorphous transformation of HgS. Solubility of HgS in Tl$_2$Se reaches 33 mol.% at the eutectic temperature and the solubility of Tl$_2$Se in HgS is insignificant. This system was investigated by DTA, XRD, and measuring of microhardness.

HgSe–Tl$_2$S. This section is nonquasibinary section of the Hg–Tl–Se–S quaternary system (Asadov 1983a). It was investigated by DTA, XRD, and measuring of micro-hardness and emf of concentrated chains.

HgS + Tl$_2$Se $\Leftrightarrow$ HgSe + Tl$_2$S. Four fields of primary crystallization exist on the liquidus surface of this system (Figure 7.17) (Asadov 1983a): HgS$_{1-x}$Se$_x$ and Hg$_3$Tl$_2$S$_{4(1-x)}$Se$_{4x}$ (γ) solid solutions and solid solutions based on Tl$_2$Se (α) and Tl$_2$S (δ). These fields are divided by the next lines of monovariant equilibria: e_1E (L $\Leftrightarrow$ δ + γ; 245°C–235°C), e_2E (L $\Leftrightarrow$ α + γ; 320°C–235°C), and e_3E (L $\Leftrightarrow$ α + δ; 385°C–235°C). Isothermal section of the HgS + Tl$_2$Se $\Leftrightarrow$ HgSe + Tl$_2$S ternary mutual system at 200°C is shown in Figure 7.18 (Asadov 1983a). This system was investigated by DTA, XRD, and measuring of microhardness and emf of concentrated chains. The ingots were annealed at 200°C for 800 h.

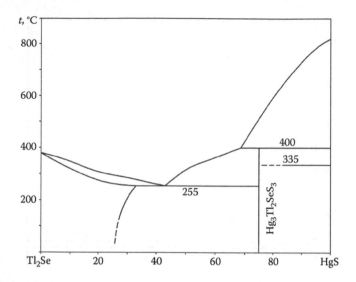

FIGURE 7.16 The HgS–Tl$_2$Se phase diagram. (From Kuliev, A.A. et al., *Zhurn. neorgan. khimii*, 23(3), 854, 1978.)

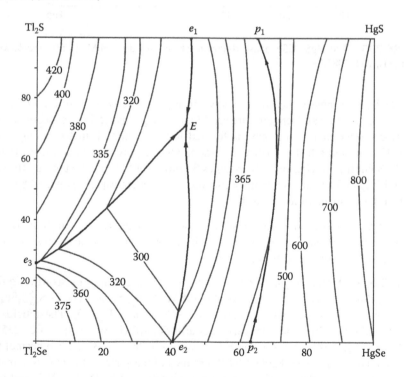

FIGURE 7.17 Liquidus surface of the HgS + Tl$_2$Se ⇔ HgSe + Tl$_2$S ternary mutual system. (From Asadov, M.M, *Zhurn. neorgan. khimii*, 28(7), 1812, 1983a.)

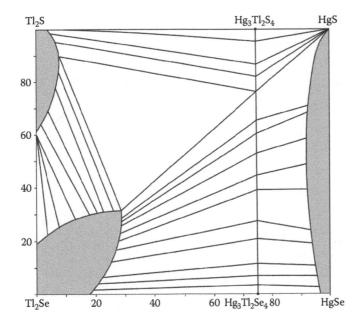

FIGURE 7.18 Isothermal section of the HgS + Tl$_2$Se $\Leftrightarrow$ HgSe + Tl$_2$S ternary mutual system at 200°C. (From Asadov, M.M., *Zhurn. neorgan. khimii*, 28(7), 1812, 1983a.)

7.44 MERCURY–THALLIUM–TELLURIUM–SULFUR

HgS–Tl$_2$Te. This section is a nonquasibinary section of the Hg–Tl–Se–S quaternary system (Asadov 1983b). It was investigated by DTA, XRD, and measuring of microhardness and emf of concentrated chains.

HgS + Tl$_2$Te $\Leftrightarrow$ HgTe + Tl$_2$S. Isothermal section of this system at 200°C is shown in Figure 7.19 (Asadov 1983b). This system was investigated by DTA, XRD, and measuring of microhardness and emf of concentrated chains.

7.45 MERCURY–LANTHANUM–OXYGEN–SULFUR

HgS–La$_2$O$_2$S. According to the data of XRD (Baranov et al. 1996), this section is a nonquasibinary section of the Hg–La–O–S quaternary system. The ingots were annealed at 730°C–830°C for 250 h.

7.46 MERCURY–URANIUM–CHLORINE–SULFUR

HgS–HgCl$_2$–UCl$_4$. The phase diagram is not constructed. [Hg$_3$S$_2$][UCl$_6$] quaternary compound is formed in this system, which crystallizes in the monoclinic structure with the lattice parameters $a = 686.5 \pm 0.5$, $b = 736.8 \pm 0.5$, $c = 1314.3 \pm 0.8$ pm, and $\beta = 91.339° \pm 0.008°$ (Bugaris and Ibers 2008).

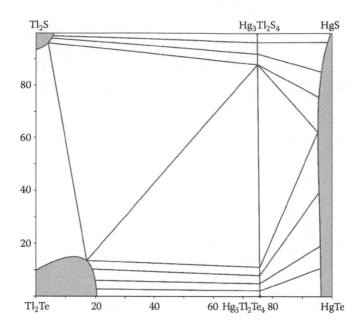

FIGURE 7.19 Isothermal section of the HgS + Tl$_2$Te $\Leftrightarrow$ HgTe + Tl$_2$S ternary mutual system at 200°C. (From Asadov, M.M., *Zhurn. neorgan. khimii*, 28(8), 2100, 1983b.)

This compound was synthesized using the mixture of HgS, UCl$_4$, and HgCl$_2$, which was heated to 850°C in 24 h, kept at this temperature for 96 h, cooled at 6.8°C/h to 200°C, and then cooled rapidly to 20°C.

7.47 MERCURY–CARBON–FLUORINE–SULFUR

Hg(SCF$_3$)$_2$ and Hg(SC$_6$F$_5$)$_2$ quaternary compounds are formed in this system. Hg(SCF$_3$)$_2$ [bis(trifluoromethylthiolato)mercury] was prepared by the reaction between HgF$_2$ and CS$_2$ and was purified by repeated vacuum sublimation at room temperature and dried in vacuo over P$_2$O$_5$ (Man et al. 1959, Downs et al. 1961). It could be also obtained by the interaction of (SCF$_3$)$_2$ with Hg under illumination (Haszeldine and Kidd 1953). Hg(SCF$_3$)$_2$ tends to crystallize in plates, which show birefringence under the microscope. This compound melts at 37°C–40°C and decomposes at 350°C (Haszeldine and Kidd 1953, Man et al. 1959, Downs et al. 1961). Heated at 220°C (48 h) in a silica tube, it gives volatile products (Haszeldine and Kidd 1955). Its experimental density is 2.911 g · cm^{-3} (Man et al. 1959).

To obtain Hg(SC$_6$F$_5$)$_2$ [bis(pentafluorobenzenethiolato)mercury], a slight excess, maximum 5%, of C$_6$F$_5$SH was added to approximately 3 mmol of Hg^{+2} ion in 25 mL of water or dilute acid (Peach 1968). The mixture was stirred magnetically until a solid was formed and no more C$_6$F$_5$SH appeared to be present. The product was filtered off, washed, and dried. Using Hg(NO$_3$)$_2$, Hg and Hg(SC$_6$F$_5$)$_2$ were formed, which were separated by dissolving Hg(SC$_6$F$_5$)$_2$ in MeOH from which it could be obtained on removal of the solvent. This compound melts at 188°C.

7.48   MERCURY–CARBON–CHLORINE–SULFUR

$Hg(SC_6Cl_5)_2$ [bis(pentachlorobenzenethiolato)mercury] is formed in this system. It was formed by shaking together an aqueous solution of $Hg(NO_3)_2 \cdot H_2O$ and a benzene solution of C_6Cl_5SH (Lucas and Peach 1969) or by the interaction of 95% ethanol solution of C_6Cl_5SH with suspension of HgO (Lucas and Peach 1970). This compound melts at the temperature higher than 300°C (Lucas and Peach 1969, 1970).

7.49   MERCURY–SILICON–FLUORINE–SULFUR

$HgS–HgF_2–SiF_4$. The phase diagram is not constructed. $Hg_3S_2SiF_6$ quaternary compound is formed in this system. This compound exists in two polymorphic modifications. α-$Hg_3S_2SiF_6$ crystallizes in the orthorhombic structure with the lattice parameters $a = 1221.9$, $b = 1490.2$, and $c = 2233.4$ pm and calculation and experimental densities 6.607 and 6.57 $g \cdot cm^{-3}$, respectively. β-$Hg_3S_2SiF_6$ crystallizes in the hexagonal structure with the lattice parameters $a = 695.7$ and $c = 458.4$ pm and calculation and experimental densities 6.82 and 6.73 $g \cdot cm^{-3}$, respectively (Puff et al. 1969).

α-$Hg_3S_2SiF_6$ was obtained at the temperatures below 120°C by the interaction of HgS solution in $Hg(CH_3COO)_2/CH_3COOH$ and H_2SiF_6, or by the interaction of HgS solution in H_2SiF_6 and H_2S (thioacetamide, HgS), or by the interaction of the solution of $Hg_2S(CH_3COO)_2$ and H_2SiF_6, or by the interaction of HgO solution in H_2SiF_6 and CS_2. β-$Hg_3S_2SiF_6$ was prepared by the heating of α-$Hg_3S_2SiF_6$ at 140°C or by the interaction of HgS with the HgO solution in H_2SiF_6 at 120°C (Puff et al. 1969).

7.50   MERCURY–GERMANIUM–SELENIUM–SULFUR

$HgS–GeSe_2$. The phase diagram is not constructed. $Hg_2GeSe_2S_2$ and $Hg_4GeSe_2S_4$ quaternary compounds, which melt incongruently at 707°C and 862°C, respectively, are formed in this system (Asadov 2006). The solubility of HgS in $GeSe_2$ and $GeSe_2$ in HgS at 600°C reaches 18 and 5 mol.%, respectively. This system was investigated by DTA, XRD, metallography, and measuring of microhardness and emf of concentration chains.

7.51   MERCURY–TIN–BROMINE–SULFUR

$HgS–SnBr_2$. The phase diagram is not constructed. $Hg_2SnS_2Br_2$ quaternary compound is formed in this system. It melts incongruently at 292°C and crystallizes in the monoclinic structure with the lattice parameters $a = 935.6$, $b = 802.8$, $c = 1063.0$ pm, and $\beta = 103.06°$ and calculated density 6.532 $g \cdot cm^{-3}$ (Blachnik et al. 1996) [$a = 950.3$, $b = 815.2$, $c = 1071.8$ pm, and $\beta = 102.68°$ according to ab initio calculation (Ruiz and Payne 1998)].

α-HgS reacts with $SnBr_2$ at 220°C to form the 1D intercalation compound $2HgS \cdot SnBr_2$ (Ruiz and Payne 1998). In this compound, the helical structure of the HgS chains remains, but the turns are modified to give tetragonal symmetry. The formation of $2HgS \cdot SnBr_2$ can be understood as a process of miscibility between the two solid phases.

7.52 MERCURY–LEAD–BROMINE–SULFUR

HgS–PbBr$_2$. The phase diagram is not constructed. Hg$_2$PbS$_2$Br$_2$ quaternary compound is formed in this system. It crystallizes in the monoclinic structure with the lattice parameters $a = 971.8$, $b = 813.1$, $c = 1088.8$ pm, and $\beta = 105.60°$ according to ab initio calculation (Ruiz and Payne 1998).

α-HgS reacts with PbBr$_2$ to form the 1D intercalation compound 2HgS·PbBr$_2$ (Ruiz and Payne 1998). In this compound, the helical structure of the HgS chains remains but the turns are modified to give tetragonal symmetry. The formation of 2HgS·PbBr$_2$ can be understood as a process of miscibility between the two solid phases.

7.53 MERCURY–LEAD–IODINE–SULFUR

HgS–PbI$_2$. The phase diagram is not constructed. Hg$_2$PbS$_2$I$_2$ quaternary compound is formed in this system. It crystallizes in the tetragonal structure with the lattice parameters $a = 1350.1$ and $c = 459.3$ pm (Blachnik et al. 1986).

7.54 MERCURY–ZIRCONIUM–CHLORINE–SULFUR

HgS–HgCl$_2$–ZrCl$_4$. The phase diagram is not constructed. Hg$_3$S$_2$[ZrCl$_6$] quaternary compound is formed in this system. It crystallizes in the monoclinic structure with the lattice parameters $a = 1290.83 \pm 0.05$, $b = 734.97 \pm 0.03$, $c = 662.18 \pm 0.02$ pm, and $\beta = 91.755° \pm 0.002°$ and calculated density 5.13 g·cm^{-3} (Beck and Hedderich 2003). To obtain this compound, the mixture of HgS, HgCl$_2$, and ZrCl$_4$ in the molar ratio 2:1:1 was used as starting materials. With a rate of 40°C/h, a temperature of 360°C was reached and kept for 5 days. After that, the ampoule was taken out of the furnace and air quenched.

7.55 MERCURY–ZIRCONIUM–BROMINE–SULFUR

HgS–HgBr$_2$–ZrBr$_4$. The phase diagram is not constructed. Hg$_3$S$_2$[ZrBr$_6$] quaternary compound is formed in this system. It crystallizes in the monoclinic structure with the lattice parameters $a = 1350.99 \pm 0.06$, $b = 756.79 \pm 0.03$, $c = 701.97 \pm 0.03$ pm, and $\beta = 92.16° \pm 0.01°$ (Beck and Hedderich 2003). To obtain this compound, the mixture of HgS, HgBr$_2$, and ZrBr$_4$ in the molar ratio 2:1:1 was used as starting materials. With a rate of 40°C/h, a temperature of 400°C was reached and kept for 5 days. After that, the ampoule was taken out of the furnace and air quenched.

7.56 MERCURY–HAFNIUM–CHLORINE–SULFUR

HgS–HgCl$_2$–HfCl$_4$. The phase diagram is not constructed. Hg$_3$S$_2$[HfCl$_6$] quaternary compound is formed in this system. It crystallizes in the monoclinic structure with the lattice parameters $a = 1291.37 \pm 0.05$, $b = 734.84 \pm 0.02$, $c = 661.24 \pm 0.01$ pm, and $\beta = 91.78° \pm 0.01°$ (Beck and Hedderich 2003). To obtain this compound, the mixture of HgS, HgCl$_2$, and HfCl$_4$ in the molar ratio 2:1:1 was used as starting materials. With a rate of 40°C/h, a temperature of 360°C was reached and kept for 5 days. After that, the ampoule was taken out of the furnace and air quenched.

7.57 MERCURY–NITROGEN–OXYGEN–SULFUR

HgS–Hg(NO$_3$)$_2$. The phase diagram is not constructed. At the interaction of Hg$_3$S$_2$Cl$_2$ with HNO$_3$, the Hg$_3$S$_2$(NO$_3$)$_2$ quaternary compound is formed in this system (Puff and Kohlschmidt 1962).

7.58 MERCURY–ARSENIC–ANTIMONY–SULFUR

Hg$_3$AsSbS$_3$ quaternary compound (mineral tvalchrelidzeite) is formed in this system (Yange et al. 2007). It crystallizes in the monoclinic structure with the lattice parameters $a = 1155.26 \pm 0.04$, $b = 438.52 \pm 0.01$, $c = 1563.73 \pm 0.05$ pm, and $\beta = 91.845° \pm 0.002°$ and calculated density 7.505 g·cm^{-3}.

7.59 MERCURY–ARSENIC–CHLORINE–SULFUR

HgS–"AsSCl." The phase diagram is not constructed. The Hg$_3$AsS$_4$Cl quaternary compound is formed in this system. It crystallizes in the hexagonal structure with the lattice parameters $a = 745.1 \pm 0.5$ and $c = 900.3 \pm 0.8$ pm (Beck et al. 2000). Hg$_3$AsS$_4$Cl was synthesized by the chemical transport reaction using the mixture of HgS, As, HgCl$_2$, and S at the temperatures 250°C–310°C.

7.60 MERCURY–ARSENIC–BROMINE–SULFUR

(HgBr$_2$)$_3$(As$_4$S$_4$) quaternary compound is formed in this system. It crystallizes in the monoclinic structure with the lattice parameters $a = 959.3 \pm 0.5$, $b = 1139.5 \pm 0.5$, $c = 1340.2 \pm 0.5$ pm, and $\beta = 107.27° \pm 0.03°$ (Bräu and Pfitzner 2007). It was obtained by melting a mixture of HgBr$_2$, As, and S at 400°C in an evacuated silica ampoules and subsequent annealing at 190°C for 2 weeks. Higher yields are obtained by the reaction of a mixture of HgBr$_2$ and As$_4$S$_4$ in CS$_2$ at 160°C for 2 weeks.

HgS–AsSBr. The phase diagram is not constructed. The Hg$_3$AsS$_4$Br quaternary compound is formed in this system. It crystallizes in the hexagonal structure with the lattice parameters $a = 743.0 \pm 0.5$ and $c = 936.4 \pm 0.5$ pm (Beck et al. 2000). Hg$_3$AsS$_4$Br was synthesized by the chemical transport reaction using the mixture of HgS, As, HgBr$_2$, and S at the temperatures 250°C–310°C.

7.61 MERCURY–ARSENIC–IODINE–SULFUR

HgI$_2$·As$_4$S$_4$ quaternary compound is formed in this system. It crystallizes in the monoclinic structure with the lattice parameters $a = 943.3 \pm 0.3$, $b = 1498.6 \pm 0.9$, $c = 1162.4 \pm 0.5$ pm, and $\beta = 127.72° \pm 0.02°$ and calculated density 4.509 g·cm^{-3} (Bräu and Pfitzner 2006a,b). It was obtained by heating HgI$_2$, gray As, and S in the molar ratio 1:4:4 in an evacuated quartz ampoules. The reactants were molten at 400°C and then annealed at 200°C for 2 weeks. HgI$_2$·As$_4$S$_4$ decomposes peritectically at 212°C and could not be obtained directly from the melt.

7.62   MERCURY–BISMUTH–CHLORINE–SULFUR

HgS–HgCl$_2$–BiCl$_3$. The phase diagram is not constructed. The Hg$_3$Bi$_2$S$_2$Cl$_8$ quaternary compound is formed in this system. It melts incongruently and crystallizes in the monoclinic structure with the lattice parameters a = 1293.81 ± 0.09, b = 738.28 ± 0.06, c = 926.06 ± 0.06 pm, and β = 116.647° ± 0.005°; calculated density 5.744 g·cm^{-3}; and energy gap ~3.26 eV (Wibowo et al. 2013). To obtain this compound, a mixture of HgS, HgCl$_2$, and BiCl$_3$ in the molar ratio 2:1:6 was loaded into a fused silica tube. Then the tube was flame sealed under vacuum, put inside a programmable furnace, heated to 300°C over 24 h, and held isothermally for 12 h, followed by a shutdown of the furnace. The excess of BiCl$_3$ was washed away using MeOH. The reaction with a 1:1 mixture of HgS and BiCl$_3$ provided single crystals of Hg$_3$Bi$_2$S$_2$Cl$_8$. To obtain single crystals of this compound, the thoroughly mixed powders of HgS and BiCl$_3$ were loaded into a fused silica tube, which was flame sealed under vacuum and heated to 450°C over 12 h and held isothermally for 1 day, followed by cooling slowly to 200°C over 36 h and then to room temperature within 1 h.

7.63   MERCURY–OXYGEN–SELENIUM–SULFUR

HgS–SeO$_2$. This section is nonquasibinary section of the Hg–O–Se–S quaternary system (Smirnova and Markovski 1964). Chemical interaction of HgS and SeO$_2$ begins at 200°C and runs with great velocity and big exothermal effect. Onset temperature of the reaction depends on the humidity of binary components and decreases from 200°C to 95°C–100°C at the increase in moisture from 0.5% to 5%. The interaction of equimolar quantities of HgS and SeO$_2$ at 210°C leads to the formation of HgSe, containing less than 0.1 mol.% HgSO$_4$, HgO, HgS, and Se. HgSe yield can reach 90%–95% from calculation possible. The heating of reaction products, obtained at 210°C up to 700°C, doesn't cause any change. Interaction of HgS with SeO$_2$ can be described by the following reaction: HgS + SeO$_2$ = HgSe + SO$_2$ + 46.5 kJ.

HgSe–HgSO$_4$. The phase diagram is not constructed. The Hg$_3$Se$_2$SO$_4$ quaternary compound is formed in this system (Puff 1963).

7.64   MERCURY–OXYGEN–TELLURIUM–SULFUR

HgTe–HgSO$_4$. The phase diagram is not constructed. The Hg$_3$Te(SO$_4$)$_2$ quaternary compound is formed in this system. It crystallizes in the cubic structure with the lattice parameter a = 1007 pm (Puff 1963).

7.65   MERCURY–OXYGEN–CHROMIUM–SULFUR

The Hg$_5$CrO$_5$S$_2$ (mineral deanesmithite) and Hg$_3$CrO$_4$S$_2$ (mineral edoylerite) quaternary compounds are formed in this system. Hg$_5$CrO$_5$S$_2$ crystallizes in the triclinic structure with the lattice parameters a = 812.87 ± 0.08, b = 949.16 ± 0.07, c = 689.40 ± 0.04 pm, α = 100.356° ± 0.006°, β = 110.163° ± 0.007°, and γ = 82.981° ± 0.008°

(Szymanski and Groat 1997, Pervukhina et al. 1999) [$a = 811.6 \pm 0.6$, $b = 950.1 \pm 0.8$, $c = 681.9 \pm 0.9$ pm, $\alpha = 100.43° \pm 0.08°$, $\beta = 110.24° \pm 0.08°$, and $\gamma = 82.80° \pm 0.08°$ and calculated density 8.14 g·cm^{-3} (Roberts et al. 1993, Jamber et al. 1994)].

Hg$_3$CrO$_4$S$_2$ crystallizes in the monoclinic structure with the lattice parameters $a = 752.83 \pm 0.04$, $b = 1483.25 \pm 0.08$, $c = 746.29 \pm 0.04$ pm, and $\beta = 118.746° \pm 0.001°$ and calculated density 7.11 g·cm^{-3} (Burna 1999) [$a = 752.4 \pm 0.7$, $b = 1481.9 \pm 0.8$, $c = 744.3 \pm 0.5$ pm, and $\beta = 118.72° \pm 0.05°$ and calculated density 7.14 g·cm^{-3} (Erd et al. 1993, Jamber et al. 1994)].

7.66 MERCURY–OXYGEN–CHLORINE–SULFUR

HgS–Hg(ClO$_4$). The phase diagram is not constructed. The Hg$_3$S$_2$(ClO$_4$)$_2$ quaternary compound is formed in this system. It crystallizes in the hexagonal structure with the lattice parameters $a = 715$ and $c = 1700$ pm (Puff 1963).

7.67 MERCURY–OXYGEN–IRON–SULFUR

HgS–Fe$_2$(SO$_4$)$_3$. The phase diagram is not constructed. Interaction of HgS (cinnabar) with the solution of Fe$_2$(SO$_4$)$_3$ leads to the formation of HgSO$_4$, which partly hydrolyzes and partly dissolves in the water (Saukov and Aydin'yan 1940). The presence of NaCl in the solutions significantly accelerates the oxidation of HgS and increases the solubility of mercury. Raising the temperature accelerates the interaction.

7.68 MERCURY–SELENIUM–TELLURIUM–SULFUR

HgS–HgSe–HgTe. Isothermal section of this system at 300°C is shown in Figure 7.20 (Malevski and Chzhun 1965). There are two two-phase regions in the HgS–HgSe–HgTe quasiternary system at this temperature. One of them is conditioned by the phase transformation of the solid solutions based on HgS, and the second is the immiscibility gap of the solid solutions with the sphalerite structure.

Figure 7.21 shows the position of the calculated miscibility gap in the HgS–HgSe–HgTe system at 400°C–800°C (Ohtani et al. 1992). It is seen that the surface of the miscibility gap extends from HgS–HgTe quasibinary system into the ternary.

7.69 MERCURY–FLUORINE–MANGANESE–SULFUR

HgS–HgF$_2$–MnF$_2$. The phase diagram is not constructed. The Hg$_2$Mn$_2$SF$_6$ quaternary compound is formed in this system. It crystallizes in the cubic structure with the lattice parameter $a = 1095.49 \pm 0.31$ pm (Bernard et al. 1975).

7.70 MERCURY–FLUORINE–COBALT–SULFUR

HgS–HgF$_2$–CoF$_2$. The phase diagram is not constructed. The Hg$_2$Co$_2$SF$_6$ quaternary compound is formed in this system. It crystallizes in the cubic structure with the lattice parameter $a = 1080.63 \pm 0.05$ pm (Bernard et al. 1975).

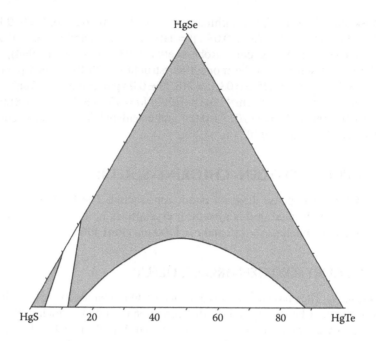

FIGURE 7.20 Isothermal section of the HgS–HgSe–HgTe quasiternary system at 300°C. (From Malevski, A.Yu. and Chzhun, T.-zhun, in *Eksperiment.-metodich. issled. rudnyh mineralov*, M.: Nauka Publish., 223, 1965.)

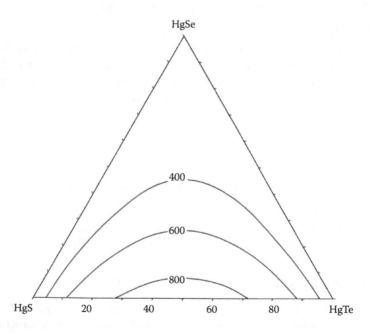

FIGURE 7.21 Calculated miscibility gap in the HgS–HgSe–HgTe quasiternary system. (From Ohtani, H. et al., *J. Alloys Compd.*, 182(1), 103, 1992.)

7.71   MERCURY–FLUORINE–NICKEL–SULFUR

HgS–HgF$_2$–NiF$_2$. The phase diagram is not constructed. The Hg$_2$Ni$_2$SF$_6$ quaternary compound is formed in this system. It crystallizes in the cubic structure with the lattice parameter a = 1095.49 ± 0.31 pm [1070.75 ± 0.13 pm (Champlon et al. 1974)] and begins to decompose at 250°C (Bernard et al. 1975). This compound was obtained by the interaction of HgS, HgF$_2$, and NiF$_2$ at 400°C–500°C for 12 h (Champlon et al. 1974).

7.72   MERCURY–CHLORINE–SELENIUM–SULFUR

HgS–HgSe–HgCl$_2$. Isothermal section of this quasiternary system at 400°C is shown in Figure 7.22 (Hudoliy et al. 1993). The limit composition of HgS$_x$Se$_{1-x}$ solid solutions in the three-phase region is HgS$_{0.8}$Se$_{0.2}$. Isothermal sections at the temperatures below 355°C include two two-phase regions where HgCl$_2$ and Hg$_3$S$_{2-2x}$Se$_{2x}$Cl$_2$ solid solutions and HgS$_x$Se$_{1-x}$ and Hg$_3$S$_{2-2x}$Se$_{2x}$Cl$_2$ solid solutions are in equilibrium.

7.73   MERCURY–CHLORINE–TELLURIUM–SULFUR

HgS–HgTe–HgCl$_2$. Isothermal section of this quasiternary system at 250°C is characterized by a wide one-phase region (Figure 7.23) (Voroshilov et al. 1996b).

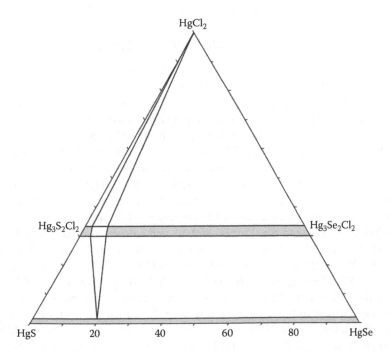

FIGURE 7.22   Isothermal section of the HgS–HgSe–HgCl$_2$ quasiternary system at 400°C. (From Hudoliy, V.A. et al., *Zhurn. neorgan. khimii*, 38(9), 1584, 1993.)

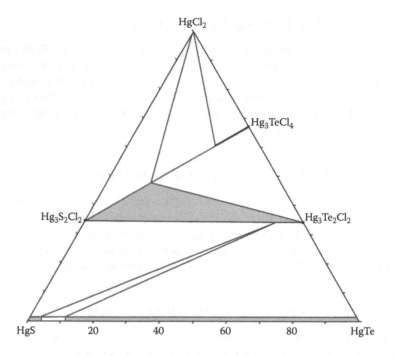

FIGURE 7.23 Isothermal section of the HgS–HgTe–HgCl₂ quasiternary system at 250°C. (From Voroshilov, Yu.V. et al., *Zhurn. neorgan. khimii*, 41(2), 287, 1996b.)

7.74 MERCURY–CHLORINE–BROMINE–SULFUR

HgS–HgCl₂–HgBr₂. Isothermal section of this quasiternary system at 200°C is shown in Figure 7.24 (Pan'ko et al. 1996). α-Solid solutions based on $Hg_3S_2Cl_2$ are in equilibrium with solid solutions based on $HgCl_2$, γ-phase of the $HgCl_2$–$HgBr_2$ system, and HgS at this temperature and $Hg_3S_2(Br_{1-x}Cl_x)_2$ solid solutions are in equilibrium with γ-phase, solid solutions based on $HgBr_2$ and HgS.

In the field of the $Hg_3S_2(Br_{1-x}Cl_x)_2$ solid solutions, the growth of the single crystals with $x \approx 0.5$ from a gaseous phase was carried out at 540°C in an evacuated silica ampoule filled with a mixture of HgS (cinnabar), $HgCl_2$, and $HgBr_2$. Two polymorphic modifications have been prepared. One of them crystallizes in the monoclinic structure with the lattice parameters $a = 1782.4 \pm 0.4$, $b = 923.8 \pm 0.2$, $c = 1026.9 \pm 0.2$ pm, and $\beta = 115.69° \pm 0.01°$ and calculated density 7.005 g·cm⁻³ (Pervukhina et al. 2006a,b) [$a = 1684.1 \pm 0.2$, $b = 912.8 \pm 0.2$, $c = 943.5 \pm 0.4$ pm, and $\beta = 90.08° \pm 0.01°$ and calculated density 6.952 g·cm⁻³ (Pervukhina et al. 2005, 2006c); $a = 1685.2 \pm 0.3$, $b = 913.6 \pm 0.2$, $c = 945.7 \pm 0.2$ pm, and $\beta = 90.09° \pm 0.03°$ and calculated density 6.925 g·cm⁻³ (Borisov et al. 1999)]. Other polymorphic modification crystallizes in the cubic structure with the lattice parameter $a = 1824.8 \pm 0.2$ and calculated density 7.026 g·cm⁻³ (Pervukhina et al. 2006a, b) [$a = 1804.8 \pm 0.2$ (Pervukhina et al. 2006c); $a = 1800.6 \pm 0.3$ and calculated density 6.909 g·cm⁻³ (Borisov et al. 1999, Pervukhina et al. 2005)].

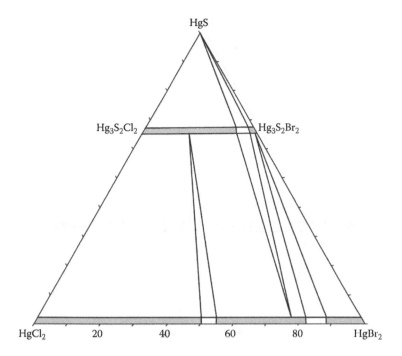

FIGURE 7.24 Isothermal section of the HgS–HgCl$_2$–HgBr$_2$ quasiternary system at 200°C. (From Pan'ko, V.V. et al., *Zhurn. neorgan. khimii*, 41(10), 1731, 1996.)

Two minerals—lavrentievite (Hg$_{3.11}$S$_{1.97}$Br$_{1.16}$Cl$_{0.76}$) and arzakite (Hg$_{3.08}$S$_{1.96}$Br$_{1.22}$Cl$_{0.75}$)— exist in this quasiternary system (Vasil'ev et al. 1984, 1986). Lavrentievite crystallizes in the monoclinic [$a = 894 \pm 2$, $b = 519.4 \pm 0.7$, $c = 1833 \pm 4$ pm, and $\beta = 92.44° \pm 0.08°$ and calculated density 7.26–7.51 g·cm^{-3} (Vasil'ev et al. 1984)] or the triclinic [$a = 890 \pm 1$, $b = 520.7 \pm 0.6$, $c = 1834 \pm 3$ pm, $\alpha = 89.67° \pm 0.07°$, $\beta = 88.88° \pm 0.07°$, and $\gamma = 92.28° \pm 0.09°$ and calculated density 7.27–7.52 g·cm^{-3} (Vasil'ev et al. 1984)] structure. Arzakite also crystallizes in the monoclinic [$a = 899 \pm 4$, $b = 524 \pm 1$, $c = 1845 \pm 8$ pm, and $\beta = 92.28° \pm 0.15°$ and calculated density 7.64–7.69 g·cm^{-3} (Vasil'ev et al. 1984, 1986)] or the triclinic [$a = 895 \pm 4$, $b = 525 \pm 2$, $c = 1849 \pm 9$ pm, $\alpha = 89.57° \pm 0.24°$, $\beta = 88.74° \pm 0.24°$, and $\gamma = 92.16° \pm 0.32°$ and calculated density 7.63 g·cm^{-3} (Vasil'ev et al. 1984)] structure.

7.75 MERCURY–CHLORINE–IODINE–SULFUR

HgS–HgCl$_2$–HgI$_2$. The phase diagram is not constructed. The Hg$_3$S$_2$ClI quaternary compound (mineral radtkeite) is formed in this system. It crystallizes in the monoclinic structure with the lattice parameters $a = 1682.7 \pm 0.4$, $b = 911.7 \pm 0.1$, $c = 1316.5 \pm 0.5$ pm, and $\beta = 130.17° \pm 0.02°$ and calculated density 7.130 g·cm^{-3} (Pervukhina et al. 2004a,b) [in the orthorhombic structure with the lattice parameters $a = 1685 \pm 1$, $b = 2027 \pm 2$, and $c = 913.3 \pm 0.2$ pm and calculation and experimental densities 7.05 ± 0.1 and 7.0 ± 0.1 g·cm^{-3}, respectively (McCormack et al. 1991)].

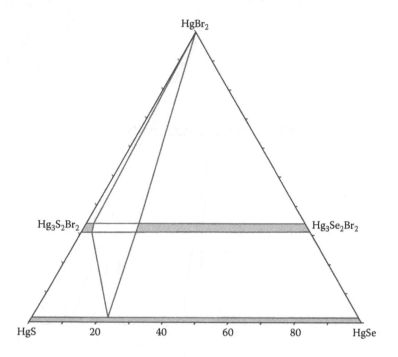

FIGURE 7.25   Isothermal section of the HgS–HgSe–HgBr$_2$ quasiternary system at 400°C. (From Hudoliy, V.A., et al., *Zhurn. neorgan. khimii*, 38(9), 1584, 1993.)

7.76 MERCURY–BROMINE–SELENIUM–SULFUR

HgS–HgSe–HgBr$_2$. Isothermal section of this quasiternary system at 400°C is shown in Figure 7.25 (Hudoliy et al. 1993). The limit composition of HgS$_x$Se$_{1-x}$ solid solutions in the three-phase region is HgS$_{0.76}$Se$_{0.24}$.

7.77 MERCURY–BROMINE–TELLURIUM–SULFUR

HgS–HgTe–HgBr$_2$. Isothermal section of this system at 200°C is shown in Figure 7.26 (Voroshilov et al. 1996a). A wide region of solid solutions based on β-Hg$_3$S$_2$Br$_2$ exists in the HgS–HgTe–HgBr$_2$ quasiternary system at this temperature.

7.78 MERCURY–BROMINE–IODINE–SULFUR

HgS–HgBr$_2$–HgI$_2$. The phase diagram is not constructed. The Hg$_3$S$_2$(Br$_{0.67}$I$_{0.33}$)$_2$ quaternary compound is formed in this system. It has a commensurately modulated structure and crystallizes in the tetragonal structure with the lattice parameters $a = 1332 \pm 1$ and $c = 446.5 \pm 0.5$ pm and calculated density 6.98 ± 0.02 g·cm^{-3} (Minets et al. 2004).

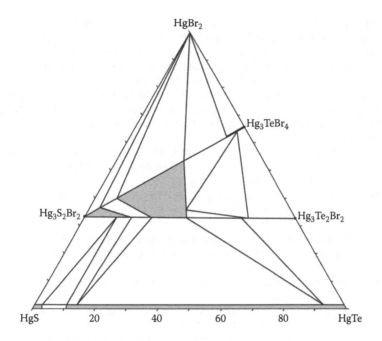

FIGURE 7.26 Isothermal section of the HgS–HgTe–HgBr$_2$ quasiternary system at 200°C. (From Voroshilov, Yu.V. et al., *Neorgan. materialy*, 32(12), 1466, 1996a.)

REFERENCES

Asadov M.M. HgS + Tl$_2$Se ⇔ HgSe + Tl$_2$S mutual system [in Russian], *Zhurn. neorgan. khimii*, **28**(7), 1812–1816 (1983a).

Asadov M.M. HgS + Tl$_2$Te ⇔ HgTe + Tl$_2$S ternary mutual system [in Russian], *Zhurn. neorgan. khimii*, **28**(8), 2100–2103 (1983b).

Asadov M.M. Dependences of physico-chemical properties of solid solutions in the systems GeSe$_2$–A^{2}B^6 (A^2 = Hg, Cd); B^6 = S, Te) versus composition [in Russian], *Azerb. khim. zhurn.*, (2), 77–81 (2006).

Asadov M.M., Babanly M.B., Kuliev A.A. Investigation of the TlSe–HgS system [in Russian], *Azerb. khim. zhurn.*, (1), 102–103 (1982).

Axtell E.A. (III), Park Y., Chondroudis K., Kanatzidis M.G. Incorporation of A$_2$Q into HgQ and dimensional reduction to A$_2$Hg$_3$Q$_4$ and A$_2$Hg$_6$Q$_7$ (A = K, Rb, Cs; Q = S, Se), *J. Am. Chem. Soc.*, **120**(1), 124–136 (1998).

Bach R.D., Weibel A.T. Nuclear magnetic resonance studies on anion-exchange reactions of alkylmercury mercaptides, *J. Am. Chem. Soc.*, **98**(20), 6241–6249 (1976).

Bach R.D., Weibel A.T., Schmonsees W., Glick M.D. Benzyltriphenylmethylthiomercury; crystal structure determination, *J. Chem. Soc. Chem. Commun.*, (23), 961–962 (1974).

Baranov I.Yu., Dolgih V.A., Popovkin B.A. Investigation of phase correlations in the La$_2$O$_2$X–MX (M = Zn, Cd, Hg; X = S, Se) systems [in Russian], *Zhurn. neorgan. khimii*, **41**(11), 1916–1919 (1996).

Beck J., Hedderich S. Expanded polycationic mercury–chalcogen networks in the layered compounds Hg$_3$E$_2$[MX$_6$] (E = S, Se; M = Zr, Hf; X = Cl, Br), *J. Solid State Chem.*, **172**(1), 12–16 (2003).

Beck J., Hedderich S., Köllisch K. Hg_3AsE_4X (E = S, Se; X = Cl, Br, I), a family of isotypic compounds with an acentric layered structure, *Inorg. Chem.*, **39**(25), 5847–5850 (2000).

Beck J., Keller H.-L., Rompel M., Wimbert L. Hydrothermalsynthese und Kristalstruktur der Münzmetall–Quecksilber–Chalkogenidhalogenide CuHgSeBr, AgHgSBr und AgHgSI, *Z. anorg. und allg. Chem.*, **627**(9), 2289–2294 (2001).

Beck J., Keller H.-L., Rompel M., Wimbert L. Ewald B. Über Münzmetal–Quecsilber–Chalkogenidehalogenide: V. Solvothermalsynthese und Kristallstrukture der Hochtemperatur-Modufikation von AgHgSI, *Z. anorg. und allg. Chem.*, **630**(7), 1031–1035 (2004).

Beck J., Rompel M. Über Münzmetal–Quecsilber–Chalkogenidehalogenide: II. Hydrothermalsynthese, Kristallstrukture und Phasenumwandlung von CuHgSCl und CuHgSBr, *Z. anorg. und allg. Chem.*, **629**(3), 421–428 (2003).

Bernard D., Pannetier J., Lucas J. $Hg_2M_2F_6S$ et $Hg_2M_2F_6O$: Deux nouvelles familles de pyrochlores contenant de mercure et des métaux M de transition divalents, *J. Solid State Chem.*, **14**(4), 328–334 (1975).

Biscarini P., Fusina L., Nivellini G. Organothiometallic compounds. Part I. Infrared and Raman spectra of some di(alkylthio)mercury(II) compounds, *J. Chem. Soc., Dalton Trans.*, (20), 2140–2144 (1974).

Björnlund G. The crystal structure of $Hg(OH)_2·2HgSO_4·H_2O$, *Acta Chem. Scand. A*, **28**(2), 169–174 (1974).

Blachnik R., Buchmeier W., Dreisbach H.A. Structure of lead (II) dimercury (II) diiodide disulfide, *Acta Crystallogr.*, **C42**(5), 515–517 (1986).

Blachnik R., Dreisbach H.A. Neue quaternäre Chalkogenohalogenide, *Monatsh. Chem.*, **117**(3), 305–311 (1986).

Blachnik R., Lytze K. Quaternary chalcogenide–halide systems, *Thermochim. Acta*, **160**(1), 79–86 (1990).

Blachnik R., Lytze K., Reuter H. A new quaternary chalcogenide halide: synthesis and structure of $Hg_2SnS_2Br_2$, *J. Solid State Chem.*, **126**(1), 95–98 (1996).

Block E., Brito M., Gernon M., McGowty D., Kang H., Zubieta J. Mercury(II) and methylmercury(II) complexes of novel sterically hindered thiolates: ^{13}C and [199]Hg NMR studies and the crystal and molecular structures of [MeHg(SC_6H_2-2,4,6-Pr_3^i)], [Hg(SC_6H_4-2-SiMe_3)_2], [Hg(2-SC_5H_3N-3-SiMe_3)_2], and [Hg{(2-SC_6H_4)_2SiMe_2}]_2, *Inorg. Chem.*, **29**(17), 3172–3181 (1990).

Bochmann M., Webb K. Novel precursors for the deposition of II-VI semiconductor films, *Mater. Res. Symp. Proc.*, **204**, 149–154 (1991a).

Bochmann M., Webb K. Sterically hindered thiolato, selenolato and tellurolato complexes of mercury(II), *J. Chem. Soc., Dalton Trans.*, (9), 2325–2329 (1991b).

Bochmann M., Webb K.J., Malik M.A., O'Brien P. Arene thiolato, selenolato, and tellurolato complexes of mercury, *Inorg. Synthes.*, **31**, 24–28 (1997).

Borisov S.V., Magarill S.A., Pervukhina N.V. To crystal chemistry of compounds $MHgYX$ (M = Cu, Ag; Y = S, Se; X = Cl, Br, I), *Z. Kristallogr.*, **220**(11), 946–953 (2005).

Borisov S.V., Magarill S.A., Romanenko G.V., Pervukhina N.V. Features of the crystal chemistry of the rare mercury minerals of a supergene origin—Possible sources of mercury in surface water and air basins [in Russian], *Khimiya v interesah ustoychivoho razvitiya*, **7**(5), 497–503 (1999).

Bowmaker G.A., Dance I.G., Dobson B.C., Rogers D.A. Syntheses and vibrational spectra of some tris(alkanethiolato)mercurate(II) complexes, and crystal structure of the hexakis(methanethiolato)dimercurate(II) dianion, *Aust. J. Chem.*, **37**(8), 1607–1618 (1984).

Bradley D.C., Kunchur N.R. Structures of mercury mercaptides. I. X-ray structural analysis of mercury methylmercaptide, *J. Chem. Phys.*, **40**(8), 2258–2261 (1964).

Bradley D.C., Kunchur N.R. Structures of mercury mercaptides: Part II. X-ray structural analysis of mercury ethylmercaptide, *Can. J. Chem.*, **43**(10), 2786–3792 (1965).

Bräu M.F., Pfitzner A. HgI$_2$·As$_4$S$_4$: An adduct from HgI$_2$ molecules and undistorted As$_4$S$_4$ cages, *Angew. Chem. Int. Ed.*, **45**(27), 4464–4467 (2006a).

Bräu M.F., Pfitzner A. HgI$_2$·As$_4$S$_4$: ein Adukt aus HgI$_2$-Molekülen und unverzerrten As$_4$S$_4$-Käfige, *Angew. Chem.*, **118**(27), 4576–4578 (2006b).

Bräu M.F., Pfitzner A. (HgBr$_2$)$_3$(As$_4$S$_4$)$_2$: An adduct of HgBr$_2$ molecules and undistorted As$_4$S$_4$ cages, *Z. anorg. und allg. Chem.*, **633**(7), 935–937 (2007).

Brown K.L., Dickson F.W. The crystal structure of synthetic christite, HgTlAsS$_3$, *Z. Kristallogr.*, **144**(1–6), 367–376 (1976b).

Bugaris D.E., Ibers J.A. Synthesis, structures and magnetic and optical properties of the compounds [Hg$_3$Te$_2$][UCl$_6$] and [Hg$_4$As$_2$][UCl$_6$], *J. Solid State Chem.*, **181**(12), 3189–3193 (2008).

Burna P.C. The structure of edoylerite determined from a microcrystal, *Can. Mineral.*, **37**(1), 113–118 (1999).

Canty A.J., Kishimoto R. Mercury(II) and organomercury(II) complexes of thiols and dithiols, including british anti-lewisite, *Inorg. Chim. Acta*, **24**, 109–122 (1977).

Canty A.J., Tyson R.K. Interaction of MeSHgO$_2$CMe with imidazole and related ligands, *Inorg. Chim. Acta*, **29**, 227–230 (1978).

Carrillo-Cabrera W., Menzel F., Brockner W. Crystal structure of copper mercury tetrathophosphate, CuHgPS$_4$, *Z. Kristallogr.*, **202**(1–2), 152–153 (1992).

Champlon F., Bernard D., Pannetier J., Lucas J. Nouveaux pyrochlores thio- et oxyfluorés contenant de mercure divalent: Hg$_2$Ni$_2$F$_6$S et Hg$_2$Ni$_2$F$_6$O, *C. R. Acad. Sci.*, **C278**(19), 1185–1187 (1974).

Downs A.J., Ebsworth E.A.V., Emeléus H.J. The Raman spectra of bis(trifluoromethylthio) mercury and derived compounds, *J. Chem. Soc.*, 3187–3192 (1961).

Dunn P.J., Cabri L.J., Chao G.Y., Fleischer M., Francis C.A., Grice L.J., Jambor J.L., Pabst A. New mineral names, *Am. Mineralog.*, **69**(3–4), 406–412 (1984).

Engel P., Nowacki W., Balić-Žunić T., Šćavničar S. The crystal structure of simonite, TlHgAs$_3$S$_6$, *Z. Kristallogr.*, **161**(3–4), 159–166 (1982).

Erd R.C., Roberts A.C., Bonardi M., Criddle A.J., Le Page Y., Gabe E.J. Edoylerite, Hg$_3^{2+}$Cr^{6+}O$_4$S$_2$, a new mineral from the Clear Creek Claim, San Benito County, California, *Mineral. Rec.*, **24**(6), 471–475 (1993).

Fleischer M., Mandarino J.A. New mineral names, *Am. Mineralog.*, **60**(9–10), 945–947 (1975).

Fraser K.A., Clegg W., Craig D.C., Scudder M.L., Dance I.G. New crystal data for [Hg(SC$_2$H$_5$)$_2$], *Acta Crystallogr.*, **C51**(3), 406–408 (1995).

Gregg D.C., Iddles H.A., Stearns, Jr. P.W. Triphenylmethyl aryl sulfides. I. Hydrogenolysis with Raney nickel. Reactions with mercuric chloride, *J. Org. Chem.*, **16**(2), 246–252 (1951).

Guillo M., Mercey B., Deschanvres A. Systèmes CuX(X = Cl, Br)–HgS. Étude des composés CuHgSX (X = Cl, Br), *Mater. Res. Bull.*, **14**(7), 947–954 (1979).

Gulay L.D., Olekseyuk I.D., Parasyuk O.V. Crystal structures of the Ag$_4$HgGe$_2$S$_7$ and Ag$_4$CdGe$_2$S$_7$ compounds, *J. Alloys Compd.*, **340**(1–2), 157–166 (2002a).

Gulay L.D., Parasyuk O.V. Crystal structure of the Ag$_6$Hg$_{0.82}$GeS$_{5.82}$ compound, *J. Alloys Compd.*, **327**(1–2), 100–103 (2001).

Gulay L.D., Parasyuk O.V., Olekseyuk I.D. Crystal structures of the Cu$_6$Hg$_{0.973}$SiS$_{5.973}$ and Ag$_6$Hg$_{0.897}$SiS$_{5.897}$ compounds, *J. Alloys Compd.*, **335**(1–2), 111–114 (2002b).

Haeuseler H., Cansiz A., Nimmrich M., Jung M. Materials with layered structures: crystal structure of Zn$_{1.25}$In$_{2.5}$S$_3$Se$_2$, a new polytype of Zn$_2$In$_2$S$_5$ and the isotypic compounds Cd$_{0.5}$Ga$_2$InS$_5$ and Hg$_{0.8}$Ga$_{1.6}$In$_{1.2}$S$_5$, *J. Solid State Chem.*, **74**(1), 171–175 (1988).

Haeuseler H., Himmrich M. Neue Verbindungen Ag$_2$HgMX$_4$ mit Wurtzstannitstruktur, *Z. Naturforsch.*, **B44**(9), 1035–1036 (1989).

Hahn H., Schulze H. Über quaternäre Chalkogenide des Germanium und Zinns, *Naturwissenschaften*, **52**(14), 426 (1965).

Harris D.C., Roberts A.C. Vaughanite, TlHgSb$_4$S$_7$, a new mineral from Hemlo, Ontario, Canada, *Mineral. Mag.*, **53**, 79–83 (1989).

Haszeldine R.N., Kidd J.M. Reactions of fluorocarbon radicals. Part XI. Synthesis and some reactions of trifluoromethanethiol and trifluoromethanesulphenyl chloride, *J. Chem. Soc.*, 3219–3225 (1953).

Haszeldine R.N., Kidd J.M. Perfluoroalkyl derivatives of sulphur. Part III. Some reactions of trifluoromethanethiol, and spectroscopic properties of the >C:S group, *J. Chem. Soc.*, 3871–3880 (1955).

Hudoliy V.A., Pan'ko V.V., Shelemba M.S., Lopit L.I., Fedor A.S., Voroshilov Yu.V. Phase equilibria in the HgS–HgSe–HgCl$_2$(Br$_2$) systems [in Russian], *Zhurn. neorgan. khimii*, **38**(9), 1584–1586 (1993).

Il'yasheva N.A., Godovikov A.A., Nenashev B.G., Popov S.P. Synthesis of laffittite AgHgAsSe$_3$ [in Russian], *Izv. AN SSSR. Neorgan. materialy*, **15**(10), 1748–1751 (1979).

Imafuku M., Nakai I., Nagashima K. Synthesis of new sulfosalt, KHgSbS$_3$, [in Japanese], *J. Chem. Soc. Jpn. Chem. Ind. Chem.*, (7), 1498–1500 (1985).

Imafuku M., Nakai J., Nagashima K. The crystal structure of a new synthetic sulfosalt, KHgSbS$_3$, *Mater. Res. Bull.*, **21**(4) 493–501 (1986).

Jamber J.L., Roberts A.C., Vanko D.A. New mineral names, *Am. Mineralog.*, **79**(9–10), 1009–1014 (1994).

Johan Z., Mantienne J., Picot P. La routhiérite, TlHgAsS$_3$, et la laffittite, AgHgAsS$_3$, deux nouvelles espèces minérales, *Bull. soc. fr. minéral., cristallogr.*, **97**, 48–53 (1974).

Kabalov Yu.K., Yevstigneeva T.L., Spiridinov E.M. Crystal structure of Cu$_2$HgSnS$_4$—Synthetic analogue of velikite [in Russian], *Kristallografiya*, **43**(1), 21–25 (1998).

Kanatzidis M.G., Liao J.H., Marking G.A. Alkali metal quaternary chalcogenides and process for the preparation of thereof, USA Patent 5 614 128. Appl. № 606565. Filed 26.02.96; Data of Patent 25.03.1997 (1997a).

Kanatzidis M.G., Liao J.-H., Marking G.A. Alkali metal quaternary chalcogenides and process for the preparation of thereof, USA Patent 5 618 471. Appl. № 606886. Filed 26.02.96; Data of Patent 08.04.1997 (1997b).

Kaplunnik L.N., Pobedimskaya E.A., Belov N.V. Crystal structure of velikite Cu$_{3.75}$Hg$_{1.75}$Sn$_2$S$_8$ [in Russian], *Kristallografiya*, **22**(1), 175–177 (1977).

Karataeva N.P., Sviridov V.V. About a product of the interaction between Ag$_2$S and HgBr$_2$ [in Russian], *Izv. AN SSSR. Neorgan. materialy*, **6**(5), 996–997 (1970).

Kawabata S., Yasui Y., Kobayashi Y., Sato M., Igawa N., Kakurai K. Studies of magnetic behavior of spin-1/2 frustrated system Hg$_2$Cu$_2$F$_6$S, *J. Magn. Magn. Mater.*, **310**(2, Pt. 2), 1295–1296 (2007).

Keller H.-L., Wimbert L. Über Münzmetal–Quecksilber–Chalkogenidehalogenide: III. Zur Kristallstruktur von Ag$_2$HgSI$_2$. *Z. anorg. und allg. Chem.*, **629**(12–13), 2337–2340 (2003).

Keller H.-L., Wimbert L. Über Münzmetal–Quecksilber–Chalkogenidehalogenide: IV. Hydrothermalsynthese und Kristallstrukture von CuHgSI und CuHg$_2$S$_2$I. *Z. anorg. und allg. Chem.*, **630**(2), 331–336 (2004).

Koton M.M., Moskvina E.P., Florinskiy F.S. Reactions of metalloorganic compounds with thiophenol [in Russian], *Zhurn. obshch. khimii*, **20**(11), 2093–2095 (1950).

Kuliev A.A., Asadov M.M., Kuliev R.A., Babanly M.B. Phase equilibria in the Tl$_2$S–HgS and Tl$_2$Se–HgS systems [in Russian], *Zhurn. neorgan. khimii*, **23**(3), 854–856 (1978).

Kunchur N.R. Polymeric structure of mercury *tert*-butyl mercury mercaptide, *Nature*, **204**(4957), 468 (1964).

Kuz'mina L.G., Bokiy N.G., Struchkov Yu.T., Kravtsov D.N., Rokhliga E.M. X-ray diffraction analysis of non-bonded interactions and coordination in organometallic compounds. IV. The crystal and molecular structure of (2,6-dimethylphenylthioato)phenylmercury [in Russian], *Zhurn. strukt. khimii*, **15**(3), 491–496 (1974).

Lecher H. Beiträge zum Valenzproblem des Schwefels. II. Über das Thiophenol-quecksilber, *Ber. deutsch. chem. Gesell.*, **48**(2), 1425–1432 (1915).

Lecher H. Über das Phenylmercapto-quecksilberchlorid, *Ber. deutsch. chem. Gesell.*, **53**(4), 568–577 (1920).

Li H., Malliakas C.D., Liu Z., Peters J.A., Jin H., Morris C.D., Zhao L., Wessels B.W., Freeman A.J., Kanatzidis M.G. CsHgInS$_3$: A new quaternary semiconductor for γ-ray detection, *Chem. Mater.*, **24**(22), 4434–4441 (2012).

Liao J.-H., Marking G.M., Hsu K.F., Matsushita Y., Ewbank M.D., Borwick R., Cunningham P., Rosker M.J., Kanatzidis M.G. α- and β-A$_2$Hg$_3$M$_2$S$_8$ (A = K, Rb; M = Ge, Sn): Polar quaternary chalcogenides with strong nonlinear optical response, *J. Am. Chem. Soc.*, **125**(31), 9484–9493 (2003).

Liu Y., Kanhere P.D., Hoo Y.S., Ye K., Yan Q., Rawat R.S., Chen Z., Ma J., Zhang Q. Cationic quaternary chalcohalide nanobelts: Hg$_4$In$_2$Q$_3$Cl$_8$ (Q = S, Se, Te), *RSC Adv.*, **2**(16), 6401–6403 (2012a).

Liu Y., Wei F., Yeo S.N., Lee F.M., Kloc C., Yan Q., Hng H.H., Ma J., Zhang Q. Synthesis, crystal structure, and optical properties of a three-dimensional quaternary Hg–In–S–Cl chalcohalide: Hg$_7$InS$_6$Cl$_5$, *Inorg. Chem.*, **51**(8), 4414–4416 (2012b).

Lucas C.R., Peach M.E. Metal derivatives of penthachlorothiophenol, *Inorg. Nucl. Chem. Lett.*, **5**(2), 73–76 (1969).

Lucas C.R., Peach M.E. Reactions of pentachlorothiophenol. I. Preparation of some simple metallic and non-metallic derivatives, *Can. J. Chem.*, **48**(12), 1869–1875 (1970).

Lutz H.D., Bertram W.W., Oft B., Hauseler H. Phase relationships in the systems MS–Cr$_2$S$_3$–In$_2$S$_3$ (M = Co, Cd, Hg), *J. Solid State Chem.*, **46**(1), 56–63 (1983).

Malevski A.Yu., Chzhun, T.Z. About isomorphous substitution of sulfur by selenium and tellurium in the mercury sulfides [in Russian]. In: *Eksperiment.-metodich. issled. rudnyh mineralov*. M.: Nauka Publish., pp. 223–226 (1965).

Mandarino J.A. Abstracts of new mineral descriptions (department), *Mineral. Rec.*, **29**(5), 476–479 (1998).

Man E.H., Coffman D.D., Muetterties E.L. Synthesis and properties of bis(trifluoromethylthio)-mercury, *J. Am. Chem. Soc.*, **81**(14), 3575–3577 (1959).

Marchuk O.V., Gulay L.D., Parasyuk O.V. The Cu$_2$S–HgS–GeS$_2$ system at 670 K and the crystal structure of the Cu$_6$Hg$_{0.92}$GeS$_{5.92}$ compound, *J. Alloys Compd.*, **333**(1–2), 143–146 (2002).

Märcker C. Ueber einige schwefelhaltige Derivate des Toluols, *Justus Liebigs Annal. Chem.*, **136**(1), 75–95 (1865).

Marking G.A., Hanko J.A., Kanatzidis M.G. New quaternary thiostannates and thiogermanates A$_2$Hg$_3$M$_2$S$_8$ (A = Cs, Rb, M = Sn, Ge) through molten A$_2$S$_x$. Reversible glass formation in Cs$_2$Hg$_3$M$_2$S$_8$, *Chem. Mater.*, **10**(4), 1191–1199 (1998).

Martan H., Weiss J. Metall-Schwefelstickstoff-Verbindungen. 17. Die Verbindungen HgN$_2$S·NH$_3$ und 2Hg(NH$_3$)$_2$I$_2$·S$_4$N$_4$, *Z. anorg. und allg. Chem.*, **515**(8), 225–229 (1984).

McCormack J.K., Dickson F.W., Leshendok M.P. Radtkeite, Hg$_3$S$_2$ClI, a new mineral from the McDermit mercury deposit, Humboldt County, Nevada, *Am. Mineralog.*, **76**(9–10), 1715–1721 (1991).

Minets Yu.V., Voroshilov Yu.V., Pan'ko V.V. The structure of mercury chacogen halogenides Hg$_3$X$_2$Hal$_2$, *J. Alloys Compd.*, **367**(1–2), 109–114 (2004).

Moldovyan N.A., Chebotaru V.Z. Growth and characteristics of HgGaInS$_4$ single crystals, *Cryst. Res. Technol.*, **25**(9), 997–1005 (1990).

Moldovyan I.A., Metlinski P.N. New HgInGaS$_4$ layered compound and its electrophysical properties [in Russian], *Izv. AN SSSR. Neorgan. materialy*, **23**(10), 1623–1625 (1987).

Moldovyan N.A., Pyshnaya N.B., Radautsan S.I. New multinary layered chalcogenides with octahedral and tetrahedral cation coordination, *Jpn. J. Appl. Phys.*, **32**(Pt. 1, Suppl.), 781–783 (1993).

Moro'oka M., Ohki H., Yamada K., Okuda T. Crystal structure of high temperature phase and ionic conductivity mechanism of CuHgSX (X = Cl, Br), *Bull. Chem. Soc. Jpn.*, **76**(11), 2111–2115 (2003).

Moro'oka M., Ohki H., Yamada K., Okuda T. Crystal structure and ionic conduction mechanism of AgHgSX (X = Br, I), *Bull. Chem. Soc. Jpn.*, **77**(5), 975–980 (2004a).

Moro'oka M., Ohki H., Yamada K., Okuda T. The crystal structure and electric conductivity of new quaternary chalcogenide halide CuHgSI, *J. Solid State Chem.*, **177**(4–5), 1401–1404 (2004b).

Morris C.D., Li H., Jin H., Malliakas C.D., Peters J.A., Trikalitis P.N., Freeman A.J., Wessels B.W., Kanatzidis M.G. $Cs_2M^{II}M_3^{IV}Q_8$ (Q = S, Se, Te): An extensive family of layered semiconductors with diverse band gaps, *Chem. Mater.*, **25**(16), 3344–3356 (2013).

Mozolyuk M.Yu., Olekseyuk I.D., Piskach L.V., Parasyuk O.V. X-ray analysis of the Tl_2GeS_3–{Zn, Cd, Hg}S sections [in Ukrainian], *Nauk. visnyk Volyn. nats. un-tu im. Lesi Ukrainky. Khim. nauky*, [(17)242], 75–78 (2012).

Nakai I., Appleman D.E. Laffittite, $AgHgAsS_3$: Crystal structure and second occurrence from the Getcholl mine, Nevada, *Am. Mineralog.*, **68**(1–2), 235–244 (1983).

Novatski V. Isotypie of aktashite $Cu_6Hg_3As_4S_{12}$ and novatskite $Cu_6Zn_3As_4S_{12}$ [in Russian], *Kristallografiya*, **27**(1), 49–50 (1982).

Nyquist R.A., Mann J.R. The vibrational spectrum of CH_3–Hg–S–CH_3 *Spectrochim. Acta Part A*, **28**(3), 511–515 (1972).

Ohtani H., Kojima K., Ishida K., Nishizawa T. Miscibility gap in II-VI semiconductor systems, *J. Alloys Compd.*, **182**(1), 103–114 (1992).

Olekseyuk I.D., Kogut Yu.M., Yurchenko O.M., Parasyuk O.V., Volkov S.V., Pekhnyo V.I. Glass formation and optical properties of the glasses in the Ag_2S–HgS–GeS_2 system, *Chem. Met. Alloys*, **2**(1–2), 49–54 (2009).

Olekseyuk I., Marchuk O., Dudchak I., Parasyuk O., Piskach L. Phase equilibria in the Cu_2SnS_3–Zn(Hg)S systems [in Ukrainian], *Visnyk L'viv. un-tu. Ser. khim.*, (39), 48–52 (2000).

Olekseyuk I.D., Mazurets I.I., Parasyuk O.V. Phase equilibria in the HgS–Ga_2S_3–GeS_2 system, *J. Alloys Compd.*, **417**(1–2), 131–137 (2006).

Olekseyuk I.D., Mozolyuk M.Yu., Piskach L.V., Litvinchuk M.B., Parasyuk O.V. Component interaction in the systems formed by the Tl(I), Hg(II), Pb(II), Si(IV) chalcogenides [in Ukrainian], *Nauk. visnyk Volyn. nats. un-tu im. Lesi Ukrainky. Khim. nauky*, [(17)242], 62–69 (2012).

Olekseyuk I.D., Mozolyuk M.Yu., Piskach L.V., Parasyuk O.V. Phase equilibria in the $Tl_2S(Se)$–HgS(Se)–SnS(Se)$_2$ systems at 520 K [in Ukrainian], *Nauk. visnyk Volyn. nats. un-tu im. Lesi Ukrainky. Khim. nauky*, (30), 19–21 (2010).

Olekseyuk I.D., Petrus' I.I., Parasyuk O.V. Isothermal section of the HgS–Ga_2S_3–GeS_2 system at 670 K [in Ukrainian], *Nauk. visnyk Volyn. derzh. un-tu. Khim. nauky*, (6), 38–40 (2001).

Pan'ko V.V., Voroshilov Yu.V., Hudoliy V.A., Shelemba M.S. Phase equilibria in the HgS(Se)–$HgCl_2$–$HgBr_2$ systems [in Russian], *Zhurn. neorgan. khimii*, **41**(10), 1731–1733 (1996).

Parasyuk O.V. Phase relations of the Ag_2SnS_3–HgS and $Ag_{33.3}Sn_{16.7}Se(Te)_{50}$–HgSe(Te) sections in Ag–Hg–Sn–S(Se, Te) systems, *J. Alloys Compd.*, **291**(1–2), 215–219 (1999).

Parasyuk O.V., Chykhrij S.I., Bozhko V.V., Piskach L.V., Bogdanyuk M.S., Olekseyuk I.D., Bulatetska L.V., Pekhnyo V.I. Phase diagram of the Ag_2S–HgS–SnS_2 system and single crystal preparation, crystal structure and properties of Ag_2HgSnS_4, *J. Alloys Compd.*, **399**(1–2), 32–37 (2005).

Parasyuk O.V., Gulay L.D., Piskach L.V., Gagalowska O.P. The Ag_2S–HgS–GeS_2 system at 670 K and the crystal structure of the Ag_2HgGeS_4 compound, *J. Alloys Compd.*, **336**(1–2), 213–217 (2002a).

Parasyuk O.V., Gulay L.D., Romanyuk Ya.E., Olekseyuk I.D. Phase diagram of the quasibinary Cu_2GeS_3–HgS system and crystals structure of the LT-modification of the Cu_2HgGeS_4 compound, *J. Alloys Compd.*, **334**(1–2), 143–146 (2002b).

Peach M.E. Some reactions of pentafluorothiophenol. Preparation of some pentafluorophenylthio metal derivatives, *Can. J. Chem.*, **46**(16), 2699–2706 (1968).

Pervukhina N.V., Borisov S.V., Magarill S.A., Naumov D.Yu., Vasil'ev V.I., Nenashev B.G. The crystal structure of synthetic analogue of the mineral radtkeite $Hg_3S_2Cl_{1.00}I_{1.00}$ [in Russian], *Zhurn. strukt. khimii*, **45**(4), 755–758 (2004a).

Pervukhina N.V., Magarill S.A., Naumov D.Yu., Borisov S.V., Vasil'ev V.I., Nenashev B.G. Crystal chemistry of mercury chalcogenide-halides. III. Crystal structure of two polymorphic modifications of the mercury sulfide-halide $Hg_3S_2Cl_{1.5}Br_{0.5}$ [in Russian], *Zhurn. strukt. khimii*, **46**(4), 663–668 (2005).

Pervukhina N.V., Magarill S.A., Naumov D.Yu., Borisov S.V., Vasil'ev V.I., Nenashev B.G. Crystal chemistry of mercury chalcogenide-halides. IV. Crystalline state of two polymorphic modifications of the mercury sulfide-halide $Hg_3S_2Br_{1.5}Cl_{0.5}$ [in Russian], *Zhurn. strukt. khimii*, **47**(2), 318–323 (2006a).

Pervukhina N.V., Romanenko G.V., Borisov S.V., Magarill S.A., Pal'chik N.A. Crystal chemistry of mercury (I)- and mercury (I, II)-containing minerals [in Russian], *Zhurn. strukt. khimii*, **40**(3), 561–581 (1999).

Pervukhina N.V., Vasil'ev V.I., Magarill S.A., Borisov S.V., Naumov D.Yu. Crystal chemistry of mercury sulfohalides of composition $Hg_3S_2Hal_2$ (Hal: Cl, Br). II. Crystal structures of two polymorphic modifications of $Hg_3S_2Br_{2-x}Cl_x$ ($x \approx 0.5$), *Can. Mineral.*, **44**(5), 1247–1255 (2006b).

Pervukhina N.V., Vasil'ev V.I., Magarill S.A., Borisov S.V., Naumov D.Yu. Crystal chemistry of mercury sulfohalides of composition $Hg_3S_2Hal_2$ (Hal: Cl, Br). I. Crystal structures of two polymorphic modifications of $Hg_3S_2Cl_{2-x}Br_x$ ($x \approx 0.5$), *Can. Mineral.*, **44**(5), 1239–1246 (2006c).

Pervukhina N.V., Vasil'ev V.I., Naumov D.Yu., Borisov S.V., Magarill S.A. The crystal structure of synthetic radtkeite, Hg_3S_2ClI, *Can. Mineral.*, **42**(1), 87–94 (2004b).

Puff H. Zur Kenntnis einiger Quecksilber-Chalkonium-Verbindungen, *Angew. Chem.*, **75**(14), 681 (1963).

Puff H., Kohlschmidt R. Quecksilberchalkogenid-halogenide, *Naturwissenschaften*, **49**(14), 299 (1962).

Puff H., Lorbacher G., Heine D. Quecksilberchalkogen-Fluorosilicate, *Naturwissenschaften*, **56**(9), 461 (1969).

Radtke A.S., Dickson F.W., Slack J.F., Brown K.L. Christite, a new thallium mineral from the Carlin gold deposit, Nevada, *Am. Mineralog.*, **62**(5–6), 421–425 (1977).

Roberts A.C., Szymanski J.T., Erd R.C., Criddle A.J., Bonardi M. Deanesmithite, $Hg_2^{1+}Hg_3^{2+}Cr^{6+}O_5S_2$, a new mineral from the Clear Creek Claim, San Benito County, California, *Can. Mineral.*, **31**(4), 787–793 (1993).

Ruiz E., Payne M.C. One-dimensional intercalation compound $2HgS \cdot SnBr_2$: Ab initio electronic structure calculations and molecular dynamics simulation, *Chem. Eur. J.*, **4**(12), 2485–2492 (1998).

Said F.F., Tuck D.G. The direct electrochemical synthesis of some thiolates of zinc, cadmium and mercury, *Inorg. Chim. Acta*, **59**, 1–4 (1982).

Saukov A.A., Aydin'yan N.H. About cinnabar oxidation [in Russian], *Tr. In-ta geol. nauk AN SSSR*, (39), 37–40 (1940).

Schäfer W., Nitsche R. Tetrahedral quaternary chalcogenides of the type Cu_2–II–IV–$S_4(Se_4)$, *Mater. Res. Bull.*, **9**(5), 645–654 (1974).

Schäfer W., Nitsche R. Zur Systematik tetraedrischer Verbindungen vom Typ $Cu_2Me^{II}Me^{IV}Me_4^{VI}$ (Stannite und Wurtzstannite), *Z. Kristallogr.*, **145**(5–6), 356–370 (1977).

Smirnova R.I., Markovski L.Ya. About interaction of mercury (II) sulfide with selenitical anhydride and selenitic acid [in Russian], *Zhurn. neorgan. khimii*, **9**(5), 1129–1133 (1964).

Steed G.M. Gortdrumite, a new sulphide mineral containing copper and mercury, from Ireland, *Mineral. Mag.*, **47**(1), 35–36 (1983).

Suchow L., Pond G.R. Photosensitive and phototropic products of solid state reaction between silver sulfide and mercuric iodide, *J. Phys. Chem.*, **58**(3), 240–242 (1954).

Sytsma L.F., Kline R.J. Mercury-protonspin-spin coupling constants of some methyl mercury compounds, *J. Organomet. Chem.*, **54**, 15–21 (1973).

Szymanski J.T., Groat L.A. The crystal structure of deanesmithite, $Hg_2Hg_3CrO_5S_2$, *Can. Mineral.*, **35**(3), 765–772 (1997).

Teske Chr.L. Darstellung und Kristallstruktur von Barium–Quecksilber–Thiostannat(IV), $BaHgSnS_4$, *Z. Naturforsch.*, **B35**(1), 7–11 (1980).

Vasil'ev V.I. New ore minerals of mercury depositions of the Mountain Altai and their paragenesis [in Russian]. In: *Voprosy metallogenii rtuti*. M.: Nauka Publish., pp. 111–129 (1968).

Vasil'ev V.I., Lavrentyev Yu.G., Pal'chik N.A. New data about arzakite and lavrentievite [in Russian], *Dokl. AN SSSR*, **290**(4), 948–951 (1986).

Vasil'ev V.I., Pal'chik N.A., Grechishchev O.K. Lavrentievite and arzakite, the new natural sulfohalogenides of mercury [in Russian], *Geologiya i geofizika*, (7), 54–63 (1984).

Vogt C. XVII. Ueber Benzylmercaptan und Zweifach–Schwefel–Benzyl, *Justus Liebigs Annal. Chem.*, **119**(2), 142–153 (1861).

Voroshilov Yu.V., Hudoliy V.A., Pan'ko V.V. Phase equilibria in the $HgS-HgTe-HgCl_2$ system and crystal structure of β-$Hg_3S_2Cl_2$ and Hg_3TeCl_4 compounds [in Russian], *Zhurn. neorgan. khimii*, **41**(2), 287–293 (1996b).

Voroshilov Yu.V., Hudoliy V.A., Pan'ko V.V., Minets Yu.V. Phase equilibria in the $HgS-HgTe-HgBr_2$ system and crystal structure of $Hg_3S_2Br_2$ and Hg_3TeBr_4 compounds [in Russian], *Neorgan. materialy*, **32**(12), 1466–1472 (1996a).

Wertheim E. Derivatives for the identification of mercaptan, *J. Am. Chem. Soc.*, **51**(12), 3661–3664 (1929).

Wibowo A.C., Malliakas C.D., Chung D.Y., Im J., Freeman A.J., Kanatzidis M.G. Mercury bismuth chalcohalides, $Hg_3Q_2Bi_2Cl_8$ (Q = S, Se, Te): Synthesis, crystal structures, band structures, and optical properties, *Inorg. Chem.*, **52**(6), 2973–2979 (2013).

Yange H., Downs R.T., Costin G., Eichler C.M. The crystal structure of tvalchrelidzeite, Hg_3SbAsS_3, and a revision of its chemical formula, *Can. Mineral.*, **45**(6), 1529–1533 (2007).

8 Systems Based on HgSe

8.1 MERCURY–HYDROGEN–CARBON–SELENIUM

Some compounds are formed in this quaternary system.

Hg(MeSe)$_2$ or C$_2$H$_6$Se$_2$Hg, bis(methaneselenolato)mercury. To obtain this compound, an excess metallic Hg was stirred with Me$_2$Se$_2$ (15 mmol) in pyridine (50 mL) for 2 days (Arnold and Canty 1981). The resulting yellow and black suspension was extracted with hot pyridine until the extract was colorless (200 mL), the black HgSe and metallic Hg being removed by filtration through cellulose powder and a fine sinter under slight positive pressure. Yellow leaflets of C$_2$H$_6$Se$_2$Hg were formed on cooling. This compound crystallizes in the monoclinic structure with the lattice parameters $a = 844.0 \pm 0.4$, $b = 1073.2 \pm 0.3$, $c = 668.1 \pm 0.3$ pm, and $\beta = 96.14° \pm 0.04°$ and calculated density 4.29 g·cm^{-3} (Arnold and Canty 1982). Its crystals were grown from pyridine.

(HgMe)$_2$Se or C$_2$H$_6$SeHg$_2$. This compound was obtained on the reaction of H$_2$Se with CH$_3$HgBr in MeOH under N$_2$ atmosphere (Breitinger and Morell 1974). The colorless precipitate was filtered and recrystallized from benzene. It melts at 130°C with decomposition.

Hg(EtSe)$_2$ or C$_4$H$_{10}$Se$_2$Hg, bis(ethaneselenolato)mercury. To obtain this compound, an excess metallic Hg was stirred with Et$_2$Se$_2$ (ca. 2 mL) in chloroform (50 mL) for 1 day (Arnold and Canty 1981). The yellow powder that formed was dissolved by the addition of hot chloroform (100 mL) and metallic Hg removed by decantation and filtration through a fine sinter under slight positive pressure. Chloroform was removed under vacuum and the yellow powder recrystallized from pyridine to form yellow prisms of C$_4$H$_{10}$Se$_2$Hg.

MeHgSeBu^t or C$_5$H$_{12}$SeHg, t-butaneselenolato(methyl)mercury. To obtain this compound, a portion of the ether extract from preparation of t-butaneselenol (ca. 20 mL) was distilled into an ice-cold solution of MeHgOH [from MeHgNO$_3$ (3.34 mmol) and NaOH (2 M, 6 mmol)] in MeOH (Arnold and Canty 1981). Water (10 mL) was added, and after standing for 12 h, the solution was filtered through a sinter containing cellulose powder and washed with MeOH (20 mL). The solution was extracted twice with hexane (25 mL), NaOH (2 M, 3 mL), and water (10 mL) added to the aqueous layer and was extracted three more times with hexane (25 mL), and the extracts were dried under low vacuum (ca. 3.3 kPa) at 0°C. The white solid C$_5$H$_{12}$SeHg sublimes at 50°C–60°C (2 kPa) and melts at 57°C–58.5°C.

Hg(Bu^{t}Se)$_2$ or C$_8$H$_{18}$Se$_2$Hg, bis(t-butaneselenolato)mercury. To obtain this compound, a portion of the ether extract from preparation of t-butaneselenol (ca. 80 mL) was distilled into a stirred solution of mercuric cyanide (24 mmol) in MeOH (50 mL)

(Arnold and Canty 1981). The precipitate formed was collected and dissolved in boiling chloroform (250 mL), and on cooling, white needles were collected.

MeHg(SeC$_6$H$_2$Pr$_3^i$-2,4,6) or **C$_{16}$H$_{26}$SeHg, tri-*i*-propylbenzeneselenolato(methyl) mercury.** This compound melts at 83°C and crystallizes in the triclinic structure with the lattice parameters $a = 591.6 \pm 0.3$, $b = 970.0 \pm 0.3$, $c = 1597.1 \pm 1.1$ pm, $\alpha = 98.23° \pm 0.04°$, $\beta = 99.12° \pm 0.5°$, and $\gamma = 95.08° \pm 0.3°$ and calculated density 1.858 g·cm^{-3} (Bochmann et al. 1992). For obtaining it to a solution of MeHgNO$_3$ (5.22 mmol) and NEt$_3$ (1 mmol) in methanol (20 mL) was added at room temperature HSeC$_6$H$_2$Me$_3$-2,4,6 (2.82 mmol). After stirring for 2 h, the mixture was filtered to remove a small amount of Hg precipitate and concentrated to give off-white crystals that were recrystallized from light petroleum. All operations were carried out under Ar using standard vacuum-line techniques.

EtHg(SeC$_6$H$_2$Pr$_3^i$-2,4,6) or **C$_{17}$H$_{28}$SeHg, tri-*i*-propylbenzeneselenolato(ethyl) mercury.** The same procedure as for obtaining C$_{16}$H$_{26}$SeHg was used to prepare C$_{17}$H$_{28}$SeHg using EtHgNO$_3$ instead of MeHgNO$_3$ (Bochmann et al. 1992).

Hg(SeC$_6$H$_2$Me$_3$-2,4,6)$_2$ or **C$_{18}$H$_{22}$Se$_2$Hg, bis(trimethylbenzeneselenolato)- mercury.** This compound sublimes at 200°C (0.13 Pa) (Bochmann and Webb 1991a,b). It was prepared according to equation Hg + (2,4,6-Me$_3$C$_6$H$_2$Se)$_2$ by stirring the mixture in toluene (Bochmann and Webb 1991b).

Pr^{n}Hg(SeC$_6$H$_2$Pr$_3^i$-2,4,6) and **Pr^{i}Hg(SeC$_6$H$_2$Pr$_3^i$-2,4,6)** or **C$_{18}$H$_{30}$SeHg, tri-*i*-propylbenzeneselenolato(*n*-propyl)mercury and tri-*i*-propylbenzeneselenolato(*i*-propyl)mercury.** The same procedure as for obtaining C$_{16}$H$_{26}$SeHg was used to prepare C$_{18}$H$_{30}$SeHg using Pr^{n}HgNO$_3$ or Pr^{i}HgNO$_3$ instead of MeHgNO$_3$ (Bochmann et al. 1992).

Hg(SeC$_6$H$_2$Pr$_3^i$-2,4,6)$_2$ or **C$_{30}$H$_{46}$Se$_2$Hg, bis(tri-*i*-propylbenzeneselenolato)- mercury.** This compound sublimes at 93°C (1.3 Pa) and melts at 133°C (Bochmann and Webb 1991a,b). It was prepared by oxidative addition of Se(C$_6$H$_2$Pr$_3^i$-2,4,6)$_2$ to Hg in light petroleum and from HgCl$_2$ and LiSeC$_6$H$_2$Pr$_3^i$-2,4,6 in tetrahydrofurane C$_{30}$H$_{46}$Se$_2$Hg (Bochmann and Webb 1991b). This compound could be also obtained if a solution of Hg[N(SiMe$_3$)$_2$]$_2$ (0.9 mmol) in light petroleum (5 mL) was added to a solution of 1.8 mmol of HSC$_6$H$_2$Pr$_3^i$-2,4,6 in light petroleum (20 mL) via syringe at room temperature. The mixture was stirred for 2 h and filtered and the residue recrystallized from toluene at −20°C to obtain C$_{30}$H$_{46}$Se$_2$Hg as white microcrystals.

Hg(SeC$_6$H$_2$Bu$_3^t$-2,4,6)$_2$ or **C$_{36}$H$_{58}$Se$_2$Hg, bis(tris-*t*-butylselenolato)mercury.** This compound sublimes at 150°C (1.3 Pa) and melts at 330°C with decomposition (Bochmann and Webb 1991a,b, Bochmann et al. 1997). For obtaining C$_{36}$H$_{58}$Se$_2$Hg, a solution of LiSeC$_6$H$_2$Bu$_3^t$-2,4,6 in tetrahydrofurane was prepared by treating (C$_6$H$_2$Bu$_3^t$-2,4,6)$_2$Se$_2$ (0.77 mmol) with two equivalents Li[BHEt$_3$] at room temperature (Bochmann and Webb 1991b, Bochmann et al. 1997). The mixture was stirred for 30 min, after which the residual Li[BHEt$_3$] was destroyed by injecting a small amount of EtOH (ca. 0.5 mL). Then, HgCl$_2$ (0.77 mmol) at room temperature was added to this mixture. Stirring was continued for 2 h. The solvent was removed in vacuo and the residue was extracted with toluene (2 × 10 mL). Concentrating in vacuo to about

10 mL and cooling to $-20°C$ overnight give pale yellow crystals, which should be protected from strong sunlight and is best stored in the dark below $10°C$.

[$Hg_{32}Se_{34}(SePh)_{36}$] or $C_{216}H_{180}Se_{70}Hg_{32}$. α-$C_{216}H_{180}Se_{70}Hg_{32}$ crystallizes in the cubic structure with the lattice parameter $a = 2224.5 \pm 0.3$ pm at 180 K, and β-$C_{216}H_{180}Se_{70}Hg_{32}$ crystallizes in the trigonal structure with the lattice parameters $a = 2275.3 \pm 0.3$ and $c = 5302.0 \pm 1.1$ pm (Behrens et al. 1996a,b). α-$C_{216}H_{180}Se_{70}Hg_{32}$ was obtained if $PhSeSiMe_3$ (1.50 mmol) was added to a suspension of $Fe(CO)_4 \cdot HgCl_2$ (0.75 mmol) in toluene (25 mL). From the resulting clear red solution, a yellow precipitate was quickly formed. After a few days, red cubes of this compound crystallize. To prepare β-$C_{216}H_{180}Se_{70}Hg_{32}$, $PhSeSiMe_3$ (1.48 mmol) was added dropwise to a suspension of $HgCl_2$ (0.74 mmol) in toluene (25 mL). The resulting clear red solution was layered with heptanes, and within a few hours, the solution lightened in color and yellow crystals of $Hg(SePh)_2$ and red needlelike crystals of β-$C_{216}H_{180}Se_{70}Hg_{32}$ were formed. All preparations were performed under N_2 atmosphere and absolute solvents were used.

8.2 MERCURY–HYDROGEN–OXYGEN–SELENIUM

$Hg_7Se_3O_{13}H_2$ and $Hg_8Se_4O_{17}H_2$ quaternary compounds were found in the Hg–H–O–Se quaternary system (Weil 2004). These compounds crystallize in the trigonal structure with the lattice parameters $a = 592.39 \pm 0.03$ and $c = 3709.6 \pm 0.3$ for $Hg_7Se_3O_{13}H_2$ and $a = 589.08 \pm 0.06$ and $c = 3104.8 \pm 0.3$ for $Hg_8Se_4O_{13}H_2$. In comparison with $Hg_7Se_3O_{13}H_2$, the structure of $Hg_8Se_4O_{17}H_2$ contains an additional and disordered $HgSeO_4$ layer that leads to a significant distortion of parts of the resulting structure. Single crystals of the mixed-valent compounds $Hg_7Se_3O_{13}H_2$ and $Hg_8Se_4O_{17}H_2$ were grown under hydrothermal conditions at the interaction of $Hg(NO_3)_2$ and H_2SeO_4 at $250°C$.

8.3 MERCURY–POTASSIUM–GERMANIUM–SELENIUM

HgSe–K_2Se–GeSe$_2$. The phase diagram is not constructed. The $K_2Hg_3Ge_2Se_8$ quaternary compound is formed in this system (Kanatzidis et al. 1997, Jin et al. 2002). This compound crystallizes in the orthorhombic structure with the lattice parameters $a = 1985.1 \pm 0.2$, $b = 1000.40 \pm 0.07$, and $c = 860.97 \pm 0.07$ pm and calculated density 5.659 g$\cdot$cm^{-3} (Jin et al. 2002). It was synthesized using a molten salt K_2Se_4 flux reaction.

8.4 MERCURY–POTASSIUM–TIN–SELENIUM

HgSe–K_2Se–SnSe$_2$. The phase diagram is not constructed. $K_2Hg_3Sn_2Se_8$ and $K_2HgSnSe_4$ quaternary compounds are formed in this system (Kanatzidis et al. 1997, Brandmayer et al. 2004). $K_2HgSnSe_4$ crystallizes in the tetragonal structure with the lattice parameters $a = 806.81 \pm 0.11$ and $c = 694.97 \pm 0.14$ pm and calculated density 5.237 g$\cdot$cm^{-3} (Brandmayer et al. 2004). To obtain this compound, $K_4SnSe_4 \cdot 1.5MeOH$ (0.15 mmol) was suspended in MeOH (5 mL) and added to a solution of $Hg(CH_3COO)_2$ (0.15 mmol) in a mixture of MeOH (4.5 mL) and H_2O (0.5 mL). After stirring overnight, the precipitate was removed by filtration, and toluene (10 mL) was allowed to flow under the filtrate. After 9 weeks, crystallization of $K_2HgSnSe_4$ started.

8.5 MERCURY–POTASSIUM–PHOSPHORUS–SELENIUM

$K_2HgP_2Se_6$ quaternary compound exists in this system (Chondroudis and Kanatzidis 1998). This compound melts congruently at 541°C, is a wide-gap semiconductor ($E_g = 2.25$ eV), and crystallizes in the monoclinic structure with the lattice parameters $a = 1303.1 \pm 0.2$, $b = 730.8 \pm 0.2$, $c = 1416.7 \pm 0.2$ pm, and $\beta = 110.63° \pm 0.01°$. To obtain this compound, a mixture of HgSe, P_2Se_5, K_2Se, and Se was sealed under vacuum in a Pyrex tube and heated to 500°C for 4 days followed by cooling to 150°C at 2°C/h.

8.6 MERCURY–POTASSIUM–ANTIMONY–SELENIUM

$HgSe–K_2Se–Sb_2Se_3$. The phase diagram is not constructed. The $KHgSbSe_3$ quaternary compound is formed in this system (Chen and Wang 2000). It crystallizes in the orthorhombic structure with the lattice parameters $a = 1948.2 \pm 0.4$, $b = 852.3 \pm 0.2$, and $c = 983.0 \pm 0.2$ pm and energy bandgap 1.85 eV. The solvothermal technique was used for the synthesis of $KHgSbSe_3$.

8.7 MERCURY–RUBIDIUM–TIN–SELENIUM

$HgSe–Rb_2Se–SnSe_2$. The phase diagram is not constructed. The $Rb_2Hg_3Sn_2Se_8$ quaternary compound is formed in this system (Kanatzidis et al. 1997).

8.8 MERCURY–RUBIDIUM–PHOSPHORUS–SELENIUM

$Rb_2HgP_2Se_6$ and $Rb_8[Hg_4(Se_2)_2(PSe_4)_4]$ quaternary compounds exist in this system (Chondroudis and Kanatzidis 1997, 1998, 2002). $Rb_2HgP_2Se_6$ melts congruently at 578°C and crystallizes in the monoclinic structure with energy gap 2.32 eV. To obtain this compound, a mixture of HgSe, P_2Se_5, Rb_2Se, and Se was sealed under vacuum in a Pyrex tube and heated to 500°C for 4 days followed by cooling to 150°C at 2°C/h (Chondroudis and Kanatzidis 1998).

$Rb_8[Hg_4(Se_2)_2(PSe_4)_4]$ melts incongruently at 413°C and crystallizes in the tetragonal structure with the lattice parameters $a = 1765.4 \pm 0.2$ and $c = 722.6 \pm 0.2$ pm, calculated density 4.699 g · cm^{-3}, and energy gap 2.32 eV (Chondroudis and Kanatzidis 1997, 2002). This compound was synthesized from a mixture of HgSe, P_2Se_5, Rb_2Se, and Se that was mixed thoroughly and transferred to a Pyrex tube inside a N_2-filled glovebox. This tube was evacuated and heated from 50°C to 495°C during 24 h, held at this temperature for 110 h with next cooling to 300°C during 130 h, then to 250°C during 24 h, and to 50°C during 24 h.

8.9 MERCURY–RUBIDIUM–ANTIMONY–SELENIUM

HgSe–Rb$_2$Se–Sb$_2$Se$_3$. The phase diagram is not constructed. The RbHgSbSe$_3$ quaternary compound is formed in this system (Chen et al. 1999, Chen and Wang 2000). It crystallizes in the monoclinic structure with the lattice parameters $a = 775.8 \pm 0.2$, $b = 1123.4 \pm 0.2$, $c = 884.9 \pm 0.2$ pm, and $\beta = 106.60° \pm 0.03°$ and energy bandgap 1.75 eV. The solvothermal technique was used for the synthesis of RbHgSbSe$_3$.

8.10 MERCURY–CESIUM–GALLIUM–SELENIUM

HgSe–Cs$_2$Se–Ga$_2$Se$_3$. The phase diagram is not constructed. The CsHg$_4$Ga$_5$Se$_{12}$ quaternary compound is formed in this system. It crystallizes in the trigonal structure with the lattice parameters $a = 1438.0 \pm 0.4$ and $c = 975.8 \pm 0.6$ pm (Krauß et al. 1996). Single crystals of this compound were prepared by flux growth from a solution of HgGa$_2$Se$_4$ in CsBr.

8.11 MERCURY–CESIUM–YTTRIUM–SELENIUM

HgSe–Cs$_2$Se–Y$_2$Se$_3$. The phase diagram is not constructed. The CsHgYSe$_3$ quaternary compound is formed in this system (Mitchell et al. 2003). It crystallizes in the orthorhombic structure with the lattice parameters $a = 422.26 \pm 0.18$, $b = 1583.9 \pm 0.7$, and $c = 1099.7 \pm 0.5$ pm at 153 ± 2 K; energy gap 2.58 cV for (010) crystal face and 2.54 eV for (001) crystal face; and calculated density 5.954 g·cm^{-3}. CsHgYSe$_3$ was prepared from the mixture of Cs$_2$Se$_3$, Y, Hg, Se, and CsI as a flux. The sample was heated to 850°C in 24 h, kept at this temperature for 96 h, and cooled to 200°C in 96 h, and then the furnace was turned off.

8.12 MERCURY–CESIUM–LANTHANUM–SELENIUM

HgSe–Cs$_2$Se–La$_2$Se$_3$. The phase diagram is not constructed. The CsHgLaSe$_3$ quaternary compound is formed in this system (Mitchell et al. 2003). It crystallizes in the orthorhombic structure with the lattice parameters $a = 440.02 \pm 0.04$, $b = 1572.79 \pm 0.14$, and $c = 1140.52 \pm 0.10$ pm at 153 ± 2 K; energy gap 2.51 eV for (010) crystal face and 2.46 eV for (001) crystal face; and calculated density 5.969 g·cm^{-3}. CsHgLaSe$_3$ was prepared from the mixture of Cs$_2$Se$_3$, La, Hg, Se, and CsI as a flux. The sample was heated to 850°C in 24 h, kept at this temperature for 96 h, and cooled to 200°C in 96 h, and then the furnace was turned off.

8.13 MERCURY–CESIUM–CERIUM–SELENIUM

HgSe–Cs$_2$Se–Ce$_2$Se$_3$. The phase diagram is not constructed. The CsHgCeSe$_3$ quaternary compound is formed in this system (Mitchell et al. 2003). It crystallizes in the orthorhombic structure with the lattice parameters $a = 437.19 \pm 0.05$, $b = 1576.58 \pm 0.17$, and $c = 1131.66 \pm 0.12$ pm at 153 ± 2 K; energy gap 1.94 eV for (010) crystal face; and calculated density 6.050 g·cm^{-3}. CsHgCeSe$_3$ was prepared from the mixture of Cs_2Se_3, Ce, Hg, Se, and CsI as a flux. The sample was heated to 850°C in 24 h, kept at this temperature for 96 h, and cooled to 200°C in 96 h, and then the furnace was turned off.

8.14   MERCURY–CESIUM–PRASEODYMIUM–SELENIUM

$HgSe–Cs_2Se–Pr_2Se_3$. The phase diagram is not constructed. The $CsHgPrSe_3$ quaternary compound is formed in this system (Mitchell et al. 2003). It crystallizes in the orthorhombic structure with the lattice parameters $a = 433.86 \pm 0.04$, $b = 1574.04 \pm 0.15$, and $c = 1124.41 \pm 0.10$ pm at 153 ± 2 K and calculated density 6.153 g·cm^{-3}. $CsHgPrSe_3$ was prepared from the mixture of Cs_2Se_3, Pr, Hg, Se, and CsI as a flux. The sample was heated to 850°C in 24 h, kept at this temperature for 96 h, and cooled to 200°C in 96 h, and then the furnace was turned off.

8.15   MERCURY–CESIUM–NEODYMIUM–SELENIUM

$HgSe–Cs_2Se–Nd_2Se_3$. The phase diagram is not constructed. The $CsHgNdSe_3$ quaternary compound is formed in this system (Mitchell et al. 2003). It crystallizes in the orthorhombic structure with the lattice parameters $a = 432.02 \pm 0.03$, $b = 1577.71 \pm 0.11$, and $c = 1119.81 \pm 0.08$ pm at 153 ± 2 K and calculated density 6.219 g·cm^{-3}. $CsHgNdSe_3$ was prepared from the mixture of Cs_2Se_3, Nd, Hg, Se, and CsI as a flux. The sample was heated to 850°C in 24 h, kept at this temperature for 96 h, and cooled to 200°C in 96 h, and then the furnace was turned off.

8.16   MERCURY–CESIUM–SAMARIUM–SELENIUM

$HgSe–Cs_2Se–Sm_2Se_3$. The phase diagram is not constructed. The $CsHgSmSe_3$ quaternary compound is formed in this system (Mitchell et al. 2003). It crystallizes in the orthorhombic structure with the lattice parameters $a = 429.4 \pm 0.2$, $b = 1587.8 \pm 0.6$, and $c = 1114.3 \pm 0.4$ pm at 153 ± 2 K; energy gap 2.36 eV for (010) crystal face and 2.37 eV for (001) crystal face; and calculated density 6.301 g·cm^{-3}. $CsHgSmSe_3$ was prepared from the mixture of Cs_2Se_3, Sm, Hg, Se, and CsI as a flux. The sample was heated to 850°C in 24 h, kept at this temperature for 96 h, and cooled to 200°C in 96 h, and then the furnace was turned off.

8.17   MERCURY–CESIUM–GADOLINIUM–SELENIUM

$HgSe–Cs_2Se–Gd_2Se_3$. The phase diagram is not constructed. The $CsHgGdSe_3$ quaternary compound is formed in this system (Mitchell et al. 2003). It crystallizes in the orthorhombic structure with the lattice parameters $a = 427.63 \pm 0.14$, $b = 1587.7 \pm 0.5$, and $c = 1111.4 \pm 0.4$ pm at 153 ± 2 K; energy gap 2.54 eV for (010) crystal face and 2.53 eV for (001) crystal face; and calculated density 6.405 g·cm^{-3}. $CsHgGdSe_3$ was prepared from the mixture of Cs_2Se_3, Gd, Hg, Se, and CsI as a flux. The sample was heated to 850°C in 24 h, kept at this temperature for 96 h, and cooled to 200°C in 96 h, and then the furnace was turned off.

8.18 MERCURY–CESIUM–TERBIUM–SELENIUM

HgSe–Cs₂Se–Tb₂Se₃. The phase diagram is not constructed. The $CsHgTbSe_3$ quaternary compound is formed in this system (Mitchell et al. 2003). It was prepared from the mixture of Cs_2Se_3, Tb, Hg, Se, and CsI as a flux. The sample was heated to 850°C in 24 h, kept at this temperature for 96 h, and cooled to 200°C in 96 h, and then the furnace was turned off.

8.19 MERCURY–CESIUM–DYSPROSIUM–SELENIUM

HgSe–Cs₂Se–Dy₂Se₃. The phase diagram is not constructed. The $CsHgDySe_3$ quaternary compound is formed in this system (Mitchell et al. 2003). It was prepared from the mixture of Cs_2Se_3, Dy, Hg, Se, and CsI as a flux. The sample was heated to 850°C in 24 h, kept at this temperature for 96 h, and cooled to 200°C in 96 h, and then the furnace was turned off.

8.20 MERCURY–CESIUM–URANIUM–SELENIUM

HgSe–Cs₂Se–USe₂. The phase diagram is not constructed. The $Cs_2Hg_2USe_5$ quaternary compound is formed in this system, which is moderately stable in air and crystallizes in the monoclinic structure with the lattice parameters $a = 1027.6 \pm 0.6$, $b = 429.9 \pm 0.2$, $c = 1543.2 \pm 0.9$ pm, and $\beta = 101.857° \pm 0.006°$ and calculated density 6.470 g·cm⁻³ at 153 ± 2 K (Bugaris et al. 2009). This compound exhibits semiconducting behavior in the (010) direction with the energy gap approximately 0.3 eV.

$Cs_2Hg_2USe_5$ compound was obtained from the solid-state reaction of Cs_2Se_3, HgSe, U, Se, and CsI as a flux. The reaction mixture was placed in a computer-controlled furnace where it was heated to 850°C in 24 h, kept at this temperature for 96 h, cooled to 200°C in 96 h, and then quickly cooled to room temperature.

8.21 MERCURY–CESIUM–TIN–SELENIUM

HgSe–Cs₂Se–SnSe₂. The phase diagram is not constructed. The $Cs_2HgSn_3Se_8$ quaternary compound is formed in this system. It crystallizes in the orthorhombic structure with the lattice parameters $a = 791.96 \pm 0.03$, $b = 1266.57 \pm 0.07$, and $c = 1853.79 \pm 0.09$ pm; calculated density 5.194 g·cm⁻³; and energy gap 1.96 eV (Morris et al. 2013). To obtain $Cs_2HgSn_3Se_8$, a mixture of Cs_2Se_2, HgSe, Sn, and Se in the appropriate ratios was added to a fused silica tube inside an N_2-filled glovebox. The tube was evacuated, flame sealed, and used in a flame-melting reaction. After cooling in air, a phase pure ingot containing dark red platelike crystals was obtained.

8.22 MERCURY–CESIUM–ANTIMONY–SELENIUM

HgSe–Cs₂Se–Sb₂Se₃. The phase diagram is not constructed. The $CsHgSbSe_3$ quaternary compound is formed in this system (Chen et al. 1999, Chen and Wang 2000). It crystallizes in the orthorhombic structure with the lattice parameters $a = 444.4 \pm 0.1$, $b = 1551.4 \pm 0.6$, and $c = 1126.1 \pm 0.7$ pm and energy bandgap 1.65 eV. The solvothermal technique was used for the synthesis of $CsHgSbSe_3$.

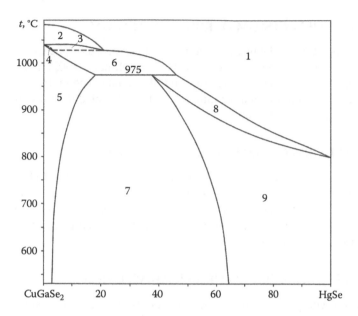

FIGURE 8.1 Phase relations in the HgSe–CuGaSe$_2$ system: 1, L; 2, L + (Ga$_2$Se$_3$); 3, L + (CuGaSe$_2$) + (Ga$_2$Se$_3$); 4, (CuGaSe$_2$) + (Ga$_2$Se$_3$); 5, (CuGaSe$_2$); 6, L + (CuGaSe$_2$); 7, (CuGaSe$_2$) + (HgSe); 8, L + (HgSe); and 9, (HgSe). (From Parasyuk, O.V. et al., *Fizyka kondens. vysokomolek. system. Nauk. zap. Rinens'kogo pedinstytutu*, (3), 158, 1997; Olekseyuk, I.D. et al., *Pol. J. Chem.*, 72(1), 49, 1998.)

8.23 MERCURY–COPPER–GALLIUM–SELENIUM

HgSe–CuGaSe$_2$. This section is not a quasibinary one as CuGaSe$_2$ melts incongruently at 1040°C (Figure 8.1) (Parasyuk et al. 1997, Olekseyuk et al. 1998). The solubility of HgSe in CuGaSe$_2$ at 600°C is equal to 3 mol.% and the solubility of CuGaSe$_2$ in HgSe at this temperature reaches 37 mol.%. The peritectic process with the formation of the solid solution based on HgSe takes place at 975°C. This system was investigated by differential thermal analysis (DTA), x-ray diffraction (XRD), and metallography, and the ingots were annealed at 600°C for 250 h (Olekseyuk et al. 1998).

8.24 MERCURY–COPPER–INDIUM–SELENIUM

HgSe–CuInSe$_2$. A continuous range of solid solution exists between HgSe and β-CuInSe$_2$ (Figure 8.2) (Gallardo 1992a, Parasyuk et al. 1997, Olekseyuk et al. 1998). CuInSe$_2$ melts congruently at 994°C and exhibits a polymorphic transformation at 804°C. Solid solution based on α-CuInSe$_2$ contains up to 7 mol.% HgSe at 600°C (Parasyuk et al. 1997, Olekseyuk et al. 1998) and 9 mol.% HgSe at 400°C (Halka et al. 2000a). Within the limits of experimental error (±0.5 pm), the variation of the lattice parameter *a* versus composition follows Vegard's rule (Gallardo 1992a).

This system was investigated by DTA, XRD, and metallography, and the ingots were annealed at 600°C for 250 h (Olekseyuk et al. 1998) [at 450°C during 720 h (Gallardo 1992a)].

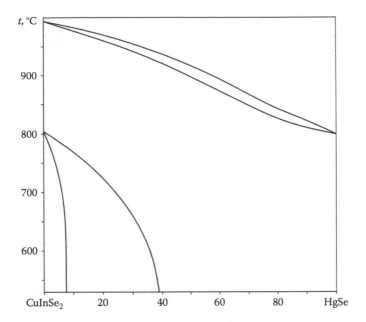

FIGURE 8.2 Phase diagram of the HgSe–CuInSe$_2$ system. (From Parasyuk, O.V. et al., *Fizyka kondens. vysokomolek. system. Nauk. zap. Rinens'kogo pedinstytutu*, (3), 158, 1997; Olekseyuk, I.D. et al., *Pol. J. Chem.*, 72(1), 49, 1998.)

HgSe–Cu$_3$InSe$_3$. This section is not a quasibinary one as Cu$_3$InSe$_3$ melts incongruently (Figure 8.3) (Halka et al. 2000b). Its liquidus consists of the curves of primary crystallization of the solid solutions based on HgSe and Cu$_3$InSe$_3$. This system was investigated by DTA, XRD, and metallography, and the ingots were annealed at 400°C for 250 h whereupon they were quenched in cold water.

HgSe–Cu$_2$Se–In$_2$Se$_3$. The liquidus surface of the Cu$_2$Se–CuInSe$_2$–HgIn$_2$Se$_4$–HgSe subsystem of the Cu$_2$Se–HgSe–In$_2$Se$_3$ quasiternary system is shown in Figure 8.4 (Halka et al. 2000b).

It consists of four fields of primary crystallization, which belong to the solid solutions based on Cu$_2$Se, Cu$_3$InSe$_3$, HgSe, and HgIn$_2$Se$_4$. The fields of primary crystallization are divided by four monovariant lines and six nonvariant points, five of which are binary and only one is ternary. The transition point U corresponds to the reaction L + Cu$_3$InSe$_3$ ⇔ Cu$_2$Se + HgSe, which takes place at 899°C. This point has the coordinates 73 mol.% Cu$_2$Se, 8 mol.% HgSe, and 19 mol.% In$_2$Se$_3$.

The isothermal section of the Cu$_2$Se–HgSe–In$_2$Se$_3$ quasiternary system at 400°C is shown in Figure 8.5 (Halka et al. 2000a). The Cu$_{1.4}$HgIn$_{16.6}$Se$_{26.6}$ quaternary compound is formed in this system. Solid solutions based on CuInSe$_2$ extend between 0 and 22 mol.% HgIn$_2$Se$_4$ in the CuInSe$_2$–HgIn$_2$Se$_4$ section. The solubility of CuInSe$_2$ in HgIn$_2$Se$_4$ is less than 5 mol.% and the maximum content of In$_2$Se$_3$ in Cu$_2$Se is less than 4 mol.%. The samples were annealed at 400°C for 250 h and the single crystals of the solid solutions based on HgSe were grown using crystallization from the solution in the melt.

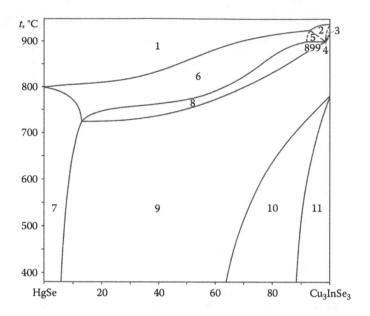

FIGURE 8.3 Phase relations in the $HgSe–Cu_3InSe_3$ system: 1, L; 2, L + (Cu_3InSe_3); 3, (Cu_3InSe_3); 4, (Cu_3InSe_3) + (HgSe); 5, L + (Cu_3InSe_3) + (HgSe); 6, L + (HgSe); 7, (HgSe); 8, L + ($CuInSe_2$) + (HgSe); 9, ($CuInSe_2$) + (HgSe); 10, ($CuInSe_2$) + (HgSe) + (Cu_2Se); and 11, ($CuInSe_2$) + (Cu_2Se). (From Halka, V.O. et al., *J. Alloys Compd.*, 309(1–2), 165, 2000b.)

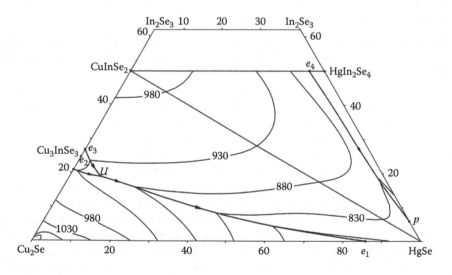

FIGURE 8.4 Liquidus surface of the $Cu_2Se–CuInSe_2–HgIn_2Se_4–HgSe$ subsystem. (From Halka, V.O. et al., *J. Alloys Compd.*, 309(1–2), 165, 2000b.)

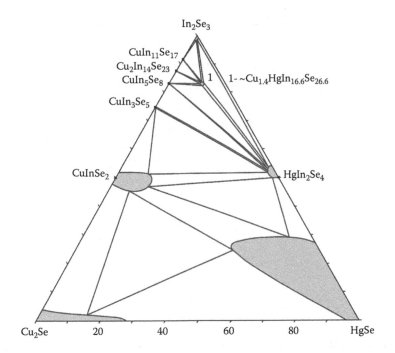

FIGURE 8.5 Isothermal section of the $Cu_2Se-HgSe-In_2Se_3$ quasiternary system at 400°C. (From Halka, V.O. et al., *J. Alloys Compd.*, 302(1–2), 173, 2000a.)

8.25 MERCURY–COPPER–SILICON–SELENIUM

HgSe–Cu₂SiSe₃. This section is a nonquasibinary section of the Hg–Cu–Si–Se quaternary system since Cu_8SiSe_6 primarily crystallizes from the Cu_2SiSe_3 side (Figure 8.6) (Parasyuk et al. 1997, 1998). The eutectic composition and temperature are 93 mol.% HgSe and 762°C, respectively. The $Cu_2HgSiSe_4$ quaternary compound is formed in this system. It melts congruently at 928°C [incongruently at 813°C ± 5°C (Schäfer and Nitsche 1977)], has polymorphous transformation at 685°C (Parasyuk et al. 1997, 1998), and crystallizes in the orthorhombic structure with the lattice parameters $a = 796.2$, $b = 682.3$, and $c = 656.9$ pm (Schäfer and Nitsche 1974, 1977). Cu_2SiSe_3 has a polymorphous transformation at 615°C and the thermal effects at 658°C correspond to the polymorphous transformation of the solid solutions based on this compound.

The solubility of HgSe in the high-temperature modification of Cu_2SiSe_3 is equal to 7 mol.% at 873°C, and the solubility of Cu_2SiSe_3 in HgSe at 400°C is not higher than 1 mol.% (Parasyuk et al. 1997, 1998).

This system was investigated by DTA, XRD, and metallography, and the ingots were annealed at 400°C for 250 h (Parasyuk et al. 1997, 1998).

HgSe–Cu₂Se–SiSe₂. The phase diagram is not constructed. $Cu_{5.976}Hg_{0.972}SiSe_6$ quaternary compound that crystallizes in the cubic structure with the lattice parameter $a = 1030.13 \pm 0.02$ pm and calculated density 6.5408 ± 0.0004 g·cm⁻³ is formed in this system (Gulay et al. 2004).

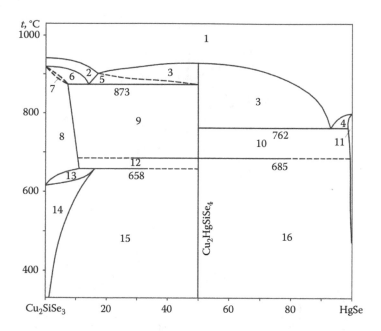

FIGURE 8.6 Phase relations in the $HgSe–Cu_2SiSe_3$ system: 1, L; 2, L + Cu_8SiSe_6; 3, L + β-$Cu_2HgSiSe_4$; 4, L + (HgSe); 5, L + Cu_8SiSe_6 + β-$Cu_2HgSiSe_4$; 6, L + (β-Cu_2SiSe_3) + Cu_8SiSe_6; 7, (β-Cu_2SiSe_3) + Cu_8SiSe_6; 8, (β-Cu_2SiSe_3); 9, (β-Cu_2SiSe_3) + β-$Cu_2HgSiSe_4$; 10, β-$Cu_2HgSiSe_4$ + (HgSe); 11, (HgSe); 12, α-$Cu_2HgSiSe_4$ + (β-Cu_2SiSe_3); 13, (α-Cu_2SiSe_3) + (β-Cu_2SiSe_3); 14, (α-Cu_2SiSe_3); 15, α-$Cu_2HgSiSe_4$ + (α-Cu_2SiSe_3); and 16, α-$Cu_2HgSiSe_4$ + (HgSe). (From Parasyuk, O.V. et al., *Fizyka kondens. vysokomolek. system. Nauk. zap. Rinens'kogo pedinstytutu*, (3), 158, 1997; Parasiuk, O.V. et al., *Ukr. khim. zhurn.*, 64(9), 20, 1998.)

8.26 MERCURY–COPPER–GERMANIUM–SELENIUM

HgSe–Cu₂GeSe₃. The phase diagram is shown in Figure 8.7 (Parasyuk et al. 1997, 1998). The eutectic compositions and temperatures are 15 and 86 mol.% HgSe and 753°C and 716°C [667°C and 597°C (Mkrtchian et al. 1988a)], respectively. The $Cu_2HgGeSe_4$ quaternary compound is formed in this system, which melts congruently at 764°C (Parasyuk et al. 1997, 1998) [at 737°C ± 5°C (Mkrtchian et al. 1988a), at 754°C ± 5°C (Schäfer and Nitsche 1977), at 760°C (Hirai et al. 1967)]. Low-temperature modification of this compound crystallizes in the tetragonal structure with the lattice parameters $a = 574.56 \pm 0.02$ and $c = 1108.34 \pm 0.04$ pm (Olekseyuk et al. 2005) [$a = 569.4$ and $c = 1102$ pm (Parasyuk et al. 1997, 1998), $a = 574.4$ and $c = 1110.3$ pm (Schäfer and Nitsche 1977), $a = 567$ and $c = 1100$ pm (Mkrtchian et al. 1988a), $a = 949$ and $c = 1796$ pm (Hirai et al. 1967)]; calculated and experimental densities 6.4993 (Olekseyuk et al. 2005) [6.66 (Hahn and Schulze 1965)] and 6.48 g·cm⁻³ (Mkrtchian et al. 1988a) [6.52 (Hahn and Schulze 1965)], respectively; and energy gap $E_g = 0.16$ eV (Mkrtchian et al. 1988b).

High-temperature modification of $Cu_2HgGeSe_4$ crystallizes in the orthorhombic structure with the lattice parameters $a = 768.22 \pm 0.03$, $b = 655.61 \pm 0.02$, and $c = 631.44 \pm 0.03$ pm and calculated density 5.5185 g·cm⁻³ (Olekseyuk et al. 2005).

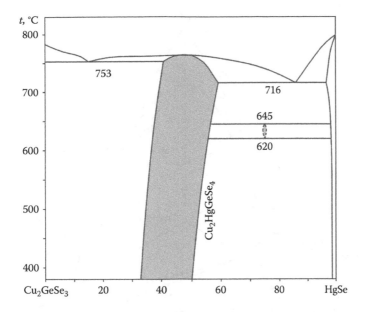

FIGURE 8.7 The HgSe–Cu$_2$GeSe$_3$ phase diagram. (From Parasyuk, O.V. et al., *Fizyka kondens. vysokomolek. system. Nauk. zap. Rinens'kogo pedinstytutu*, (3), 158, 1997; Parasiuk, O.V. et al., *Ukr. khim. zhurn.*, 64(9), 20, 1998.)

The homogeneity region of this compound at 400°C is within the interval of 32–43 mol.% HgSe (Parasyuk et al. 1997, 1998, Olekseyuk et al. 2005) [35–42 mol.% HgSe (Olekseyuk et al. 2001b)]. A new phase is formed within the interval of 620°C–645°C, but its composition was not determined (the ratio of HgSe and Cu$_2$GeSe$_3$ in this phase is approximately 3/1) (Parasyuk et al. 1997, 1998).

The solubility of Cu$_2$GeSe$_3$ in HgSe is equal to 1.5 mol.% [15 mol.% at room temperature (Mkrtchian et al. 1988a)], and the solubility of HgSe in Cu$_2$GeSe$_3$ is not higher than 1 mol.% (Parasyuk et al. 1997, 1998) [9 mol.% (Mkrtchian et al. 1988a)].

This system was investigated by DTA, XRD, metallography, and measuring of microhardness and density (Hirai et al. 1967, Mkrtchian et al. 1988a, Parasyuk et al. 1997, 1998). The ingots were annealed at 400°C for 250 h [for 500 h (Olekseyuk et al. 2001b)] (Parasyuk et al. 1997, 1998, Olekseyuk et al. 2005) [at 430°C–530°C for 500 h (Mkrtchian et al. 1988a)]. Single crystals of the Cu$_2$HgGeSe$_4$ compound were grown by the directional crystallization of the melt (Olekseyuk et al. 2001b).

HgSe–Cu$_2$Se–GeSe$_2$. The liquidus surface of this quasiternary system (Figure 8.8) consists of seven fields of primary crystallization of the phases Cu$_2$Se, Cu$_8$GeSe$_6$, Cu$_2$GeSe$_3$, HgSe, Hg$_2$GeSe$_4$, GeSe$_2$, and Cu$_2$HgGeSe$_4$ (Marchuk et al. 2008). The fields of primary crystallization are separated by 16 monovariant lines and 16 nonvariant points. The modes and temperatures of the occurring ternary nonvariant processes are given in Table 8.1.

The isothermal section of the HgSe–Cu$_2$Se–GeSe$_2$ quasiternary system at 400°C is shown in Figure 8.9 (Olekseyuk et al. 2005). Seven single-phase regions,

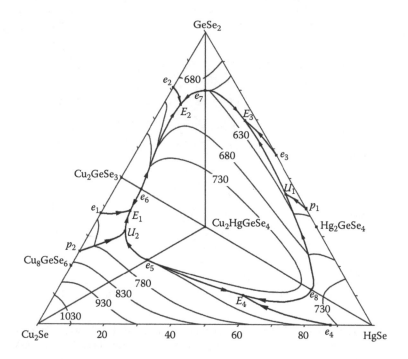

FIGURE 8.8 Liquidus surface of the HgSe–Cu$_2$Se–GeSe$_2$ quasiternary system. (From Marchuk, O.V. et al., *J. Alloys Compd.*, 457(1–2), 337, 2008.)

TABLE 8.1

Temperatures and Compositions of the Ternary Invariant Points in the HgSe–Gu$_2$Se–GeSe$_2$ Quasiternary System

Invariant Points	Reaction	Composition, mol.%			t, °C
		Cu$_2$Se	HgSe	GeSe$_2$	
E_1	L $\Leftrightarrow$ Cu$_8$GeSe$_6$ + Cu$_2$HgGeSe$_4$ + Cu$_2$GeS$_3$	53	8	39	737
E_2	L $\Leftrightarrow$ GeSe$_2$ + Cu$_2$HgGeSe$_4$ + Cu$_2$GeS$_3$	20	5	75	622
E_3	L $\Leftrightarrow$ GeSe$_2$ + Cu$_2$HgGeSe$_4$+ Hg$_2$GeSe$_4$	3	27	70	575
E_4	L $\Leftrightarrow$ β-Cu$_2$Se + HgSe + Cu$_2$HgGeSe$_4$	34	56	10	680
U_1	L + HgSe $\Leftrightarrow$ Cu$_2$HgGeSe$_4$+ Hg$_2$GeSe$_4$	4	51	45	600
U_2	L + β-Cu$_2$Se $\Leftrightarrow$ Cu$_8$GeSe$_6$ + Cu$_2$HgGeSe$_4$	58	10	32	747

Source: Marchuk, O.V. et al., *J. Alloys Compd.*, 457(1–2), 337, 2008.

12 two-phase regions, and 6 three-phase regions exist in the HgSe–Cu$_2$Se–GeSe$_2$ system at this temperature. The solubility of binary and ternary phases is less than 2–3 mol.% of the respective component.

The glass-forming region in the HgSe–Cu$_2$Se–GeSe$_2$ quasiternary system is elongated along the HgSe–GeSe$_2$ quasibinary system (Figure 8.10) (Olekseyuk et al. 2001a).

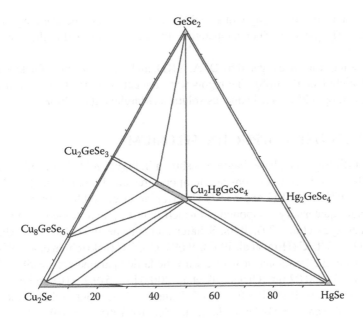

FIGURE 8.9 Isothermal section of the HgSe–Cu$_2$Se–GeSe$_2$ quasiternary system at 400°C. (From Olekseyuk, I.D. et al., *J. Alloys Compd.*, 398(1–2), 80, 2005.)

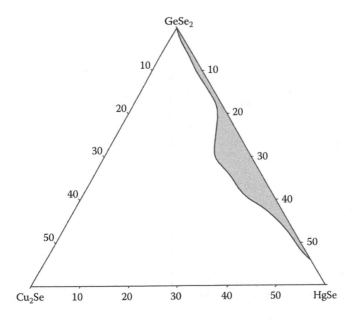

FIGURE 8.10 Glass-forming region in the HgSe–Cu$_2$Se–GeSe$_2$ quasiternary system. (From Olekseyuk, I.D. et al., *Fiz. i khim. tv. tila*, 2(1), 69, 2001a.)

The maximum content of Cu_2Se that could be incorporated in the glass ingots is equal to 7 mol.%. The glass transition temperature for such alloys is within the interval from 218°C to 316°C.

This system was investigated by DTA, XRD, and metallography (Olekseyuk et al. 2001a, Marchuk et al. 2008). The ingots were annealed at 400°C for 500 h with the next quenching in 25% NaCl aqueous solution (Marchuk et al. 2008).

8.27   MERCURY–COPPER–TIN–SELENIUM

HgSe–Cu$_2$SnSe$_3$. The phase diagram is shown in Figure 8.11 (Parasyuk et al. 1997, 1998). The eutectic compositions and temperatures are 11 and 84 mol.% HgSe and 682°C and 677°C [627°C and 652°C (Mkrtchian et al. 1988a)], respectively. The $Cu_2HgSnSe_4$ quaternary compound is formed in this system. It melts congruently at 708°C (Parasyuk et al. 1997, 1998, Schäfer and Nitsche 1977) [at 712°C (Parasyuk et al. 1999), at 710°C (Hirai et al. 1967), at 697°C ± 5°C (Mkrtchian et al. 1988a)] and crystallizes in the tetragonal structure with the lattice parameters $a = 582.88 \pm 0.01$ and $c = 1141.79 \pm 0.02$ pm (Olekseyuk et al. 2002) [$a = 579.2 \pm 0.3$ and $c = 1138.9 \pm 0.5$ (Parasyuk et al. 1997, 1998), $a = 582.5$ and $c = 1141.3$ pm (Schäfer and Nitsche 1977), $a = 581.4$ and $c = 1147$ pm (Hahn and Schulze 1965), $a = 579$ and $c = 1145$ pm (Mkrtchian et al. 1988a), $a = 949$ and $c = 1715$ pm (Hirai et al. 1967)]; calculated and experimental densities 6.514 (Olekseyuk et al. 2002) and 6.45 g·cm^{-3} (Mkrtchian et al. 1988a) (6.51 and 6.47 (Hahn and Schulze 1965)], respectively; and energy gap $E_g = 0.17$ eV (Mkrtchian et al. 1988b)].

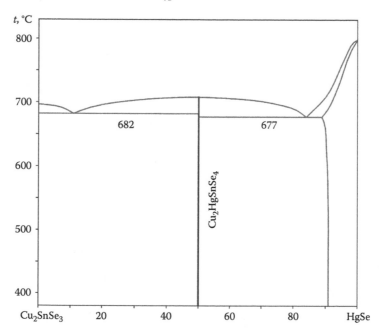

FIGURE 8.11   The HgSe–Cu$_2$SnSe$_3$ phase diagram. (From Parasiuk, O.V. et al., *Ukr. khim. zhurn.*, 64(9), 20, 1998.)

The solubility of Cu_2SnSe_3 in HgSe at 400°C is equal to 9 mol.% [15 mol.% at room temperature (Mkrtchian et al. 1988a)], and the solubility of HgSe in Cu_2SnSe_3 is insignificant (Parasyuk et al. 1997, 1998) [17 mol.% (Mkrtchian et al. 1988a)].

This system was investigated by DTA, metallography, XRD, and measuring of microhardness and density (Mkrtchian et al. 1988a, Parasyuk et al. 1997, 1998). The ingots were annealed at 400°C for 250 h (Parasyuk et al. 1998) [at 430°C–530°C for 500 h (Mkrtchian et al. 1988a)]. The single crystals of the $Cu_2HgSnSe_4$ compound were grown by the directional crystallization of the melt (Olekseyuk et al. 2001b) or using a solution–fusion method (Olekseyuk et al. 2002).

HgSe–Cu₂Se–SnSe₂. The liquidus surface of this quasiternary system consists of six surfaces of primary crystallization: Cu_2Se, Cu_2SnSe_3, HgSe, $SnSe_2$, β-Hg_2SnSe_4, and $Cu_2HgSnSe_4$ (Figure 8.12) (Parasyuk et al. 1999). The regions of primary crystallization are separated by 14 monovariant lines and by 14 invariant points; 9 of these points (8 eutectics and 1 peritectic) belong to three-phase invariant processes, and 5 of them belong to four-phase invariant processes. The parameters, temperatures, and compositions of the ternary invariant points are given in Table 8.2.

The isothermal section of the HgSe–Cu₂Se–SnSe₂ quasiternary system at 400°C is shown in Figure 8.13 (Parasyuk et al. 1999). $Cu_2HgSnSe_4$ quaternary compound is

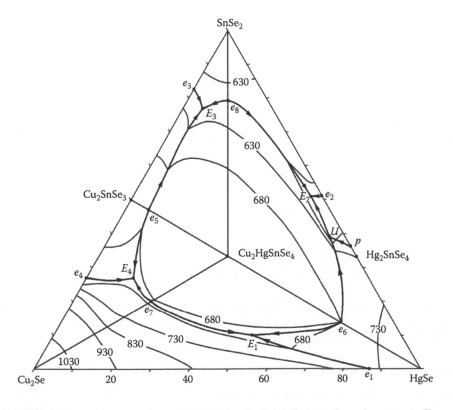

FIGURE 8.12 Liquidus surface of the HgSe–Cu₂Se–SnSe₂ quasiternary system. (From Parasyuk, O.V. et al., *J. Alloys Compd.*, 287(1–2), 197, 1999.)

TABLE 8.2

Temperatures and Compositions of the Ternary Invariant Points in the HgSe–Cu₂Se–SnSe₂ Quasiternary System

Invariant Points	Reaction	Composition, mol.%			t, °C
		Cu_2Se	HgSe	$SnSe_2$	
E_1	$L \Leftrightarrow Cu_2Se + HgSe + Cu_2HgSnSe_4$	40	50	10	645
E_2	$L \Leftrightarrow Cu_2HgSnSe_4 + \beta\text{-}Hg_2SnSe_4 + SnSe_2$	3	66	41	546
E_3	$L \Leftrightarrow Cu_2HgSnSe_4 + Cu_2SnSe_3 + SnSe_2$	18	5	77	915
E_4	$L \Leftrightarrow Cu_2Se + Cu_2SnSe_3 + Cu_2HgSnSe_4$	61	12	17	680
U	$L + HgSe \Leftrightarrow Cu_2HgSnSe_4 + \beta\text{-}Hg_2SnSe_4$	4	57	39	667

Source: Parasyuk, O.V. et al., *J. Alloys Compd.*, 287(1–2), 197, 1999.

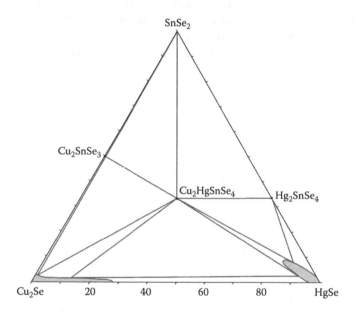

FIGURE 8.13 Isothermal section of the HgSe–Cu₂Se–SnSe₂ quasiternary system at 400°C. (From Parasyuk, O.V. et al., *J. Alloys Compd.*, 287(1–2), 197, 1999.)

in thermal equilibrium with all phases of this system. This system was investigated by DTA, XRD, metallography, and microhardness measurements. The alloys were annealed at 400°C for 250 h and quenched in cold water.

8.28 MERCURY–COPPER–CHLORINE–SELENIUM

HgSe–CuCl. The phase diagram is not constructed. The CuHgSeCl quaternary compound is formed in this system that crystallizes in the orthorhombic structure with the lattice parameters $a = 694.44 \pm 0.06$, $b = 1275.61 \pm 0.19$, and $c = 425.26 \pm 0.05$ pm and calculated and experimental densities 6.67 and 6.71 ± 0.03 g·cm^{-3}, respectively (Guillo et al. 1980). The crystal chemistry of this compound was discussed in Borisov et al. (2005), and it was shown that the degree of anion ordering in this compound substantially exceeds (by 15%–20%) the degree of cation ordering in the same interval of interplanar spacings d_{hkl}, which determines the structure arrangement.

8.29 MERCURY–COPPER–BROMINE–SELENIUM

HgSe–CuBr. The phase diagram is not constructed. The CuHgSeBr quaternary compound is formed in this system that crystallizes in the orthorhombic structure with the lattice parameters a = 1020.1 ± 0.3, b = 431.2 ± 0.1, and c = 925.6 ± 0.3 pm and calculated density 6.90 g·cm^{-3} (Beck et al. 2001). The crystal chemistry of this compound was discussed in Borisov et al. (2005), and it was shown that the degree of anion ordering in this compound substantially exceeds (by 15%–20%) the degree of cation ordering in the same interval of interplanar spacings d_{hkl}, which determines the structure arrangement.

CuHgSeBr has been synthesized by the hydrothermal reaction of HgSe and CuBr in concentrated HBr as a solvent at 285°C (Beck et al. 2001).

8.30 MERCURY–SILVER–GALLIUM–SELENIUM

HgSe–AgGaSe$_2$. The phase diagram of this system is a eutectic type (Figure 8.14) (Parasyuk et al. 1997). The eutectic contains 73 mol.% HgSe and crystallizes at 696°C. The solubility of HgSe in AgGaSe$_2$ at the eutectic temperature reaches 44 mol.%

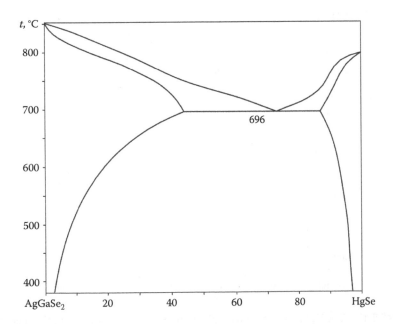

FIGURE 8.14 The HgSe–AgGaSe$_2$ phase diagram. (From Parasyuk, O.V. et al., *Fizyka kondens. vysokomolek. system. Nauk. zap. Rinens'kogo pedinstytutu*, (3), 158, 1997.)

and is equal to 5 mol.% at 400°C. The solubility of $AgGaSe_2$ in HgSe increases from about 5 mol.% at 400°C to 13 mol.% at the eutectic temperature. This system was investigated by DTA, XRD, and metallography, and the ingots were annealed at 400°C during 250 h.

8.31 MERCURY–SILVER–INDIUM–SELENIUM

$HgSe–AgInSe_2$. The phase diagram of this system is a eutectic type (Figure 8.15) (Kozer et al. 2010, Parasyuk et al. 1997). The eutectic contains 54 mol.% HgSe and crystallizes at 720°C. The solubility of HgSe in β-$AgInSe_2$ at the eutectic temperature reaches 18 mol.%, and the solubility of HgSe in α-$AgInSe_2$ is equal to 9 mol.% [7 mol.% (Parasyuk et al. 1997)] at the temperature of a eutectoid reaction (502°C) and decreases to 8 mol.% [5 mol.% (Parasyuk et al. 1997)] at 400°C (Kozer et al. 2010). The solubility of $AgInSe_2$ in HgSe at 400°C is equal to about 39 mol.% and practically does not change at the increasing temperature up to solidus temperature (Kozer et al. 2010) [according to the data of Parasyuk et al. (1997), the solubility of $AgInSe_2$ in HgSe at the eutectic temperature reaches 14 mol.% and decreases up to about 5 mol.% at 400°C]. It was indicated (Gallardo 1992b) that it is possible to predict the value at which the chalcopyrite phase in this system disappears (x_c) and $T_c(x)$ behavior for the $Hg_{2x}(AgIn)_{1-x}Se_2$ alloys for which a chalcopyrite–sphalerite-like disordering transition with extensive terminal solid solution formation exists.

This system was investigated by DTA, XRD, and metallography, and the ingots were annealed at 400°C during 250 h (Kozer et al. 2010, Parasyuk et al. 1997).

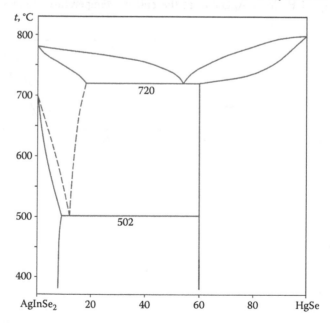

FIGURE 8.15 The $HgSe–AgInSe_2$ phase diagram. (From Kozer, V.R. et al., *Neorgan. materialy*, 46(6), 686, 2010.)

8.32   MERCURY–SILVER–SILICON–SELENIUM

HgSe–Ag$_2$Se–SiSe$_2$. The isothermal section of the HgSe–Ag$_2$Se–SiSe$_2$ quasiternary system at room temperature is shown in Figure 8.16 (Parasyuk et al. 2003b). The existence of several intermediate phases located in the Ag$_8$SiSe$_6$–Hg$_4$SiSe$_6$ section was established. The Ag$_6$HgSiSe$_6$ quaternary compound is formed in this system with the homogeneity range from 23 to 43 mol.% Hg$_4$SiSe$_6$ (Ag$_{6.16–4.56}$Hg$_{0.92–1.72}$SiSe$_6$, γ-phase). It crystallizes in the orthorhombic structure with the lattice parameters $a = 767.52 \pm 0.09$, $b = 767.72 \pm 0.09$, and $c = 1085.4 \pm 0.1$ pm and calculated density 6.929 ± 0.002 g·cm^{-3} (Gulay et al. 2002).

The formation of three new intermediate phases Ag$_{-7.04--6.32}$Hg$_{-0.48--0.84}$SiSe$_6$ (β-phase), Ag$_{-3.44--2.96}$Hg$_{-2.28--2.52}$SiSe$_6$ (δ-phase), and Ag$_2$Hg$_3$SiSe$_6$ (ε-phase) was also found (Parasyuk et al. 2003b). The β-phase (of Ag$_{6.4}$Hg$_{0.8}$SiSe$_6$ composition) and δ-phase (of Ag$_{3.2}$Hg$_{2.4}$SiSe$_6$ composition) crystallize in the cubic structure with the lattice parameters $a = 1088.06 \pm 0.04$ and 1071.08 ± 0.02 pm and calculated densities 6.9212 ± 0.0008 and 7.1479 ± 0.0004 g·cm^{-3}, respectively.

The alloys were annealed at 400°C for 500 h (Parasyuk et al. 2003b). After that, the ampoules were cooled to 100°C at the rate of 10°C/h, and at this temperature, the furnace was turned off and the ampoules with the samples were cooled to room temperature.

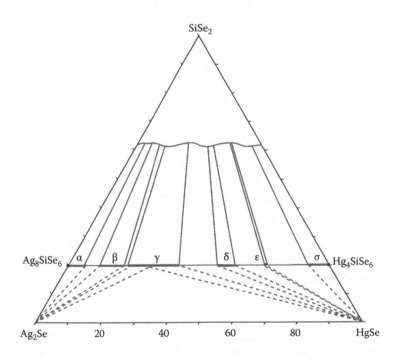

FIGURE 8.16   Isothermal section of the HgSe–Ag$_2$Se–SiSe$_2$ quasiternary system at room temperature. (From Parasyuk, O.V., et al., *J. Alloys Compd.*, 348(1–2), 157, 2003b.)

8.33 MERCURY–SILVER–GERMANIUM–SELENIUM

$HgSe–Ag_2Se–GeSe_2$. The isothermal section of this quasiternary system at room temperature is shown in Figure 8.17 (Parasyuk et al. 2003a). The formation of five intermediate phases β $(Ag_{-7.12-6.32}Hg_{-0.44-0.82}GeSe_6)$, γ $(Ag_{-6.08-4.00}Hg_{-0.96-2.00}GeSe_6)$, δ $(Ag_{3.4}Hg_{2.3}GeSe_6)$, ε $(Ag_{-2.24-2.00}Hg_{-2.88-3.00}GeSe_6)$, and $~Ag_{1.4}Hg_{1.3}GeSe_6$ was established.

β-Phase crystallizes in the cubic structure with the lattice parameter $a = 1090.26 ± 0.04$ pm and calculated density $7.3330 ± 0.0008$ g·cm^{-3} for the composition $Ag_{6.504}Hg_{0.912}GeSe_6$ (Parasyuk et al. 2003a). γ-Phase crystallizes in the orthorhombic structure with the lattice parameters $a = 770.65 ± 0.09$, $b = 770.73 ± 0.07$, and $c = 1089.8 ± 0.1$ pm and calculated density $7.192 ± 0.003$ g·cm^{-3} for the composition $Ag_6HgGeSe_6$ (Gulay et al. 2002); $a = 769.61 ± 0.07$, $b = 769.58 ± 0.06$, and $c = 1088.4 ± 0.1$ pm and calculated density $7.172 ± 0.002$ g·cm^{-3} for the composition $Ag_{5.6}Hg_{1.2}GeSe_6$; $a = 767.22 ± 0.08$, $b = 767.26 ± 0.07$, and $c = 1085.0 ± 0.1$ pm and calculated density $7.241 ± 0.002$ g·cm^{-3} for the composition $Ag_{4.8}Hg_{1.6}GeSe_6$; and $a = 764.25 ± 0.07$, $b = 764.14 ± 0.07$, and $c = 1080.6 ± 0.1$ pm and calculated density $7.285 ± 0.002$ g·cm^{-3} for the composition $Ag_4Hg_2GeSe_6$ (Parasyuk et al. 2003a). $Ag_{3.4}Hg_{2.3}GeSe_6$ (δ-phase) crystallizes in the cubic structure with the lattice parameter $a = 1077.67 ± 0.08$ pm and calculated density $7.500 ± 0.002$ g·cm^{-3}. $Ag_2HgGeSe_4$ that crystallizes in the cubic structure with the lattice parameter $a = 1079.6$ pm (Haeuseler and Himmrich 1989) was not observed at 25°C by Parasyuk et al. (2003a).

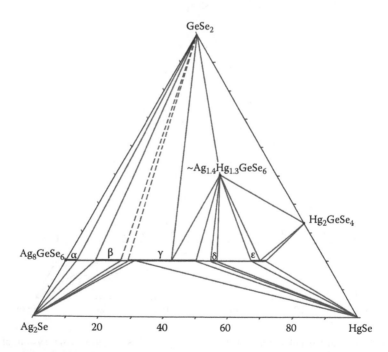

FIGURE 8.17 Isothermal section of the $HgSe–Ag_2Se–GeSe_2$ quasiternary system at room temperature. (From Parasyuk, O.V. et al., *J. Alloys Compd.*, 351, 135, 2003a.)

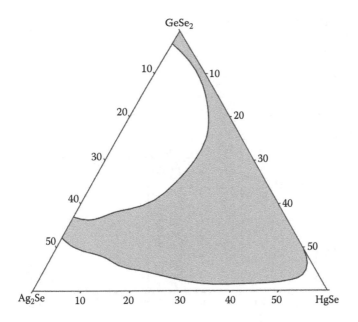

FIGURE 8.18 Glass-formation region in the HgSe–Ag$_2$Se–GeSe$_2$ quasiternary system. (From Olekseyuk, I.D. et al., *Chem. Met. Alloys*, 2(3–4), 146, 2009.)

Glass-formation region in the HgSe–Ag$_2$Se–GeSe$_2$ quasiternary system crosses the concentration triangle (Figure 8.18) (Olekseyuk et al. 2009). The minimum content of the glass-forming component is 43 mol.% GeSe$_2$. For GeSe$_2$ concentration over 80 mol.%, the glass-formation region narrows along the HgSe–GeSe$_2$ side, and the content of the modifier Ag$_2$Se does not exceed 3 mol.%. The characteristic temperatures of the glassy alloys, namely, the glass transition temperature (T_g), the crystallization temperature (T_c), and the melting temperature (T_m) of the crystallized alloys, have been measured.

This system was investigated by DTA, XRD, and metallography (Parasyuk et al. 2003a, Olekseyuk et al. 2009). The alloys were annealed at 400°C for 500 h (Parasyuk et al. 2003a). After that, the ampoules were cooled to 100°C at the rate of 10°C/h, and at this temperature, the furnace was turned off and the ampoules with the samples were cooled to room temperature. Ag$_2$HgGeSe$_4$ was obtained by the interaction of chemical elements at 700°C (Haeuseler and Himmrich 1989).

8.34 MERCURY–SILVER–TIN–SELENIUM

HgSe–"Ag$_2$SnSe$_3$." The phase relations in this section are given in Figure 8.19 (Parasyuk 1999). The Ag$_2$HgSnSe$_4$ quaternary compound is formed in this system. It melts congruently at 557°C, is characterized by a homogeneity region within 47–52 mol.% HgSe at 400°C (Parasyuk 1999), and crystallizes in the orthorhombic structure with the lattice parameters $a = 846.1 \pm 0.1$, $b = 734.0 \pm 0.1$, and $c = 699.01 \pm 0.06$ pm and calculated density 6.523 ± 0.002 g·cm^{-3} (Parasyuk et al. 2002) [$a = 849.4$, $b = 731.4$, and $c = 697.2$ pm (Haeuseler and Himmrich 1989)].

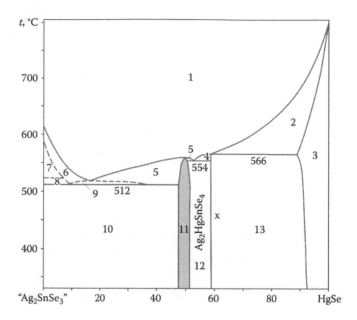

FIGURE 8.19 Phase relations in the HgSe–"Ag$_2$SnSe$_3$" section: 1, L; 2, L + (HgSe); 3, (HgSe); 4, L + X; 5, L + Ag$_2$HgSnSe$_4$; 6, L + Ag$_8$SnSe$_6$; 7, L + Ag$_8$SnSe$_6$ + Ag$_x$Sn$_{1-x}$Se; 8, L + Ag$_8$SnSe$_6$ + SnSe$_2$; 9, L + Ag$_8$SnSe$_6$ + Ag$_2$HgSnSe$_4$; 10, Ag$_8$SnSe$_6$ + SnSe$_2$ + Ag$_2$HgSnSe$_4$; 11, Ag$_2$HgSnSe$_4$; 12, Ag$_2$HgSnSe$_4$ + X; and 13, X + (HgSe). (From Parasyuk, O.V., *J. Alloys Compd.*, 291(1–2), 215, 1999.)

In the homogeneity region, the lattice parameters decrease from $a = 850.2 \pm 0.3$, $b = 736.2 \pm 0.3$, and $c = 700.7 \pm 0.3$ pm to $a = 848.0 \pm 0.3$, $b = 732.3 \pm 0.3$, and $c = 699.6 \pm 0.4$ pm (Parasyuk 1999).

A new quaternary phase X with composition lying in the interval 58–60 mol.% HgSe was found in Parasyuk (1999). This phase melts incongruently at 566°C and crystallizes in the tetragonal structure. A eutectic with a melting temperature of 554°C and a composition of 53 mol.% HgSe is situated between quaternary compounds. HgSe forms a solid solution that exceeds 10 mol.% at the peritectic temperature and equals 8.5 mol.% at 400°C.

This system was investigated by DTA, XRD, and metallography, and the ingots were annealed at 250°C during 500 h with further quenching in cold water (Parasyuk 1999). The Ag$_2$HgSnSe$_4$ compound was obtained by the interaction of chemical elements at 700°C (Haeuseler and Himmrich 1989).

HgSe–Ag$_2$Se–SnSe$_2$. The isothermal section of this quasiternary system at 400°C is shown in Figure 8.20 (Parasyuk et al. 2002). Three intermediate phases Ag$_2$HgSnSe$_4$, Ag$_{2.66}$Hg$_2$Sn$_{1.34}$Se$_6$, and ~Ag$_6$HgSnSe$_6$ exist at this temperature. Ag$_{2.66}$Hg$_2$Sn$_{1.34}$Se$_6$ crystallizes in the orthorhombic structure with the lattice parameters $a = 1279.5 \pm 0.2$, $b = 426.31 \pm 0.06$, and $c = 582.07 \pm 0.04$ pm and calculated density 6.914 ± 0.003 g·cm^{-3} (Gulay and Parasyuk 2002).

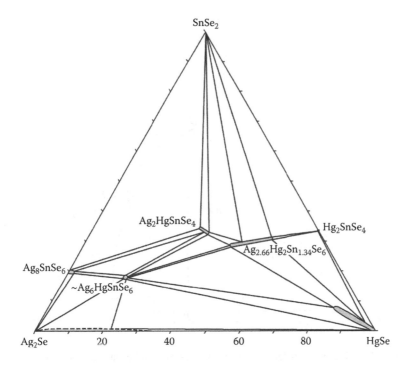

FIGURE 8.20 Isothermal section of the HgSe–Ag$_2$Se–SnSe$_2$ quasiternary system at 400°C. (From Parasyuk, O.V. et al., *J. Alloys Compd.*, 339(1–2), 140, 2002.)

This system was investigated by XRD and metallography, and the ingots were annealed at 400°C during 250–500 h with the next quenching in air (Gulay and Parasyuk 2002, Parasyuk et al. 2002).

8.35 MERCURY–GALLIUM–GERMANIUM–SELENIUM

HgSe–Ga$_2$Se$_3$–GeSe$_2$. The liquidus surface of this system (Figure 8.21) consists of six regions of primary crystallization: HgSe, GeSe$_2$, Ga$_2$Se$_3$, α-HgGa$_2$Se$_4$, β-HgGa$_2$Se$_4$, and Hg$_2$GeSe$_4$ (Olekseyuk et al. 1996). They are separated by 10 invariant lines and 5 invariant points. The parameters, temperatures, and compositions of the ternary invariant points are given in Table 8.3.

The isothermal section of the HgSe–Ga$_2$Se$_3$–GeSe$_2$ quasiternary system at 500°C is shown in Figure 8.22 (Olekseyuk et al. 1996). Solid solutions based on HgSe and Ga$_2$Se$_3$ exist at this temperature. The solubility on the basis of other compounds is not significant.

The glass-forming region occupies almost all the field of GeSe$_2$ primary crystallization and is elongated along the HgSe–GeSe$_2$ quasibinary system (Figure 8.23) (Olekseyuk et al. 1997). The characteristic temperatures of the glassy alloys, namely, the glass transition temperature (T_g), the crystallization temperature (T_c), and the melting temperature (T_m) of the crystallized alloys, have been measured.

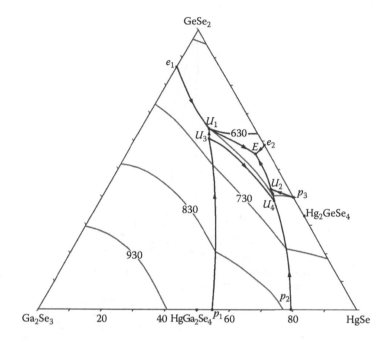

FIGURE 8.21 Liquidus surface of the $HgSe$–Ga_2Se_3–$GeSe_2$ quasiternary system. (From Olekseyuk, I.D. et al., *J. Alloys Compd.*, 238(1–2), 141, 1996.)

TABLE 8.3

Temperatures and Compositions of the Ternary Invariant Points in the $HgSe$–Ga_2Se_3–$GeSe_2$ Quasiternary System

Invariant Points	Reaction	Composition, mol.%			t, °C
		HgSe	Ga₂Se₃	GeSe₂	
E	$L \Leftrightarrow \beta\text{-}HgGa_2Se_4 + Hg_2GeSe_4 + GeSe_2$	40	6	54	572
U_1	$L + Ga_2Se_3 \Leftrightarrow \beta\text{-}HgGa_2Se_4 + GeSe_2$	21	16	63	631
U_2	$L + HgSe \Leftrightarrow \beta\text{-}HgGa_2Se_4 + Hg_2GeSe_4$	51	6	43	592
U_3	$L + Ga_2Se_3 \Leftrightarrow \alpha\text{-}HgGa_2Se_4 + \beta\text{-}HgGa_2Se_4$	23	17	60	652
U_4	$L + HgSe \Leftrightarrow \alpha\text{-}HgGa_2Se_4 + Hg_2GeSe_4$	56	8	36	652

Source: Olekseyuk, I.D. et al., *J. Alloys Compd.*, 238(1–2), 141, 1996.

This system was investigated by DTA, XRD, metallography, and microhardness measurements, and the alloys were annealed at 500°C for 250°C and quenched in cold water (Olekseyuk et al. 1996). The glass alloys were obtained by the annealing at 1000°C during 6 h with the next quenching (Olekseyuk et al. 1997).

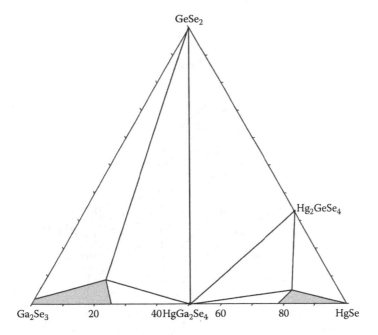

FIGURE 8.22 Isothermal section of the HgSe–Ga$_2$Se$_3$–GeSe$_2$ quasiternary system at 500°C. (From Olekseyuk, I.D. et al., *J. Alloys Compd.*, 238(1–2), 141, 1996.)

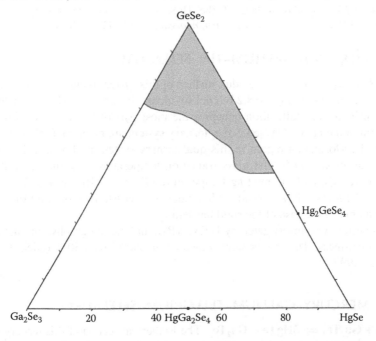

FIGURE 8.23 Glass-formation region in the HgSe–Ga$_2$Se$_3$–GeSe$_2$ quasiternary system. (From Olekseyuk, I.D. et al., *Fizyka kondens. vysokomolek. system. Nauk. zap. Rinens'kogo pedinstytutu*, (3), 148, 1997.)

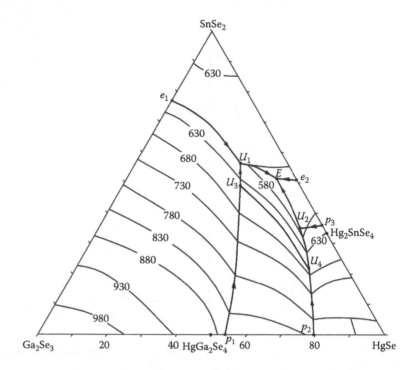

FIGURE 8.24 Liquidus surface of the $HgSe–Ga_2Se_3–SnSe_2$ ternary system. (From Olekseyuk, I.D. and Parasyuk, O.V., *Zhurn. neorgan. khimii*, 42(5), 838, 1997.)

8.36 MERCURY–GALLIUM–TIN–SELENIUM

$HgSe–Ga_2Se_3–SnSe_2$. The liquidus surface of this system includes six fields of primary crystallizations (Figure 8.24) (Olekseyuk and Parasyuk 1997). The field of the Ga_2Se_3 primary crystallization occupies the most part of this surface. Nonvariant equilibria in the $HgSe–Ga_2Se_3–SnSe_2$ ternary system are given in Table 8.4.

The glass-formation region in this quasiternary system is shown in Figure 8.25 (Olekseyuk et al. 1999). The glass transition temperature (T_g), the crystallization temperature (T_c), and the melting temperature (T_m) were determined for obtained glasses. The glasses have considerable tendency to crystallization, and they can be obtained only in the case of the rigid hardening.

This system was investigated by DTA, XRD, and metallography, and measuring of microhardness. The ingots were annealed at 450°C for 250 h (Olekseyuk and Parasyuk 1997).

8.37 MERCURY–GALLIUM–TELLURIUM–SELENIUM

$3HgSe + Ga_2Te_3 \Leftrightarrow 3HgTe + Ga_2Te_3$. The isothermal section of this ternary mutual system at 630°C is shown in Figure 8.26 (Kerkhoff and Leute 2005). The cubic one-phase field extends only along the edges of the phase square. It is interrupted by miscibility gaps caused by the tetragonal superstructure near 75 mol.% $Ga_2Te_3(Ga_2Se_3)$. Moreover,

TABLE 8.4

Nonvariant Equilibria in the HgSe–Ga$_2$Se$_3$–SnSe$_2$ Ternary System

		Composition, mol.%			
Symbol	Reaction	HgSe	Ga$_2$Se$_3$	SnSe$_2$	t, °C
E	L $\Leftrightarrow$ β-HgGa$_2$Se$_4$ + Hg$_2$SnSe$_4$ + SnSe$_2$	42	6	52	552
U_1	L + Ga$_2$Se$_3$ $\Leftrightarrow$ β-HgGa$_2$Se$_4$ + SnSe$_2$	50	13	57	582
U_2	L + HgSe $\Leftrightarrow$ β-HgGa$_2$Se$_4$ + Hg$_2$SnSe$_4$	58	7	35	567
U_3	L + Ga$_2$Se$_3$ $\Leftrightarrow$ α-HgGa$_2$Se$_4$ + β-HgGa$_2$Se$_4$	34	18	49	652
U_4	L + HgSe $\Leftrightarrow$ α-HgGa$_2$Se$_4$ + β-HgGa$_2$Se$_4$	67	11	22	652

Source: Olekseyuk, I.D. and Parasyuk, O.V., *Zhurn. neorgan. khimii*, 42(5), 838, 1997.

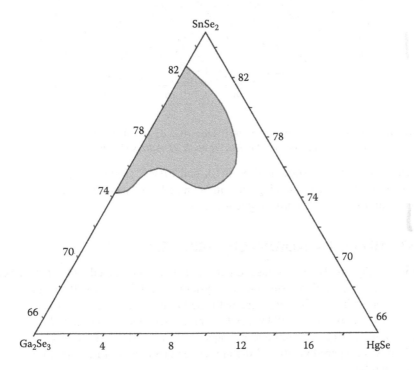

FIGURE 8.25 The glass-formation region in the HgSe–Ga$_2$Se$_3$–SnSe$_2$ quasiternary system. (From Olekseyuk, I.D. et al., *Funct. Mater.*, 6(3), 474, 1999.)

there are two two-phase fields between two cubic regions. The three three-phase fields represent always two cubic phases and one tetragonal phase in equilibrium.

The diffusion profiles and reaction paths have revealed that thermodynamics plays an important part in diffusion experiments in this ternary mutual system (Leute 2007). Typical of this system is that there are ordering tendencies at special composition in addition to the usual thermodynamic interactions. These ordering tendencies complicate the landscape of the unstable and metastable regions.

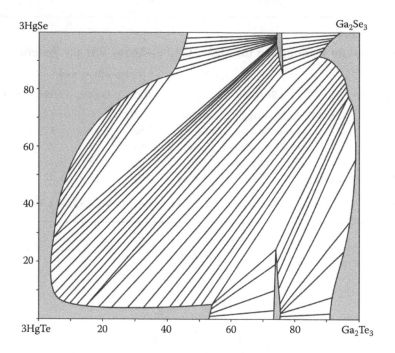

FIGURE 8.26 Isothermal section of the $3HgSe + Ga_2Te_3 \Leftrightarrow 3HgTe + Ga_2Te_3$ ternary mutual system at 630°C. (From Kerkhoff, M. and Leute, V., *J. Alloys Compd.*, 391(1–2), 42, 2005.)

This system was investigated by DTA and EPMA (Kerkhoff and Leute 2005). It was shown that the phase diagram can be modeled by a Gibbs energy function for a subregular system with ordering tendencies.

8.38 MERCURY–INDIUM–CHLORINE–SELENIUM

HgSe–HgCl₂–InCl₃. The phase diagram is not constructed. The $Hg_4In_2Se_3Cl_8$ quaternary compound is formed in this system. It is stable up to 300°C and crystallizes in the orthorhombic structure with the lattice parameters $a = 1371.13 \pm 0.11$, $b = 746.16 \pm 0.06$, and $c = 1843.05 \pm 0.15$ pm; calculated density 5.469 g·cm⁻³; and calculation and experimental energy gaps 2.25 and 2.34 eV, respectively (Liu et al. 2012). It was prepared via the solid-state reaction from a mixture containing Hg_2Cl_2, In_2Se_3, and Se.

8.39 MERCURY–THALLIUM–SILICON–SELENIUM

HgSe–Tl₂SiSe₃. The phase diagram is a eutectic type (Figure 8.27) (Olekseyuk et al. 2012). The $HgTl_2SiSe_4$ quaternary compound is formed in this system. It melts incongruently at 430°C and crystallizes in the tetragonal structure with the lattice parameters $a = 800.32 \pm 0.03$ and $c = 668.79 \pm 0.04$ pm and calculated density 7.390 ± 0.001 g·cm⁻³ (Mozolyuk et al. 2012, Olekseyuk et al. 2012). The eutectic composition and temperature are 35 mol.% HgSe and 380°C, respectively.

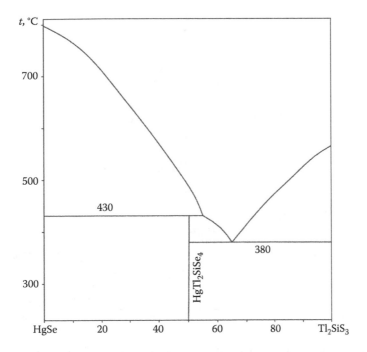

FIGURE 8.27 The HgSe–Tl$_2$SiSe$_3$ phase diagram. (From Olekseyuk, I.D. et al., *Nauk. visnyk Volyn. nats. un-tu im. Lesi Ukrainky. Khim. nauky*, [(17)242], 62, 2012.)

This system was investigated by DTA, XRD, and measuring of microhardness. The ingots were annealed at 250°C for 250 h (Mozolyuk et al. 2012, Olekseyuk et al. 2012).

HgSe–Tl$_2$Se–SiSe$_2$. The isothermal section of this system at 250°C is shown in Figure 8.28 (Mozolyuk et al. 2012, Olekseyuk et al. 2012). HgTl$_2$SiSe$_4$ and HgTl$_2$Si$_2$Se$_6$ quaternary compounds exist in the HgSe–Tl$_2$Se–SiSe$_2$ quasiternary system at this temperature.

8.40 MERCURY–THALLIUM–TIN–SELENIUM

HgSe–Tl$_2$Se–SnSe$_2$. The isothermal section of this system at 250°C is shown in Figure 8.29 (Olekseyuk et al. 2010, Mozolyuk et al. 2012). HgTl$_2$SnSe$_4$ quaternary compound exists in the HgSe–Tl$_2$Se–SnSe$_2$ quasiternary system at this temperature. It crystallizes in the tetragonal structure with the lattice parameters a = 804.07 ± 0.01 and c = 688.52 ± 0.02 pm and calculated density 7.7871 ± 0.0005 g·cm^{-3}. The ingots were annealed at 250°C for 250 h.

8.41 MERCURY–THALLIUM–TELLURIUM–SELENIUM

HgSe–Tl$_2$Te. This section is a nonquasibinary section of the Hg–Tl–Te–Se quaternary system (Asadov 1984b). It crosses the fields of the primary crystallization of solid solutions based on Tl$_2$Te and HgTe$_x$Se$_{1-x}$ and Hg$_3$Tl$_2$Te$_{4x}$Se$_{4-x}$ solid solutions.

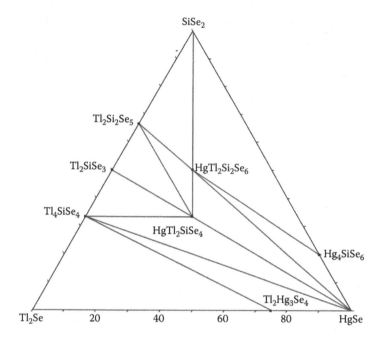

FIGURE 8.28 Isothermal section of the HgSe–Tl$_2$Se–SiSe$_2$ quasiternary system at 250°C. (From Olekseyuk, I.D. et al., *Nauk. visnyk Volyn. nats. un-tu im. Lesi Ukrainky. Khim. nauky*, [(17)242], 62, 2012; Mozolyuk, M.Yu. et al., *Mater. Res. Bull.*, 47(11), 3830, 2012.)

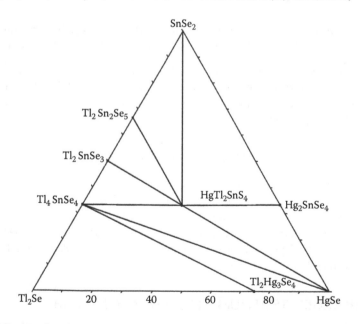

FIGURE 8.29 Isothermal section of the HgSe–Tl$_2$Se–SnSe$_2$ quasiternary system at 250°C. (From Olekseyuk, I.D. et al., *Nauk. visnyk Volyn. nats. un-tu im. Lesi Ukrainky. Khim. nauky*, (30), 19, 2010; Mozolyuk, M.Yu. et al., *Mater. Res. Bull.*, 47(11), 3830, 2012.)

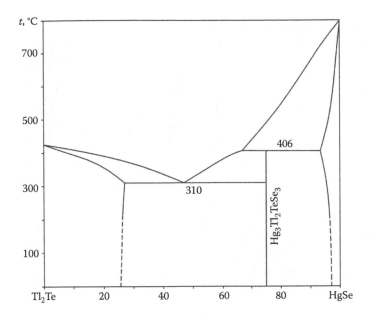

FIGURE 8.30 The HgSe–Tl₂Te phase diagram. (From Asadov, M.M. et al., *Izv. AN SSSR. Neorgan. materialy*, 14(5), 960, 1978.)

According to the data of Asadov et al. (1978), the phase diagram of this system is shown in Figure 8.30. The eutectic composition and temperature are 47 mol.% HgSe and 310°C, respectively. The $Hg_3Tl_2TeSe_3$ quaternary compound is formed in this system, which melts incongruently at 406°C, but later this author (Asadov 1984a,b) concluded that this compound is one of the compositions of the $Hg_3Tl_2Te_{4-x}Se_{4x}$ solid solutions. The peritectic composition is 67 mol.% HgSe (Asadov et al. 1978). The solubility of HgSe in Tl_2Te and Tl_2Te in HgSe reaches 27 and 12 mol.% and decreases to 25 and 6 mol.% at 200°C, respectively. This system was investigated by DTA, XRD, and measuring of microhardness, and the ingots were annealed at 200°C for 300 h.

HgTe–Tl₂Se. This section is a nonquasibinary section of the Hg–Tl–Te–Se quaternary system (Asadov 1984a). The solubility of HgTe in Tl_2Se at 300°C is equal to 25 mol.%. This system was investigated by DTA, XRD, and measuring of microhardness and emf of concentrated chains. The ingots were annealed at 230°C for 600 h.

HgSe + Tl₂Te ⇔ HgTe + Tl₂Se. The liquidus surface of this ternary mutual system includes five fields of primary crystallization (Figure 8.31) (Asadov 1984a): solid solutions based on Tl_2Se and Tl_2Te and $HgTe_xSe_{1-x}$, $Hg_3Tl_2Te_{4-x}Se_{4x}$, and $Hg_3Tl_2Te_{4x}Se_{4-x}$ solid solutions. Nonvariant equilibria are given in Table 8.5.

The isothermal section of the HgSe + Tl₂Te ⇔ HgTe + Tl₂Se ternary mutual system at 200°C is shown in Figure 8.32 (Asadov 1984b). This system was investigated by DTA, XRD, and measuring of microhardness and emf of concentrated chains (Asadov 1984a,b). The ingots were annealed at 200°C [230°C (Asadov 1984a)] for 600 h (Asadov 1984b).

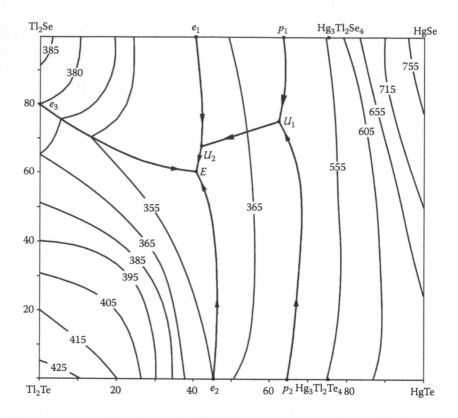

FIGURE 8.31   Liquidus surface of the HgSe + $Tl_2Te \Leftrightarrow$ HgTe + Tl_2Se ternary mutual system. (From Asadov, M.M., *Izv. AN SSSR. Neorgan. materialy*, 20(1), 38, 1984a.)

TABLE 8.5

Nonvariant Equilibria in the HgSe + $Tl_2Te \Leftrightarrow$ HgTe + Tl_2Se Ternary Mutual System

Symbol	Reaction	Composition, mol.%			t, °C
		Tl_2Te	HgSe	Tl_2Se	
E	$L \Leftrightarrow (Tl_2Se) + (Tl_2Te) + Hg_3Tl_2Te_{4-x}Se_{4x}$	40	40	20	302
U_1	$L + HgTe_xSe_{1-x} \Leftrightarrow Hg_3Tl_2Te_{4-x}Se_{4x} + Hg_3Tl_2Te_{4x}Se_{4-x}$	22	68	10	407
U_2	$L + Hg_3Tl_2Te_{4-x}Se_{4x} \Leftrightarrow (Tl_2Se) + Hg_3Tl_2Te_{4x}Se_{4-x}$	30	45	25	312

Source: Asadov, M.M., *Izv. AN SSSR. Neorgan. materialy*, 20(1), 38, 1984.

8.42   MERCURY–LANTHANUM–OXYGEN–SELENIUM

HgSe–La$_2$O$_2$Se. According to the data of Baranov et al. (1996), this section is a nonquasibinary section of the Hg–La–O–Se quaternary system. This system was investigated by XRD. The ingots were annealed at 730°C–830°C for 250 h.

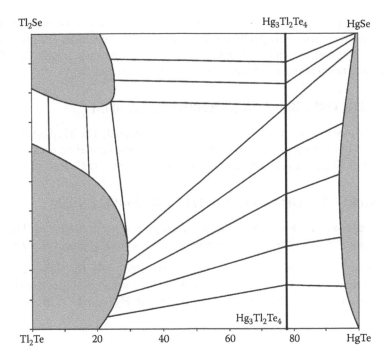

FIGURE 8.32   Isothermal section of the HgSe + Tl$_2$Te $\Leftrightarrow$ HgTe + Tl$_2$Se ternary mutual system at 200°C. (From Asadov, M.M., *Izv. AN SSSR. Neorgan. materialy*, 20(10), 1621, 1984b.)

8.43   MERCURY–URANIUM–CHLORINE–SELENIUM

HgSe–HgCl$_2$–UCl$_4$. The phase diagram is not constructed. [Hg$_3$Se$_2$][UCl$_6$] quaternary compounds are formed in this system that crystallizes in the monoclinic structure with the lattice parameters $a = 683.0 \pm 0.1$, $b = 752.0 \pm 0.2$, and $c = 1330.0 \pm 0.8$ pm and $\beta = 91.88° \pm 0.03°$ (Bugaris and Ibers 2008). This compound was synthesized using the mixture of HgSe, UCl$_4$, and HgCl$_2$, which was heated to 850°C in 24 h, kept at this temperature for 96 h, cooled at 6.8°C/h to 200°C, and then cooled rapidly to 20°C.

8.44   MERCURY–CARBON–FLUORINE–SELENIUM

C$_2$F$_6$Se$_2$Hg [bis(trifluoromethylselenolato)mercury] and C$_6$F$_{14}$Se$_2$Hg [bis(heptafluoropropylselenolato)mercury] quaternary compounds are formed in this system. C$_2$F$_6$Se$_2$Hg was obtained if C$_2$F$_6$Se$_2$ was sealed in a quartz tube with Hg and shaken (20 h) at 4 cm from the mercury arc lamp (Dale et al. 1958, Clase and Ebsworth 1965). The residue in the tube was extracted with ether, and the product was sublimed in a vacuum at 50°C to give this compound as yellow needles. The resublimed crystals melt sharply at 51°C–52°C.

To obtain C$_6$F$_{14}$Se$_2$Hg, dried silver heptafluorobutyrate (25 mmol) and Se (60 mmol) were heated in a Carius tube at 280°C for 2 h (Eméleus and Welcman 1963).

Preliminary separation of the products gave a mixture of the mono- and diselenides, CO_2, and small amounts of heptafluorobutyric acid and its anhydride. The mixture was shaken for 48 h with a fivefold–sixfold excess of Hg in an evacuated tube. $C_6F_{14}Se_2Hg$ formed white needles and was purified by vacuum sublimation.

8.45 MERCURY–SILICON–FLUORINE–SELENIUM

HgSe–HgF$_2$–SiF$_4$. The phase diagram is not constructed. The $Hg_3Se_2SiF_6$ quaternary compound is formed in this system. This compound could be obtained by the interaction of HgSe with the solution of HgO in H_2SiF_6 at 120°C (Puff et al. 1969).

8.46 MERCURY–GERMANIUM–TELLURIUM–SELENIUM

HgSe + GeTe $\Leftrightarrow$ HgTe + GeSe. The phase diagram is not constructed. The glass-forming region in this system was determined (Figure 8.33) (Feltz et al. 1976).

HgTe–GeSe$_2$. The phase diagram is not constructed. The $Hg_2GeSe_2Te_2$ quaternary compound that melts incongruently at 477°C is formed in this system (Asadov 2006). The solubility of HgTe in $GeSe_2$ and $GeSe_2$ in HgTe at 477°C is equal to 20

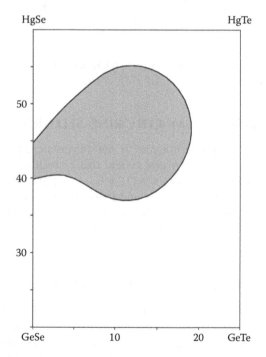

FIGURE 8.33 The glass-forming region in the HgSe + GeTe $\Leftrightarrow$ HgTe + GeSe ternary mutual system. (From Feltz, A. et al., New vitreous semiconductors, *Tr. 6th Mezhdunar. konf. po amorf. i zhidkim poluprovodn.: Struktura i sv-va nekristal. poluprovodn.* L., Nov. 18–24, 1975, L.: Nauka Publish., 24, 1976.)

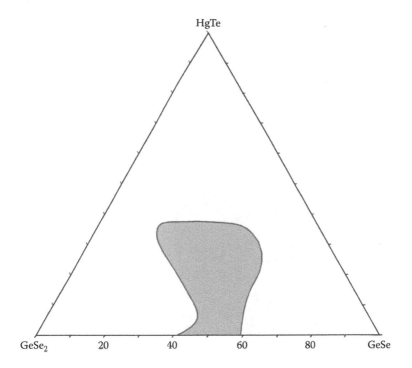

FIGURE 8.34 The glass-forming region in the HgTe–GeSe–GeSe$_2$ quasiternary system. (From Feltz, A. et al. New vitreous semiconductors, *Tr. 6th Mezhdunar. konf. po amorf. i zhidkim poluprovodn.: Struktura i sv-va nekristal. poluprovodn.* L., Nov. 18–24, 1975, L.: Nauka Publish., 24, 1976.)

and 20 mol.%, respectively. This system was investigated through DTA, XRD, metallography, and measuring of microhardness and emf of concentrated chains.

HgTe–GeSe–GeSe$_2$. The phase diagram is not constructed. The glass-forming region in this system was determined (Figure 8.34) (Feltz et al. 1976).

HgTe–GeSe$_2$–GeTe. The phase diagram is not constructed. The glass-forming region in this system was determined (Figure 8.35) (Feltz et al. 1976).

8.47 MERCURY–LEAD–TELLURIUM–SELENIUM

HgSe + PbTe ⇔ HgTe + PbSe. The phase diagram is not constructed. Solid solutions with sphalerite structure based on mercury chalcogenides and with rock salt structure based on lead chalcogenides exist in this ternary mutual system (Leute and Köller 1986). The solubility of mercury chalcogenides in lead chalcogenides at 630°C reaches 5 mol.%, and the solubility of lead chalcogenides in mercury chalcogenides is insignificant. Between the temperatures 430°C and 630°C, equilibrium is shifted to the formation of HgTe and PbSe.

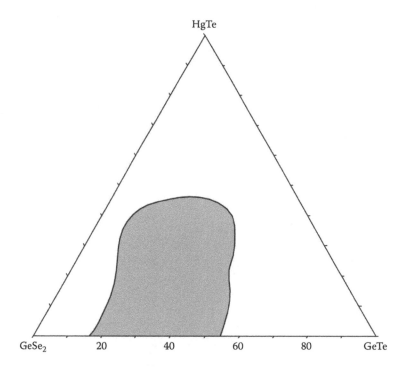

FIGURE 8.35 The glass-forming region in the HgTe–GeSe₂–GeTe quasiternary system. (From Feltz, A. et al., New vitreous semiconductors, *Tr. 6th Mezhdunar. konf. po amorf. i zhidkim poluprovodn.: Struktura i sv-va nekristal. poluprovodn.* L., Nov. 18–24, 1975, L.: Nauka Publish., 24, 1976.)

8.48 MERCURY–LEAD–IODINE–SELENIUM

HgSe–PbI₂. The phase diagram of this system is a eutectic type (Figure 8.36) (Odin et al. 2004). The eutectic crystallizes at $386°C \pm 2°C$ and contains 14 ± 1 mol.% HgSe. The solubility of PbI₂ in HgSe is less than 0.5 mol.%, and the solubility of HgSe in PbI₂ is not higher than 1 mol.%. A solid solution based on PbI₂ has a "mixed" structure in which the layers of 2H-PbI₂ and 6R-PbI₂ polytypes alternate. A new metastable phase μ, which contains 20 mol.% HgSe and 80 mol.% PbI₂, is formed in this system. It crystallizes in the rhombohedral structure with the lattice parameters $a = 746.3 \pm 0.6$ pm and $\alpha = 35.31° \pm 0.08°$ or in the hexagonal structure with the lattice parameters $a = 455.9 \pm 0.4$ and $b = 2103 \pm 2$ pm.

This system was investigated by DTA and metallography (Odin et al. 2004). Metastable phase μ was obtained by the quenching of the melts from the temperatures 810°C–947°C.

HgSe–PbSe–PbI₂. The liquidus surface of this system (Figure 8.37) consists of the fields of the primary crystallization of HgSe, PbSe, and PbI₂. Ternary eutectic *E* crystallizes at 371°C. The isothermal section of this system at 360°C is given in Figure 8.38 (Odin et al. 2004).

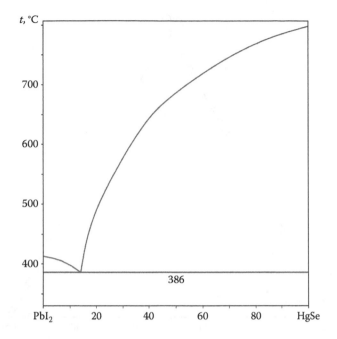

FIGURE 8.36   Phase diagram of the HgSe–PbI$_2$ quasibinary system. (From Odin, I.N. et al., *Zhurn. neorgan. khimii*, 49(9), 1562, 2004.)

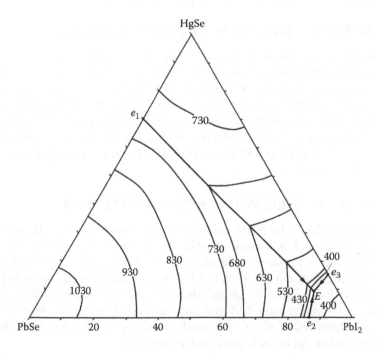

FIGURE 8.37   Liquidus surface of the HgSe–PbSe–PbI$_2$ quasiternary system. (From Odin, I.N. et al., *Zhurn. neorgan. khimii*, 49(9), 1562, 2004.)

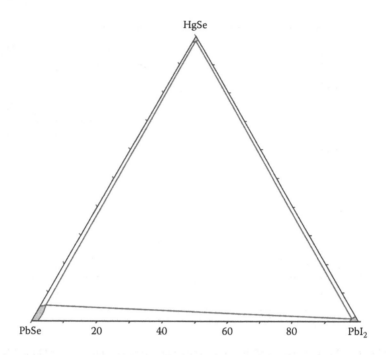

FIGURE 8.38 Isothermal section of the HgSe–PbSe–PbI$_2$ quasiternary system at 360°C. (From Odin, I.N. et al., *Zhurn. neorgan. khimii*, 49(9), 1562, 2004.)

8.49 MERCURY–ZIRCONIUM–CHLORINE–SELENIUM

HgSe–HgCl$_2$–ZrCl$_4$. The phase diagram is not constructed. The Hg$_3$Se$_2$[ZrCl$_6$] quaternary compound is formed in this system. It crystallizes in the monoclinic structure with the lattice parameters $a = 1311.33 \pm 0.02$, $b = 749.75 \pm 0.01$, $c = 663.35 \pm 0.01$ pm, and $\beta = 91.207° \pm 0.003°$ and calculated density 5.42 g·cm^{-3} (Beck and Hedderich 2003). To obtain this compound, the mixture of HgSe, HgCl$_2$, and ZrCl$_4$ in the molar ratio 2:1:1 was used as starting materials. With a rate of 40°C/h, a temperature of 360°C was reached and kept for 5 days. After that, the ampoule was taken out of the furnace and air quenched.

8.50 MERCURY–ZIRCONIUM–BROMINE–SELENIUM

HgSe–HgBr$_2$–ZrBr$_4$. The phase diagram is not constructed. The Hg$_3$Se$_2$[ZrBr$_6$] quaternary compound is formed in this system. It crystallizes in the monoclinic structure with the lattice parameters $a = 1350.99 \pm 0.06$, $b = 756.79 \pm 0.03$, $c = 701.97 \pm 0.03$ pm, and $\beta = 92.164° \pm 0.003°$ and calculated density 6.16 g·cm^{-3} (Beck and Hedderich 2003). To obtain this compound, the mixture of HgSe, HgBr$_2$, and ZrBr$_4$ in the molar ratio 2:1:1 was used as starting materials. With a rate of 40°C/h, a temperature of 400°C was reached and kept for 5 days. After that, the ampoule was taken out of the furnace and air quenched.

8.51 MERCURY–HAFNIUM–CHLORINE–SELENIUM

HgSe–HgCl$_2$–HfCl$_4$. The phase diagram is not constructed. The Hg$_3$Se$_2$[HfCl$_6$] quaternary compound is formed in this system. It crystallizes in the mono-clinic structure with the lattice parameters $a = 1309.50 \pm 0.06$, $b = 749.90 \pm 0.04$, $c = 662.72 \pm 0.03$ pm, and $\beta = 91.992° \pm 0.003°$ and calculated density 5.88 g·cm^{-3} (Beck and Hedderich 2003). To obtain this compound, the mixture of HgS, HgCl$_2$, and HfCl$_4$ in the ratio 2:1:1 was used as starting materials. With a rate of 40°C/h, a temperature of 360°C was reached and kept for 5 days. After that, the ampoule was taken out of the furnace and air quenched.

8.52 MERCURY–ARSENIC–BROMINE–SELENIUM

HgSe–AsSeBr. The phase diagram is not constructed. The Hg$_3$AsSe$_4$Br quaternary compound is formed in this system. It crystallizes in the hexagonal structure with the lattice parameters $a = 707 \pm 8$ and $c = 947 \pm 2$ pm (Beck et al. 2000). Hg$_3$AsSe$_4$Br was synthesized by the chemical transport reaction using the mixture of HgSe, As, HgBr$_2$, and Se at the temperatures 250°C–310°C

8.53 MERCURY–ARSENIC–IODINE–SELENIUM

HgSe–AsSeI. The phase diagram is not constructed. The Hg$_3$AseS$_4$I quaternary compound is formed in this system. It crystallizes in the hexagonal structure with the lattice parameters $a = 769.02 \pm 0.07$ and $c = 996.8 \pm 0.1$ pm (Beck et al. 2000). Hg$_3$AsSe$_4$I was synthesized by the chemical transport reaction using the mixture of HgSe, As, HgI$_2$, and Se at the temperatures 250°C–310°C.

8.54 MERCURY–BISMUTH–CHLORINE–SELENIUM

HgSe–HgCl$_2$–BiCl$_3$. The phase diagram is not constructed. The Hg$_3$Bi$_2$Se$_2$Cl$_8$ quaternary compound is formed in this system. It melts incongruently and crys-tallizes in the monoclinic structure with the lattice parameters $a = 697.1 \pm 0.1$, $b = 863.4 \pm 0.2$, $c = 812.0 \pm 0.2$ pm, and $\beta = 106.87° \pm 0.03°$ (Wibowo et al. 2013). To obtain this compound, a mixture of HgSe, HgCl$_2$, and BiCl$_3$ in the molar ratio 2:1:6 was loaded into a fused silica tube. Then, the tube was flame sealed under vacuum, put inside a programmable furnace, heated to 300°C over 24 h, and held isothermally for 12 h, followed by a shutdown of the furnace. The excess of BiCl$_3$ was washed away using MeOH. The reaction with a 1:1 mixture of HgSe and BiCl$_3$ provided single crystals of Hg$_3$Bi$_2$Se$_2$Cl$_8$. To obtain the single crystals of this com-pound, the thoroughly mixed powders of HgSe and BiCl$_3$ were loaded into a fused silica tube, which was flame sealed under vacuum and heated to 600°C over 16 h and held isothermally for 1 day, followed by cooling slowly to 200°C over 36 h and then to room temperature within 1 h.

8.55 MERCURY–OXYGEN–MANGANESE–SELENIUM

HgSe–MnSe–O. The phase diagram is not constructed. According to the data of thermodynamic simulations, an oxidation of the $Hg_xMn_{1-x}Se$ solid solutions begins from the oxidation of MnSe with formation of MnO and Se. And then, HgSe is oxidized to form $HgSeO_3$ (Medvedev and Berchenko 1994).

8.56 MERCURY–TELLURIUM–CHLORINE–SELENIUM

HgSe–HgTe–HgCl$_2$. The isothermal section of this quasiternary system at 230°C is shown in Figure 8.39 (Pan'ko et al. 1989). The part of this section adjoining to the HgSe–HgTe quasibinary system is a two-phase region of $HgSe_xTe_{1-x}$ and $Hg_3Se_{2-x}Te_{2x}Cl_2$ solid solutions. The $HgCl_2$–$Hg_3Te_2Cl_2$–$Hg_3Se_2Cl_2$ subsystem includes two two-phase regions and one three-phase region. A narrow band of two-phase equilibria is situated along the HgTe–HgCl$_2$ quasibinary system between $HgCl_2$ and Hg_3TeCl_4. The ingots were annealed at 230°C for 1080 h.

8.57 MERCURY–TELLURIUM–BROMINE–SELENIUM

HgSe–HgTe–HgBr$_2$. The isothermal section of this quasiternary system at 200°C is shown in Figure 8.40 (Hudoliy et al. 1990). The limit of three-phase region in the quasibinary system HgSe–HgTe corresponds to the $HgTe_{0.18}Se_{0.82}$ composition.

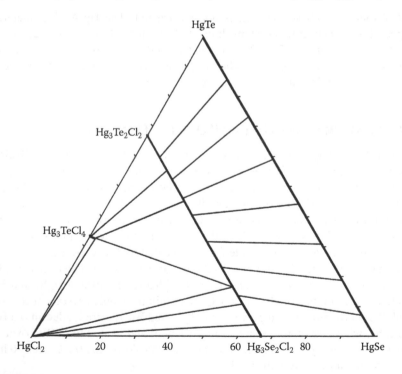

FIGURE 8.39 Isothermal section of the HgSe–HgTe–HgCl$_2$ quasiternary system at 230°C. (From Pan'ko, V.V. et al., *Ukr. khim. zhurn.*, 55(3), 232, 1989.)

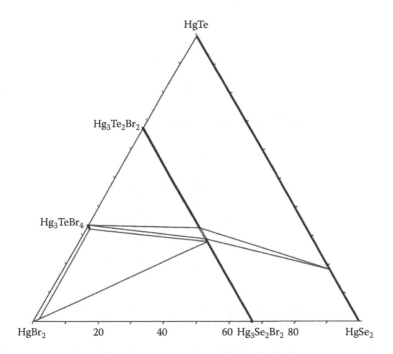

FIGURE 8.40 Isothermal section of the HgSe–HgTe–HgBr$_2$ quasiternary system at 200°C. (From Hudoliy, V.A. et al., *Izv. AN SSSR. Neorgan. materialy*, 26(11), 2425, 1990.)

Two-phase regions are situated between Hg$_3$TeBr$_4$ and Hg$_3$Te$_{2-x}$Se$_{2x}$Br$_2$ solid solutions, HgBr$_2$ and Hg$_3$Se$_{2-x}$Te$_{2x}$Br$_2$ solid solutions, Hg$_3$Te$_{2-x}$Se$_{2x}$Br$_2$ and HgTe$_{1-x}$Se$_x$ solid solutions, and Hg$_3$Se$_{2-x}$Te$_{2x}$Br$_2$ and HgSe$_{1-x}$Te$_x$ solid solutions. The Hg$_3$TeBr$_4$ + Hg$_3$Te$_{2-x}$Se$_{2x}$Br$_2$ + Hg$_3$Se$_{2-x}$Te$_{2x}$Br$_2$ and HgBr$_2$ + Hg$_3$TeBr$_4$ + Hg$_3$Se$_{2-x}$Te$_{2x}$Br$_2$ three-phase regions are divided by the narrow band of Hg$_3$TeBr$_4$ + Hg$_3$Se$_{2-x}$Te$_{2x}$Br$_2$ two-phase region. The ingots were annealed at 200°C for 720 h.

8.58 MERCURY–CHLORINE–BROMINE–SELENIUM

HgSe–HgCl$_2$–HgBr$_2$. The isothermal section of this quasiternary system at 200°C is shown in Figure 8.41 (Pan'ko et al. 1996). At this temperature, α-solid solutions based on Hg$_3$Se$_2$Cl$_2$ are in equilibrium with solid solutions based on HgCl$_2$ and γ-phase of the HgCl$_2$–HgBr$_2$ system, and HgSe and Hg$_3$Se$_2$(Br$_{1-x}$Cl$_x$)$_2$ solid solutions are in equilibrium with γ-phase, solid solutions based on HgBr$_2$ and HgSe.

8.59 MERCURY–BROMINE–IODINE–SELENIUM

HgSe–HgBr$_2$–HgI$_2$. The isothermal section of this quasiternary system at 200°C is shown in Figure 8.42 (Minets 1997, Minets et al. 2004). The phase equilibria are determined by the solid solutions in the Hg$_3$Se$_2$Br$_2$–Hg$_3$Se$_2$I$_2$. In the top part of the diagram, there are two two-phase regions, involving the equilibria of HgSe

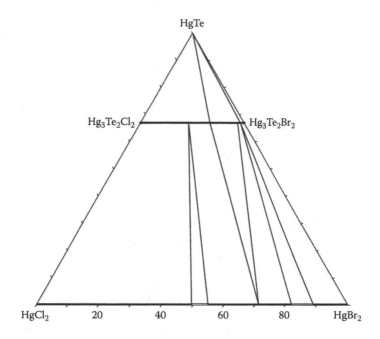

FIGURE 8.41 Isothermal section of the HgSe–HgCl$_2$–HgBr$_2$ quasiternary system at 200°C. (From Pan'ko, V.V. et al., *Zhurn. neorgan. khimii*, 41(10), 1731, 1996.)

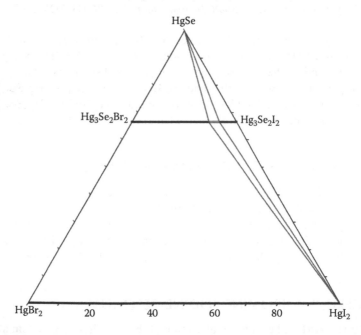

FIGURE 8.42 Isothermal section of the HgSe–HgBr$_2$–HgI$_2$ quasiternary system at 200°C. (From Minets, Yu.V., *Nauk. visnyk Uzhhorod. un-tu. Ser. Khimiya*, (2), 45, 1997; Minets, Yu.V. et al., *J. Alloys Compd.*, 365(1–2), 121, 2004.)

with $Hg_3Se_2Br_2$ and $Hg_3Se_2I_2$, respectively. They are separated by the three-phase field ($HgSe + Hg_3Se_2Br_2 + Hg_3Se_2I_2$). A similar situation exists in the bottom part: the two two-phase fields formed by $HgBr_2$ and HgI_2 with the corresponding selenide halides are separated by a narrow three-phase region based on the two-phase equilibrium in the $Hg_3Se_2Br_2$–$Hg_3Se_2I_2$ system. The alloys were annealed at 350°C and 200°C for 350–400 h.

REFERENCES

Arnold A.P., Canty A.J. Synthesis, structure and spectroscopic studies of mercury(II)seleno-lates and MeHgSeBuᵗ, *Inorg. Chim. Acta*, **55**, 171–176 (1981).

Arnold A.P., Canty A.J. Mercury(II)selenolates. Crystal structures of polymeric $Hg(SeMe)_2$ and the tetrameric pyridinates $[\{HgCl(py)(SeEt)\}_4]$ and $[\{HgCl(py)_{0.5}(SeBu^t)\}_4]$, *J. Chem. Soc., Dalton Trans.*, (3), 607–613 (1982).

Asadov M.M. Homogeneity region of phases in the HgSe + Tl_2Te ⇔ HgTe + Tl_2Se system [in Russian], *Izv. AN SSSR. Neorgan. materialy*, **20**(10), 1621–1623 (1984b).

Asadov M.M. The HgSe + Tl_2Te ⇔ HgTe + Tl_2Se ternary mutual system [in Russian], *Izv. AN SSSR. Neorgan. materialy*, **20**(1), 38–40 (1984a).

Asadov M.M. Dependences of physico-chemical properties of solid solutions in the systems $GeSe_2$–A^2B^6 (A^2 = Hg, Cd); B^6 = S, Te) versus composition [in Russian], *Azerb. khim. zhurn.*, (2), 77–81 (2006).

Asadov M.M., Babanly M.B., Kuliev A.A. Phase equilibria in the Tl_2Te–HgTe(HgSe) systems [in Russian], *Izv. AN SSSR. Neorgan. materialy*, **14**(5), 960–961 (1978).

Baranov I.Yu., Dolgih V.A., Popovkin B.A. Investigation of phase correlations in the La_2O_2X–MX (M = Zn, Cd, Hg; X = S, Se) systems [in Russian], *Zhurn. neorgan. khimii*, **41**(11), 1916–1919 (1996).

Beck J., Hedderich S. Expanded polycationic mercury–chalcogen networks in the layered compounds $Hg_3E_2[MX_6]$ (E = S, Se; M = Zr, Hf; X = Cl, Br), *J. Solid State Chem.*, **172**(1), 12–16 (2003).

Beck J., Hedderich S., Köllisch K. Hg_3AsE_4X (E = S, Se; X = Cl, Br, I), a family of isotypic compounds with an acentric layered structure, *Inorg. Chem.*, **39**(25), 5847–5850 (2000).

Beck J., Keller H.-L., Rompel M., Wimbert L. Hydrothermalsynthese und Kristalstruktur der Münzmetall–Quecksilber–Chalkogenidhalogenide CuHgSeBr, AgHgSBr und AgHgSI, *Z. anorg. allg. Chem.*, **627**(9), 2289–2294 (2001).

Behrens S., Bettenhausen M., Deveson A.C., Eichhöfer, A., Fenske D., Lohde A., Woggon U. Synthese und Struktur der Nanocluster $[Hg_{32}Se_{14}(SePh)_{36}]$, $[Cd_{32}Se_{14}(SePh)_{36}(PPh_3)_4]$, $[Et_2P(Ph)C_4H_8OSiMe_3]_5[Cd_{18}I_{17}(PSiMe_3)_{12}]$ und $[Et_3NC_4H_8OSiMe_3]_5[Cd_{18}I_{17}(PSiMe_3)_{12}]$, *Angew. Chem.*, **108**(19), 2360–2363 (1996a).

Behrens S., Bettenhausen M., Deveson A.C., Eichhöfer, A., Fenske D., Lohde A., Woggon U. Synthesis and structure of the nanoclusters $[Hg_{32}Se_{14}(SePh)_{36}]$, $[Cd_{32}Se_{14}(SePh)_{36}(PPh_3)_4]$, $[P(Et)_2(Ph)C_4H_8OSiMe_3]_5[Cd_{18}I_{17}(PSiMe_3)_{12}]$, and $[N(Et)_3C_4H_8OSiMe_3]_5[Cd_{18}I_{17}(PSiMe_3)_{12}]$, *Angew. Chem. Int. Ed. Engl.*, **35**(19), 2215–2218 (1996b).

Bochmann M, Coleman A.P., Powell A.K. Synthesis of some alkyl metal selenolato complexes of zinc, cadmium and mercury. X-ray crystal structure of Me, Hg, $Se(2,4,6-Pr^i_3C_6H_2)$, *Polyhedron*, **11**(5), 507–512 (1992).

Bochmann M., Webb K. Novel precursors for the deposition of II-VI semiconductor films, *Mater. Res. Symp. Proc.*, **204**, 149–154 (1991a).

Bochmann M., Webb K. Sterically hindered thiolato, selenolato and tellurolato complexes of mercury(II), *J. Chem. Soc., Dalton Trans.*, (9), 2325–2329 (1991b).

Bochmann M., Webb K.J., Malik M.A., O'Brien P. Arene thiolato, selenolato, and tellurolato complexes of mercury, *Inorg. Synthes.*, **31**, 24–28 (1997).

Borisov S.V., Magarill S.A., Pervukhina N.V. To crystal chemistry of compounds MHgYX (M = Cu, Ag; Y = S, Se; X = Cl, Br, I), Z. Kristallogr., 220(11), 946–953 (2005).

Brandmayer M.K., Clérac R., Weigend F., Dehnen S. Ortho-chalcogenostannates as ligands: Syntheses, crystal structures, electronic properties, and magnetism of novel compounds containing ternary anionic substructures [M₄(μ₄-Se)(SnSe₄)₄]¹⁰⁻ (M = Mn, Zn, Cd, Hg), {[Hg₄(μ₄-Se)(SnSe₄)₃]⁶⁻}, or {[HgSnSe₄]²⁻}, Chem. Eur. J., 10(20), 5147–5157 (2004).

Breitinger D., Morell W. Tris(methylmercurio) selenonium salts [Se(HgCH₃)₃]X and bis(methylmercuric) chalcogenides Ch(HgCH₃)₂ (Ch = Se and Te), Inorg. Nucl. Chem. Lett., 10(5), 409–411 (1974).

Bugaris D.E., Ibers J.A. Synthesis, structures and magnetic and optical properties of the compounds [Hg₃Te₂][UCl₆] and [Hg₄As₂][UCl₆], J. Solid State Chem., 181(12), 3189–3193 (2008).

Bugaris D.E., Wells D.M., Ibers J.A. Synthesis, structure and magnetic and electronic properties of Cs₂Hg₂USe₅, J. Solid State Chem., 182(5), 1017–1020 (2009).

Chen Z., Wang R.-J. Crystal structures and semiconductor properties of alkaline metal selenides MHgSbSe₃ (M = K, Rb, Cs), Acta Chim. Sin., 58(3), 326–331 (2000).

Chen Z., Wang R.-J., Li J. New quaternary selenoantimonates AHgSbSe₃ (A = Rb, Cs): Synthesis, structure and optical properties, Mater. Res. Soc. Symp. Proc., 547, 419–424 (1999).

Chondroudis K., Kanatzidis M.G. [M₄(Se₂)₂(PSe₄)₄]⁸⁻: A novel, tetranuclear, cluster anion with a stellane-like core, J. Chem. Soc., Chem. Commun., (4), 401–402 (1997).

Chondroudis K., Kanatzidis M.G. Group-10 and 12 one-dimensional selenodiphosphates: A₂MP₂Se₆ (A = K, Rb, Cs; M = Pd, Zn, Cd, Hg), J. Solid State Chem., 138(2), 321–328 (1998).

Chondroudis K., Kanatzidis M.G. Synthesis of quaternary selenophosphates using molten salt fluxes: Rb₈Hg₄(Se₂)₂(PSe₄)₄, K₄In₂(PSe₅)₂(P₂Se₆), Rb₄Ti₂(P₂Se₇)(P₂Se₉)₂, Rb₄U₄(Se)₂(Se₂)₄(PSe₄)₄, Inorg. Synthes., 33, 122–130 (2002).

Clase H.J., Ebsworth E.A.V. Some reactions and derivatives of bis(trifluoromethylseleno)mercury, J. Chem. Soc., 940–947 (1965).

Dale J.W., Emeléus H.J., Haszeldine R.N. Organometallic and organometalloidal fluorine compounds. Part XIV. Trifluoromethyl derivatives of selenium, J. Chem. Soc., 2939–2945 (1958).

Emeléus H.J., Welcman N. Heptafluoropropyl derivatives of selenium, J. Chem. Soc., 1268–1271 (1963).

Feltz A., Burckhardt W., Senf L. New vitreous semiconductors, Tr. 6th Mezhdunar. konf. po amorf. i zhidkim poluprovodn.: Struktura i sv-va nekristal. poluprovodn. L., Nov. 18–24, 1975. L.: Nauka Publish., pp. 24–31 (1976).

Gallardo P.G. Hg₂ₓ(CuIn)₁₋ₓSe₂ alloys: Phase diagram and lattice parameter values, Phys. Status Solidi (a), 130(1), 39–44 (1992a).

Gallardo P.G. Order-disorder phase transitions in D²ᴵᴵ₂ₓ(AᴵBᴵᴵᴵ)₁₋ₓC₂ᵛᴵ alloys systems, Phys. Status Solidi (a), 134(1), 119–125 (1992b).

Guillo M., Mercey B., Labbé P., Deschanvres A. The structure of copper (I) mercury (II) chloride selenide, Acta Crystallogr. B, 36(11), 2520–2523 (1980).

Gulay L.D., Olekseyuk I.D., Parasyuk O.V. Crystal structures of the Ag₆HgGeSe₆ and Ag₆HgSiSe₆ compounds, J. Alloys Compd., 343(1–2), 116–121 (2002).

Gulay L.D., Parasyuk O.V. Crystal structure of the Ag₂.₆₆Hg₂Sn₁.₃₄Se₆ and Hg₂SnSe₄ compounds, J. Alloys Compd., 337(1–2), 94–98 (2002).

Gulay L.D., Parasyuk O.V., Romanyuk Ya.E., Olekseyuk I.D. Crystal structure of the Cu₅.₉₇₆Hg₀.₉₇₂SiSe₆ compound, J. Alloys Compd., 367(1–2), 121–125 (2004).

Haeuseler H., Himmrich M. Neue Verbindungen Ag₂HgMX₄ mit Wurtzstannitstruktur, Z. Naturforsch., B44(9), 1035–1036 (1989).

Hahn H., Schulze H. Über quaternäre Chalcogenide des Germanium und Zinns, Naturwissenschaften, 52(14), 426 (1965).

Halka V.O., Olekseyuk I.D., Parasyuk O.V. The $Cu_2Se-HgSe-In_2Se_3$ system at 670 K, *J. Alloys Compd.*, **302**(1–2), 173–176 (2000a).

Halka V.O., Olekseyuk I.D., Parasyuk O.V. The $Cu_2Se-HgSe-In_2Se_3$ quasiternary system. I. Description of the quadrangle $Cu_2Se-CuInSe_2-HgIn_2Se_4-HgSe$, *J. Alloys Compd.*, **309**(1–2), 165–171 (2000b).

Hirai T., Kurata K., Takeda Y. Deviation of new semiconducting compounds by cross substitution for group IV semiconductors and their semiconducting and thermal properties, *Solid State Electron.*, **10**(10), 975–981 (1967).

Hudoliy V.A., Pan'ko V.V., Voroshilov Yu.V., Shelemba M.S. Phase equilibria in the $HgTe-HgSe-HgBr_2$ system [in Russian], *Izv. AN SSSR. Neorgan. materialy*, **26**(11), 2425–2428 (1990).

Jin X., Zhang L., Shu G., Wang R., Guo H. Synthesis and characterization of a novel quaternary metal selenide, $K_2Hg_3Ge_2Se_8$, *J. Alloys Compd.*, **347**(1–2), 67–71 (2002).

Kanatzidis M.G., Liao J.-H., Marking G.A. Alkali metal quaternary chalcogenides and process for the preparation of thereof, USA Patent 5618 471. Appl. № 606886. Filed 26.02.96; Data of Patent 08.04.1997 (1997).

Kerkhoff M., Leute V. The phase diagram of the quasiternary system $Ga_{2k}Hg_{3-3k}Se_{3l}Te_{3-3l}$, *J. Alloys Compd.*, **391**(1–2), 42–48 (2005).

Kozer V.R., Parasyuk O.V., Olekseyuk I.D. Phase equilibria in the $AgInSe_2-HgIn_2Se_4$ and $AgInSe_2-HgSe$ systems [in Russian], *Neorgan. materialy*, **46**(6), 686–690 (2010).

Krauß G., Keller E., Krämer V. Crystal structure of caesium tetramercury pentagallium dodecaselenide, $CsHg_4Ga_5Se_{12}$, *Z. Kristallogr.*, **211**(3), 188 (1996).

Leute V. Computer simulations of interdiffusion processes in the quasiternary system $Ga_{2k}Hg_{3-3k}Se_{3l}Te_{3-3l}$, *J. Alloys Compd.*, **433**(1–2), 154–161 (2007).

Leute V., Köller H.-J. Thermodynamic properties of the quasiternary system $Hg_kPb_{(1-k)}Se_lTe_{(1-l)}$, *Z. Phys. Chem. (BRD)*, **150**(2), 227–243 (1986).

Liu Y., Kanhere P.D., Hoo Y.S., Ye K., Yan Q., Rawat R.S., Chen Z., Ma J., Zhang Q. Cationic quaternary chalcohalide nanobelts: $Hg_4In_2Q_3Cl_8$ (Q = S, Se, Te), *RSC Adv.*, **2**(16), 6401–6403 (2012).

Marchuk O.V., Olekseyuk I.D., Grebenyuk A.G. Phase equilibria in the system $Cu_2Se-HgSe-GeSe_2$, *J. Alloys Compd.*, **457**(1–2), 337–343 (2008).

Medvedev Yu.V., Berchenko N.N. Analysis of the phase compositions at the interface of $Mn_{1-x}A_x^{II}B^{VI}(A^{II} = Zn, Cd, Hg; B^{VI} = Te, Se)$ solid solution with own oxides [in Russian], *Zhurn. neorgan. khimii*, **39**(5), 846–848 (1994).

Minets Yu.V. Interaction in the $HgSe-HgBr_2-HgI_2$ ternary system [in Ukrainian], *Nauk. visnyk Uzhhorod. un-tu. Ser. Khimiya*, (2), 45–47 (1997).

Minets Yu.V., Voroshilov Yu.V., Pan'ko V.V., Khudoliy V.A. Phase equilibria in the $HgSe-HgBr_2-HgI_2$ system and crystal structure of $Hg_3Se_2Br_2$ and $Hg_3Se_2I_2$, *J. Alloys Compd.*, **365**(1–2), 121–125 (2004).

Mitchell K., Huang F.Q., McFarland A.D., Haynes Ch.L., Somers R.C., Van Duyne R.P., Ibers J.A. The $CsLnMSe_3$ semiconductors (Ln = rare-earth element, Y; M = Zn, Cd, Hg), *Inorg. Chem.*, **42**(13), 4109–4116 (2003).

Mkrtchian S.A., Dovletov K., Zhukov E.G., Melikdzhanian A.G., Nuryiev S. Interaction in the $Cu_2Ge(Sn)Se_3-HgSe$ systems [in Russian], *Zhurn. neorgan. khimii*, **33**(9), 2379–2384 (1988a).

Mkrtchian S.A., Dovletov K., Zhukov E.G., Melikdzhanian A.G., Nuryiev S. Electrophysical properties of the $Cu_2A^{II}B^{IV}Se_4$ (A^{II} – Cd, Hg: B^{IV} – Ge, Sn) compounds [in Russian], *Izv. AN SSSR. Neorgan. materialy*, **24**(7), 1094–1096 (1988b).

Morris C.D., Li H., Jin H., Malliakas C.D., Peters J.A., Trikalitis P.N., Freeman A.J., Wessels B.W., Kanatzidis M.G. $Cs_2M^{II}M_3^{IV}Q_8$ (Q = S, Se, Te): An extensive family of layered semiconductors with diverse band gaps, *Chem. Mater.*, **25**(16), 3344–3356 (2013).

Mozolyuk M.Yu., Piskach L.V., Fedorchuk A.O., Olekseyuk I.D., Parasyuk O.V. Physicochemical interaction in the $Tl_2Se-HgSe-D^{IV}Se_2$ systems (D^{IV} – Si, Sn), *Mater. Res. Bull.*, **47**(11), 3830–3834 (2012).

Odin I.N., Grin'ko V.V., Kozlovskiy V.F., Safronov E.V., Gapanovich M.V. Formation of the stable and metastable phases in the PbSe + MI$_2$ ⇔ MSe + PbI$_2$ (M = Hg, Mn, Sn) mutual systems [in Russian], *Zhurn. neorgan. khimii*, **49**(9), 1562–1567 (2004).

Olekseyuk I.D., Gulay L.D., Dudchak I.V., Piskach L.V., Parasyuk O.V., Marchuk O.V. Single crystal preparation and crystal structure of the Cu$_2$Zn(Cd,Hg)SnSe$_4$ compounds, *J. Alloys Compd.*, **340**(1–2), 141–145 (2002).

Olekseyuk I.D., Kogut Yu.M., Parasyuk O.V., Piskach L.V., Gorgut G.P., Kus'ko O.P., Pekhnyo V.I., Volkov S.V. Glass-formation in the Ag$_2$Se–Zn(Cd, Hg)Se–GeSe$_2$, *Chem. Met. Alloys*, **2**(3–4), 146–150 (2009).

Olekseyuk I.D., Marchuk O.V., Bozhko V.V., Trofymchuk L.V. Electrophysical properties of the Cu$_2$HgC^{VI}Se$_4$ (C – Sn, Ge) quaternary compounds [in Ukrainian], *Nauk. visnyk Volyn. derzh. un-tu. Khim. Nauky*, (6), 34–37 (2001b).

Olekseyuk I.D., Marchuk O.V., Gulay L.D., Zhbankov O.Ye. Isothermal section of the Cu$_2$Se–HgSe–GeSe$_2$ system at 670 K and crystal structure of the compounds Cu$_2$HgGeSe$_4$ and HT-modification of Cu$_2$HgGeSe$_4$, *J. Alloys Compd.*, **398**(1–2), 80–84 (2005).

Olekseyuk I.D., Marchuk O.V., Parasyuk O.V., Bozhko V.V., Galyan V.V. Physico-chemical and physical properties of glasses in the Cu$_2$Se–HgSe–GeSe$_2$ system [in Ukrainian], *Fiz. i khim. tv. tila*, **2**(1), 69–76 (2001a).

Olekseyuk I.D., Mozolyuk M.Yu., Piskach L.V., Litvinchuk M.B., Parasyuk O.V. Component interaction in the systems formed by the Tl(I), Hg(II), Pb(II), Si(IV) chalcogenides [in Ukrainian], *Nauk. visnyk Volyn. nats. un-tu im. Lesi Ukrainky. Khim. nauky*, [(17)242], 62–69 (2012).

Olekseyuk I.D., Mozolyuk M.Yu., Piskach L.V., Parasyuk O.V. Phase equilibria in the Tl$_2$S(Se)–HgS(Se)–SnS(Se)$_2$ systems at 520 K [in Ukrainian], *Nauk. visnyk Volyn. nats. un-tu im. Lesi Ukrainky. Khim. nauky*, (30), 19–21 (2010).

Olekseyuk I.D., Parasyuk O.V. Phase equilibria in the HgSe–Ga$_2$Se$_3$–SnSe$_2$ system [in Russian], *Zhurn. neorgan. khimii*, **42**(5), 838–844 (1997).

Olekseyuk I.D., Parasyuk O.V., Bozhko V.V., Galyan V.V., Petrus' I.I. Glass-forming in the Zn(Cd,Hg)Se–Ga$_2$Se$_3$–GeSe$_2$ systems [in Ukrainian], *Fizyka kondens. vysokomolek. system. Nauk. zap. Rinens'kogo pedinstytutu*, (3), 148–152 (1997).

Olekseyuk I.D., Parasyuk O.V., Bozhko V.V., Petrus' I.I., Galyan V.V. Formation and properties of the quasiternary Zn(Cd,Hg)Se–Ga$_2$Se$_3$–SnSe$_2$ system glasses, *Funct. Mater.*, **6**(3), 474–477 (1999).

Olekseyuk I.D., Parasyuk O.V., Galka V.O. The CuGaSe$_2$–HgSe and CuInSe$_2$–HgSe systems, *Pol. J. Chem.*, **72**(1), 49–54 (1998).

Olekseyuk I.D., Parasyuk O.V., Salamakha P.S., Prots' Yu.M. The phase equilibria in the quasiternary HgSe–Ga$_2$Se$_3$–GeSe$_2$ system, *J. Alloys Compd.*, **238**(1–2), 141–148 (1996).

Pan'ko V.V., Hudoliy V.A., Stetsovich M.Yu., Voroshilov Yu.V. Phase equilibria in the HgSe–HgTe–HgCl$_2$ system [in Russian], *Ukr. khim. zhurn.*, **55**(3), 232–233 (1989).

Pan'ko V.V., Voroshilov Yu.V., Hudoliy V.A., Shelemba M.S. Phase equilibria in the HgS(Se)–HgCl$_2$–HgBr$_2$ systems [in Russian], *Zhurn. neorgan. khimii*, **41**(10), 1731–1733 (1996).

Parasyuk O.V. Phase relations of the Ag$_2$SnS$_3$–HgS and Ag$_{33.3}$Sn$_{16.7}$Se(Te)$_{50}$–HgSe(Te) sections in Ag–Hg–Sn–S(Se, Te) systems, *J. Alloys Compd.*, **291**(1–2), 215–219 (1999).

Parasyuk O.V., Gulay L.D., Piskach L.V., Kumanska Yu.O. The Ag$_2$Se–HgSe–SnSe$_2$ system and the crystal structure of the Ag$_2$HgSnSe$_4$ compound, *J. Alloys Compd.*, **339**(1–2), 140–143 (2002).

Parasyuk O.V., Gulay L.D., Romanyuk Ya.E., Olekseyuk I.D. The Ag$_2$Se–HgSe–SiSe$_2$ system in the 0–60 mol.% SiSe$_2$ region, *J. Alloys Compd.*, **348**(1–2), 157–166 (2003b).

Parasyuk O.V., Gulay L.D., Romanyuk Ya.E., Olekseyuk I.D., Piskach L.V. The Ag$_2$Se–HgSe–GeSe$_2$ system and crystal structures of the compounds, *J. Alloys Compd.*, **351**, 135–144 (2003a).

Parasyuk O.V., Olekseyuk I.D., Galka V.O., Marchuk O.V. Phase equilibria in the $A^I B^{III} Se_2$–HgSe and $A_2^I C^{IV} Se_3$–HgSe (A^I = Ag, Cu; B^{III} = Ga, In; C^{IV} = Si, Ge, Sn) systems [in Ukrainian], *Fizyka kondens. vysokomolek. system. Nauk. zap. Rinens'kogo pedinstytutu*, (3), 158–162 (1997).

Parasyuk O.V., Olekseyuk I.D., Marchuk O.V. Phase equilibria in the $Cu_2Si(Ge, Sn)Se_3$–HgSe systems [in Ukrainian], *Ukr. khim. zhurn.*, **64**(9), 20–23 (1998).

Parasyuk O.V., Olekseyuk I.D., Marchuk O.V. The Cu_2Se–HgSe–$SnSe_2$ system, *J. Alloys Compd.*, **287**(1–2), 197–205 (1999).

Puff H., Lorbacher G., Heine D. Quecksilberchalkogen-Fluorosilicate, *Naturwissenschaften*, **56**(9), 461 (1969).

Schäfer W., Nitsche R. Tetrahedral quaternary chalcogenides of the type Cu_2–II–IV–$S_4(Se_4)$, *Mater. Res. Bull.*, **9**(5), 645–654 (1974).

Schäfer W., Nitsche R. Zur Systematik tetraedrischer Verbindungen vom Typ $Cu_2Me^{II}Me^{IV}Me_4^{VI}$ (Stannite und Wurtzstannite), *Z. Kristallogr.*, **145**(5–6), 356–370 (1977).

Weil M. The crystal structure of $Hg_7Se_3O_{13}H_2$ and $Hg_8Se_4O_{17}H_2$—Two mixed-valent mercury oxoselenium compounds with a multivarious crystal chemistry, *Z. Kristallogr.*, **219**(10), 621–629 (2004).

Wibowo A.C., Malliakas C.D., Chung D.Y., Im J., Freeman A.J., Kanatzidis M.G. Mercury bismuth chalcohalides, $Hg_3Q_2Bi_2Cl_8$ (Q = S, Se, Te): Synthesis, crystal structures, band structures, and optical properties, *Inorg. Chem.*, **52**(6), 2973–2979 (2013).

9 Systems Based on HgTe

9.1 MERCURY–HYDROGEN–CARBON–TELLURIUM

Some compounds are formed in this quaternary system.

(HgMe)$_2$Te or C$_2$H$_6$TeHg$_2$. This compound was obtained on the reaction of H$_2$Te with CH$_3$HgBr in MeOH under N$_2$ atmosphere (Breitinger and Morell 1974). The colorless precipitate was filtered and recrystallized from benzene. It melts at 130°C with decomposition.

HgTeBu$_2^t$ or C$_8$H$_{18}$TeHg. To obtain this compound, 11.8 mL of *t*-BuTeLi solution (~3.7 mmol) with stirring at −5°C was added to 3.65 mmol of *t*-butylmercuric chloride in 11 mL of tetrahydrofuran (Harris et al. 1987). After 10 min, the green-yellow solution was placed in a CCl$_4$/dry ice slush bath (−23°C) and kept at this temperature while being concentrated to dryness under vacuum. The yellow residues were extracted for 3 min with 30 mL of toluene at −5°C and centrifuged, after which the solid was discarded. The yellow liquid was diluted with 15 mL pentane and left to crystallize at −78°C. The resultant yellow solid was washed twice with pentane at −78°C and dried under a stream of Ar for 7 h in the dark at 20°C. The mother liquor was concentrated to 5 mL under vacuum at −23°C, warmed to room temperature for less than 1 min to dissolve the solid, and diluted with 5 mL of pentane, and second crop of crystals was collected at −78°C. Recrystallization with toluene–pentane or from neat pentane gave fine yellow prisms with irreproducible partial melting in the 80°C–90°C range and gradual darkening above 110°C. The reaction and all handling of this compound were carried out in dim light.

The yellow light-sensitive C$_8$H$_{18}$TeHg is stable for at least 8 h at 20°C under Ar in the dark. It sublimes at 0.1 Pa at 70°C.

Hg(BuTe)$_2$ or C$_8$H$_{18}$Te$_2$Hg, bis(butyltellurolato)mercury. To obtain this compound, dibutyl ditelluride (1.67 mmol) was added to a suspension of Hg (3.29 mmol) in toluene/heptane (10 mL/10 mL) (Brennan et al. 1990). The mixture was allowed to stir in the dark for 16 h, while a yellow precipitate was formed. The precipitate was collected by filtration, washed with heptane (5 mL), and dried under vacuum to give an insoluble orange-yellow solid.

Hg(TePhMe)$_2$ or C$_{14}$H$_{14}$Te$_2$Hg, bis(methylbenzenetellurato)mercury. To obtain this compound, a solution of 0.55 equivalent of HgCl$_2$ was added to a solution of PhTeLi in tetrahydrofuran (Steigerwald and Sprinkle 1987). Upon completion of addition, the mixture was stirred for 30 min. Filtration and washing (pentane) gave the crude red-orange product. It was purified by extraction with toluene/PMe$_3$ (10:1 by volume). C$_{14}$H$_{14}$Te$_2$Hg is an amorphous powder. When this compound is sealed under vacuum and heated at 120°C for 24 h, HgTe is isolated. This is a very mild route to polycrystalline mercury telluride.

Hg(TeC$_6$H$_2$Me$_3$-2,4,6)$_2$ or **C$_{18}$H$_{22}$Te$_2$Hg, bis(trimethylbenzenetellurolato) mercury.** To obtain this compound, Hg (2.0 mmol) was added to a stirred solution of (2,4,6-Me$_3$C$_6$H$_2$)$_2$Te$_2$ (2.2 mmol) in toluene (20 mL) at room temperature (Bochmann and Webb 1991, Bochmann et al. 1997). The reaction was carried out in a 100 mL three-necked flask connected via a stopcock adaptor to the inert gas supply of a vacuum line. The mixture was stirred rigorously until Hg has disappeared (ca. 24 h). The pale yellow precipitate was filtered off, washed with toluene and diethyl ether, dissolved in hot pyridine (5 mL), and filtered hot to remove Hg residues. The filtrate was left to crystallize at –20°C to give yellow crystals of the product. A second fraction can be obtained by the addition of light petroleum to the mother liquor. The crystals of C$_{18}$H$_{22}$Te$_2$Hg darken on heating above 100°C, sublime with some decomposition at 184°C, and decompose rapidly at 215°C.

C$_{18}$H$_{22}$Te$_2$Hg could be also prepared if a solution of Hg[N(SiMe$_3$)$_2$]$_2$ (1 mmol) in light petroleum (5 mL) was added to a solution of 2,4,6-Me$_3$C$_6$H$_2$TeH (2 mmol) at –20°C via syringe or if a solution of Hg[N(SiMe$_3$)$_2$]$_2$ (1 mmol) in light petroleum at room temperature was treated with (2,4,6-Me$_3$C$_6$H$_2$)$_2$Te$_2$ (1.03 mmol) and stirred for 15 min (Bochmann and Webb 1991). A yellow precipitate was formed immediately. The mixture was warmed to room temperature and filtered, and the residue recrystallized from hot toluene.

Hg(TeC$_6$H$_2$Pr$_3^i$-2,4,6)$_2$ or C$_{30}$H$_{46}$Te$_2$Hg, bis(tri-i-propylbenzenetelluronolato)mercury. This compound melts at 75°C with decomposition (Bochmann and Webb 1991). To obtain C$_{30}$H$_{46}$Te$_2$Hg, 1.5 mmol of Hg was stirred with a suspension of (C$_6$H$_2$Pr$_3^i$-2,4,6)$_2$Te$_2$ (1.51 mmol) in MeOH (20 mL) at room temperature for 24 h. A yellow-orange powder precipitated, which was filtered off, washed with MeOH, and dried in vacuo. Recrystallization from acetonitrile gave orange needlelike crystals, contaminated with a small quantity of (C$_6$H$_2$Pr$_3^i$-2,4,6)$_2$Te$_2$, which were separated manually.

9.2 MERCURY–POTASSIUM–TIN–TELLURIUM

K$_2$HgSnTe$_4$ quaternary compound, which crystallizes in the tetragonal structure with the lattice parameters $a = 858.0 \pm 0.2$ and $c = 735.8 \pm 0.4$ pm, is formed in this system (Dhingra and Haushalter 1994). To obtain this compound, a mixture of 0.255 mmol of K$_4$SnTe$_4$ and 0.254 mmol of HgCl$_2$ was treated with 0.3 mL of ethylenediamine and sealed in a quartz tube under vacuum. After heating for 48 h at 100°C, the solution was cooled and the small, metallic gray, cubic-shaped crystals were filtered and washed with ethylenediamine. It is possible to prepare K$_2$HgSnTe$_4$ at high temperature by the reaction of 0.972 mmol of K$_2$Te, 0.972 mmol of Hg, 0.969 mmol of Sn, and 2.915 mmol of Te at 500°C for 10 h in an evacuated quartz tube. Single crystals of this compound were prepared by the treatment of K$_4$SnTe$_4$, which was prepared from the reaction of KSn and Te, with 1 equivalent HgCl$_2$ in ethylenediamine at 100°C.

9.3 MERCURY–RUBIDIUM–ANTIMONY–TELLURIUM

HgTe–Rb$_2$Te–Sb$_2$Te$_3$. The phase diagram is not constructed. The RbHgSbTe$_3$ quaternary compound is formed in this system. It crystallizes in the orthorhombic structure with the lattice parameters $a = 459.0 \pm 0.2$, $b = 1574.5 \pm 0.4$, and $c = 1173.7 \pm 0.2$ pm;

calculated density 6.191 g·cm^{-3}; and energy bandgap 0.2 eV (Li et al. 1997). RbHgSbTe$_3$ was obtained by the interaction of Rb$_2$Te, Hg$_2$Cl$_2$, Sb$_2$Te$_3$, and Te as a starting material, and ethylenediamine was added to the mixture. The reaction took place in the tube sealed under vacuum at 180°C for 7 days. This compound could be synthesized also by the solid-state reaction of HgTe, Rb$_2$Te, and Sb$_2$Te$_3$ in a 1:1:2 ratio at 400°C–600°C.

9.4 MERCURY–COPPER–INDIUM–TELLURIUM

HgTe–CuInTe$_2$. The phase diagram is not constructed. It was indicated (Gallardo 1992) that it is possible to predict the value at which the chalcopyrite phase in this system disappears (x_c) and $T_c(x)$ behavior for the Hg$_{2x}$(CuIn)$_{1-x}$Te$_2$ alloys for which a chalcopyrite–sphalerite-like disordering transition with extensive terminal solid solution formation exists.

9.5 MERCURY–COPPER–SILICON–TELLURIUM

HgTe–Cu$_2$SiTe$_3$. The phase diagram is not constructed. Cu$_2$HgSiTe$_4$ quaternary compound is formed in this system (Haeuseler et al. 1991). It crystallizes in the tetragonal structure with the lattice parameters $a = 609.2 \pm 0.1$ and $c = 1183.1 \pm 0.4$ pm.

9.6 MERCURY–COPPER–GERMANIUM–TELLURIUM

HgTe–Cu$_2$GeTe$_3$. The phase diagram of this quasibinary system is shown in Figure 9.1 (Hirai et al. 1967). The eutectic crystallizes at 512°C and contains 35 mol.% Cu$_2$GeTe$_3$ (from Figure 9.1). Cu$_2$HgGeTe$_4$ quaternary compound is formed in this system. It melts congruently at 520°C (Hirai et al. 1967) and crystallizes in the

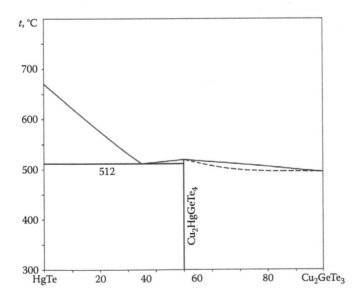

FIGURE 9.1 Phase diagram of the HgTe–Cu$_2$GeTe$_3$ quasibinary system. (From Hirai, T. et al., *Solid State Electron.*, 10(10), 975, 1967.)

tetragonal structure with the lattice parameters $a = 611.4 \pm 0.1$ and $c = 1192.8 \pm 0.3$ pm (Haeuseler et al. 1991) [$a = 793$ and $c = 1715$ pm (Hirai et al. 1967)].

9.7 MERCURY–COPPER–TIN–TELLURIUM

HgTe–Cu$_2$SnTe$_3$. The phase diagram is not constructed. The Cu$_2$HgSnTe$_4$ quaternary compound is formed in this system. It crystallizes in the tetragonal structure with lattice parameters $a = 616.2$ and $c = 1228$ pm [$a = 619.1 \pm 0.1$ and $c = 1226.3 \pm 0.3$ pm (Haeuseler et al. 1991)] and calculated and experimental densities 6.81 and 6.66 g·cm^{-3}, respectively (Hahn and Schulze 1965).

9.8 MERCURY–SILVER–GALLIUM–TELLURIUM

HgTe–AgGaTe$_2$. The phase relations in this system are shown in Figure 9.2 (Galka et al. 1999). The eutectic crystallizes at 641°C and contains 57 mol.% HgTe. The solubility of HgTe in AgGaTe$_2$ reaches 20 mol.% at the eutectic temperature and decreases to 4 mol.% at 400°C. The solubility of AgGaTe$_2$ in HgTe is equal to 33 mol.% at 400°C. This system was investigated by differential thermal analysis (DTA), x-ray diffraction (XRD), and metallography, and the alloys were annealed at 400°C for 250 h.

9.9 MERCURY–SILVER–INDIUM–TELLURIUM

HgTe–AgInTe$_2$. This section is a nonquasibinary section of the Hg–Ag–In–Te quaternary system (Figure 9.3) (Gallardo et al. 1988, Galka et al. 1999). The liquidus is represented by the lines of the primary crystallization of AgIn$_5$Te$_8$ and the solid solution based on HgTe (Galka et al. 1999). The solubility of HgTe in α-AgInTe$_2$

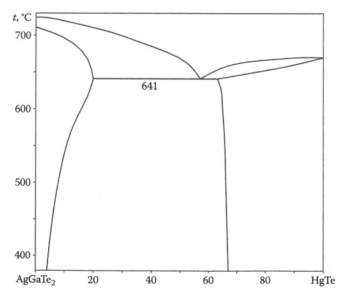

FIGURE 9.2 Phase relations in the HgTe–AgGaTe$_2$ quasibinary system. (From Galka, V.O. et al., *Pol. J. Chem.*, 73(4), 743, 1999.)

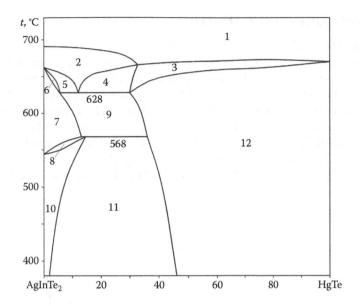

FIGURE 9.3 Phase relations in the HgTe–AgInTe$_2$ system: 1, L; 2, L + AgIn$_5$Te$_8$; 3, L + (HgTe); 4, L + AgIn$_5$Te$_8$ + (HgTe); 5, L + (β-AgInTe$_2$) + AgIn$_5$Te$_8$; 6, (β-AgInTe$_2$) + AgIn$_5$Te$_8$; 7, (β-AgInTe$_2$); 8, (β-AgInTe$_2$) + (α-AgInTe$_2$); 9, (β-AgInTe$_2$) + (HgTe); 10, (α-AgInTe$_2$); 11, (α-AgInTe$_2$) + (HgTe); and 12, (HgTe). (From Galka, V.O. et al., *Pol. J. Chem.*, 73(4), 743, 1999.)

at 400°C is equal to 2 mol.% and increases with increasing temperature. The solubility of HgTe in β-AgInTe$_2$ at 628°C does not exceed 6 mol.% and increases with decreasing temperature. Peritectic transformation takes place at 568°C. Solid solution based on HgTe contains at 400°C up to 54 mol.% AgInTe$_2$, and the solubility increases at the increasing temperature. It was indicated (Gallardo 1992) that it is possible to predict the value at which the chalcopyrite phase in this system disappears (x_c) and $T_c(x)$ behavior for the Hg$_{2x}$(AgIn)$_{1-x}$Te$_2$ alloys for which a chalcopyrite–sphalerite-like disordering transition with extensive terminal solid solution formation exists.

According to the data of Voytovich (1986) and Voytovich et al. (1986), solid solutions with sphalerite structure are formed within the interval of 0–60 mol.% AgInTe$_2$.

This system was investigated by DTA, XRD, metallography, and microradioscopic analysis and measuring of microhardness (Voytovich 1986, Voytovich et al. 1986, Gallardo et al. 1988, Galka et al. 1999). The alloys were annealed at 400°C for 250 h (Galka et al. 1999). Single crystals of solid solutions were grown by the Bridgman method or by the solid-state recrystallization (Andruhiv et al. 1980, Voytovich 1986, Voytovich et al. 1980, 1986).

HgTe–Ag$_2$Te–In$_2$Te$_3$. The phase diagram is not constructed. The Ag$_4$Hg$_3$In$_6$Te$_{14}$ quaternary compound is formed in this system. It melts congruently at 670°C and crystallizes in the tetragonal structure with lattice parameters $a = 1035.2 \pm 2.8$ and $c = 2990.0 \pm 18.1$ pm (Mayet and Roubin 1980).

9.10   MERCURY–SILVER–THALLIUM–TELLURIUM

HgTe–AgTlTe. The phase diagram of this quasibinary system is shown in Figure 9.4 (Gardes et al. 1987, 1988). It is characterized by an intermediate quaternary compound $Ag_3HgTl_3Te_4$ and a high-temperature phase δ for $t > 324°C$ in the range from 30 to 60 mol.% HgTe. $Ag_3HgTl_3Te_4$ and the high-temperature phase melt incongruently at 404°C and 449°C, respectively. The eutectic crystallizes at 413°C and contains 28 mol.% HgTe (from Figure 9.4). The solubility of HgTe in α-AgTlTe and in β-AgTlTe reaches 13 mol.% at 404°C and 18 mol.% at 429°C, respectively. The solubility of AgTlTe in HgTe is equal to 26 mol.% at 449°C. This system was investigated by DTA and XRD, and the alloys were annealed at 300°C for 1–3 months.

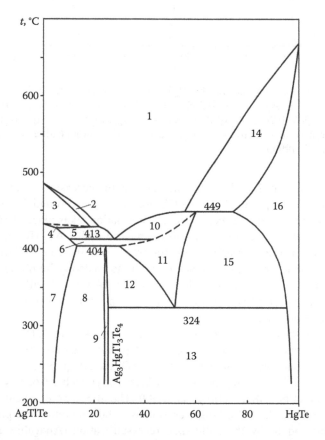

FIGURE 9.4   Phase diagram of the HgTe–AgTlTe quasibinary system: 1, L; 2, L + (β-AgTlTe); 3, (β-AgTlTe); 4, (β-AgTlTe) + (α-AgTlTe); 5, L + (α-AgTlTe); 6, (α-AgTlTe) + δ; 7, (α-AgTlTe); 8, (α-AgTlTe) + $Ag_3HgTl_3Te_4$; 9, $Ag_3HgTl_3Te_4$; 10, L + δ; 11, δ; 12, δ + $Ag_3HgTl_3Te_4$; 13, $Ag_3HgTl_3Te_4$ +(HgTe); 14, L + (HgTe); 15, δ + (HgTe); and 16, (HgTe). (From Gardes, B. et al., *Port. Electrochim. Acta*, 5(Dez.), 285, 1987; Gardes, B. et al., *J. Solid State Chem.*, 73(2), 502, 1988.)

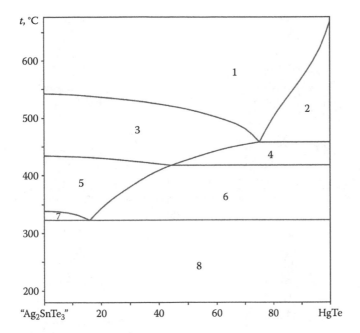

FIGURE 9.5 Phase relations in the HgTe–"Ag_2SnTe_3" section: 1, L; 2, L + HgTe; 3, L + SnTe; 4, L + HgTe + SnTe; 5, L + $AgSnTe_2$ + SnTe; 6, L + $AgSnTe_2$ + HgTe + SnTe; 7, L + $AgSnTe_2$ + SnTe + Te; and 8, $AgSnTe_2$ + HgTe + SnTe + Te. (From Parasyuk, O.V., *J. Alloys Compd.*, 291(1–2), 215, 1999.)

9.11 MERCURY–SILVER–TIN–TELLURIUM

HgTe–"Ag_2SnTe_3." This section is a nonquasibinary section of the Hg–Ag–Sn–Te quaternary system (Figure 9.5) (Parasyuk 1999). Quaternary phases were not discovered in this section. The liquidus consists of the two fields corresponding to the primary crystallization of HgTe and SnTe. Under the solidus, four phases, HgTe, SnTe, $AgSnTe_2$, and Te, are present. This system was investigated by DTA, XRD, and metallography, and the ingots were annealed at 250°C during 500 h with further quenching in cold water.

9.12 MERCURY–ALUMINUM–ANTIMONY–TELLURIUM

HgTe–AlSb. The phase diagram is not constructed. According to the data of metallography and XRD, the solubility of AlSb in HgTe is not higher than 5 mol.% (Burdiyan and Georgitse 1970).

9.13 MERCURY–GALLIUM–INDIUM–TELLURIUM

3HgTe–Ga_2Te_3–In_2Te_3. Isothermal sections of this quasiternary system at 630°C, 480°C, 430°C, and 380°C as derived from XRD and EPMA data are shown in Figure 9.6 (von Wensierski et al. 1997a, Leute et al. 1999). It was shown that ordered

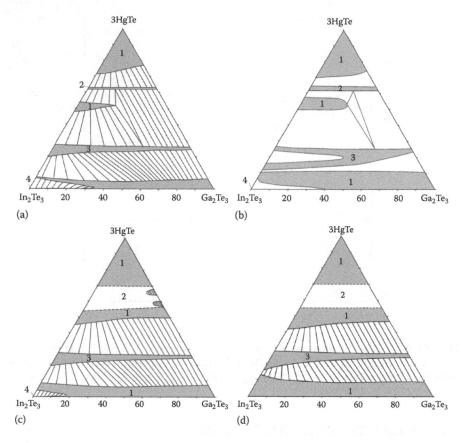

FIGURE 9.6 Isothermal sections of the $3HgTe–Ga_2Te_3–In_2Te_3$ quasiternary system at (a) 380°C, (b) 430°C, (c) 480°C, and (d) 630°C: 1, cubic phase; 2, ordered phase; 3, chalcopyrite phase; and 4, orthorhombic phase. (From von Wensierski, H. et al., *J. Alloys Compd.*, 255 (1–2), 169, 1997a; Leute, V. et al., *J. Alloys Compd.*, 289(1–2), 233, 1999.)

structures are formed at special concentrations of structural vacancies. Depending on the temperature and concentration, continuous order–disorder transitions could be detected. Besides structural miscibility gaps and a three-phase region, there occurs also a miscibility gap within an ordered region.

At 630°C (Figure 9.6d), there are only two miscibility gaps. Within the HgTe-rich alloy region, there is a composition region yielding additional superstructure reflections of the low-temperature phase. But in this case, a miscibility gap between this region and the surrounding cubic phase does not exist. With decreasing temperature, the phase diagram becomes more complicated. At 480°C (Figure 9.6c) near In_2Te_3 corner of the compositional triangle, there occurs a narrow quasiternary region. There must be an additional disordered cubic phase coexisting with the orthorhombic phase. Within the chalcopyrite phase (3), a miscibility gap occurs that extends at 480°C from $3HgTe–In_2Te_3$ edge up to 40 mol.% Ga_2Te_3. On the $3HgTe–Ga_2Te_3$ edge at 62.5 mol.% 3HgTe, a new totally ordered phase (2) occurs (von Wensierski et al. 1997a, Leute et al. 1999).

At decreasing temperature to 430°C (Figure 9.6b), the ordered phase (2) becomes more and more stable causing an extension of the adjacent miscibility gaps. As the critical line for the order–disorder transitions reaches $3HgTe-In_2Te_3$ at 450°C, the ordered phase (2) extends at 430°C along the whole quasibinary section $Hg_5Ga_2Te_8-Hg_5In_2Te_8$. At 380°C (Figure 9.6a) compared to 430°C, the extension of the disordered (1) regions has decreased and that of the heterogeneous regions has increased (von Wensierski et al. 1997a, Leute et al. 1999). The existence region of superstructure phase (2) is somewhat broader, whereas that of chalcopyrite phase (3) remained nearly unchanged.

Interdiffusion experiments along the quasibinary section $(Hg_{3n}Ga_{2m}In_{2(1-m-n)}V_n)Te_3$ have shown that the structural vacancies (V) determine the diffusion properties of such alloys in different ways (von Wensierski et al. 1997b). At low vacancy densities, the structural vacancies behave like usual defects introduced by doping. At higher vacancy densities, the diffusion coefficient increases much less than the concentration of the structural vacancies because a great part of them are bound in associates. At special stoichiometric conditions ($n = 5/8$ and $1/4$), the structural vacancies take strictly ordered positions and behave no more as defects, but as interstitial sites of an ordered superstructure of the zinc-blende lattice. Near the points of transition from the ordered phases to the more or less disordered alloys, the diffusion coefficient increases continuously.

Annealing times for preparing the ingots of the $3HgTe-Ga_2Te_3-In_2Te_3$ quasiternary system ranged between 20 days and 4 months for the low annealing temperatures (von Wensierski et al. 1997a, Leute et al. 1999). To preserve the high-temperature equilibrium composition, the ampoules were quenched in ice water.

9.14   MERCURY–GALLIUM–ANTIMONY–TELLURIUM

HgTe–GaSb. The phase diagram is not constructed. According to the data of metallography and XRD, the solubility of GaSb in HgTe reaches 10 mol.% (Ambros and Burdiyan 1970, Burdiyan 1970, 1971). Energy gap of forming solid solutions changes near linearly with composition (Burdiyan 1970, 1971).

9.15   MERCURY–INDIUM–ARSENIC–TELLURIUM

HgTe–InAs. The phase diagram is shown in Figure 9.7 (Kuz'mina 1976). Solid solutions over entire range of concentrations are formed in this system (Goriunova et al. 1962, Iniutkin et al. 1964, Kuz'mina 1976). Small quantities of inclusions were found from the InAs-rich side (Goriunova et al. 1962). Lattice parameters of forming solid solutions change linearly with concentration. The region of possible solid solution degradation was found near the liquidus. Two-phase region was determined at 500°C by the XRD of ingots annealed for 250 h.

This system was investigated by DTA, metallography, and measuring of electroconductivity (Goriunova et al. 1962, Iniutkin et al. 1964, Kuz'mina 1976). The ingots were annealed at 610°C for 250 h and then at 650°C–850°C (Kuz'mina 1976) [at 570°C–600°C for 550–600 h (Goriunova et al. 1962)].

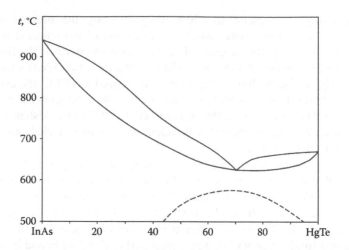

FIGURE 9.7 The HgTe–InAs phase diagram. (From Kuz'mina, G.A., *Izv. AN SSSR. Neorgan. materialy*, 12(6), 1121, 1976.)

9.16 MERCURY–INDIUM–ANTIMONY–TELLURIUM

HgTe–InSb. The phase diagram is a eutectic type (Figure 9.8) (Pashaev et al. 2006). The eutectic composition and temperature are 27 mol.% HgTe and 350°C, respectively. The solubility of HgTe in InSb is equal to 10 mol.% at room temperature.

According to the data of Belotski et al. (1978), the solubility of HgTe in InSb at 500°C reaches 10 mol.% and the solubility of InSb in HgTe at the same temperature is equal to 5 mol.%. The ingots within the interval of 10–80 mol.% InSb contain elemental mercury.

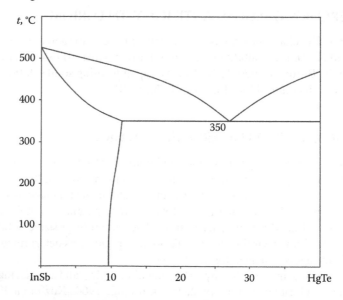

FIGURE 9.8 Part of the HgTe–InSb phase diagram. (From Pashaev, A.M. et al., *Zhurn. khim. problem*, (1), 119, 2006.)

This system was investigated by DTA, XRD, metallography, and measuring of microhardness and density (Pashaev et al. 2006). The ingots were annealed at 340°C for 500 h and the ingots containing 8, 9, 10, 12, and 15 mol.% HgTe were additionally annealed at 100°C, 200°C, and 300°C during 250 h with the next quenching.

9.17 MERCURY–INDIUM–CHLORINE–TELLURIUM

$HgTe–HgCl_2–InCl_3$. The phase diagram is not constructed. $Hg_4In_2Te_3Cl_8$ quaternary compound is formed in this system. It is stable up to 300°C and crystallizes in the orthorhombic structure with the lattice parameters $a = 1505.7 \pm 0.3$, $b = 767.45 \pm 0.15$, and $c = 1887.9 \pm 0.4$ pm; calculated density 5.171 $g \cdot cm^{-3}$; and calculation and experimental energy gaps 2.21 and 1.89 eV, respectively (Liu et al. 2012). It was prepared via the solid-state reaction from a mixture containing Hg_2Cl_2, In_2Te_3, and Te.

9.18 MERCURY–THALLIUM–GERMANIUM–TELLURIUM

$HgTl_2GeTe_4$ quaternary compound is formed in this system. It crystallizes in the tetragonal structure with the lattice parameters $a = 835.71 \pm 0.11$ and $c = 706.84 \pm 0.14$ pm (McGuire et al. 2005). This compound was synthesized by the reaction of Tl, Te, HgTe, and Ge in 2:1:1:3 molar ratio.

9.19 MERCURY–THALLIUM–TIN–TELLURIUM

$HgTl_2SnTe_4$ quaternary compound is formed in this system. It crystallizes in the tetragonal structure with the lattice parameters $a = 839.7 \pm 0.4$ and $c = 715.7 \pm 0.6$ pm (McGuire et al. 2005). This compound was synthesized by the reaction of Tl, Te, HgTe, and Sn in 2:1:1:3 molar ratio.

9.20 MERCURY–DYSPROSIUM–MANGANESE–TELLURIUM

$HgTe–DyTe–MnTe$. The phase diagram is not constructed. $Hg_{1-x-y}Mn_xDy_yTe$ $(0 < x \leq 0.09, y = 0.01)$ single crystals were prepared by the Bridgman method (Kovalyuk et al. 2013). At 27°C, the energy gap of the crystals with $x + y = 0.063$ is equal to 0.092 eV.

9.21 MERCURY–URANIUM–CHLORINE–TELLURIUM

$HgTe–HgCl_2–UCl_4$. The phase diagram is not constructed. $[Hg_3Te_2][UCl_6]$ quaternary compound is formed in this system that crystallizes in the monoclinic structure with the lattice parameters $a = 683.5 \pm 0.1$, $b = 786.4 \pm 0.1$, $c = 1363.2 \pm 0.2$ pm, and $\beta = 91.530° \pm 0.02°$; calculated density 5.929 $g \cdot cm^{-3}$ at 153 ± 2 K; and energy gap ~2.6 eV (Bugaris and Ibers 2008).

This compound was synthesized using the mixture of HgTe, UCl_4, and $HgCl_2$, which was heated to 850°C in 24 h, kept at this temperature for 96 h, cooled at 6.8°C/h to 200°C, and then cooled rapidly to 20°C (Bugaris and Ibers 2008). In an alternative heating procedure, the reaction mixture was heated to 600°C in 15 h and kept at this temperature for 120 h, and then the furnace was turned off.

9.22   MERCURY–SILICON–FLUORINE–TELLURIUM

HgTe–HgF$_2$–SiF$_4$. The phase diagram is not constructed. Hg$_3$Te$_2$SiF$_6$ quaternary compound is formed in this system. This compound could be obtained by the interaction of HgTe with the solution of HgO in H$_2$SiF$_6$ at 120°C (Puff et al. 1969).

9.23   MERCURY–BISMUTH–CHLORINE–TELLURIUM

HgTe–HgCl$_2$–BiCl$_3$. The phase diagram is not constructed. The Hg$_3$Bi$_2$Te$_2$Cl$_8$ quaternary compound is formed in this system. It melts incongruently and crystallizes in the monoclinic structure with the lattice parameters a = 1748.3 ± 0.4, b = 768.4 ± 0.2, c = 1341.5 ± 0.3 pm, and β = 104.72° ± 0.03°; calculated density 5.939 g·cm^{-3}; and energy gap 2.80 eV (Wibowo et al. 2013). To obtain this compound, a mixture of HgTe, HgCl$_2$, and BiCl$_3$ in the molar ratio 2:1:6 was loaded into a fused silica tube, which was flame sealed under vacuum, put inside a programmable furnace, heated to 200°C over 14 h, and held isothermally for 48 h, followed by a shutdown of the furnace. Single crystals of Hg$_3$Bi$_2$Te$_2$Cl$_8$ were prepared by a reaction of HgTe and BiCl$_3$ at 1:1 ratio. The mixture in a sealed quartz tube was heated to 600°C over 16 h and held isothermally for 24 h, followed by cooling slowly to 200°C over 36 h and then to room temperature within 1 h.

9.24   MERCURY–OXYGEN–MANGANESE–TELLURIUM

HgTe–MnTe–O. The phase diagram is not constructed. According to the data of thermodynamic calculations, an oxidation of the Hg$_x$Mn$_{1-x}$Te solid solutions begins from the oxidation of MnTe with formation of MnO and Te. And then, HgTe is oxidized to form elemental Hg (Medvedev and Berchenko 1994).

9.25   MERCURY–CHLORINE–BROMINE–TELLURIUM

HgTe–HgCl$_2$–HgBr$_2$. Isothermal section of this quasiternary system at 200°C is shown in Figure 9.9 (Hudoliy et al. 1995). There are some two-phase regions in this section: the first of them exists between HgTe and Hg$_3$Te$_2$(Cl$_{1-x}$Br$_x$)$_2$ solid solutions, the second between Hg$_3$Te$_2$(Cl$_{1-x}$Br$_x$)$_2$ and Hg$_3$Te(Cl$_{1-x}$Br$_x$)$_4$ solid solutions, and the third in the HgCl$_2$–HgBr$_2$–Hg$_3$Te$_2$(Cl$_{1-x}$Br$_x$)$_2$ subsystem, which are divided by the γ-phase of the HgCl$_2$–HgBr$_2$ quasibinary system into three subsystems. This system was investigated by DTA and XRD, and the ingots were annealed at 200°C for 600 h.

9.26   MERCURY–BROMINE–IODINE–TELLURIUM

HgTe–HgBr$_2$–HgI$_2$. The phase diagram is not constructed. The Hg$_3$Te$_2$BrI quaternary compound is formed in this system. It melts incongruently at 522°C, has the homogeneity range along the section Hg$_3$Te$_2$Br$_2$–Hg$_3$Te$_2$I$_2$ within the interval of 44–65 mol.% Hg$_3$Te$_2$I$_2$ at 200°C (Minets et al. 1998), and crystallizes in the monoclinic structure with lattice parameters a = 1837.6 ± 0.5, b = 958.7 ± 0.2, c = 1057.5 ± 0.2 A, and β = 90.12° ± 0.02° and calculated density 7.585 ± 0.005 g·cm^{-3} (Voroshilov and Minets 1998, Minets et al. 2004).

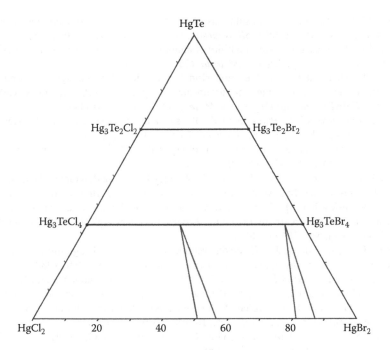

FIGURE 9.9 Isothermal section of the HgTe–HgCl$_2$–HgBr$_2$ quasiternary system at 200°C. (From Hudoliy, V.A. et al., *Neorgan. materialy*, 31(3), 396, 1995.)

REFERENCES

Ambros V.P., Burdiyan I.I. Investigation of mercury telluride solubility in gallium antimonide [in Russian], *Uchen. zap. Tirasp. ped. in-ta*, (21, Pt. 1), 31–38 (1970).

Andruhiv M.V., Voytovich O.E., Konnikov S.G., Matiushkin S.F., Tsiutsiura D.I., Shuptar D.D. Electrical and photoelectrical properties of narrow gap (AgInTe$_2$)$_x$–(HgTe)$_{1-x}$ solid solutions [in Russian]. In: *Poluprovodn. s uzkoy zapreshchen. zonoy i polumet. Materialy 5th Vses. simpoz.* P.2, L'vov, Ukraine, pp. 252–253 (1980).

Belotski D.P., Babiuk P.F., Cherveniuk G.I. Investigation of solid solutions in the InSb–HgTe system [in Russian], *Izv. AN SSSR. Neorgan. materialy*, **14**(3), 589–590 (1978).

Bochmann M., Webb K. Sterically hindered thiolato, selenolato and tellurolato complexes of mercury(II), *J. Chem. Soc., Dalton Trans.*, (9), 2325–2329 (1991).

Bochmann M., Webb K.J., Malik M.A., O'Brien P. Arene thiolato, selenolato, and tellurolato complexes of mercury, *Inorg. Synthes.*, **31**, 24–28 (1997).

Breitinger D., Morell W. Tris(methylmercurio) selenonium salts [Se(HgCH$_3$)$_3$]X and bis(methylmercuric) chalcogenides Ch(HgCH$_3$)$_2$ (Ch = Se and Te), *Inorg. Nucl. Chem. Lett.*, **10**(5), 409–411 (1974).

Brennan J.G., Siegrist T., Carroll P.J., Stuczynski S.M., Reynders P., Brus L.E., Steigerwald M.L. Bulk and nanostructure group II-VI compounds from molecular organometallic precursors, *Chem. Mater.*, **2**(4), 403–409 (1990).

Bugaris D.E., Ibers J.A. Synthesis, structures and magnetic and optical properties of the compounds [Hg$_3$Te$_2$][UCl$_6$] and [Hg$_4$As$_2$][UCl$_6$], *J. Solid State Chem.*, **181**(12), 3189–3193 (2008).

Burdiyan I.I. The solid solutions of gallium antimonide with cadmium, zinc and mercury tellurides [in Russian]. In: *Nekot. voprosy khimii i fiz. poluprovodn. slozhn. sostava.* Uzhgorod, Ukraine: Uzhgorod. Un-t Publish., pp. 190–195 (1970).

Burdiyan I.I. Investigation of solid solutions of GaSb with cadmium, zinc and mercury tellurides [in Russian], *Izv. AN SSSR. Neorgan. materialy*, **7**(3), 414–416 (1971).

Burdiyan I.I., Georgitse E.I. About solubility of aluminium antimonide in mercury telluride [in Russian], *Uchen. zap. Tirasp. ped. in-ta*, (21, P.1), 3–5 (1970).

Dhingra S.S., Haushalter R.C. One dimensional inorganic polymers: Synthesis and structural characterization of the main group metal polymers $K_2HgSnTe_4$, $(Et_4N)_2HgSnTe_4$, $(Ph_4P)GeInTe_4$, and $RbInTe_2$, *Chem. Mater.*, **6**(12), 2376–2381 (1994).

Galka V.O., Krykhovets O.V., Parasyuk O.V., Olekseyuk I.D. Phase equilibria in the $AgGaTe_2$–HgTe and $AgInTe_2$–HgTe, *Pol. J. Chem.*, **73**(4), 743–748 (1999).

Gallardo P.G. Order-disorder phase transitions in $D_{2x}^{II}(A^IB^{III})_{1-x}C_2^{VI}$ alloys systems, *Phys. Status Solidi (a)*, **134**(1), 119–125 (1992).

Gallardo P.G. Quintero V., Perez G.S., Tovar R., Woolley J.C. Phase diagram and lattice parameter values for the $Hg_{2x}(AgIn)_{1-x}Te_2$, *Phys. Status Solidi (a)*, **107**(1), 165–169 (1988).

Gardes B., Ayral-Marin R.M., Brun G., Tedenac J.C. Contribution a l'étude du système quaternaire argent–mercure–thallium–tellure: Étude de la coupe AgTlTe–HgTe, *J. Solid State Chem.*, **73**(2), 502–506 (1988).

Gardes B., Brun G., Tedenac J.C. Étude du système AgTlTe–HgTe. Diagramme d'équilibre et caracterisation physique, *Port. Electrochim. Acta*, **5**(Dez.), 285–288 (1987).

Goriunova N.A., Grigor'eva V.S., Sharavski P.V., Osnach L.A. Solid solutions in the InAs–HgTe system [in Russian]. In: *Fizika*, L.: Publish. of Leningr. inzh.-stroit. in-t, pp. 7–10 (1962).

Haeuseler H., Ohrendori F., Himmrich M. Zur Kenntnis quaternärer Telluride $Cu_2MM'Te_4$ mit Tetrahederstrukturen, *Z. Naturforsch.*, **B46**(8), 1049–1052 (1991).

Hahn H., Schulze H. Über quaternäre Chalkogenide des Germanium und Zinns, *Naturwissenschaften*, **52**(14) 426 (1965).

Harris D.C., Nissan R.A., Higa K.T. Synthesis and characterization of tert-butyl(tert-butyltelluro)mercury(II), *Inorg. Chem.*, **26**(5), 765–768 (1987).

Hirai T., Kurata K., Takeda Y. Deviation of new semiconducting compounds by cross substitution for group IV semiconductors and their semiconducting and thermal properties, *Solid State Electron.*, **10**(10), 975–981 (1967).

Hudoliy V.A., Pan'ko V.V., Voroshilov Yu.V. Phase equilibria in the $HgTe$–$HgCl_2$–$HgBr_2$ system [in Russian], *Neorgan. materialy*, **31**(3), 396–398 (1995).

Iniutkin A., Kolosov E., Osnach L., Habarova V., Habarov E., Sharavski P. Some investigations of solid solutions based on III–V and II–VI compounds [in Russian], *Izv. AN SSSR. Ser. fiz.*, **28**(6), 1010–1016 (1964).

Kovalyuk T.T., Maistruk E.V., Maryanchuk P.D. Magnetic, optical, and kinetic properties of $Hg_{1-x-y}Mn_xDy_yTe$ crystals, *Inorg. Mater. (Engl. Trans.)*, **49**(5), 445–449 (2013).

Kuz'mina G.A. Investigation of the AlSb–CdTe and InAs–HgTe phase diagram [in Russian], *Izv. AN SSSR. Neorgan. materialy*, **12**(6), 1121–1122 (1976).

Leute V., Wetze D., Zeppenfeld A. Phase diagrams of II-VI/III-VI solid solutions with ordering tendencies, *J. Alloys Compd.*, **289**(1–2), 233–243 (1999).

Li J., Chen Zh., Wang X., Proseprio D.M. A novel two-dimensional mercury antimony telluride: low temperature synthesis and characterization of $RbHgSbTe_3$, *J. Alloys Compd.*, **262–263**, 28–33 (1997).

Liu Y., Kanhere P.D., Hoo Y.S., Ye K., Yan Q., Rawat R.S., Chen Z., Ma J., Zhang Q. Cationic quaternary chalcohalide nanobelts: $Hg_4In_2Q_3Cl_8$ (Q = S, Se, Te), *RSC Adv.*, **2**(16), 6401–6403 (2012).

Mayet F., Roubin M. Contribution a l'étude des système ternaire Ag_2Te–In_2Te_3–HgTe. Mise en evidence de deux phases nouvelles, *C. R. Acad. Sci. C*, **291**(13), 291–294 (1980).

McGuire M.A., Scheidemantel T.J., Badding J.V., DiSalvo F.J. Tl_2AXTe_4 (A = Cd, Hg, Mn; X = Ge, Sn): Crystal structure, electronic structure, and thermoelectric properties, *Chem. Mater.*, **17**(24), 6186–6191 (2005).

Medvedev Yu.V., Berchenko N.N. Analysis of the phase compositions at the interface of $Mn_{1-x}A_x^{II}B^{VI}$ (A^{II} = Zn, Cd, Hg; B^{VI} = Te, Se) solid solution with own oxides [in Russian], *Zhurn. neorgan. khimii*, **39**(5), 846–848 (1994).

Minets Yu.V., Hudoliy V.A., Pan'ko V.V., Voroshilov Yu.V. The $Hg_3Te_2Br_2$–$Hg_3Te_2I_2$ phase diagram [in Ukrainian], *Nauk. visnyk Uzhgorod. un-tu. Ser. Khimia*, (3), 43–44 (1998).

Minets Yu.V., Voroshilov Yu.V., Pan'ko V.V. The structure of mercury chalcogenhalogenides $Hg_3X_2Hal_2$, *J. Alloys Compd.*, **367**(1–2), 109–114 (2004).

Parasyuk O.V. Phase relations of the Ag_2SnS_3–HgS and $Ag_{33.3}Sn_{16.7}Se(Te)_{50}$–HgSe(Te) sections in Ag–Hg–Sn–S(Se, Te) systems, *J. Alloys Compd.*, **291**(1–2), 215–219 (1999).

Pashaev A.M., Ismailov A.M., Aliev I.I., Kuliev K.G. Synthesis and physico-chemical investigation of the $(InSb)_{1-x}(HgTe)_x$ solid solutions [in Russian], *Zhurn. khim. problem*, (1), 119–121 (2006).

Puff H., Lorbacher G., Heine D. Quecksilberchalkogen-Fluorosilicate, *Naturwissenschaften*, **56**(9), 461 (1969).

Steigerwald M.L., Sprinkle C.R. Organometallic synthesis of II-VI semiconductors. 1. Formation and decomposition of bis(organotelluro)mercury and bis(organotelluro)cadmium compounds, *J. Am. Chem. Soc.*, **109**(23), 7200–7201 (1987).

Voroshilov Yu.V., Minets Yu.V. Crystal structure of Hg_3Te_2BrI [in Ukrainian], *Nauk. visnyk Uzhgorod. un-tu. Ser. Khimia*, (3), 39–42 (1998).

Voytovich O.E. Solid solutions based on HgTe in the Ag–In–Te–Hg system [in Russian], *Fiz. elektronika (L'vov)*, (33), 77–83 (1986).

Voytovich O.E., Liubchenko A.V., Tsiutsiura D.I., Shuptar D.D. Obtaining and investigation of electrical properties of $(AgInTe_2)_x(HgTe)_{1-x}$ solid solutions [in Russian], *Fiz. elektronika (L'vov)*, (20), 63–65 (1980).

Voytovich O.E., Tsiutsiura D.I., Pashkovski M.V. The $(AgInTe_2)_x(2HgTe)_{1-x}$ solid solutions [in Russian], *Izv. AN SSSR. Neorgan. materialy*, **22**(11), 1918–1920 (1986).

von Wensierski H., Bolwin H., Zeppenfeld A., Leute V. Ordering phenomena and demixing in the quasiternary system $Ga_2Te_3/Hg_3Te_3/In_2Te_3$, *J. Alloys Compd.*, **255**(1–2), 169–177 (1997a).

von Wensierski H., Weitze D., Leute V. Ordering and diffusion in II-VI/III-VI alloys with structural vacancies, *Solid State Ionics*, **101–103**(Pt. 1), 479–487 (1997b).

Wibowo A.C., Malliakas C.D., Chung D.Y., Im J., Freeman A.J., Kanatzidis M.G. Mercury bismuth chalcohalides, $Hg_3Q_2Bi_2Cl_8$ (Q = S, Se, Te): Synthesis, crystal structures, band structures, and optical properties, *Inorg. Chem.*, **52**(6), 2973–2979 (2013).

Index

A